Forensic Law Casebook

While there are several texts that focus on forensic science techniques and applications, there are few to no quality books that adequately address the judicial interpretation of forensic legal and scientific principles. The field of forensic science and law has long been in need of a historic casebook.

Forensic Law Casebook: Judicial Reasoning and the Application of Forensic Science in Criminal Cases fills the current void by reviewing actual case law and translating the practical application of science to the courtroom. Each chapter represents a unique forensic discipline, providing a short introduction to the subject matter, the relevant case law and court cases that pertain to that subject area and posing a variety of questions and issues to the student. All cases provided contain a sufficient portion of the legal decision—and its implications to the evidence and analytical practices of that discipline—in order to then pose critical and analytical questions to the student, once they have fully read the case material and the decision and considered its implications.

Each chapter ends its theoretical examination with real-world experience encountered by those laboring in the investigative and collection processes—as well as problems or challenges encountered by those employed in the office of the prosecutor, public defender, medical examiner or other aligned office. This last section of each chapter gives true meaning and impact as to how forensic law decision-making impacts forensic practitioners, and a true understanding of the responsibility placed on law enforcement, investigators and scientists tasked with collecting, preserving and analyzing the evidence.

Forensic Law Casebook provides the reader with an array of legal cases and decisions that lay out the parameters of forensic law and its evidentiary value. In the end, what emerges from this are the bedrock principles that guide current forensic evidence and the admissibility of various practices common to the field applications of forensic science. Practitioners, law students, undergraduate and graduate students in compatible majors—as well as law and university libraries—will benefit from this essential reference and adjunct to anyone studying forensic science, criminalistics and the law.

Forensic Law Casebook

Judicial Reasoning and the Application of Forensic Science in Criminal Cases

Charles P. Nemeth, JD, PhD, LLM

CRC Press
Taylor & Francis Group
Boca Raton London New York

CRC Press is an imprint of the
Taylor & Francis Group, an **informa** business

Designed cover image: © Shutterstock

First edition published 2024
by CRC Press
2385 NW Executive Center Drive, Suite 320, Boca Raton FL 33431

and by CRC Press
4 Park Square, Milton Park, Abingdon, Oxon, OX14 4RN

CRC Press is an imprint of Taylor & Francis Group, LLC

Library of Congress Cataloging-in-Publication Data
Names: Nemeth, Charles P., 1951- author.
Title: Forensic law casebook : judicial reasoning and the application of forensic science in criminal cases / Charles P. Nemeth, JD, PhD, LLM.
Description: Boca Raton : CRC Press, 2024. | Includes bibliographical references and index.
Identifiers: LCCN 2023034209 (print) | LCCN 2023034210 (ebook) | ISBN 9781032211749 (hardback) | ISBN 9781032207056 (paperback) | ISBN 9781003267126 (ebook)
Subjects: LCSH: Evidence, Expert. | Evidence, Expert--Cases. | Criminal investigation. | Evidence, Criminal. | Admissible evidence. | Documentary evidence. | Forensic sciences. | LCGFT: Casebooks (Law)
Classification: LCC K5485 .N46 2024 (print) | LCC K5485 (ebook) | DDC 347.73/6--dc23/eng/20231002
LC record available at https://lccn.loc.gov/2023034209
LC ebook record available at https://lccn.loc.gov/2023034210

ISBN: 978-1-032-21174-9 (hbk)
ISBN: 978-1-032-20705-6 (pbk)
ISBN: 978-1-003-26712-6 (ebk)

DOI: 10.4324/9781003267126

Typeset in Times
by MPS Limited, Dehradun

Access the Support Material: www.routledge.com/9781032211749

Dedication

To all my children - Eleanor, Stephen, Anne Marie, John, Joseph, Mary Claire and Michael

To my spouse and closest friend, Jean Marie - 52 Years and Counting!

To St. Thomas Aquinas who remarked:

Thus there may be two causes for the just change of human law: one on the part of reason; the other on the part of man whose acts are regulated by law. The cause on the part of reason is that it seems natural to human reason to advance gradually from the imperfect to the perfect. Hence, in speculative sciences, we see that the teaching of the early philosophers was imperfect, and that it was afterwards perfected by those who succeeded them. So also in practical matters: for those who first endeavored to discover something useful for the human community, not being able by themselves to take everything into consideration, set up certain institutions which were deficient in many ways; and these were changed by subsequent lawgivers who made institutions that might prove less frequently deficient in respect of the common weal.

Summa Theologiae *I-II, Q. 97, art. 1*

Contents

About the Author

Dr. Charles P. Nemeth has spent much of his professional life in the study and practice of law and justice.

He is a recognized expert on professional ethics and the justice system, private-sector justice, law and private security systems. He is professor and program director at Franciscan University of Steubenville, leaving behind an impressive record of program leadership and curricular development for nearly four decades. Before joining Franciscan, he was Chair and Professor of Security, Fire and Emergency Management and Director of the Center of Private Security and Safety at John Jay College in New York City.

He is a prolific writer, having published numerous texts and articles on law and justice throughout his career. His text, *Private Security and the Law 5th*-(CRC Press/Taylor Routledge, 2018) is considered the foremost treatise on the subject matter. His private security expertise is further buttressed by: *Private Security and Investigative Process 4th edition* (CRC Press, 2019); *Private Security: Principles and Practice, 2nd edition* (CRC Press, 2019, 2023); *Criminal Law 2nd* (CRC Press, 2012) with a new edition planned for 2022, and *Law and Evidence 3rd* (CRC Press/Taylor Francis, 2022). In the area of Homeland Security, he has published *Introduction to Homeland Security: Practices and Principles* 4th edition (CRC Press, 2010, 2014, 2017, 2021).

He has also authored a series of philosophical works on Thomas Aquinas including *Aquinas on Crime* (St. Augustine's Press, 2010); *Aquinas in the Courtroom* (Praeger/Greenwood Press, 2001); *Aquinas and King: A Discourse on Civil Disobedience* (Carolina Academic Press, 2011) and *Cicero and Aquinas: A Comparative Study of Nature and the Natural Law* (Bloomsbury Publishing, 2018). More recently he has completed two manuscripts: *Natural Law and the U.S Supreme Court since Roe v. Wade* (Anthem Press, 2020) and *Happiness in a Complex World: Rules from Aristotle and Aquinas* (Anthem Press, 2020). A second edition shall be published by Sophia Institute Press. He has also served as Chief Editor to a peer reviewed journal, *The Homeland Security Review,* since 2005, and currently serves as Chief Editor for the *Law, Ethics and Jurisprudence* series to be published by Anthem Press. In addition, Dr. Nemeth is founding Editor of a new journal dedicated to the study and analysis of natural law jurisprudence, *Natural Law and Justice*. The journal is a major initiative of a newly established Center for Criminal Justice, Law and Ethics at Franciscan University. He has been an educator for more than 40 years. He holds memberships in the New York, North Carolina and Pennsylvania Bars and has been active as a general practitioner. Dr. Nemeth was previously a Chair at the State University of New York at Brockport and California University of Pennsylvania, one of Pennsylvania's 14 State Universities. He is a much-sought-after legal consultant for security companies and a recognized scholar on issues involving law, professional ethics and morality and the impact of privatization on public justice models.

Dr. Nemeth resides in Pittsburgh, Pennsylvania, with his spouse, Jean Marie, together for 50 years and blessed with seven children, all of whom are accomplished personally and professionally.

Preface

The field of forensic science and law has long been deprived of a historic casebook. While the marketplace is replete with forensic science "application" texts, it has seen few produced in the "Casebook" style where the emphasis is placed on the judicial interpretation of forensic legal and scientific principles. In this way, the text's legalistic emphasis compliments the real-world application of forensics, scientific analysis and method. It has always been an alliance subject to two worlds: namely the science of things and its corresponding admissibility. That is the central aim of this work—to provide the reader with an array of legal cases and decisions that lay out the parameters of forensic law and its evidentiary value. In the end, what emerges from this examination are the bedrock principles that guide forensic evidence and the admissibility of various practices common to the field applications of forensic science.

Chapter 1 lays out the "Trilogy" of case decisions which broadly set forth the various tests for forensic science admissibility. Commencing with the Frye decision—an almost revered test still applicable in dozens of American jurisdictions—the reader shall discover the importance of scientific rigor and the concept of scientific consensus. Frye is a test that demands a majoritarian view on the questioned practice in forensic science. From Frye, the attention turns to how the Federal Rules of Evidence gave birth to a new test of "helpfulness" in Daubert. Here, the rules become a bit more flexible, and the potentiality of admission expanded. In addition, Daubert has the effect of transferring judgments about admission from the scientific community to the judge as the "gatekeeper."

Finally, in the last of the three tests, the reader encounters Kumho, where the concept of forensic science and its various fields is liberalized to include more forms of "technical" knowledge. Each of these tests dictates how particular jurisdictions can and may deal with forensic law questions.

Chapter 2's coverage closely follows the test noted earlier but looks more keenly at what party may give testimony about the forensic field and its application in court. At its base, there are two choices—either the expert witness or the lay witness. In matters of forensic science and law, the preference is almost uniformly for the expert witness, whose training, preparation, education and research mirror the field under scrutiny. The chapter fully assesses what qualities and qualifications are needed to be an expert witness. In contrast, the chapter lays out the contrast between the expert and the lay witness and also delivers practical suggestions on how to impeach or attack the veracity, credibility and viability of the expert.

The evidentiary proof of identity is crucial in nearly every form of civil and criminal investigation. Who are the parties? Who is the perpetrator? Who are the defendants? In Chapter 3, how identity can be established through various forensic means and methods is fully examined. From proof by fingerprint and palmprint to skeletal analysis and anthropological methods, forensic practitioners seek to identify the parties. Throughout Chapter 3, readers are exposed to not only historic techniques of identity but also the cutting-edge scientific methods involving trace evidence, forensic dental evidence and the emergent and increasing reliance on DNA evidence.

Chapter 4 is a natural follow up in matters of identity since the collection of various fluids leads to claims and conclusions on identity. In matters of sexual assault, rape and other sexual offenses, the forensic collection of seminal fluid and spermatozoa are conclusive about person and identity. Blood typing can eliminate or provide an inclusive basis for targeting a particular individual as other forms of fluids such as spit and saliva, sweat, urine and human excrement.

The chapter often deals with questions relating to evidence and its sufficiency and whether forensic evidence of this sort meets the burden of proof.

Chapter 5 covers the field of forensic toxicology, which is a unique blend of both chemistry and biological science. In toxicology, the chemist or medical examiner often seeks to identify substances that may or may not be the cause of death or injury. Hence, the forensic law settings for toxicology usually deal with proof of intoxication or drug usage. Just as critically, the role of poisons and their proof are essential in homicide cases, or even civil cases where wrongful death is at issue. Finally, the complexities of toxicology are discoverable in the chapter's cases where defendants seek to break the chain of causation in the alleged substance, triggering death with an alternative explanation. Indeed, the Daubert decision, despite is landmark impact, deals with this very notion—whether a drug caused birth defects or whether there were other possible and plausible explanations.

Chapter 6 zeroes in on the idea and conceptualization of what "trace" evidence is and how it leads to a variety of legal challenges. Trace evidence is sometimes referred to as microscopic evidence due to its size and its incapability to be self-explanatory. Hence, hair, fiber, glass shards, chemical residue and paint chips, to name a few, cannot speak for themselves but need to be presented in a way that demonstratively or graphically illustrates the evidence. Cases presented throughout the chapter include arson, accident investigation, proof of presence or agency as well as a host of identifiers which objectively match the perpetrator.

Chapter 7 highlights the various documentary evidence forms as litigated and disputed. Commencing with coverage of the "Best Evidence Rule," the chapter stresses how functional and practical the Federal Rules of Evidence are when it comes to usage and admissibility. The concept of what constitutes a writing or document receives a great deal of scrutiny. Then, the Chapter examines a host of cases which weigh and evaluate whether documents and other writings have been altered and modified and whether or not the composition of the writing or document is consistent with various impressions, types of ink or typeface match—or contrasts. The Chapter includes a bevy of cases which challenge the veracity and integrity of a document used for proof of a right or privileged status.

Chapter 8's coverage of ballistics targets the usual legal and evidentiary challenges to the quality and consistency of evidence relating to projectiles, weapons and their imprint and impressions. Starting with the initial residue left on hands, clothing and the victim, courts make every attempt to reach certain deductive conclusions about type of weaponry, distance from weapon to target, caliber and other variables. Ample attention is given to how markings, impressions, as well as striations, can be matched to a particular weapon.

In Chapter 9, the emphasis is on pathological evidence relating to death, trauma and injury. Special emphasis is given to the time, cause and manner of death and the diverse elements that factor into such determinations. A whole array of causes are closely scrutinized from a legal perspective and measured against commonly agreed formulas for timing and method of death and injury. Other time is spent on the instruments of death, from blunt force to the types of projectiles inflicting damage on the human body.

Chapter 10 is a unique blend of both the "hard" and "soft" sciences involving psychiatry and psychology. While the former has a harder edge, especially when involving neurological explanations for crime and criminality, the behavioral essence of psychology is less rigorous. Emotion is always tougher to compute and quantify than a medical diagnosis involving the brain. Throughout the chapter, readers encounter theories and explanations as to criminality that are less than objectively measured, and, instead, these are diagnoses that venture into subjective territory. Even the full coverage of the three tests for legal insanity cannot erase the confusion and subjectivity inherent to these fields. The chapter distinguishes legal insanity from medical insanity and delivers a full array of theories and explanations that may or may not qualify.

In the final chapter, the reader encounters a prophetic view on where forensic science and forensic law may be headed. What sorts of innovation are likely to occur? What new fields and avenues may emerge in the years ahead? On the other hand, will some credited science and law see a future of general disbelief? A historical review of forensic law and science makes clear that the discipline is not fixed nor static. Chapter 11 prods the reader to see where the shortfalls are—to critique the shortcomings of our present understanding of forensic science and law. Most practitioners engaged in this area of law are quite certain to encounter calls for a bit more rigor and stronger efforts to reign in the "junk" side of forensic science and law. To be sure, the future of the field will depend on a bit more quality control than is currently witnessed. Readers should think critically about how things can improve.

Acknowledgments

The idea of an accessible and useful Casebook for Forensic Law has long been on my radar. The need has been acutely obvious over the last decade or so. And with the rise of forensic programs and growing interest in criminalistics and related subject matter, it was clear that students were missing out on this sort of legal approach—something essential to a full understanding of the field. It was, and has been, far more daunting than I ever envisioned when I commenced the project. The sheer size, volume and expansive subject matter has made this a formidable challenge. I owe a debt of gratitude to many and realize this project would have been merely a pipedream without their support and assistance.

First at Taylor & Francis my work with Senior Editor Mark Listewnik continues to be professional in every way, and I am deeply grateful to Mark for not only being an editor who has developed an extraordinary library in forensic science but also an editor willing to a take a risk here and there. Mark has always seen the need for this subject matter.

At Franciscan University, the support is beyond charitable. From Father Dave Pivonka to our Academic Vice President Daniel Kempton, who is soon to retire be replaced by Dr. Steve Hilldebrand, I am fortunate to work in such a wonderful environment. Most especially worthy of praise is Dean Chrissy Jungers of the School of Professional Studies, whose generous nature and endless support fostered this project and many others. It is a delight to work with such decent people. Finally, our Criminal Justice Department was blessed to have a research assistant who mightily contributed to this project, namely Luz Piedra. Luz manifested a research skill level well beyond all expectations. I am forever grateful for her excellent work. On the graphics side of the work, as well as high interest insertions and web exercises, the project relied heavily on another remarkable student from Franciscan—Samantha Matthews. Here, too, we encounter a student displaying skill levels more commensurate with a doctoral student.

On the editorial and manuscript side, I again cannot find suitable words which capture the talent level of Hope Haywood. Nearly 35 years ago, I encountered Hope when she was a student and all these years later she still organizes every production and does so without any rival. And in this particular case, she has taken on a gargantuan task. There is not much Hope cannot tackle, and, without her, the project would have never been completed.

Charles P. Nemeth JD, PhD, LL.M
Professor of Criminal Justice
Director of the Center for Criminal Justice, Law and Ethics
Pittsburgh, Pennsylvania
6/8/2023

1 Forensic Science, Expert Evidence and Standards of Admissibility

1.1 INTRODUCTION

Any determination on the admissibility of forensic science is inexorably tied to a larger community of scholarship, research, collegial agreement and a majoritarian view of the propriety of the evidence offered. Forensic law encompasses this very large question and assesses whether or not the evidence offered up meets the criteria for admission. Forensic law determinations are guided by the various state and federal rules on evidence but even more critically, the case law conclusions as these rules and common law principles relating to the relevance, materiality and scientific consensus on the forensic topic. Both the expert witness and the expert subject matter must meet specified guidelines regarding admission.

Whether or not the expert possesses the necessary knowledge to testify as an expert is a central factor, though not the exclusive basis, in any admissibility determination. More is required. Whether the subject matter of the expert's testimony is reasonably reliable, has achieved scientific consensus, generated a body of research and literature and has been tested in the scientific marketplace are equally crucial considerations.[1] Has the expert witness' subject matter, the scientific topic proffered, met with a generally favorable review in the scientific community and, at the same time, do courts, either by current decision or long-standing precedent, admit the same? Courts primarily consider whether or not the scientific field has gained acceptance and consensus—a standard essentially followed since the 1923 case of *Frye v. U.S.*[2]

In *Frye*, a decision of the DC Circuit Court of Appeals, the court's consideration of forensic science will come to depend upon a variety of factors—mostly a majoritarian, consensus driven and provable proposition. *Frye* will not be a welcoming home for the proponents of *avante garde* or questionable science. In subsequent cases, soon to be discussed, courts have wider discretion in making the judgment on scientific legitimacy even in the absence of a consensus. Today, judges see scientific and forensic evidence as being part of the mix and as long as it is "helpful" to the trier of fact, admission is more likely than not.[3] Both state and federal courts establish minimal legal thresholds for expert admission, for there is no uniform, mandatory test or measure for expert evidence for the 50 states. At the federal level, the litigants and judiciary are guided by the Federal Rules of Evidence and a series of cases that interpret and mirror the FRE. More on the central cases will be assessed shortly (Figure 1.1).

1.2 THE FRYE DOCTRINE: "CROSSING THE BARRIER OF JUDICIAL ACCEPTABILITY"

In making any judgment on the admissibility of expert and forensic evidence, courts have historically looked to a long-standing precedent, *Frye v. U.S.*[4] In Frye, the question of whether systolic blood pressure could be measured by a polygraph machine was a rather innovative scientific question for the time. The Court was hesitant to admit scientific evidence unless it was convinced that the evidence and its corresponding scientific experts had "crossed the barrier of

DOI: 10.4324/9781003267126-1

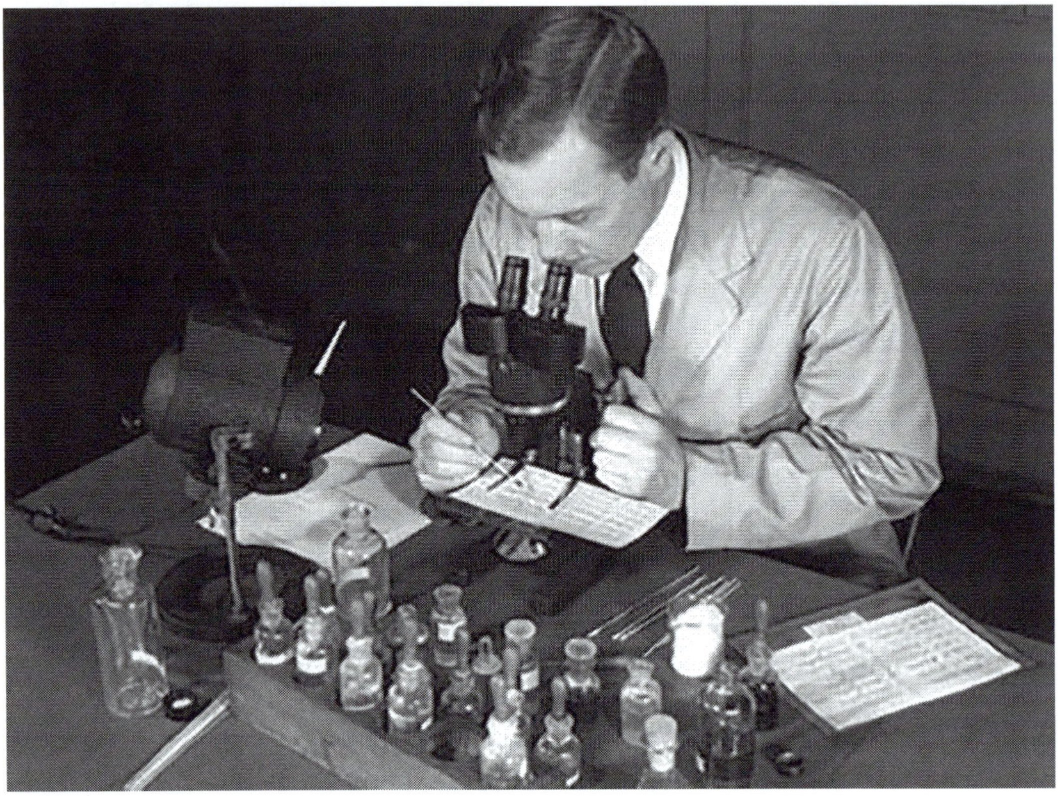

FIGURE 1.1 A laboratory technician examines ink on questioned documents ca. 1950s, FBI Media at https://multimedia.fbi.gov/item?type=image&id=674.

judicial acceptability."[5] The *Frye* decision is much deliberated and oft-quoted. Its critical message conveys the following advice:

> Just when a scientific principle or discovery crosses the line between the experimental and demonstrable stages is difficult to define. Somewhere in this twilight zone, the evidential force of the principle must be recognized, and while the courts will go a long way in admitting expert testimony deducted from a well recognized scientific principle or discovery, the thing from which the deduction is made must be sufficiently established to have gained general acceptance in the particular field in which it belongs.[6]

Read about the case of serial killer Wayne Williams failing his polygraph test multiple times at https://www.fbi.gov/news/stories/-serial-killers-part-5-wayne-williams-and-the-atlanta-child-murders

Whether evidence is reliable and thus admissible under the Frye doctrine will depend on a variety of factors, including:

1. The validity of the underlying scientific principle.
2. The validity of the technique or process that applies the principle.
3. The condition of any instrumentation used in the process.

4. Adherence to proper procedures.
5. The qualifications of the person who performs the test.
6. The qualifications of a person who interprets the results.[7]

Under this guidance, it is fair to conclude that the *Pre-Menstrual Syndrome (PMS) defense* has crossed *Frye's* barrier of judicial acceptability. Do the majority of psychologists and psychiatrists concur on this judgment or are there wide-ranging opinions on the legitimacy of this defense? What about the fields of *astrology* and *para-sensory perception*? Do either of these fields have a reliable science to undergird the theories? Will the "Twinkie defense," argued by murdering felons claiming sugar imbalance, suffice under the *Frye* criteria? It is clear that under Frye, a majoritarian view of the scientific conclusion must be proven. *Frye* is not a case precedent displaying sweeping tolerance for any new scientific theory but instead is a cautious case that urges the advocates to wait for a consensus in the scientific community before proceeding.

> There are many legal scholars and practitioners who long for a return to *Frye*. The longstanding case had a sort of admirable rigor to it that is no longer evident in more modern case decisions. Visit: https://web.ornl.gov/sci/techresources/Human_Genome/hg5yp/goal.shtml

Under the *Frye* doctrine, an expert's testimony and the forensic subject matter must be a field or intellectual endeavor that is generally credible and scientifically sound. Some legal interpreters extend these notions to non-scientific forms of expert testimony—or put another way—the distinction between the "hard" and "soft" sciences. And in more recent case law, such as *Kumho Tire*,[8] the definition of scientific knowledge now encompasses "technical" or specialized knowledge.

While ballistics depends on physics and chemistry, typically hard sciences, other fields have little or no scientific substantiality in the traditional sense, e.g., economics, accounting, psychology, security law and so on. Hard science and technical knowledge now appear closely knit and even under historic principles, the courts entertained fields that were hardly hard by design such as blood spatter analysis, handwriting and graphology as a host of other fields that do not have the rigor of chemistry, biology and physics. In nonscientific evidence or better stated, technical knowledge, the judicial test is not one of substance but method. *Frye*, as has already been shown, worries first about the field. Is it a scientific conclusion that has majoritarian support in the scientific community or is there a consensus? Do most scientists reach the same conclusion? Does the field engage in quantitative and qualitative research regarding its content and at the same time produce scholarly and peer-reviewed research? "To evaluate the validity and accuracy of the expert's conclusions, *Frye* requires trial courts to survey relevant scholarship. Admission is predicated on a showing that the expert's conclusions command an independent, supporting consensus. Under *Frye*, trial courts therefore assess expert evidence in its native context, a process that does not discriminate against different nonscientific fields."[9]

Frye v. United States
293 F. 1013 (D.C. Cir. 1923)
December 3, 1923

Appellant, defendant below, was convicted of the crime of murder in the second degree, and from the judgment prosecutes this appeal.

A single assignment of error is presented for our consideration. In the course of the trial counsel for defendant offered an expert witness to testify to the result of a deception test made upon defendant. The test is described as the systolic blood pressure deception test. It is asserted that blood pressure is influenced by change in the emotions of the witness, and that the systolic blood pressure rises are brought about by nervous impulses sent to the sympathetic branch of the autonomic nervous system. Scientific experiments, it is claimed, have demonstrated that fear, rage, and pain always produce a rise of systolic blood pressure, and that conscious deception or falsehood, concealment of facts, or guilt of crime, accompanied by fear of detection when the person is under examination, raises the systolic blood pressure in a curve, which corresponds exactly to the struggle going on in the subject's mind, between fear and attempted control of that fear, as the examination touches the vital points in respect of which he is attempting to deceive the examiner.

In other words, the theory seems to be that truth is spontaneous, and comes without conscious effort, while the utterance of a falsehood requires a conscious effort, which is reflected in the blood pressure. The rise thus produced is easily detected and distinguished from the rise produced by mere fear of the examination itself. In the former instance, the pressure rises higher than in the latter, and is more pronounced as the examination proceeds, while in the latter case, if the subject is telling the truth, the pressure registers highest at the beginning of the examination, and gradually diminishes as the examination proceeds.

Prior to the trial defendant was subjected to this deception test, and counsel offered the scientist who conducted the test as an expert to testify to the results obtained. The offer was objected to by counsel for the government, and the court sustained the objection. Counsel for defendant then offered to have the proffered witness conduct a test in the presence of the jury. This also was denied.

Counsel for defendant, in their able presentation of the novel question involved, correctly state in their brief that no cases directly in point have been found. The broad ground, however, upon which they plant their case, is succinctly stated in their brief as follows:

"The rule is that the opinions of experts or skilled witnesses are admissible in evidence in those cases in which the matter of inquiry is such that inexperienced persons are unlikely to prove capable of forming a correct judgment upon it, for the reason that the subject-matter so far partakes of a science, art, or trade as to require a previous habit or experience or study in it, in order to acquire a knowledge of it. When the question involved does not lie within the range of common experience or common knowledge, but requires special experience or special knowledge, then the opinions of witnesses skilled in that particular science, art, or trade to which the question relates are admissible in evidence."

Numerous cases are cited in support of this rule. Just when a scientific principle or discovery crosses the line between the experimental and demonstrable stages is difficult to define. Somewhere in this twilight zone the evidential force of the principle must be recognized, and while courts will go a long way in admitting expert testimony deduced from a well-recognized scientific principle or discovery, the thing from which the deduction is made must be sufficiently established to have gained general acceptance in the particular field in which it belongs.

We think the systolic blood pressure deception test has not yet gained such standing and scientific recognition among physiological and psychological authorities as would justify the courts in admitting expert testimony deduced from the discovery, development, and experiments thus far made.

The judgment is affirmed.

Questions:

1. What does the Court measure relative to systolic blood pressure?
2. Frye has long been accused of inhibiting novel and innovative scientific claims in the law. Is this a fair critique?
3. Frye has often been labeled a case of both substance and process. How so?

1.3 FROM FRYE'S CROSSING THE BARRIER TO DAUBERT'S HELPFULNESS TEST FOR FORENSIC SCIENCE

In the end, *Frye* is more restrictive as to admissibility than the subsequent *Daubert* and *Kumho* decisions announced nearly 75 years later. Not all were happy with *Frye's* restrictiveness. The federal courts, by and through a pivotal ruling, *Daubert v. Merrell Dow Pharmaceuticals*,[10] have effectively made the *Frye* rule moot in their jurisdiction even though nearly two dozen states continue to adhere to *Frye*. In *Daubert*,[11] which is crudely known as the "*Helpfulness*" test, the U.S. Supreme Court rejects the *Frye* standard at the federal level primarily because the language of the Federal Rules of Evidence does not encompass the *Frye* standards.[12] The FRE tends to be more discretionary than *Frye* could ever be since its emphasis is aiding or helping the trier of fact in the evidentiary arena. The language and aim of FRE 702, which guides expert opinion, stand in contradiction to the narrower Frye doctrine. FRE Rule 702 calls for *scientific, technical* and *specialized* knowledge as a pre-requisite to expert evidence admissibility. The uniqueness of the knowledge being presented before the court, that is, knowledge above and beyond the normal competency of a lay witness, is what castes the evidence in an expert mode.[13] Succinctly stated, experts exist because lay witnesses are both incapable of giving testimony about technical subject matter and are necessary because of the practical benefits they bring to the judicial arena. Ordinary lay persons lack the intellectual, occupational or educational background essential in the resolution of complex cases. However, there are cases in which the lines between expert and lay witness testimony are often blurred.[14]

Daubert, by most accounts, has liberalized the doorway or threshold of admittance for various reasons. First, strict reliance on the general consensus test of *Frye* is being replaced with a "more searching and flexible inquiry about the reliability and relevance of the offered evidence."[15] *Frye's* majoritarian view within the scientific community has been abrogated, although the proponent of said evidence must still prove the legitimacy of the theory, argue its dependability and reasonableness, and at the same time show that scientific analysis and critiques have been published. Most critically, the *Frye* consensus in the scientific community has been replaced with judges now being the screeners of the evidence. Under *Daubert*, judges are "gatekeepers" performing dual functions—screening "expert scientific testimony not merely to assure that the expert is qualified, but also to assure that the expert methodology is 'reliable.'"[16] Judges were to determine whether "the scientific evidence had sufficient testing, peer review and publication."[17] But those views need not be majoritarian and if the judge, in his or her gatekeeping function, finds that the evidence is "helpful" to court's determination, it will cross the once more demanding threshold. To give judges this extraordinary power—evaluating the quality and content of scientific evidence has caused a good bit of consternation. It is questionable whether judges are even capable of performing the latter function.[18] Like it or not, until and when *Daubert* modified *Frye*, the consensus argument was firmly entrenched. Since *Daubert*, the net result is that many more forms of alleged science have been admitted since a singular view on a particular topic may be all that is necessary.[19]

> Visit the American Board of Criminalistics at https://www.criminalistics.com/ and learn about the certification process.

Daubert v. Merrell Dow Pharmaceuticals, Inc.
509 U.S. 579 (1993)

In this case, we are called upon to determine the standard for admitting expert scientific testimony in a federal trial.

I

Petitioners Jason Daubert and Eric Schuller are minor children born with serious birth defects. They and their parents sued respondent in California state court, alleging that the birth defects had been caused by the mothers' ingestion of Bendectin, a prescription antinausea drug marketed by respondent. Respondent removed the suits to federal court on diversity grounds.

After extensive discovery, respondent moved for summary judgment, contending that Bendectin does not cause birth defects in humans and that petitioners would be unable to come forward with any admissible evidence that it does. In support of its motion, respondent submitted an affidavit of Steven H. Lamm, physician and epidemiologist, who is a well-credentialed expert on the risks from exposure to various chemical substances. Doctor Lamm stated that he had reviewed all the literature on Bendectin and human birth defects—more than 30 published studies involving over 130,000 patients. No study had found Bendectin to be a human teratogen (i.e., a substance capable of causing malformations in fetuses). On the basis of this review, Doctor Lamm concluded that maternal use of Bendectin during the first trimester of pregnancy has not been shown to be a risk factor for human birth defects.

Petitioners did not (and do not) contest this characterization of the published record regarding Bendectin. Instead, they responded to respondent's motion with the testimony of eight experts of their own, each of whom also possessed impressive credentials. These experts had concluded that Bendectin can cause birth defects. Their conclusions were based upon "in vitro" (test tube) and "in vivo" (live) animal studies that found a link between Bendectin and malformations; pharmacological studies of the chemical structure of Bendectin that purported to show similarities between the structure of the drug and that of other substance known to cause birth defects; and the "reanalysis" of previously published epidemiological (human statistical) studies.

The District Court granted respondent's motion for summary judgment. The court stated that scientific evidence is admissible only if the principle upon which it is based is "'sufficiently established to have general acceptance in the field to which it belongs.'" 727 F. Supp. 570, 572 (S.D. Cal. 1989), quoting United States v. Kilgus, 571 F.2d 508, 510 (CA9 1978). The court concluded that petitioners' evidence did not meet this standard. Given the vast body of epidemiological data concerning Bendectin, the court held, expert opinion which is not based on epidemiological evidence is not admissible to establish causation. 727 F. Supp., at 575. Thus, the animal cell studies, live animal studies, and chemical structure analyses on which petitioners had relied could not raise, by themselves, a reasonably disputable jury issue regarding causation. Ibid. Petitioners' epidemiological analyses, based as they were on recalculations of data in previously published studies that had found no causal link between the drug and birth defects, were ruled to be inadmissible because they had not been published or subjected to peer review. Ibid.

The United States Court of Appeals for the Ninth Circuit affirmed. 951 F.2d 1128 (1991). Citing Frye v. United States, 54 App.D.C. 46, 47, 293 F. 1013, 1014 (1923), the court stated that expert opinion based on a scientific technique is inadmissible unless the technique is "generally accepted" as reliable in the relevant scientific community. 951 F.2d, at 1129-1130. The court declared that expert opinion based on a methodology that diverges "significantly from the procedures accepted by recognized authorities in the field … cannot be shown to be 'generally accepted as a reliable technique.'" Id., at 1130, quoting United States v. Solomon, 753 F.2d 1522, 1526 (CA9 1985).

The court emphasized that other Courts of Appeals considering the risks of Bendectin had refused to admit reanalyses of epidemiological studies that had been neither published nor subjected to peer review. 951 F.2d, at 1130-1131. Those courts had found unpublished reanalyses "particularly problematic in light of the massive weight of the original published studies supporting [respondent's] position, all of which had undergone full scrutiny from the scientific community."

Id., at 1130. Contending that reanalysis is generally accepted by the scientific community only when it is subjected to verification and scrutiny by others in the field, the Court of Appeals rejected petitioners' reanalyses as "unpublished, not subjected to the normal peer review process, and generated solely for use in litigation." Id., at 1131. The court concluded that petitioners' evidence provided an insufficient foundation to allow admission of expert testimony that Bendectin caused their injuries and, accordingly, that petitioners could not satisfy their burden of proving causation at trial.

We granted certiorari, 506 U.S. 914 (1992), in light of sharp divisions among the courts regarding the proper standard for the admission of expert testimony. Compare, e.g., United States v. Shorter, 257 U.S.App.D.C. 358, 363-364, 809 F.2d 54, 59-60 (applying the "general acceptance" standard), cert. denied, 484 U.S. 817 (1987), with DeLuca v. Merrell Dow Pharmaceuticals, Inc., 911 F.2d 941, 955 (CA3 1990) (rejecting the "general acceptance" standard).

II A

In the 70 years since its formulation in the Frye case, the "general acceptance" test has been the dominant standard for determining the admissibility of novel scientific evidence at trial. See E. Green C. Nesson, Problems, Cases, and Materials on Evidence 649 (1983). Although under increasing attack of late, the rule continues to be followed by a majority of courts, including the Ninth Circuit.

The Frye test has its origin in a short and citation-free 1923 decision concerning the admissibility of evidence derived from a systolic blood pressure deception test, a crude precursor to the polygraph machine. In what has become a famous (perhaps infamous) passage, the then Court of Appeals for the District of Columbia described the device and its operation and declared:

> "Just when a scientific principle or discovery crosses the line between the experimental and demonstrable stages is difficult to define. Somewhere in this twilight zone the evidential force of the principle must be recognized, and while courts will go a long way in admitting expert testimony deduced from a well-recognized scientific principle or discovery, the thing from which the deduction is made must be sufficiently established to have gained general acceptance in the particular field in which it belongs." 54 App. D.C., at 47, 293 F., at 1014 (emphasis added).

Because the deception test had "not yet gained such standing and scientific recognition among physiological and psychological authorities as would justify the courts in admitting expert testimony deduced from the discovery, development, and experiments thus far made," evidence of its results was ruled inadmissible. Ibid.

The merits of the Frye test have been much debated, and scholarship on its proper scope and application is legion. Petitioners' primary attack, however, is not on the content, but on the continuing authority, of the rule. They contend that the Frye test was superseded by the adoption of the Federal Rules of Evidence. We agree.

…

Here there is a specific Rule that speaks to the contested issue. Rule 702, governing expert testimony, provides:

> "If scientific, technical, or other specialized knowledge will assist the trier of fact to understand the evidence or to determine a fact in issue, a witness qualified as an expert by knowledge, skill, experience, training, or education, may testify thereto in the form of an opinion or otherwise."

Nothing in the text of this Rule establishes "general acceptance" as an absolute prerequisite to admissibility. Nor does respondent present any clear indication that Rule 702 or the Rules as a whole were intended to incorporate a "general acceptance" standard. The drafting history makes no mention of Frye, and a rigid "general acceptance" requirement would be at odds with the "liberal thrust" of the Federal Rules and their "general approach of relaxing the traditional barriers to 'opinion' testimony." Beech Aircraft Corp. v. Rainey, 488 U.S., at 169 (citing Rules 701 to 705). See also Weinstein, Rule 702 of the Federal Rules of Evidence is Sound; It Should Not Be Amended, 138 F.R.D. 631 (1991) ("The Rules were designed to depend primarily upon lawyer-adversaries and sensible triers of fact to evaluate conflicts"). Given the Rules' permissive backdrop and their inclusion of a specific rule on expert testimony that does not mention "general acceptance," the assertion that the Rules somehow assimilated Frye is unconvincing. Frye made "general acceptance" the exclusive test for admitting expert scientific testimony. That austere standard, absent from, and incompatible with, the Federal Rules of Evidence, should not be applied in federal trials.

…

C

Faced with a proffer of expert scientific testimony, then, the trial judge must determine at the outset, pursuant to Rule 104(a), whether the expert is proposing to testify to (1) scientific knowledge that (2) will assist the trier of fact to understand or determine a fact in issue. …

Another pertinent consideration is whether the theory or technique has been subjected to peer review and publication. Publication (which is but one element of peer review) is not a sine qua non of admissibility; it does not necessarily correlate with reliability, see S. Jasanoff, The Fifth Branch: Science Advisors as Policymakers 61-76 (1990), and, in some instances, well-grounded but innovative theories will not have been published, see Horrobin, The Philosophical Basis of Peer Review and the Suppression of Innovation, 263 JAMA 1438 (1990). Some propositions, moreover, are too particular, too new, or of too limited interest to be published. But submission to the scrutiny of the scientific community is a component of "good science," in part because it increases the likelihood that substantive flaws in methodology will be detected. See J. Ziman, Reliable Knowledge: An Exploration of the Grounds for Belief in Science 130-133 (1978); Relman Angell, How Good Is Peer Review?, 321 New Eng.J.Med. 827 (1989). The fact of publication (or lack thereof) in a peer reviewed journal thus will be a relevant, though not dispositive, consideration in assessing the scientific validity of a particular technique or methodology on which an opinion is premised.

Additionally, in the case of a particular scientific technique, the court ordinarily should consider the known or potential rate of error, see, e.g., United States v. Smith, 869 F.2d 348, 353-354 (CA7 1989) (surveying studies of the error rate of spectrographic voice identification technique), and the existence and maintenance of standards controlling the technique's operation, see United States v. Williams, 583 F.2d 1194, 1198 (CA2 1978) (noting professional organization's standard governing spectrographic analysis), cert. denied, 439 U.S. 1117 (1979).

Finally, "general acceptance" can yet have a bearing on the inquiry. A "reliability assessment does not require, although it does permit, explicit identification of a relevant scientific community and an express determination of a particular degree of acceptance within that community." United States v. Downing, 753 F.2d, at 1238. See also 3 Weinstein Berger ¶ 702[03], pp. 702-41 to 702-42. Widespread acceptance can be an important factor in ruling particular evidence admissible, and "a known technique which has been able to attract only minimal support within the community," Downing, 753 F.2d, at 1238, may properly be viewed with skepticism.

…

IV

To summarize: "General acceptance" is not a necessary precondition to the admissibility of scientific evidence under the Federal Rules of Evidence, but the Rules of Evidence—especially Rule 702—do assign to the trial judge the task of ensuring that an expert's testimony both rests on a reliable foundation and is relevant to the task at hand. Pertinent evidence based on scientifically valid principles will satisfy those demands.

The inquiries of the District Court and the Court of Appeals focused almost exclusively on "general acceptance," as gauged by publication and the decisions of other courts. Accordingly, the judgment of the Court of Appeals is vacated, and the case is remanded for further proceedings consistent with this opinion.

It is so ordered.

Questions:

1. Why is the "General Acceptance" standard ended with the Daubert decision? What is the basis for shift away from "General Acceptance?"
2. How does Daubert deal with the issue of limited research or publications which have tested the theory or scientific proposition? Under Frye a body of literature and scholarly analysis was a natural expectation. How does Daubert liberalize that requirement.
3. When describing Frye, Daubert refers to the ruling as "austere." What does this mean?

1.4 KUMHO TIRE AND THE MOVE TO INHIBIT JUNK SCIENCE

Daubert's lenient and inconsistent results have generated enormous criticism. The U.S. Supreme Court's 1999 ruling, in *Kumho Tire Co. v. Carmichael,*[20] accelerated the gatekeeping role of the judge when it comes to the quality and content of expert evidence. Not only are the questions of qualification and field pertinent but also the methodology behind the results testified to.[21] But the Kumho decision does even more to liberalize things, even though the ambition is otherwise. First, it makes non-mandatory the Daubert criteria of peer review, research methodology and consensus; second, it expands the idea of what scientific knowledge is once the province of the hard sciences such as biology, chemistry and physics to be more inclusive of mere "technical" knowledge. Indeed, Kumho grapples with the notion that evaluating tire failure is less objective science and more a subjective interpretation (Figure 1.2).

Tire analysis, as posed by the *Kumho* expert, was more about patterns and tread wear than pure engineering. It is a highly interpretive technical knowledge based on experience more than materials failure. In *Kumho*, the court expanded the types of knowledge that could be helpful to the court but insisted on proper methodology. And given this pattern for tolerance of emerging forensic knowledge, few would argue that *Daubert* and *Kumho* have fostered an environment in which junk science finds a friendlier evaluator. With the consensus argument gone and the shift from science to judge as "gatekeeper," it is no wonder that novel theories proliferate. *Kumho*, some hold, is an attempt to restrain a burgeoning "junk science" industry from the courtroom, although its basis and reasoning do provide the means. Expert-forensic evidence should be primarily rooted in a "real science"[22] that only arises from "careful and controlled experimentation."[23]

Kumho Tire Co., Ltd., Et Al. v. Carmichael Et Al.
526 U.S. 137
March 23, 1999

When a tire on the vehicle driven by Patrick Carmichael blew out and the vehicle overturned, one passenger died and the others were injured. The survivors and the decedent's representative, respondents here, brought this diversity suit against the tire's maker and its

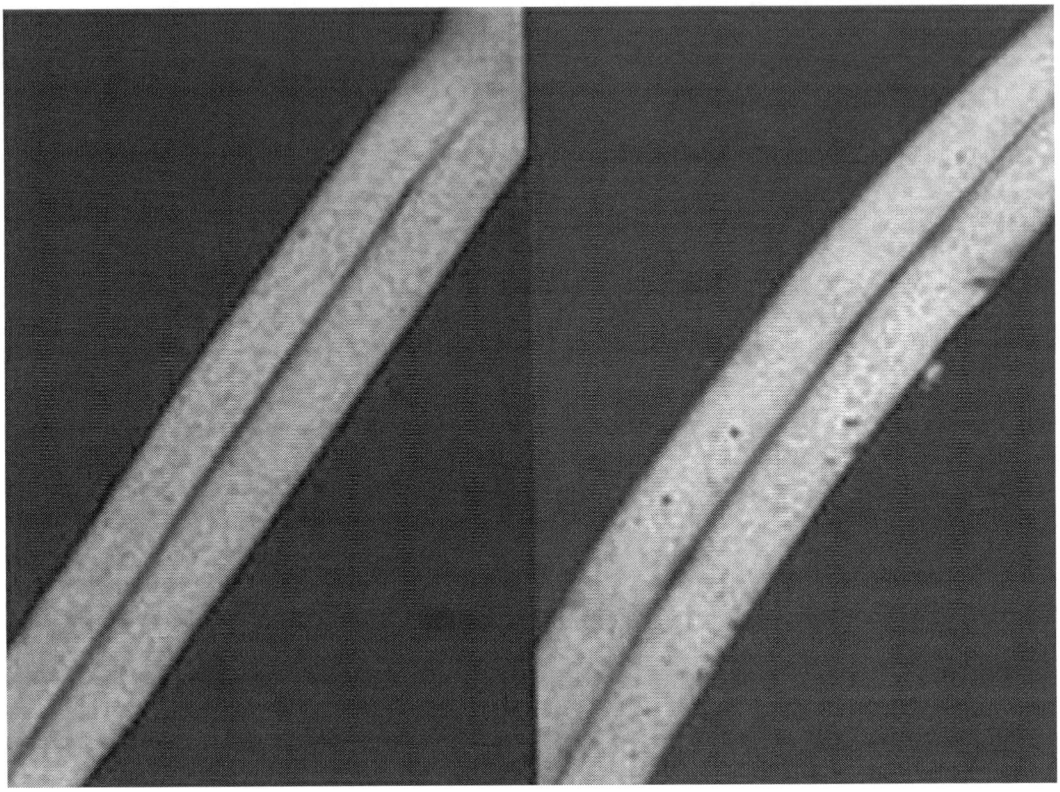

FIGURE 1.2 Fiber comparison of spandex fibers, Attorney General of South Dakota at https://atg.sd.gov/LawEnforcement/Forensics/traceevidence.aspx#gsc.tab=0.

distributor (collectively Kumho Tire), claiming that the tire that failed was defective. They rested their case in significant part upon the depositions of a tire failure analyst, Dennis Carlson, Jr., who intended to testify that, in his expert opinion, a defect in the tire's manufacture or design caused the blowout. That opinion was based upon a visual and tactile inspection of the tire and upon the theory that in the absence of at least two of four specific, physical symptoms indicating tire abuse, the tire failure of the sort that occurred here was caused by a defect. Kumho Tire moved to exclude Carlson's testimony on the ground that his methodology failed to satisfy Federal Rule of Evidence 702, which says: "If scientific, technical, or other specialized knowledge will assist the trier of fact ..., a witness qualified as an expert ... may testify thereto in the form of an opinion." Granting the motion (and entering summary judgment for the defendants), the District Court acknowledged that it should act as a reliability "gatekeeper" under Daubert v. Merrell Dow Pharmaceuticals, Inc., 509 U. S. 579, 589, in which this Court held that Rule 702 imposes a special obligation upon a trial judge to ensure that scientific testimony is not only relevant, but reliable. The court noted that Daubert discussed four factors-testing, peer review, error rates, and "acceptability" in the relevant scientific community-which might prove helpful in determining the reliability of a particular scientific theory or technique, id., at 593-594, and found that those factors argued against the reliability of Carlson's methodology. On the plaintiffs' motion for reconsideration, the court agreed that Daubert should be applied flexibly, that its four factors were simply illustrative, and that other factors could argue in favor of admissibility. However, the court affirmed its earlier order because it found insufficient indications of the reliability of Carlson's methodology. In reversing, the Eleventh Circuit held that the District Court had erred as a matter of law in applying Daubert. Believing that Daubert was limited to the scientific context, (138) the court held that the Daubert factors did not apply to Carlson's testimony, which it characterized as skill or experience based.

Held:

1. The Daubert factors may apply to the testimony of engineers and other experts who are not scientists. Pp. 147-153.
   a. The Daubert "gatekeeping" obligation applies not only to "scientific" testimony, but to all expert testimony. Rule 702 does not distinguish between "scientific" knowledge and "technical" or "other specialized" knowledge, but makes clear that any such knowledge might become the subject of expert testimony. It is the Rule's word "knowledge," not the words (like "scientific") that modify that word, that establishes a standard of evidentiary reliability. 509 U. S., at 589-590. Daubert referred only to "scientific" knowledge because that was the nature of the expertise there at issue. Id., at 590, n. 8. Neither is the evidentiary rationale underlying Daubert's "gatekeeping" determination limited to "scientific" knowledge. Rules 702 and 703 grant all expert witnesses, not just "scientific" ones, testimonial latitude unavailable to other witnesses on the assumption that the expert's opinion will have a reliable basis in the knowledge and experience of his discipline. Id., at 592. Finally, it would prove difficult, if not impossible, for judges to administer evidentiary rules under which a "gatekeeping" obligation depended upon a distinction between "scientific" knowledge and "technical" or "other specialized" knowledge, since there is no clear line dividing the one from the others and no convincing need to make such distinctions. Pp. 147-149.
   b. A trial judge determining the admissibility of an engineering expert's testimony may consider one or more of the specific Daubert factors. The emphasis on the word "may" reflects Daubert's description of the Rule 702 inquiry as "a flexible one." 509 U. S., at 594. The Daubert factors do not constitute a definitive checklist or test, id., at 593, and the gatekeeping inquiry must be tied to the particular facts, id., at 591. Those factors mayor may not be pertinent in assessing reliability, depending on the nature of the issue, the expert's particular expertise, and the subject of his testimony. Some of those factors may be helpful in evaluating the reliability even of experience-based expert testimony, and the Court of Appeals erred insofar as it ruled those factors out in such cases. In determining whether particular expert testimony is reliable, the trial court should consider the specific Daubert factors where they are reasonable measures of reliability. Pp. 149-152.
   c. A court of appeals must apply an abuse-of-discretion standard when it reviews a trial court's decision to admit or exclude expert (139) testimony. General Electric Co. v. Joiner, 522 U. S. 136, 138-139. That standard applies as much to the trial court's decisions about how to determine reliability as to its ultimate conclusion. Thus, whether Daubert's specific factors are, or are not, reasonable measures of reliability in a particular case is a matter that the law grants the trial judge broad latitude to determine. See id., at 143. The Eleventh Circuit erred insofar as it held to the contrary. Pp. 152-153.
2. Application of the foregoing standards demonstrates that the District Court's decision not to admit Carlson's expert testimony was lawful. The District Court did not question Carlson's qualifications, but excluded his testimony because it initially doubted his methodology and then found it unreliable after examining the transcript in some detail and considering respondents' defense of it. The doubts that triggered the court's initial inquiry were reasonable, as was the court's ultimate conclusion that Carlson could not reliably determine the cause of the failure of the tire in question. The question was not the reliability of Carlson's methodology in general, but rather whether he could reliably determine the cause of failure of the particular tire at issue. That tire, Carlson conceded, had

traveled far enough so that some of the tread had been worn bald, it should have been taken out of service, it had been repaired (inadequately) for punctures, and it bore some of the very marks that he said indicated, not a defect, but abuse. Moreover, Carlson's own testimony cast considerable doubt upon the reliability of both his theory about the need for at least two signs of abuse and his proposition about the significance of visual inspection in this case. Respondents stress that other tire failure experts, like Carlson, rely on visual and tactile examinations of tires. But there is no indication in the record that other experts in the industry use Carlson's particular approach or that tire experts normally make the very fine distinctions necessary to support his conclusions, nor are there references to articles or papers that validate his approach. Respondents' argument that the District Court too rigidly applied Daubert might have had some validity with respect to the court's initial opinion, but fails because the court, on reconsideration, recognized that the relevant reliability inquiry should be "flexible," and ultimately based its decision upon Carlson's failure to satisfy either Daubert's factors or any other set of reasonable reliability criteria. Pp. 153-158.

131 F.3d 1433, reversed.

Questions:

1. What is the Court's view of the consensus theory and the role of the judge in forensic evidence analysis.
2. How does the court label or characterize the expert testimony on tire failure?
3. What is the Court's holding as to the admissibility of the tire expert?

1.5 THE LEGACY OF FRYE, DAUBERT AND KUMHO: SCIENCE OR JUNK

There is little question that the proliferation of experts and novel, proclaimed fields of expertise have caused concerns about the quality and content of expert witnesses. The *Frye* doctrine seemed to anticipate the quality issue by demanding evidentiary rigor. *Daubert's* descent into "helpfulness" and far-reaching judicial discretion was bound to loosen up the standards of both the scientific field and its results. While both *Daubert* and *Kumho* talk the talk of scientific method, peer review publications and research and a generic scientific method, the words are undercut by the discretionary "gatekeeping" function which promotes laxity in admission. *Frye's* historical reputation and its influence have not been vanquished—for *Frye* is still the test in nearly two dozen states, such as New York and Pennsylvania. In time, the only way to regulate the proliferation of junk science is to return to admission criteria that assure some level of quality control and evidentiary rigor. For some, in the *Frye* camp, there is this optimism that we shall return to first principles on these questions to assure greater integrity.[24]

> The proliferation of junk science impacts not only courthouses but also the society at large. And during the COVID-19 challenge, junk science seems to grow without much restraint. Visit the top 20 junk science claims during the COVID-19 pandemic at: https://www.realclearscience.com/blog/2021/12/11/the_biggest_junk_science_of_2021_covid-19_edition_806977.html

Other legal tests since *Frye* have opened the floodgates of expertise and experts espousing the aligned theories. While *Frye* provides a quality control barrier, other legal tests display a tolerance and flexibility that enables "junk science." In *Daubert* and *Kumho*, judges become the major screeners of the scientific proposition and this has led to not only incorrect conclusions but also a sort of activism shielded by nebulous arguments based on discretion, gatekeeping and helpfulness. *Daubert*, in particular, "assumes much relative to judicial capacity by asking judges to make subtle and sophisticated determinations about scientific methodology and its relation to the conclusions offered by an expert witness ... ."[25]

Such activism has enormous support primarily because proponents of this liberality find restrictiveness worse than laxity. Argued from another angle, the anti-*Frye* crowd appears in the school of judicial activism by being comfortable with judges making a threshold determination of whether an expert's testimony will be more likely to aid or mislead the jury rather than be tied to a stricter test of scientific reliability.[26] In this environment, junk is a more likely candidate for admission than under stricter historic rules.

A "junk science" witness is not hard to detect because the expert lacks the traditional prerequisites, like:

1. He is qualified in his field, but the subject matter of his opinions is not quite in his field.
2. He has published, but not on the subject matter about which he is advancing opinions.
3. He has not offered his opinion in this case for publication in any peer-reviewed journal.
4. He is a member of numerous professional societies, but does not serve in any official capacity in any of them.
5. He can point to no literature in his field that supports or confirms his opinions.
6. His opinion is not shared by recognized authorities in his field.
7. His opinion relies upon unestablished facts, untenable inferences and/or speculation.
8. His opinion is contrary to the facts, logic and/or established science.
9. His opinion cannot be confirmed by collateral analysis.
10. He sticks to his opinion even after his fallacies and inaccuracies have been exposed.[27]

On the other side of the ocean are proponents of judicial restraint, holding that the role of judge should not be so over-weighted that the scientific world is muted and irrelevant. In addition, when experts, that lack the *Frye* pre-requisites, are elevated as more reasoned and insightful than the larger scientific community, the power of the expert is disproportionate to their role and purpose.[28] Juries become less capable of their own deliberation because the expert's pedestal placement impedes an honest critique. To some, the weight and importance granted to expert testimony are grotesquely disproportionate.[29] And isn't this over-dependence on the expertise of others a social and cultural concern as well?[30] One does not simply "buy shoes" anymore. Instead, the customer seeks a "shoe consultant." "Experts" offer classes for every imaginable form of stress and give advice on things like parenting, heightened self-esteem and enhanced assertiveness. Everyone is an expert, from real estate sales agents to the pest inspector. This form of obsessive reliance upon expertise has irrefutably weakened the role of a jury and a judge in the processes of litigation. If the expert says something, it is perceived *ex cathedra*. The judicial role and jury deliberative process are vanquished by this same overreliance. "The expanding array of scientific (as well as some not so scientific) specialties available for testimony raises hard questions. Will courts require that the witness' opinions be reasonably based upon trustworthy data? How far must judges inquire into the practice of other experts in the same field prior to allowing a trial witness to proffer an expert opinion? How much should the expert's supporting data be received in evidence?"[31]

All of us encounter junk science regularly. Today's carcinogen is tomorrow's salvation as in the case of chocolate. Esteemed scientists have long disagreed about many political issues that allegedly are scientifically based. Climate change is a most prominent example. Aside from the endless cacophony that "the science is the science," there are those that disagree in the scientific community. Visit: https://skepticalscience.com/argument.php

While some courts admit "junk" science because they believe that the trier of fact cannot be harmed by even the most vacuous of fields, others are only comfortable in admitting hard, concrete and well-established sciences like physics, chemistry, pathology and its academic counterparts. On a more liberal front, some judges hold that the legal sufficiency of evidence does not necessarily translate or equate into scientific certainty, and thus, if a field is not fully developed, its admission should be permitted regardless. Then again, nonscientific expertise, e.g., police use of excessive force, real estate valuation, etc., is more attuned to methodology than field. Although courts are leery of expert testimony when the basis and scientific foundation for the testimony have not been rigorously scrutinized, this shortcoming does not preclude admission. Indeed, there appears increasing leniency in the courts toward the admission of once unheard-of theories and propositions. Select fields, like psychology, push those boundaries out further each and every day in the forensic world. So edgy are these conclusions that some in the scientific community are raising alarms. Some have called for a period of "soul-searching" and a "broad recognition" that: (1) the reliability of published research depends crucially on methodological rigor in ways that had not been sufficiently appreciated, and (2) lack of methodological rigor can contaminate research results in a manner that is entirely undetectable to a peer examining those results.[32]

Louisiana State University monitors junk science at http://biotech.law.lsu.edu/IEEE/ieee29.htm.

To be sure, most science commences its intellectual life with a minority of scientific support, and only by subsequent research and experience is the general scientific community eventually persuaded. The legal system and law enforcement community should avoid the use of "junk" science, the stuff popularized by tabloid television and magazines. Beware of fashionable consulting groups whose services win cases at any cost. Of recent interest has been the blurred acceptance of experts who testify to the injuries inflicted on sexual abuse victims.

Sensationalized trials, such as the *McMartin Pre-school* case in California and the *Kelley* trial in North Carolina, or the *Duke Lacrosse Team Rape allegations*,[33] have highlighted the inordinate influence of so-called sexual abuse experts.[34] Emotionally reaching out to child victims of sexual abuse is a commendable response, but such empathy must be tempered by evidentiary discipline.[35]

Our natural outrage and corresponding sympathy to those alleging child sexual abuse, rape or other sexual assault needs to be subjected to the traditional scrutiny that any sort of evidence experiences. Things are not always black and white. For example, a few courts have affirmed the use of expert testimony regarding child sexual abuse syndrome,[36] while others resist. How precisely is this syndrome defined? Sexual abuse experts deduce abuse because the child suffers from sleep disruptions, nightmares, sexual dysfunction, abuse of other children, "acting out" or the exhibition of pseudo-mature seductive behavior, anxiety and guilt, runs away, experiences isolation from peers, suffers bed wetting or faces other habitual reactions. These characteristics and traits are presented as being a typical personality profile of a sexual abuse victim. Are these traits dependable predictors of a syndrome? Nightmares, anxiety and bed wetting are not conclusive of sexual abuse but only reliable indicators of something emotional in scope.[37]

At the same time, litigators should think creatively about experts and recognize that today's "junk" due to its innovation and novelty may evolve into something dependable. Uncannily, some junk evolves into substance and worth. "An open mind is also helpful in keeping up with the rapidly evolving 'technology of expert testimony.'"[38]

The influence of *Frye v. U.S.* cannot be overstated since its pronouncement nearly 100 years ago. Even the more modern scientific tests for admissibility cannot avoid the legacy of *Frye*. *Frye* set a standard for both subject matter and methodology that had to be met before the evidence was worthy of admission. Even subsequent rulings that overturned or replaced *Frye* in the federal courts and selected states can never fully escape this sort of analysis.

In Department of Human Services, the Frye ruling dominates the discussion since Pennsylvania is still a Frye jurisdiction. Here we encounter the Pennsylvania Supreme Court's insistence on the Frye standard when it comes to subject matter and the aligned scientific methodology.

Interest of M.R.
2021 PA Super 30
Superior Court of Pennsylvania.
Filed: March 1, 2021.

The Philadelphia Department of Human Services ("DHS") appeals from the June 26, 2020 orders adjudicating dependent M.R., a child, and his twin sister, J.R., a child (collectively, "Children"), but not finding child abuse as to B.R. ("Mother") and R.R. ("Father") (collectively, "Parents"). Children's guardians ad litem ("GAL") also appeal from those same orders. At issue in these appeals is whether the trial court abused its discretion by admitting the testimony of Parents' expert witness, Marvin Miller, M.D., who opined that metabolic bone disease of infancy ("MBDI") caused Children's multiple fractures, instead of abuse. Alternatively, DHS and GAL contend that, even if Dr. Miller's testimony was admissible, the trial court abused its discretion in declining to find child abuse, given the overwhelming medical evidence presented at the adjudicatory hearing that Children's injuries were caused by abuse. After careful review, we conclude that the trial court improperly admitted Dr. Miller's testimony and, therefore, we reverse its orders refusing to make a finding of child abuse against Parents in each child's case.

...

Analysis

Both DHS and GAL attack the trial court's above-stated rationale. They argue that Dr. Miller has not applied a generally accepted scientific methodology to reach his conclusion that MBDI caused Children's fractures. We agree.

To begin, the trial court emphasized in its analysis that the Journal of Pediatric Endocrinology and Metabolism published Dr. Miller's study and that other, similar studies have been published. While we recognize that scientific publications and studies are ways in which an expert's methodology can be shown to be generally accepted in the relevant field, see Hopkins , supra , we concur with DHS that "the mere fact of publication is not enough to establish general acceptance, especially where the medical establishment's reaction to those publications has been opprobrium and concern over the misuse of TBBD/MBDI in the courtroom." DHS's Brief at 22. Both the Society of Pediatric Radiology and the Diagnostic Imaging of Child Abuse textbook have pointed out flaws in Dr. Miller's scientific methodology, such as, inter alia , the bias involved and

Dr. Miller's refusal to acknowledge more reasonable explanations for a child's injuries, like abuse. As DHS observes:

> The value of the peer[-]review process is not simply in an article's being green-lit for publication by a journal's editorial board or reviewers. Rather, the purpose of publication is to stimulate scientific discourse. Publication allows other scientists to test the theory by examining and critiquing a study's methodology, or by attempting to replicate or falsify its results. It is this broader dialogue of publication and response that generates a scientific consensus about the theory's reliability. See Daubert v. Merrell Dow Pharm., Inc., 509 U.S. 579, 593, 113 S.Ct. 2786, 125 L.Ed.2d 469 (1993) ("Publication (which is but one element of peer review) is not a sine qua non of admissibility; it does not necessarily correlate with reliability…[.] But submission to the scrutiny of the scientific community is a component of 'good science,' in part because it increases the likelihood that substantive flaws in methodology will be detected.").

DHS's Brief at 30-31. In addition, the fact that other doctors have published a similar study does not automatically make Dr. Miller's methodology generally accepted in the scientific community. That study may contain the same flaws as Dr. Miller's research. DHS astutely

> Parents point to just one source used to assess general acceptance: whether any peer-reviewed, published research exists on the topic. But they ignore all of the other sources, including: how the field has reacted to peer-reviewed research supporting the expert's theory; the views of professional organizations in the field; whether the technique is taught in textbooks; the testimony of other experts; and how other courts have treated the methodology.

DHS's Reply Brief at 1-2 (citation omitted). Thus, we disagree with the trial court that publication proves that doctors and scientists in the medical field generally accept Dr. Miller's scientific methodology as a means for arriving at the conclusion that MBDI caused Children's injuries, especially when considering the plethora of other sources discrediting his methodology.

Furthermore, we reject the trial court's determination that Dr. Miller used the same scientific methodology as the CHOP and Nemours doctors, in that he examined medical records, diagnostic imaging, and the medical histories of both Children and Mother to reach his conclusion of MBDI.

…

Indeed, Dr. Miller emphasized that "[t]he diagnosis of [MBDI] is strongly inferred by reading of the x-rays[,]" and that "x-rays are often the telling study that will allow us to jump on th[e] diagnosis" of MBDI. N.T., 6/26/20, at 38, 49. However, as DHS argues, Dr. Miller's method of interpreting the x-rays was itself not generally accepted by the medical community, given the testimony of Dr. Servaes regarding the use of PowerPoint to view x-rays, the findings of other doctors at CHOP and Nemours that Children's bones were normal, and the Society for Pediatric Radiology's critique of Dr. Miller's published article that he mischaracterized normal bones as rickets. Further, Dr. Servaes testified that, in making a proper diagnosis, it is "important to think about risk factors and take that into consideration, but the evidence is what the x-rays look like and what the laboratory values are and the pattern. You have to take the entire picture, not just single aspects and decide, because this one thing is present, the conclusion is drawn. You have to look at the entire picture." Id. at 94. Here, unlike the other doctors who evaluated Children, Dr. Miller did not consider the 'entire picture,' in particular the results of Children's lab tests, in making his diagnosis of MBDI. Instead, Dr. Miller essentially ignored these tests and drew conclusions based on a purported risk factor being present. See N.T., 6/26/20, at 48 (Dr. Miller's testifying that, as long as one risk factor is present in an infant, that infant could have MBDI). Thus, the trial court abused its discretion in ascertaining that Dr. Miller applied a generally accepted scientific methodology simply because he claimed to have used medical records, diagnostic imaging, and the medical histories of both Children and Mother. Accordingly, as Parents did not meet their burden of showing that the medical community generally accepts the scientific methodology used by Dr. Miller in reaching his conclusion that MBDI caused Children's fractures, we hold that the trial court should not have admitted Dr. Miller's testimony.

The exclusion of Dr. Miller's testimony affects the trial court's finding on child abuse. In not finding child abuse as to Parents, the trial court provided:

> The petitioning party in a dependency action must demonstrate the existence of child abuse by clear and convincing evidence. 23 Pa.C.S.[] § 6381(d); [s]ee In re L.Z. [631 Pa. 343], 111 A.3d [1164,] 1179 [(2015)]. However, the identity of the abuser need only be established by prima facie evidence. (Id.). The Child Protective Services Law provides for the following evidentiary presumption:
>
> Evidence that a child has suffered child abuse of such a nature as would ordinarily not be sustained or exist except by reason of the acts or omissions of the parent or other person responsible for the welfare of the child shall be prima facie evidence of child abuse by the parent or other person responsible for the welfare of the child.
>
> 23 Pa.C.S.[] § 6381(d). In the application of Section 6381(d), evidence that a child has suffered injury that would not ordinarily be sustained but for the acts or omissions of the parent or responsible person is sufficient to establish that the parent or responsible person perpetrated that abuse unless the parent or responsible person rebuts the presumption. See In re L.Z., 111 A.3d at 1185. Here, DHS failed to establish a prima facie case of child abuse because they failed to demonstrate that … Children were victims of child abuse. Although the medical experts offered by DHS concluded that … Children's injuries resulted from child abuse, Dr. Miller's theory contradicted this testimony. (N.T.[,] 2/7/20[,] at 84, 86-[8]7; N.T.[,] 6/26/20[,] at 21). Dr. Miller's testimony presented an alternative medical explanation for the injuries … Children suffered other than child abuse. His testimony that … Children suffered from MBDI demonstrates an alternative and accidental cause for … Children's injuries. Because, according to Dr. Miller's expert testimony, MBDI could have plausibly caused the multiple rib fractures, DHS failed to establish clear and convincing evidence that … Children were victims of child abuse. (N.T.[,] 6/26/20[,] at 26). Therefore, absent clear and convincing evidence that child abuse occurred, there is no prima facie evidence Parents were the perpetrators of child abuse. For the foregoing reasons, this [c]ourt properly declined to make a child abuse finding as to Parents.

TCO at 12-13.

As the trial court abused its discretion in admitting Dr. Miller's testimony pursuant to Rule 702 and Frye, its explanation for not finding child abuse against Parents falters. Without Dr. Miller's testimony, it is evident that DHS has demonstrated child abuse by clear and convincing evidence; all of the remaining evidence in the case overwhelmingly points to abuse as the cause of Children's injuries.

"The term 'child abuse' shall mean intentionally, knowingly or recklessly doing any of the following: (1) Causing bodily injury to a child through any recent act or failure to act." 23 Pa.C.S. § 6303(b.1)(1). 'Bodily injury' is defined as "[i]mpairment of physical condition or substantial pain." 23 Pa.C.S. § 6303(a). "The requisite standard of proof for a finding of child abuse pursuant to Section 6303(b.1) is clear and convincing evidence." Interest of A.C., 237 A.3d 553, 558 (Pa. Super. 2020) (citation omitted). "Clear and convincing evidence is 'evidence that is so clear, direct, weighty, and convincing as to enable the trier of fact to come to a clear conviction, without hesitancy, of the truth of the precise facts in issue.'" Id. (citation omitted). Moreover, "[t]he standard of review in dependency cases requires an appellate court to accept the findings of fact and credibility determinations of the trial court if they are supported by the record, but does not require the appellate court to accept the lower court's inferences or conclusions of law. Accordingly, we review for an abuse of discretion." Id. at 557 (citations omitted).

As GAL aptly states:

> The trial court's [o]pinion confirms the existence of clear and convincing evidence… that Parents intentionally, knowingly[,] or recklessly caused bodily injury to … [C]hildren through their actions or failure to act; [the trial court] wrote:

- Dr. Henry testified that (1) … [C]hildren had no underlying genetic conditions or underlying bone conditions that could have caused the injuries, (2) that … [C]hildren showed paradoxical fussiness that can be indicative of pain from rib fractures, (3) that Mother denied any accidental trauma, (4) that there was no family history of bone disease, (5) the amount of force required for rib fractures is rarely seen in accidental injuries, and (6) … [C]hildren's fractures were most consistent with abuse. [TCO at 3-4.]
- Dr. Skraban concluded that [(1)] the [x]-rays were normal with no sign of an underlying genetic condition or bone demineralization, (2) Dr. Bober confirmed CHOP's findings, and (3) … [C]hildren did not have osteogenesis imperfecta. [Id. at 4].

The trial court further stated that Mother testified she was … the primary caretaker of … [C]hildren, but family members frequently visited. [Id. at 5; see also N.T., 2/7/20, at 221-24.] As such, there is prima facie evidence that Parents were responsible for … [C]hildren.

GAL's Brief at 45-46 (internal citation omitted). We agree that DHS presented clear and convincing evidence to establish child abuse by Parents. Thus, we reverse the trial court's orders, and direct it to make a finding of child abuse as to Parents for each child.

Orders reversed. Case remanded. Jurisdiction relinquished.

Questions:

1. What type of scientific inquiry is central to this case?
2. While the DHS charged child abuse due to fractures and other traumatic injury, in two children, how does the defense counter these claims?
3. What is the ruling of the Court and Why?

Visit the Mid-Atlantic Association of Forensic Scientists at https://www.maafs.org/.

Another argument often posed in the early stages of trial or even on appeal is the claim that a trial court "abused its discretion" in allowing certain scientific evidence into the litigation. Under the trilogy of *Frye*, *Daubert* and *Kumho*, the courts have a role to play as to screener, evaluator and even the oft-heard term "gatekeeper." Courts can and do make judgments each day on forensic evidence and the appellate courts give enormous deference to the trial courts faced with these decisions.

In Barton, the appeal rests on whether or not the trial court abused its inherent discretion by allowing testimony as to DNA (Figure 1.3).

United States v. Barton
909 F.3d 1323
June 12, 2018

Robert Barton was charged by a grand jury sitting in the Middle District of Florida, tried by a petit jury, and convicted in a single count of being a felon in possession of a firearm in violation of 18 U.S.C. § 922(g). The evidence adduced at trial was overwhelming, including Barton's two confessions, first at the scene of his arrest and later to a jailhouse informant, corroborating testimony from the passenger in Barton's vehicle at the time of the arrest, and deoxyribonucleic acid (DNA) evidence directly linking Barton to the firearm. The single issue raised on appeal is whether the district court abused its considerable discretion under Daubert v. Merrell Dow Pharmaceuticals, Inc.,

FIGURE 1.3 A biologist performs DNA testing at the Terrorist Explosive Device Analytical Center (TEDAC) in Huntsville, Alabama. TEDAC was formally established in 2003 to serve as the single inter-agency organization to receive, fully analyze and exploit all terrorist improvised explosive devices, or IEDs, of interest to the United States. On February 16, 2016, TEDAC celebrated the completion of its new laboratory facilities at Redstone Arsenal in Huntsville. FBI Media at https://multimedia.fbi.gov/item?type=image&id=4360.

509 U.S. 579, 113 S.Ct. 2786, 125 L.Ed.2d 469 (1993), in admitting expert testimony concerning the DNA evidence. We can discern no error in the trial court's exercise of its critical gatekeeping function and, accordingly, affirm the conviction.

I.

…

B.

Before trial, Barton moved in limine to exclude the DNA evidence and Zuleger's testimony. The district court referred the motion to a magistrate judge, who held a Daubert hearing at which Zuleger and Johnson both testified.

…

At the Daubert hearing, Zuleger explained Trinity's methodology for testing and analyzing DNA.

…

Zuleger explained that the DNA in Barton's case was tested using the Polymerase Chain Reaction /Short Tandem Repeat (PCR/STR) method—a method used by every accredited laboratory in the country. Importantly, Barton does not challenge the reliability of PCR/STR generally. … Zuleger testified, however, that even when quantitation estimates very little or no DNA at all in the sample, the lab continues with later stages of analysis because the manufacturer's validation studies revealed that it is nevertheless possible to obtain usable data. Zuleger indicated that the optimal amount of DNA is .5 nanograms, or 500 picograms, and that in this case, quantitation estimated 210 picograms in the sample obtained from the firearm.

Using PCR in the amplification step, the analyst then makes millions of copies of the DNA by adding it to a "DNA kit," a combination of chemicals that produces the reaction. Before laboratory use, a DNA kit is subjected to validation testing, which establishes interpretation guidelines for profile analysis. The interpretation guidelines from validation testing on the DNA kit used in this case, the PowerPlex Fusion, appear in Trinity's procedures manual, and include the parameters in which an analyst may conclude that there is a "major donor" in a sample that contains more than one person's DNA. Finally, Zuleger explained, the sample is analyzed by a computer, which produces an electropherogram, a visualization of the genetic data from which the analyst conducts profile analysis. The alleles at each loci appear as "peaks" on the electropherogram. Profile analysis is relatively straightforward where the DNA in the sample comes from one individual. One or two peaks will appear at each loci and a DNA profile will be readily determined. The process becomes more complicated where the DNA in the sample is a mixture from two or more individuals. In that case, more than two peaks may appear at any given locus and an individual profile may be determined only if the analyst is able to distinguish between donors, including by isolating a so-called "major donor." Major donors are determined by peak-height differences, and Trinity's procedures manual dictates the peak-height ratios and peak-height intensity considerations necessary to identify a major donor.

…

Zuleger used a modified random match probability statistical analysis, and concluded that the frequency of occurrence of the major DNA profile derived from the firearm sample was 1 in 41 million in the general population. The obvious implication of Zuleger's testimony was that it is extraordinarily unlikely that the DNA found on the firearm came from someone other than Barton.

On the second day of the Daubert hearing, defense DNA expert Dr. Elizabeth Johnson testified that there were three major problems with Zuleger's methodology and analysis: (1) that the "major donor" call was unreliable because Trinity's standards regarding peak-height ratios were inadequate; (2) that no validation studies pertaining to three-or-more person mixtures had been performed on the PowerPlex Fusion DNA kit; and (3) that the quantity of DNA material (210 picograms) was so low as to be classified as "low copy," and thus inherently unreliable. In support of her contentions, Johnson cited the procedures manuals from the Harris County crime lab and the California Department of Justice, and claimed that they are representative of what is generally accepted in the scientific community.

In his 16-page Report and Recommendation—with the benefit of over six hours of expert testimony and argument of counsel—the magistrate judge recommended that Barton's motion be denied. For starters, he concluded that Barton's argument regarding the validation studies lacked merit because the studies were done under well-accepted standards and Zuleger testified that the same principles outlined in the validation studies for mixtures applied whether the mixture contained DNA from two, three, or more donors. Meanwhile, Barton failed to present anything demonstrating the result would have differed if Trinity's validation studies were conducted with complex mixtures from three or more donors. Next, the court found that while 210 picograms is below the optimal amount of DNA for testing and may increase the risk of stochastic effects, those effects may be "appropriately considered by trained forensic biologists under acceptable standards."

…

On appeal, Barton argues only that the district court erred in admitting Zuleger's expert testimony and asks this Court to overturn his conviction and remand the case for a new trial.

II.

"We review for abuse of discretion the district court's decisions regarding the admissibility of expert testimony and the reliability of an expert opinion." United States v. Frazier, 387 F.3d 1244, 1258 (11th Cir. 2004) (en banc); see also Kumho Tire Co., Ltd. v. Carmichael, 526 U.S. 137, 152, 119 S.Ct. 1167, 143 L.Ed.2d 238 (1999) ("[A] court of appeals is to apply an abuse-of-discretion standard when it reviews a trial court's decision to admit or exclude expert testimony.") (quotations and alterations omitted).

This abuse-of-discretion standard "recognizes the range of possible conclusions the trial judge may reach," Frazier, 387 F.3d at 1259, and thus affords the district court "considerable leeway" in evidentiary rulings, id. at 1258. Accordingly, "the 'deference that is the hallmark of abuse-of-discretion review,' requires that we not reverse an evidentiary decision of a district court 'unless the ruling is manifestly erroneous.' " Id. at 1259 (first quoting Gen. Elec. Co. v. Joiner, 522 U.S. 136, 143, 118 S.Ct. 512, 139 L.Ed.2d 508 (1997) ; then quoting id. at 142, 118 S.Ct. 512). That is, this Court "must affirm unless we find that the district court has made a clear error of judgment, or has applied the wrong legal standard." Id.

Further, even an abuse of discretion will not warrant reversal where the resulting error was harmless. We "will not overturn an evidentiary ruling and order a new trial unless the objecting party has shown a substantial prejudicial effect from the ruling." Maiz v. Virani, 253 F.3d 641, 667 (11th Cir. 2001). Substantial prejudice goes to the outcome of the trial; "where an error had no substantial influence on the outcome, and sufficient evidence uninfected by error supports the verdict, reversal is not warranted." United States v. Drury, 396 F.3d 1303, 1315 (11th Cir. 2005) (quoting United States v. Hawkins, 905 F.2d 1489, 1493 (11th Cir. 1990)).

III.

A.

...

"As the Supreme Court made abundantly clear in Daubert, Rule 702 compels the district courts to perform the critical 'gatekeeping' function concerning the admissibility of expert scientific evidence." Frazier, 387 F.3d at 1260 (quoting Daubert v. Merrell Dow Pharm., Inc., 509 U.S. 579, 589 n.7, 597, 113 S.Ct. 2786, 125 L.Ed.2d 469 (1993)) (emphasis added in original). Because expert testimony can be both highly persuasive and difficult for a lay jury to evaluate, the importance of this gatekeeping function "cannot be overstated." Id. To faithfully discharge its gatekeeping duty, the trial court must engage in a rigorous three-part analysis and consider whether:

> (1) the expert is qualified to testify competently regarding the matters he intends to address; (2) the methodology by which the expert reaches his conclusions is sufficiently reliable as determined by the sort of inquiry mandated in Daubert; and (3) the testimony assists the trier of fact, through the application of scientific, technical, or specialized expertise, to understand the evidence or to determine a fact in issue.

Id. (quoting City of Tuscaloosa v. Harcros Chems., Inc., 158 F.3d 548, 562 (11th Cir. 1998)).

Only the second prong—reliability—is at issue in this case. "When evaluating the reliability of scientific expert opinion, the trial judge must assess 'whether the reasoning or methodology underlying the testimony is scientifically valid and ... whether that reasoning or methodology properly can be applied to the facts in issue.'" Id. at 1261–62 (footnote omitted) (quoting Daubert, 509 U.S. at

592–93, 113 S.Ct. 2786). Daubert sets forth a number of illustrative, though not exhaustive, factors that should be considered in determining reliability, including:

> Indeed, the first and third prongs are beyond dispute. Zuleger has extensive scientific training and decades of work in DNA testing laboratories, and Barton stipulated at the Daubert hearing that she is a qualified DNA expert. As for the third prong, the presence or absence of the Defendant's DNA on the firearm goes to the very heart of the case—whether the Defendant, a convicted felon, was in fact in possession of a firearm—and thus indisputably assists the jury "to determine a fact at issue." Frazier, 387 F.3d at 1260.
>
> --------
>
> (1) whether the expert's theory can be and has been tested; (2) whether the theory has been subjected to peer review and publication; (3) the known or potential rate of error of the particular scientific technique; and (4) whether the technique is generally accepted in the scientific community.

Quiet Tech. DC-8 v. Hurel-Dubois UK Ltd., 326 F.3d 1333, 1341 (11th Cir. 2003); see also Frazier, 387 F.3d at 1262 ("These factors are illustrative, not exhaustive; not all of them will apply in every case, and in some cases other factors will be equally important in evaluating the reliability of proffered expert opinion.").

Notwithstanding its critical gatekeeping function, the trial court is just that—a gatekeeper—and Rule 702 is a screening procedure, not an opportunity to substitute the trial court's judgment for that of a jury. In that regard, "it is not the role of the district court to make ultimate conclusions as to the persuasiveness of the proffered evidence," Quiet Tech., 326 F.3d at 1341, and "[v]igorous cross-examination, presentation of contrary evidence, and careful instruction on the burden of proof are the traditional and appropriate means of attacking shaky but admissible evidence," id. (quoting Daubert, 509 U.S. at 596, 113 S.Ct. 2786).

B.

Here, the trial court could fairly determine that Zuleger's opinions were based on reliable methods and sufficiently reliable application of those methods to satisfy Daubert and warrant admission. The reliability of the basic methodology—PCR/STR testing—is not in dispute. Barton agrees that this type of testing is reliable and generally accepted within the scientific community. He instead argues that the use of this testing with the low quantity of a complex DNA mixture present in this case, and in the absence of appropriate validation testing and interpretive thresholds for complex mixtures, rendered the data unreliable and Zuleger's opinion drawn from the data inappropriate. Thus, he argues that the trial court abused its discretion in admitting the evidence.

We remain unpersuaded. The magistrate judge had before him the competing scientific testimony of two qualified experts. One relied on Trinity's procedures manual, while the other relied on the procedures of two other laboratories. All of these procedures, though different, were reviewed and approved by a highly regarded accrediting body. Looking at the explanations, methodology, and procedures used by Zuleger, the magistrate judge concluded that the Daubert standard was met. And far from "rubber stamp[ing]" the expert testimony, as Barton claims, the magistrate judge carefully considered each of Barton's arguments against the reliability of Zuleger's proposed testimony and found the opinions reliable based on the methods she employed and in reference to Trinity's audited procedures. While it is true that "[t]he trial court's gatekeeping function requires more than simply 'taking the expert's word for it,'" Fed. R. Evid. 702 advisory committee's note (2000 amends.), Zuleger explained in detail how she formed her expert opinion based on reliable and generally accepted methods.

Moreover, the magistrate judge credited Zuleger's testimony only in light of other indicia of reliability, including the auditing and accreditation process. Barton argues for the first time in this

Court that accreditation and auditing history are not useful indicia of a lab's adherence to reliable methodology. In support of this argument, Barton cites a number of news articles regarding laboratories that were either closed or reprimanded for serious lapses despite having been accredited. This information was not presented to the trial court. But even if it had been, the failure of accreditation and auditing to uncover problems in a few different labs does not make it an abuse of discretion for the court to have found accreditation probative of reliability.

Ultimately, the trial court concluded that the arguments raised by Barton did not undermine admissibility under Daubert, but rather went to the weight of the evidence and could be presented to the jury at trial. In so holding, the trial judge relied heavily on our opinion in Quiet Technology, which is analogous to this case. In that case, the appellant argued not about the reliability of the particular kind of study or methodology, but rather that the expert "misapplied the generally valid principles underlying" the type of study he performed. Quiet Tech., 326 F.3d at 1344. Specifically, appellant argued that the expert had failed to use the proper equations in calculating certain data that was ultimately put into analysis software. Similarly here, Barton argues that Zuleger failed to perform certain calculations that would have altered the data ultimately used to determine that the frequency of occurrence of the major DNA profile derived from the firearm sample was 1 in 41 million in the general population. The Quiet Technology Court found that the "identification of such flaws in generally reliable scientific evidence is precisely the role of cross-examination." Id. at 1345; see also Rosenfeld v. Oceania Cruises, Inc., 654 F.3d 1190, 1193 (11th Cir. 2011) ("Indeed, 'in most cases, objections to the inadequacies of a study are more appropriately considered an objection going to the weight of the evidence rather than its admissibility.'") (quoting Hemmings v. Tidyman's Inc., 285 F.3d 1174, 1188 (9th Cir. 2002)). The Court concluded that "[b]ecause [the expert's] methods and results were discernible and rooted in real science—i.e., were 'intellectual[ly] rigor[ous]'—they were empirically testable. As such, they were subject to effective cross examination and, indeed, were questioned vigorously" by the appellant at trial. Quiet Tech., 326 F.3d at 1346 (quoting Kumho Tire, 526 U.S. at 152, 119 S.Ct. 1167). The same is true here: Zuleger testified at trial, Barton vigorously cross-examined Zuleger and offered the rebutting testimony of Johnson, another expert, and the jury was left to evaluate the evidence. Moreover, the district judge appropriately instructed the jury that they were to treat expert scientific testimony like "any other witness's testimony" and "decide for [themselves] whether to rely upon the opinion."

The trial court rightly reached its decision based on an evaluation of the foundations of Zuleger's testimony and the failure of the defense to rebut it with anything but the testimony of a competing expert, who employed the same general methodology. Indeed, it would have gone beyond the gatekeeping function of the trial court to exclude Zuleger's testimony on the basis of a credibility determination favoring Johnson's. See Ambrosini v. Labarraque, 101 F.3d 129, 141 (D.C. Cir. 1996) ("The district court[ ] err[ed] … [by] misconce[iving] of the limited 'gatekeeper' role envisioned in Daubert. By attempting to evaluate the credibility of opposing experts and the persuasiveness of competing scientific studies, the district court conflated the questions of the admissibility of expert testimony and the weight appropriately to be accorded such testimony by a fact finder.") (cited approvingly in Quiet Tech., 326 F.3d at 1341). The issues raised by Johnson's competing testimony went to the weight owed Zuleger's expert opinion, and were properly left to the jury. Under these circumstances, the trial court did not abuse its discretion in allowing the presentation of Zuleger's expert testimony to the jury.

…

In addition to the strength of the other evidence against Barton, the trial judge properly instructed the jury on weighing expert testimony. As we have repeatedly found, proper instruction to the jury on weighing expert testimony may render errors in admitting the evidence harmless. See United States v. Myers, 972 F.2d 1566, 1578 (11th Cir. 1992) ("This error, however, was harmless in light of the court's instructions to the jury that it was their 'duty to decide … the specific facts,' and whether to 'accept … [and] rely upon an expert witness.'"); United States v. Herring, 955 F.2d 703, 708–09 (11th Cir. 1992) (finding any possible error harmless where the district court "carefully instructed the jury" and "admonished [it] to give no special deference to the expert testimony"). Here, using this Circuit's pattern jury instruction, the district judge told the jurors to treat expert scientific testimony like "any other witness's testimony" and "decide for [themselves] whether to rely upon the opinion." After hearing the extensive testimony from dueling DNA experts, the jury was properly directed to evaluate the competing testimony of Zuleger and Johnson and decide for itself what weight, if any, to accord the DNA evidence.

In short, even if we were to conclude that the district court had abused its discretion in allowing the expert testimony of Ms. Zuleger, any claimed error would have been harmless. The evidence was overwhelming and we can find no reasonable likelihood that the claimed error affected the Defendant's substantial rights.

Accordingly, we affirm.

Questions:

1. Describe how the Court characterizes the abuse of discretion standard. Is it to be narrowly or broadly interpreted? Is there a natural deference paid to the trial court? Is it more or less likely that that a higher court would uphold an abuse of discretion case?
2. What is the Daubert gatekeeping function so crucial to a question of abuse of discretion?
3. What is the subject matter and DNA issue that gives rise to the appeal?

NOTES

1 See Miiko Kuma, "Admissibility of Expert Evidence: Proving the Basis for an Expert's Opinion," *Sydney Law Review* 33, no. 3 (September 2011): 427, http://classic.austlii.edu.au/au/journals/SydLawRw/2011/19.pdf.
2 293 F. 1013, (D.C. Cir. 1923).
3 Daubert v. Merrell Dow Pharmaceuticals, 509 U.S. 579 (1993).
4 293 F. 1013, (D.C. Cir. 1923).
5 509 US 579 (1993).
6 Frye v. U.S., 293 F. 1013, 1014 (D.C. Cir. 1923).
7 Michael Graham, *Evidence, Text, Rules, Illustrations and Problems* (Louisville, CO: National Institute for Trial Advocacy, 1983), 305–306.
8 Kumho Tire Co. v. Carmichael 526 U.S. 137 (1999).
9 Peter B. Oh, "Assessing Admissibility of Nonscientific Expert Evidence under Federal Evidence Rule 702," *Defense Counsel Journal* (October 1997): 556–567; Joelle Anne Moreno, "Beyond the Polemic Against Junk Science: Navigating the Oceans That Divide Science and Law with Justice Breyer at The Helm," *Boston University Law Review* 81, no. 5 (2001): 1033, https://papers.ssrn.com/sol3/papers.cfm?abstract_id=1440894.
10 509 US 579 (1993).
11 Ibid.
12 Edward K. Cheng & Albert H. Yoon, "Does Frye or Daubert Matter? A Study of Scientific Admissibility Standards," *Virginia Law Review* 91, no. 2 (2005): 471, https://www.virginialawreview.org/articles/does-frye-or-daubert-matter-study-scientific-admissibility-standards/.
13 Ladd states, "There is no more certain test that work for determining when experts may be used than the common sense inquiry where the untrained lay witness would be unqualified to determine intelligently, and to the best possible degree, the particular issue without enlightenment from those having a specialized understanding of the subject involved in the dispute." Mason Ladd, "Expert Testimony," *Vanderbilt Law Review* 5 (1952): 414, 418.
14 See "Expert or Law Opinion-Where is the Line?" Advocacy and Evidence Blog, Temple University Beasley School of Law, https://law.temple.edu/aer/2019/06/19/expert-or-lay-opinion-where-is-the-line.

15 Ellen Moskkowitz, "Junk Science," *Hastings Center Reporter*, 48 (1996); See also Lewis H. LaRue & David S. Caudill, "A Non-Romantic View of Expert Testimony," *Seton Hall Law Review* 35, no. 1 (2004): 1; Gary Edmond, "Supersizing Daubert Science for Litigation and Its Implications for Legal Practice and Scientific Research," *Villanova Law Review* 52, no. 4 (2007): 857.
16 Michael H. Gottesman, "Should State Courts Impose a 'Reliability' Threshold?" *Trial* (September 1997): 20–23.
17 Edward R. Cavanagh, "Decision Extends Daubert Approach to All Expert Testimony," *New York State Bar Journal* (July/Aug. 1999): 9; See also Edward J. Imwinkelried, "The Next Step After Daubert: Developing A Similarly Epistemological Approach to Ensuring the Reliability of Nonscientific Expert Testimony," *Cardozo Law Review* 15, no. 6 (1994): 2271.
18 Paul Reidinger, "They Blinded Me with Science," *ABA Journal* (September 1996): 58–62.
19 In fact, the Federal Judicial Center has formally recognized the need for judicial involvement by increasing science training for district court judges and by publishing a handbook on scientific evidence, which includes detailed "reference guides" on scientific and technical specialties frequently encountered in the courtroom. Paul S. Miller & Bert W. Rein, "Whither Daubert? Reliable Resolution of Scientifically-Based Causality Issues In Toxic Tort Cases," *Rutgers Law Review* 50, no. 2 (1998): 563–584; See also Victor E. Schwartz & Cary Silverman, "The Draining of Daubert and the Recidivism of Junk Science in Federal and State Courts," *Hofstra Law Review* 35, no. 1 (2006): 217. For an exceptional overview of these cases, see Gary Edmond, Simon Cole, Emma Cunliffe, and Andrew Roberts, "Admissibility Compared: The Reception of Incriminating Expert Evidence (i.e., Forensic Science) in Four Adversarial Jurisdictions," *University of Denver Criminal Law Review* 3 (Spring 2013): 31.
20 119 S.Ct. 1167 (1999).
21 Robert W. Littleton, "Supreme Court Dramatically Changes the Rules on Experts," *New York State Bar Journal* (July/Aug. 1999): 8, 12.
22 Ibid.
23 Ibid.
24 See: Gary L. Brown, "Who's on First? Frye, then Daubert, then Frye, and now back to Daubert? The Florida Supreme Court's Internal Battle Over the Standard for Admissibility for Expert Testimony," Kelley Kronenberg Blog, May 2019, https://www.kelleykronenberg.com/whos-on-first-frye-then-daubert-then-frye-and-now-back-to-daubert-the-florida-supreme-courts-internal-battle-over-the-standard-for-admissibility-for-expert-testimony/; See also Jill Lepore, "On Evidence: Proving *Frye* as a Matter of Law, Science, and History," *Yale Law Journal* 124, no. 4 (2015), https://www.yalelawjournal.org/essay/on-evidence-proving-frye-as-a-matter-of-law-science-and-history.
25 Thomas D. Schroeder, "Toward A More Apparent Approach to Considering the Admission of Expert Testimony," *Notre Dame Law Review* 95, no. 5 (2020): 2043, http://ndlawreview.org/wp-content/uploads/2020/06/7.-Schroeder.pdf.
26 Kimberly S. Moore, "Exploring the Inconsistencies of Scrutinizing of Expert Testimony Under the Federal Rules of Evidence," *Texas Tech Law Review* 22 (1991): 885, 891, https://hdl.handle.net/2346/87444.
27 George M. Walker, "What to Do When It Happens to You," *Alabama Lawyer* 17 (1993): 117, 118.
28 See: Michelle Garcia and Nichole C. Patton, "Experts and Opinions: The Pitfalls and Possibilities of Expert Witness Testimony," https://www.americanbar.org/content/dam/aba/publications/pass_it_on/experts_opinions_witness_testimony_PIO_F14.pdf.
29 Moore, "Exploring Inconsistencies," 891.
30 The judicial use of psychometric data measuring racist tendencies is another clear example of the Daubert influence. How does one really measure, at least scientifically, one's racial inclination? Better said, who has the expertise to formulate such findings? In *Price Waterhouse v. Hopkins*, the U.S. Supreme Court addressed this issue. In weighing the viability of psychological research on sexual stereotypes by Dr. Lisa Luke, the Court rejected defendant's characterization of her testimony as "Gossamer evidence" based only on "intuitive hunches" and of her detection of sex stereotyping as "intuitively divined." "Nor are we disposed to adopt the dissent's dismissive attitude toward Dr. Fiske's field of study and toward her own professional integrity … . Indeed, we are tempted to say that Dr. Fiske's expert testimony was merely icing on Hopkins' cake." Jane Goodman, "Evaluating Psychological Expertise on Questions of Social Fact: The Case of Price Waterhouse v. Hopkins," *Law & Human Behavior* 17, no. 2 (1993): 249, 253.
31 Carlson, "Policing the Bases," 578.
32 Edith Beerdsen, "Litigation Science After the Knowledge Crisis," *Cornell Law Review* 106, no. 3 (March 2021): 532, https://www.cornelllawreview.org/2021/03/27/litigation-science-after-the-knowledge-crisis/.

33 See Michael L. Seigel, *Race to Injustice: Lessons Learned from the Duke Lacrosse Rape Case* (Durham, NC: Carolina Academic Press, 2009), https://cap-press.com/books/isbn/9781594605147/Race-to-Injustice.

34 There is a passionate, ongoing debate about the "helpfulness" of child advocates in sexual abuse cases. "Child advocates say that such expert testimony is necessary because legal proof of sexual abuse is often unavailable, especially when a parent or family member is the alleged abuser. Defense attorneys say that there is no documentable profile of the sexual abuse victim and that juries can understand the manifestations of abuse without expert help." Note, "The Unreliability of Expert Testimony on the Typical Characteristics of Sexual Abuse Victims," *Georgetown Law Journal* 74 (1985): 429, 431; See also Joelle Anne Moreno, "Einstein on the Bench?: Exposing What Judges Do Not Know About Science and Using Child Abuse Cases to Improve How Courts Evaluate Scientific Evidence," *Ohio State Law Journal* 64, no.2 (2003): 531.

35 In a wide-ranging study conducted by Andrew Cohen for the *Georgetown Law Review*, the expert testimony in child abuse cases, while well-intentioned, was found to be suspect. "However strong are the policies of protecting victims to child abuse and of liberally allowing expert testimony, most expert testimony on the typical characteristics of sexual abuse victims has not been helpful for juries and presents serious dangers of prejudice." Ibid, 429. See also: Linda E. Carter, "Admissibility of Expert Testimony in Child Sexual Abuse Cases in California: Retire Kelly-Frye and Return to a Traditional Analysis," *Loyola of Los Angeles Law Review* 22, no. 4 (1989): 1103; See also: Mireille Cyr and Guy Bruneau, "False Allegations of Child Sexual Abuse," *Institut national de santé publique Québec*, last updated November 2012, https://www.inspq.qc.ca/en/sexual-assault/fact-sheets/false-allegations-child-sexual-abuse.

36 State v. Floody, 481 N.W. 2d 242 (S.D. 1992). (1985); See also Moreno, "Einstein."

37 Note, "Unreliability Expert Testimony," 449.

38 Nancy Hollander & Lauren M. Baldwin, "Winning with Experts," *Trial* (1993): 16, 23–24.

2 Forensic Law and the Role of Expertise

Most of the field of forensic science depends upon the research and opinion of the expert witness rather than a lay witness. This is because forensic science is highly specialized and equally in need of significant experience. Its content is not mere opinion as one perceives or encounters the world around us. Lay witnesses have the luxury of giving testimony on what the witness sees or experiences without any need for expert education or background. On the other hand, the expert witness must have specialized knowledge and training, a reliable field of study and a methodology that can withstand scientific scrutiny.

While lay witnesses are prohibited from expressing personal opinions in their testimonial function except as to what the lay witness has seen or heard or has the competence to relay a stated view, experts may testify by opinion to events, circumstances and conditions these witnesses never personally experienced. By their skill and qualification, courts not only admit, but expect such opinion commentary.[1]

Even if an expert has particularized, specialized knowledge which would greatly aid the trier of fact, the expert's testimony must have a conceptual nexus with the case's subject matter. The expert's opinion must correlate with the issue before the court. Some call this correlation a proper "fit" between the expert subject matter and the case before the tribunal.[2] FRE Rule 703 addresses this requirement, noting:

> The facts or data in the particular case upon which an expert bases an opinion or inference may be those perceived by or made known to the expert at or before the hearing. If of a type reasonably relied upon by experts in the particular field, in forming opinions or inferences upon the subject, the facts or data need not be admissible in evidence.[3]

Keeping this rule in mind, the expert provides opinion testimony that is based either on firsthand observation or, on occasion, hypothetical questions, or finally, by an examination and application of data, test results, reports, documents and other studies (Figure 2.1). In other words, the expert's testimony is rooted in the basic tenets of evidence law—relevancy and materiality. Even at common law, the expert witness' opinion had to be based on a reliable scientific, technical or other specialized knowledge which formed the basis of the expertise. Firsthand observation of the facts at trial is not required for any expert.[4]

Rule 703 also permits an expert witness to base his opinion or inference on facts, data or opinions presented in and outside of court other than by his own direct perception, without regard to whether such facts, data or opinions have or will be admitted or have or would be ruled inadmissible if offered into evidence. The facts, data or opinions, however, must be of a type reasonably relied on by experts in the field for forming opinions or inferences on the subject.[5]

Witnesses with firsthand knowledge may be asked about the facts underlying his opinion because the facts themselves are needed to satisfy the burden of proof.[6]

Commonwealth v. Rogers
250 A.3d 1209, (Pa. 2021)

DOI: 10.4324/9781003267126-2

FIGURE 2.1 Expert witness testifying in a court room. Source: Department of Defense. European Command. Office of Military Government for Germany (U.S.) Office of the Chief Counsel for War Crimes. 3/15/1947-6/20/1949, at https://catalog.archives.gov/id/10465282.

The primary issue raised in this discretionary appeal is whether the trial court properly excluded evidence that two of Appellant's rape victims had a history of prostitution convictions, where Appellant's defense included a contention that the encounters were consensual instances of prostitution.

I.

Section 3104 of Pennsylvania's Crimes Code, referred to as the Rape Shield Law (the "shield law"), is designed to protect alleged rape victims in the context of a criminal trial. Subject to limited exceptions, it excludes evidence of an alleged victim's past sexual history. Its purpose is to prevent the trial from shifting its focus from the defendant's guilt or innocence to the victim's reputation or moral virtue. See Commonwealth v. Johnson, 536 Pa. 153, 158, 638 A.2d 940, 942 (1994) (citing Commonwealth v. Majorana, 503 Pa. 602, 470 A.2d 80 (1983)). The law provides:

(a) General rule.-- Evidence of specific instances of the alleged victim's past sexual conduct, past sexual victimization, allegations of past sexual victimization, opinion evidence of the alleged victim's past sexual conduct, and reputation evidence of the alleged victim's past sexual conduct shall not be admissible in prosecutions of any offense listed in subsection (c) except evidence of the alleged victim's past sexual conduct with the defendant where consent of the alleged victim is at issue and such evidence is otherwise admissible pursuant to the rules of evidence.

18 Pa.C.S. § 3104(a). Under subsection (c), the above provision applies, inter alia, in cases where the defendant is charged under Chapter 31 of the Crimes Code, relating to sexual offenses. See id. § 3104(c).

In Philadelphia, over the course of approximately ten months beginning in May 2011, Appellant physically and sexually assaulted five women – two of whom were minors – and stole various items of personal property from them. Appellant was charged with dozens of crimes, including multiple counts of rape, robbery, and involuntary deviate sexual intercourse. The charges were consolidated for trial.

In a pretrial motion in limine, Appellant sought permission to introduce evidence that two of the three adult victims, A.P. and M.H., had a history of convictions for prostitution in the general area where the incidents occurred. He wanted to use such proofs to support the defense theory, as to those victims, that his encounters with them were consensual acts of prostitution. In his filing, labeled as a combined motion and memorandum of law, Appellant acknowledged the shield law, but maintained it did not exclude the evidence in question. He also claimed that exclusion would violate his rights under the United States and Pennsylvania Constitutions. The common pleas court, per Judge Anders, denied the motion in relevant part, noting that none of the convictions related to encounters with Appellant. See N.T., Feb. 9, 2015, at 5-6. The matter proceeded to a consolidated, four-day waiver trial before Judge Woelpper in February 2015. At trial, the victims testified and described the attacks, including the threats of violence and actual violence Appellant used to subdue his victims – including tackling, punching, and choking. Several of the victims indicated that, after Appellant assaulted them, he stole personal property from them, such as credit cards, identification cards, cell phones, and cash. With regard to the three adult victims, the Commonwealth presented evidence that DNA matching Appellant's was recovered from the victims' clothing or bodies shortly after each attack.

Appellant testified in his defense. He admitted to having sexual relations with all of the victims, but portrayed it as consensual in each instance. He denied that he physically harmed, or stole property from, any of them. Further, he described the encounters with the adult victims as sex-for-money transactions. He claimed to have propositioned A.P. and that she agreed she wanted to "make some money." N.T., Feb. 11, 2015, at 103. As for M.H., Appellant testified that she was walking back and forth in a manner suggestive of solicitation. See id. at 104.

The court, sitting as fact-finder, convicted Appellant of rape, aggravated assault, robbery, and related offenses as to all three of the adult victims. The court also found him guilty of rape and other offenses as to one of the minor victims, and corruption of minors with regard to the other minor victim. See N.T., Feb. 19, 2015, at 3-4. Appellant was classified as a sexually-violent predator and sentenced to an aggregate prison term of 55-170 years. See N.T., July 2, 2015, at 50. Appellant's post-sentence motion was denied.

...

In the latter opinion, the court relied on the Rape Shield Law as the basis for its determination that the proffered evidence was properly excluded. The court explained that, while exculpatory proofs are not barred by the statute, that precept was inapplicable in the present case:

Defendant sought to introduce evidence that [two] of the rape victims had prior convictions for prostitution. All of these convictions were for sexual conduct with persons other than Defendant. As in [Commonwealth v. Jones, 826 A.2d 900 (Pa. Super. 2003) (en banc)], these convictions did not exculpate Defendant and were inadmissible to prove that any of the victims consented to sexual intercourse with Defendant. For similar reasons, the convictions were also inadmissible to test the credibility of the [two] rape victims because the convictions were irrelevant to the defense of consent and were more prejudicial than probative.

Id. at 3-4.

On appeal, Appellant renewed, inter alia, his challenge to the exclusion of the prostitution evidence, as well as his contention that the verdicts were against the weight of the evidence.

A three-judge panel of the Superior Court affirmed the judgment of sentence in a non-precedential opinion. See Commonwealth v. Rogers, No. 342 EDA 2017, 2019 WL 4686960, at *8 (Pa. Super. Sept. 25, 2019). As to the issue involving the shield law, the panel noted that Appellant urged it to reexamine its precedent in Jones and Commonwealth v. Dear, 342 Pa. Super. 191, 492 A.2d 714 (1985), both of which had held that a complainant's history of prostitution convictions involving third persons, when offered to prove consent with regard to the defendant, is inadmissible under the Rape Shield Law. See id. at 202, 492 A.2d at 720 ; Jones, 826 A.2d at 909 (indicating that such evidence "is not probative of the complainant's willingness to commit sexual acts with [the defendant], for hire or for any other reason"). The panel stated that it lacked the power to overrule a prior decision of the Superior Court except where there is intervening authority by this Court calling that decision into question – and that Appellant did not assert any such intervening authority existed. See id. at *4 (quoting Commonwealth v. Pepe, 897 A.2d 463, 465 (Pa. Super. 2006)).

...

This Court allowed further review to determine whether the exclusion of evidence concerning A.P.'s and M.H.'s prostitution convictions denied Appellant his right to cross-examine witnesses and present a defense where Appellant claimed at trial that the encounters were "prostitution transactions." Commonwealth v. Rogers, —— Pa. ——, ——, 224 A.3d 1263, 1264 (2020) (per curiam). We also granted review as to the question of whether the Superior Court erred in holding that Appellant's evidentiary-weight claim was waived. See id.

II.

The admissibility of proffered evidence generally depends on its relevance and probative value. See Commonwealth v. Drumheller, 570 Pa. 117, 135, 808 A.2d 893, 904 (2002). Even where evidence is probative, it may be excludable for policy reasons. For example, evidence may be barred under the Pennsylvania Rules of Evidence where there is a danger of "unfair prejudice, confusing the issues, misleading the jury, undue delay, wasting time, or needlessly presenting cumulative evidence." Pa.R.E. 403. As illustrated by the present case, it may also be excluded based on policy as embodied in a legislative enactment.

Rulings on admissibility are committed to the common pleas court's discretion and will only be reversed on appeal where there is an abuse of discretion. See Commonwealth v. Maconeghy, 642 Pa. 770, 778, 171 A.3d 707, 712 (2017). An abuse of discretion occurs when the law is overridden or misapplied, or the judgment exercised was either manifestly unreasonable or the product of partiality, prejudice, bias, or ill will. See Commonwealth v. Spiewak, 533 Pa. 1, 7 n.4, 617 A.2d 696, 699 n.4 (1992). Here, the contention is that the Rape Shield statute should not have been applied under the circumstances. This raises an issue of law as to which our review is de novo and plenary. See Commonwealth v. Renchenski, 616 Pa. 608, 614-15, 52 A.3d 251, 255 (2012).

Initially, it is not entirely clear from Appellant's pretrial motion whether he sought permission to introduce documentary evidence concerning the victims' alleged prostitution convictions, or merely permission to cross-examine them about those matters. See Commonwealth v. Rogers, Nos. CP-51-CR-5861-2012, et al., Defendant's Motion & Memorandum of Law in Support Thereof to Admit Evidence Pursuant to the Pennsylvania Rape Shield Statute dated Oct. 21, 2015, at ¶6 (C.P. Phila.) (stating in general terms that "the defense seeks admission of evidence of the complainants' sexual conduct to corroborate [the defense's] contention that in the case[s] of [A.P.] and [M.H.] the sexual contact[s] were consensual acts of prostitution"). The motions court understood the request as directed to cross-examination of the complainants during the prosecution's case in chief, and Appellant did not suggest it was more expansive in response to the court's invitation to make any corrections on the record. See N.T., Feb. 9, 2015 (Motion), at 5-6. Further, there is nothing in the parties' present advocacy suggesting, based on any identified aspect of the record, that the request should be viewed as having extended beyond cross-examination. Accordingly, we will proceed from the premise that Appellant challenges the motions court's denial of his request as framed in terms of cross-examination of A.P. and M.H., as well as the Superior Court's affirmance of the denial.

The purpose of the Rape Shield Law has been explained in prior decisions and need not be repeated at length. To summarize, it is intended to "prevent a trial from shifting its focus from the culpability of the accused towards the virtue and chastity of the victim." Commonwealth v. Allburn, 721 A.2d 363, 366-67 (Pa. Super. 1998) (internal quotation marks and citations omitted). This protective measure is salient where defendants attempt to utilize evidence of the complainant's alleged promiscuity to bolster their claim of consent. ... Thus, the shield law "prevent[s] a sexual assault trial from degenerating into an attack upon the victim's reputation for chastity." Commonwealth v. Berkowitz, 537 Pa. 143, 151, 641 A.2d 1161, 1165 (1994) (citing cases). It additionally removes obstacles to the reporting of sex crimes. Accord Williams v. State, 681 N.E.2d 195, 200 (Ind. 1997).

With that said, the shield law may not be applied in a manner that violates a defendant's constitutional right to a fair trial, including his right to present evidence and cross-examine witnesses. See Spiewak, 533 Pa. at 11, 617 A.2d at 701 ("Notwithstanding these worthy legislative aims, rules excluding evidence cannot be mechanistically applied to abridge a defendant's right of confrontation by denying admission of highly reliable and relevant evidence critical to his defense."). In this regard, the Sixth Amendment and Article I, Section 9 of the state Charter both protect a defendant's right to be confronted with adverse witnesses. See U.S. CONST. amend. VI ("In all criminal prosecutions, the accused shall enjoy the right... to be confronted with the witnesses against him[.]); PA. CONST. art. I, § 9 (same). The federal right to "be confronted with" such witnesses has been incorporated to the States and includes the right to conduct reasonable cross-examination. ...

At the same time, the confrontation right is not absolute. It guarantees "an opportunity for effective cross-examination, not cross-examination that is effective in whatever way, and to whatever extent, the defense might wish." United States v. Owens, 484 U.S. 554, 559, 108 S. Ct. 838, 842, 98 L.Ed.2d 951 (1988) (internal quotation marks and emphasis omitted). Thus, trial courts "retain wide latitude insofar as the Confrontation Clause is concerned to impose reasonable limits on such cross-examination based on concerns about, among other things, harassment, prejudice, confusion of the issues, the witness' safety, or interrogation that is repetitive or only marginally relevant." Delaware v. Van Arsdall, 475 U.S. 673, 679, 106 S. Ct. 1431, 1435, 89 L.Ed.2d 674 (1986).

In terms of the present controversy, our intermediate court has noted that "Rape Shield laws, if rigidly construed, could impermissibly encroach upon a defendant's right to confront and cross-examine witnesses which is secured under the United States and Pennsylvania Constitutions." Commonwealth v. Nieves, 399 Pa. Super. 277, 287, 582 A.2d 341, 346 (1990). As a consequence, Pennsylvania courts have sought to balance the defendant's fundamental right to a fair trial, including his right to confront his accuser, against the state's interests embodied in the statute (as outlined above) and in the rules of evidence. …

On the threshold question of whether the prostitution evidence promoted by Appellant is statutorily precluded, we note initially that some of the evidence pertained to alleged instances of prostitution in the post-offense timeframe. See supra note 2. An issue thus arises whether the evidence concerns "past sexual conduct" for purposes of the shield law. The statute is ambiguous on this point because, as a textual matter, "past" could reasonably mean either prior to trial or prior to the offense. … We will therefore ascertain the meaning that best aligns with legislative intent. See 1 Pa.C.S. § 1921(a) (indicating that the object of all interpretation and construction of statutes is to ascertain and effectuate legislative intent); Commonwealth v. Cullen-Doyle, 640 Pa. 783, 787, 164 A.3d 1239, 1242 (2017) (same).

It makes little difference in terms of the shield law's purposes whether the evidence in question relates to conduct which occurred before or after the alleged offense. The shield law reflects the General Assembly's judgment that, in relation to the issue of consent with the defendant, a complainant's sexual history with third parties should not be aired. This objective would be ill served if the victim's excludable sexual history were limited to that which occurred prior to the alleged offense. This is true because evidence of post-offense sexual conduct could lead to the same types of distractions as proofs concerning pre-incident conduct – most notably, the improper shifting of the trial's focus away from the defendant's guilt or innocence and toward the complainant's character. … Additionally, although the statute appears in the Crimes Code, it is remedial rather than penal. Consequently, the textual reference to the alleged victim's "past" conduct should be read broadly. See 1 Pa.C.S. § 1928(c) (directing that, subject to certain enumerated exceptions not implicated here, statutory provisions should be "liberally construed to effect their objects and to promote justice"). Understood broadly, "past sexual conduct" signifies sexual conduct which occurred at any time in the past, meaning, prior to trial rather than solely prior to the alleged sexual assault.

Under our broad construction, there is little doubt that the proofs offered by Appellant were statutorily precluded. Assuming the complainants engaged in prostitution with third parties at any time before trial, those encounters qualify as their "past sexual conduct," and there is no suggestion that Appellant himself was involved in any of the prior acts of prostitution.

Given this state of affairs, Appellant seeks admission of the evidence on constitutional grounds by analogizing the present case to those in which courts have determined that otherwise-barred proofs may be admitted as bearing on the question of the complainant's credibility. He argues that here, the defense theory – consistent with what Appellant told the police – was that A.P. and M.H. gave their financially-induced consent to the encounter, and that he should have been allowed to put evidence before the fact-finder that they had engaged in prostitution with others. Summarizing a litany of extra-jurisdictional cases which have announced a variety of rules and applied them with varying results, see Brief for Appellant at 27-48, Appellant contends we should follow the lead of states that have adopted a liberal approach to admissibility in order to avoid injury to his constitutional rights. See id. at 48-49.

Appellant additionally maintains that the evidence's probative value outweighed its prejudicial effect in the circumstances. He emphasizes that the balancing test in reality deals, not with prejudice as such – as most relevant trial evidence is, by design, prejudicial to the opposition – but with the risk of unfair prejudice. See Pa.R.E. 403 (permitting the exclusion of relevant evidence if its probative value is outweighed by a danger of, among other things, unfair prejudice). Appellant indicates that, to embody that risk, the evidence must "inflame the mind" of the fact-finder. Brief for Appellant at 50 (quoting Eck, 413 Pa. Super. at 549, 605 A.2d at 1254 (stating that otherwise-admissible evidence may be excluded if "it would so inflame the minds of the jurors that its probative value is outweighed by unfair prejudice")). He argues that that standard was not met here as the complainants' alleged involvement in the sex-for-hire trade was "just business" and unrelated to an accusation of promiscuity. Id. at 52.

...

The question becomes, then, whether the shield law's prohibition must yield, under the Sixth Amendment and Article I, Section 9, to Appellant's ability to question A.P. and M.H. concerning their acts of prostitution with third parties. As reflected in the cases reviewed above, the circumstances in which Pennsylvania courts have admitted evidence notwithstanding the shield law involve proofs offered to demonstrate factual premises other than consent – such as that the conduct was committed by someone other than the defendant, the complainant harbored bias and hostility toward the defendant which would induce him or her to fabricate or color testimony, or that the complainant otherwise had an ulterior motive to manufacture charges.

In some limited situations, prostitution evidence may be germane to these other issues. ...

This matter is not at all like Joyce. Here, both A.P. and M.H. initiated contact with the police, and the trial evidence indicated they had been physically beaten shortly before reporting the crimes. ... Thus, Appellant's objective in promoting the evidence is limited to suggesting that, because the victims had been engaged in prostitution in the same geographic area on other occasions, the fact-finder should give more weight to his testimony that his own encounters with them entailed prostitution. The prostitution evidence, therefore, amounted to propensity evidence – that is, evidence of the complainants' prior bad acts promoted in an effort to show that, on the occasion in question, they acted in accordance with their character. While propensity evidence may have some limited probative value, it is inadmissible to demonstrate a person's character in order to show conduct conforming therewith. ...

Appellant denies that he sought introduction of the evidence solely to establish that the complainants acted in conformity with their conduct on other occasions. He couches his reason for promoting the prostitution evidence as supporting his "defense of financially-induced consent." Brief for Appellant at 17. The argument fails, however, as it is internally inconsistent: the evidence was relevant only insofar as the complainants gave their "financially-induced consent" to third parties on other occasions. Indeed, "absent [Appellant's] disavowed propensity theory, he is left with no theory of relevance at all." Amicus Brief of the Office of Attorney General, at 4.

In light of the above, we conclude, consistent with Dear, that the proofs offered by Appellant are the very types of items that the Rape Shield Law is designed to preclude. They would cast aspersions upon the moral character of the complainants and do little to prove consent at the relevant time – even if the victims had engaged in prostitution on other occasions. ... Further, the evidence was not offered to prove that another person committed the crimes or that the complainants harbored hostility toward Appellant or otherwise had a motive to fabricate or exaggerate the charges.

Finally, Appellant was not prohibited from establishing a consent defense. He testified concerning the events and, as noted, portrayed them all as consensual. On this record there is no reason to believe that the fact-finder was impeded in reconciling the conflict between Appellant's testimony and the other trial evidence based on judgments concerning the credibility or veracity of that evidence. In brief, the shield law barred the evidence and the statutory prohibition was not overridden in this case by Appellant's constitutional rights.

IV.

For the reasons given above, the Superior Court's ruling is affirmed insofar as it upheld the common pleas court's denial of Appellant's motion in limine, thereby excluding from trial any evidence concerning the alleged prostitution convictions of A.P. and M.H. The intermediate court's order is vacated, however, to the extent it found Appellant's weight-of-the-evidence claim waived, and the matter is remanded for a merits disposition of that claim.

Questions:

1. What is the court's dual analysis, in terms of competing interest for the rape shield law in Pennsylvania?
2. What does the court rule on the question of prostitution as always being an impermissible line of questioning?
3. Does the court review cases on rape shield which granted the right to introduce evidence based on the Confrontation clause of the U.S. Constitution?

> Read the press release regarding a bill passed in the state of California to raise standards for expert witness testimony at https://sd11.senate.ca.gov/news/20210318-senator-wiener%E2%80%99s-legislation-end-wrongful-convictions-resulting-faulty-expert-witness.

Thus, while courts are flexible in their general approach to permitting expert testimony, testimony given without a proper foundation will be excluded. An expert's bare conclusions unsupported by factual evidence are simply inadmissible, and courts therefore require that an expert's testimony rest on some factual predicate.[7] The modern tendency to liberally admit with the hopes that questionable evidence will aid the trier of fact has been well-intentioned but clearly naïve. The proliferation of questionable science has led to the well-known moniker-"junk science."

2.1 CONTRAST AND COMPARISON OF LAY AND EXPERT WITNESSES

In this way, it is important to compare and contrast the lay and expert witness but just as crucially to accept that the lines of these two types of witnesses may crossover. In other words, the expert may or may need to be expert if his lay experiences qualify for that status- and conversely, the lay witness may or may not evolve into an expert depending upon the scope and nature of the testimony (Figure 2.2).

In United States v. Brinson, the entwining of the lay and the expert becomes part of appeal approach for defendant challenges expertise while the prosecution reminds the court that a lay witness perception of a weapon usage can be just as compelling in proving a particular element. In addition, is the expert really an expert under Daubert simply because of experience and "helpfulness"? Is a police detective the type of "expert" that was once expected in Frye?

United States v. Brinson
Oct 21, 2019
No. 18-13185 (11th Cir. Oct. 21, 2019)

Christopher Brinson appeals from his conviction for possession of a firearm by a convicted felon, in violation of 18 U.S.C. § 922(g)(1). Brinson contends the district court plainly erred in allowing the expert testimony of Detective Stephen Barborini regarding his prior identifications of firearms, and the district court erred in denying Brinson's motion for judgment of acquittal, as the evidence was insufficient to support his conviction because the Government failed to prove an

FIGURE 2.2 Lindbergh takes the witness stand during the 1935 trial of Hauptmann in Flemington, New Jersey. Photo courtesy of the Library of Congress.

object he was holding in images on social media was, in fact, a firearm. After review, we affirm Brinson's conviction.

I. DISCUSSION

A. Prior Firearms Identifications

Rule 702 of the Federal Rules of Evidence allows "[a] witness who is qualified as an expert by knowledge, skill, experience, training, or education," to give expert testimony, so long as the following requirements are met:

> (a) the expert's scientific, technical, or other specialized knowledge will help the trier of fact to understand the evidence or to determine a fact in issue; (b) the testimony is based on sufficient facts or data; (c) the testimony is the product of reliable principles and methods; and (d) the expert has reliably applied the principles and methods to the facts of the case.

Fed. R. Evid. 702. Rule 703 of the Federal Rules of Evidence provides that "[a]n expert may base an opinion on facts or data in the case that the expert has been made aware of or personally observed." Fed. R. Evid. 703. However, "if the facts or data would otherwise be inadmissible, the proponent of the opinion may disclose them to the jury only if their probative value in helping the jury evaluate the opinion substantially outweighs their prejudicial effect." Id.

Expert testimony that meets the requirements of Rule 702 may still be excluded under Rule 403. United States v. Frazier, 387 F.3d 1244, 1263 (11th Cir. 2004) (en banc). Rule 403 of the Federal Rules of Evidence provides the court may exclude relevant evidence "if its probative value is substantially outweighed by a danger of one or more of the following: unfair prejudice, confusing the issues, misleading the jury, undue delay, wasting time, or needlessly presenting cumulative evidence." Fed. R. Evid. 403. "Evidence is relevant if: (a) it has any tendency to make a fact more or less probable than it would be without the evidence; and (b) the fact is of consequence in determining the action." Fed. R. Evid. 401.

Plain-error review applies because, as Brinson concedes, although he objected to testimony regarding the details of Barborini's 32 prior firearms identifications, he did not object to Barborini's testimony he had made 32 prior, accurate determinations. See United States v. Vereen, 920 F.3d 1300, 1312 (11th Cir. 2019) ("Objections not raised in the district court are reviewed only for plain error."). The district court did not plainly err in allowing Barborini's testimony regarding his prior, accurate firearms identifications. See id. (explaining under plain-error review, we may correct an error where (1) an error occurred; (2) the error was plain; (3) the error affects substantial rights; and (4) the error "seriously affects the fairness, integrity, or public reputation of judicial proceedings"). Barborini's testimony regarding the 32 prior determinations was not part of his expert opinion; rather, it occurred while he was being qualified as an expert witness and supported his experience in the field of firearms identification, in which he was qualified as an expert without objection. See Fed. R. Evid. 702; Frazier, 387 F.3d at 1260 (stating the burden of establishing a witness is qualified to testify as an expert, pursuant to Rule 702, rests on the proponent of the expert testimony). Thus, the testimony regarding the 32 prior determinations was relevant as it pertained to Barborini's experience and qualifications as an expert and relayed to the jury how his prior experience informed the conclusions he reached in this case. Barborini's qualification as an expert also relied on his advanced training with the ATF, which included traveling to firearms manufacturers and learning about the markings of various firearms. See Fed. R. Evid. 702. Further, even if Barborini's testimony were inadmissible, its probative value of relaying to the jury his experience in firearms identification outweighed any prejudicial effect. See Fed. R. Evid. 403; United States v. Jernigan, 341 F.3d 1273, 1284 (11th Cir. 2003) (stating we review issues related to Rule 403 "in a light most favorable to [the evidence's] admission, maximizing its probative value and minimizing its undue prejudicial impact" (quotation marks omitted)).

...

B. Sufficiency of the Evidence

"To justify a conviction for violation of § 922(g)(1), the government must have proved that: (1) [the defendant] was a convicted felon, (2) he knowingly possessed a firearm, and (3) the firearm was in or affected interstate commerce." United States v. Howard, 742 F.3d 1334, 1341 (11th Cir. 2014). A firearm includes "any weapon (including a starter gun) which will or is designed to or may readily be converted to expel a projectile by the action of an explosive." 18 U.S.C. § 921(a)(3).

We have held a conviction for brandishing a gun during the commission of a crime, under 18 U.S.C. § 924(c), "may be sustained by lay witness testimony that a defendant carried or used a gun," rejecting the defendant's argument the government failed to offer sufficient evidence the object he brandished during a robbery was a firearm capable of firing a projectile by means of an explosive. United States v. Hunt, 187 F.3d 1269, 1270-71 (11th Cir. 1999). Evidence in Hunt included witness testimony Hunt possessed a firearm during a bank robbery, including one witness's testimony Hunt "cocked the gun" in her face. Id. at 1270. We determined the government did not have to "show to a scientific certainty that a defendant is carrying a device that fires projectiles by means of an explosive" nor did it have to offer the gun into evidence or present expert witness testimony identifying the object as a firearm. Id. at 1271. In a factually similar case, we applied Hunt in affirming a defendant's § 924(c) conviction, even though the firearm was not in evidence, based upon witness testimony about the weapon. United States v. Woodruff, 296 F.3d 1041, 1049 (11th Cir. 2002).

The evidence was sufficient to support Brinson's § 922(g)(1) conviction. See United States v. Lander, 668 F.3d 1289, 1296 (11th Cir. 2012) ("We review de novo the district court's denial of a judgment of acquittal on sufficiency of the evidence grounds."). Although Brinson argues Hunt and Woodruff are distinguishable because they involved violent crimes, our holding in both cases, that the government need not show to a scientific certainty a defendant is carrying a firearm, did not depend on the violent nature of the robberies in those cases. See Woodruff, 296 F.3d at 1049; Hunt, 187 F.3d at 1270-71. While the officers who testified at Brinson's trial did not see the alleged firearm in person, they all viewed the photographs and video depicting Brinson holding what appeared to them to be a firearm. Although Brinson contrasts Barborini's testimony with that of his expert witness's, the jury had exclusive province over credibility determinations and was entitled to accept Barborini's testimony and reject Brinson's expert witness's testimony. See United States v. Feliciano, 761 F.3d 1202, 1206 (11th Cir. 2014) ("The jury has exclusive province over the credibility of witnesses … ."). Further, the officers' testimonies, based on viewing the images of Brinson and their experiences handling similar firearms, were not incredible as a matter of law. See id. (stating in conducting sufficiency review, we will not revisit the question of witness credibility "unless it is incredible as a matter of law"). Accordingly, the evidence was sufficient to support Brinson's conviction.

II. CONCLUSION

We affirm Brinson's 18 U.S.C. § 922(g) conviction.
AFFIRMED.

Questions:

1. How does FRE Rule 702 lay out the possibility for a detective or police specialists in firearms to be an expert?
2. Of the criteria for expert testimony, noted at Rule 702, what is the most relevant to the police detective? Secondly, is it fair to argue that the police detective may not meet all the stated criteria?
3. The Court also observes that even if the expert testimony comes up short, the "lay" witness testimony as to weaponry, may be sufficient in the order of proof. How does it reach this conclusion?

The crossover and complexities of expert witnesses being fully distinguishable from a lay witness remain a recurring theme in testimonial evidence. Succinctly stated it is possible to be both but more likely to be severable and distinct. In U.S. v. Brown, the merger and crossover ss completely evident. A police investigator operates in both worlds- yet even so, jury instructions must point out this duality.

United States v. Brown
Jun 6, 2016
651 F. App'x 653 (9th Cir. 2016)
Before: WARDLAW, PAEZ, and BEA, Circuit Judges.

After a jury trial, Deandre Brown was convicted of one count of conspiracy to commit sex trafficking of children in violation of 18 U.S.C. § 371, two counts of sex trafficking of children in violation of 18 U.S.C. § 1591(a)(1), and two counts of participating in a sex trafficking venture in violation of 18 U.S.C. § 1591(a)(2). He appeals his conviction and 360-month prison sentence. We affirm the conviction, but we vacate the sentence and remand for resentencing. 1. The district court did not plainly err by qualifying Detective Stigerts as an expert based on his experience and training, and by admitting his expert testimony regarding pimping and prostitution culture. See United States v. Brooks, 610 F.3d 1186, 1195-96 (9th Cir. 2010).

We held in United States v. Vera, however, that when a law enforcement officer provides both expert testimony and lay opinions based on his investigation of the defendant's wrongful conduct, "the jury must be instructed on how to appropriately evaluate each form of testimony offered by the officer." 770 F.3d 1232, 1235 (9th Cir. 2014). Because Detective Stigerts offered both expert and lay opinion, the district court plainly erred in failing to give such an instruction. See id. at 1246. The failure to clearly delineate for the jury which opinions were based on Stigerts's experience and training and which were based on his investigation of Brown compounded the error. See United States v. Torralba-Mendia, 784 F.3d 652, 658 (9th Cir. 2015).

We are unable to conclude, however, that these errors affected Brown's substantial rights. In Vera, the challenged testimony comprised nearly the "sole evidence" of a key contested issue. 770 F.3d at 1246. Here, by contrast, considerable evidence supported the verdict apart from Stigerts's challenged testimony. In closing argument, Brown's counsel emphasized that the two important questions were whether Brown used force to induce the victims into prostitution and whether Brown knew the victims were underage. But, little of Stigerts's testimony was necessary to establish the age of the victims, or that Brown used force against them. Both victims testified to Brown's violent acts, as did Brittney Beacham. Their testimony was corroborated by photographs and the testimony of medical professionals and other officers. And all three women testified that they feared Brown. On this record, Brown fails to demonstrate that the plain error prejudiced him. See United States v. Olano, 507 U.S. 725, 734 (1993). 2. Brown's other assignments of trial error do not require reversal whether considered individually or cumulatively. See United States v. Wilkes, 662 F.3d 524, 543 (9th Cir. 2011) ("[I]n contrast to this circuit's cases that reversed a defendant's convictions based on cumulative error, the government in this case presented ample evidence of Wilkes's guilt."). 3. By contrast, Brown's sentence must be vacated. As the government concedes, the district court failed to ensure that Brown read and discussed the presentence report ("PSR") with his trial counsel as required by Federal Rule of Criminal Procedure 32(i)(1)(A), an error we have called "deplorable and easily avoided." United States v. Sustaita, 1 F.3d 950, 954 (9th Cir. 1993). Because Brown represents that his trial counsel never reviewed or discussed the PSR with him, and because he identifies factual errors in the PSR that may have influenced the district court's imposition of sentence, we cannot conclude that "it is clear that no prejudice resulted" from the Rule 32 violation. Sustaita, 1 F.3d at 954.

Rule 32(i)(1)(A) states that the district court "must verify that the defendant and the defendant's attorney have read and discussed the presentence report and any addendum to the report." --------

Moreover, the term of incarceration imposed violated the so-called parsimony principle that a criminal sentence be "sufficient, but not greater than necessary" to serve the purposes codified in 18 U.S.C. § 3553(a)(2). United States v. Carty, 520 F.3d 984, 991 (9th Cir. 2008) (en banc) (quoting 18 U.S.C. § 3553(a)). In an exchange with Brown at sentencing, the district court stated, "Whether it's 300 or 360 months, sir, you're going to spend most of your adult life in prison until you're in your fifties. The difference is 20 years versus - 25 years versus 30 years, which is not significant." To the contrary, "[a]uthority does not suggest that a minimal amount of additional time in prison cannot constitute prejudice." Glover v. United States, 531 U.S. 198, 203 (2001). If the district court believed the difference between potential sentences was insignificant, parsimony required that the lesser sentence be imposed. We hold that the violations of Rule 32 and § 3553(a) require vacating Brown's sentence and remanding for resentencing. 4. Because we vacate Brown's sentence, we do not reach the other challenges to his sentence.

AFFIRMED IN PART; VACATED IN PART; and REMANDED.

Questions:

1. What does the court say about distinguishing lay from expert witnesses as a conclusion or finding in this case?
2. Which part of the police officer's activities are construed as lay- and which part expert?
3. In the final analysis, was the Court convinced the failure to instruct the jury on the distinction of lay and expert testimony should result in an acquittal or dismissal?

Just where the line crosses for an expert to be an expert and lay witness to be nothing more, is not always an easy finding. Especially true in field work for the law enforcement community, many entrenched forensic practices in fingerprints, ballistics and impressions, are the province of those heavily experienced in the practices. These experienced players can and do qualify for expert status under all the tests mentioned, yet failing to differentiate and instruct jurors of when the expert starts or ends and so too with the lay beginning and ending, can be an appeal basis. In U.S v. Clark, the court was able to differentiate but most delicately.

United States v. Clark
Oct 12, 2017
No. 16-16277 (11th Cir. Oct. 12, 2017)
Woodrow Andrew Clark appeals his conviction for conspiracy to distribute heroin. We affirm.

I. BACKGROUND

On March 24, 2015, Clark and his codefendant, Robert Elie, were charged with conspiring to possess heroin with the intent to distribute, in violation of 21 U.S.C. §§ 841 and 846 (Count 1), and possessing heroin with the intent to distribute, in violation of 21 U.S.C. § 841 (Count 2). While Elie pled guilty in January 2016, Clark proceeded to a jury trial in June 2016.

Department of Homeland Security ("DHS") Special Agent Steven Ledgerwood testified that, on March 9, 2015, he was on duty and learned of a potential drug transaction occurring at the Camden St. Clair Apartments. Agent Ledgerwood was investigating a Mexican cartel that was sending large quantities of drugs to the metro-Atlanta area. While monitoring the area where he believed the drug transaction would occur, Agent Ledgerwood observed a bright orange Chevrolet truck. He saw a black male exit the truck and walk through the breezeway leading to the suspect apartment where DHS believed the drugs were being kept. He later saw the same orange truck pull back into the complex. He also observed a black Infinity, which he believed to be carrying drugs, enter the complex. Agent Ledgerwood also read into the record a stipulation that Clark previously had been convicted of selling heroin on two prior occasions in Florida.

DHS Agent Charles Engle also testified at trial. Agent Engle testified that he was assigned to the organized crime and drug task force and had participated in physical surveillance in ongoing investigations. When the government began questioning Agent Engle about countersurveillance, Clark objected that Agent Engle was providing expert testimony and the government had failed to provide the requisite notice. The district court stated that Agent Engle could not "get into countersurveillance," but he could explain why law enforcement was watching the apartment. R. at 887. Clark argued that Agent Engle could testify regarding what he observed, but that he could not provide his opinion that, based on his experience, he believed Clark was conducting countersurveillance. The court noted that such an opinion would be based on specialized knowledge, as a person without law-enforcement training would not draw the same conclusion. The court permitted the testimony to continue but instructed the government not to get into testimony regarding what actions constitute countersurveillance.

Agent Engle resumed his testimony. He stated that he had participated in surveillance of the Camden St. Clair Apartments on March 9, 2015. He was in a parked car with heavily tinted windows on the top floor of a parking deck and could see the entrance to the breezeway leading to the suspect apartment. An orange truck parked near him, even though there were available parking spaces near the entrance of the suspect apartment. The truck faced the breezeway, just as Agent Engle's vehicle did. Agent Engle later identified Elie as the driver and Clark as the passenger. Agent Engle testified that Clark and Elie remained in the truck for about ten minutes, then Elie exited the truck and Clark handed Elie a red bag through the truck window. Elie walked into the breezeway toward the apartment. Clark remained in the truck for five minutes and then drove out of the complex. About ten minutes later, Clark returned in the truck and parked in the same location as before. Clark remained in the driver's seat and looked around. About ten minutes later, Clark again drove out of the complex.

The government asked Agent Engle why he believed the truck might be involved in the drug activity. Clark renewed his objection; the court noted that it had already ruled. Agent Engle testified that, at that point, he believed the orange truck was involved in the transaction because a person in the truck had walked to the suspect apartment, the truck had entered and left the apartment complex, and the driver had not exited the truck. Agent Engle also observed the black Infinity, which was believed to be delivering narcotics to the apartment, enter the complex. He observed a Hispanic male, carrying a package, exit the Infinity and approach the suspect apartment. Agent Engle saw the orange truck return to the complex 15 minutes later. The truck approached the apartment; Elie exited the breezeway with the red bag and entered the passenger side of the truck. Agent Engle stated that, once Elie was inside, the truck began to slowly leave the complex.

Agent Engle testified that he began to follow the truck but could not follow Clark out of the complex without being too noticeable. He later joined other agents in the pursuit of the orange truck and eventually caught up to the truck. Agent Engle watched law-enforcement officers attempt to conduct a traffic stop. The truck began to flee; later the driver and passenger began to flee on foot. When agents later searched the truck, they found the red bag containing $24,000, 2 kilograms of heroin, and 3 cell phones. Agents also seized two cell phones belonging to Clark from the truck. The agents were able to retrieve information from one phone and found evidence that Clark used a messaging app.

During a recess, the government provided the district court and defense counsel with several cases in which this court had permitted lay opinion testimony from a law-enforcement officer that included opinions formed based on law-enforcement experience. The district court clarified that, while this court has held that narcotics agents may testify about the significance of certain conduct, the government had failed to provide any notice of its intent to use Agent Engle as an expert; thus, he could not testify as to any specialized knowledge.

Agent Engle published to the jury the seized heroin and stated the amount of heroin was approximately 18,000 "hits." R. at 943. Agent Engle testified that he believed that a narcotics deal was occurring in the suspect apartment. Because Clark passed open parking spaces closer to the apartment and continued looking around the complex, Agent Engle believed that Clark had been acting as a lookout. Agent Engle testified that he believed Clark was involved in a drug transaction and this had heightened his suspicion of the truck. Additionally, he believed that the driver was driving slowly to see whether anyone was following. At the conclusion of Agent Engle's testimony, the court noted for the record that it had ruled that his testimony was lay opinion testimony because it was based on his perception.

...

Georgia State Patrol Trooper Doug Allen testified that he was called to assist DHS with a traffic stop involving narcotics on March 9, 2015. He followed the orange truck for several miles and eventually determined that the driver failed to maintain his lane and was not using his headlights while it was raining and that the truck's brake lights were not properly working. After observing the traffic violations, Trooper Allen initiated a traffic stop. Trooper Allen stated that the truck began to slow but did not come to a complete stop. He explained that, in his experience, when a person slows his vehicle but does not stop completely, he is considering fleeing from the police. He then stated that the driver fled the truck on foot and identified Clark as the driver. He chased Clark and was able to catch him, though he had to use his Taser to subdue and arrest Clark.

The jury found Clark guilty as to Count 1 and not guilty as to Count 2. The district court sentenced Clark to 156 months of imprisonment, to run concurrently to a state sentence that previously had been imposed. The court also sentenced Clark to a five-year term of supervised release. On appeal, Clark argues that the district court erred by permitting a law-enforcement agent to testify as a lay witness that, based on Clark's behavior, the agent believed Clark was acting as a lookout during a drug transaction. Clark asserts that, because the agent's testimony was based in part on his experience as a law-enforcement officer, his statements were expert testimony and required notice.

II. DISCUSSION

We review a district court's evidentiary rulings for abuse of discretion. United States v. Frediani, 790 F.3d 1196, 1199 (11th Cir. 2015). Moreover, even if a district court abused its discretion, we will only reverse if the error was not harmless. Id. at 1200. An error is harmless unless, in light of the record, there is a reasonable likelihood that the error had a substantial influence on the outcome of the proceedings. United States v. Bradley, 644 F.3d 1213, 1270 (11th Cir. 2011).

The Federal Rules of Evidence permit lay witnesses to give opinion testimony if the testimony is "(a) rationally based on the witness's perception; (b) helpful to clearly understating the witness's testimony or to determining a fact in issue; and (c) not based on scientific, technical, or other specialized knowledge within the scope of [Federal] Rule [of Evidence] 702." Fed. R. Evid. 701. Rule 702 permits opinion testimony from a witness who is "qualified as an expert by knowledge, skill, experience, training, or education." Fed. R. Evid. 702. If the government intends to call an expert witness at trial, the government must, upon request, provide the defendant with a written summary of any expert testimony the government intends to use in its case-in-chief. Fed. R. Crim. P. 16(a)(1)(G). No such notice is required for lay opinion testimony.

To be rationally based on the witness's perception and thus qualify as lay opinion, testimony must be based on "first-hand knowledge or observation." United States v. Marshall, 173 F.3d 1312, 1315 (11th Cir. 1999). We have held that the mere fact that a witness's perception is based in part on his past experiences does not render it an expert opinion. United States v. Novaton, 271 F.3d 968, 1008 (11th Cir. 2001) …

Contrary to Clark's argument on appeal, we have recognized that a law-enforcement officer may testify based on knowledge gained from his experience in a particular field, and that testimony does not necessarily qualify as "specialized knowledge" within the meaning of Rule 702. See Tampa Bay Shipbuilding, 320 F.3d at 1223 & n.17; Novaton, 271 F.3d at 1008. The district court specifically prevented Agent Engle from testifying about Clark's actions based on any specialized knowledge; the government elicited testimony only as to Agent Engle's perception based on his observation of Clark's actions on the day of the heroin transaction. Agent Engle did not testify on direct examination regarding what "countersurveillance" is or make any general observations regarding conspirators in a drug ring. Because Agent Engle's testimony was based on his perception of Clark's actions, it was admissible as lay opinion testimony under Rule 701. See Fed. R. Evid. 701. The district court did not clearly err in permitting the agent to testify as a lay witness that, based on his observations, he believed that Clark was acting as a lookout during the drug transaction.

Furthermore, even if the district court did abuse its discretion in admitting Agent Engle's opinion testimony that Clark was acting as a lookout, any error was harmless. See Bradley, 644 F.3d at 1270. Significant evidence, aside from Agent Engle's opinion that Clark was acting as a lookout, showed that Clark conspired to possess with intent to distribute heroin.

AFFIRMED.

Questions:

1. What allegedly were the "expert" areas that required expert witness testimony?
2. In this case was the court satisfied a conviction with either form of evidence- either lay or expert would have be sustained despite the error in instruction.
3. What are the parameters for lay testimony? Are these merely questions of comprehension and the ability to relay and interpret what was personally experienced?

The question of whether expert testimony and its requirements be satisfactorily adhered to, is a perennial problem in the judicial analysis. In U.S. McCleod, two major conclusions can be drawn: first, that appellate courts give wide deference and latitude to trial courts; and secondly, that even if there is an error, the appellate court often devise an alternative path which justifies the conviction. In this case, the question of child pornography and child sexual imagery could be proven by the investigative-lay witness approach of the investigating officer, or whether proof that the image was in fact extracted from the phone and thus custodially tied to the defendant weighs heavy in the opinion.

Read the following piece regarding eyewitness testimony utilized in establishing an alibi in Sacco and Vanzetti case, as well as questioning of the defendants at https://www.mass.gov/info-details/sacco-vanzetti-the-evidence.

United States v. McLeod
Aug. 28, 2018
No. 16-50013 (9th Cir.)
Before: BEA and MURGUIA, Circuit Judges, and MOLLOY, District Judge.

Tony Lee McLeod was convicted by a jury of nine counts of persuading or attempting to persuade a minor to engage in sexually explicit conduct for the purpose of producing an image of that conduct, 18 U.S.C. § 2251(a), (e), one count of traveling with the intent to engage in illicit sexual conduct, 18 U.S.C. §2423(b), and one count of transportation of a minor with the purpose of engaging in illicit sexual conduct, 18 U.S.C. § 2423(a). McLeod appeals his conviction, alleging the district court abused its discretion by: (1) failing to make a reliability finding on purported expert testimony concerning information obtained from a cell phone through a Cellebrite device, which he claims is required by Federal Rule of Evidence 702; (2) admitting testimony from one of the victims about physical contact between McLeod and the victim, over McLeod's objection based on Rule 403; and (3) failing to sever the § 2251 production counts from the § 2423 travel and transport counts. McLeod also asserts that § 2251 is unconstitutional as applied to him.

...

1. McLeod argues the district court abused its discretion when it admitted Detective Damian Jackson's testimony at trial without making a reliability finding under Rule 702. Rule 702 governs

the admission of expert testimony and requires that proposed expert testimony be reliable. Further, under Rule 702, where the testimony's "factual basis, data, principles, methods, or [its] application" is called into question, a trial judge must make a reliability determination. See Kumho Tire Co., Ltd. v. Carmichael, 526 U.S. 137, 149 (1999) (quoting Daubert v. Merrell Dow Pharm., Inc., 509 U.S. 579 (1993)).

It appears that the district court overruled McLeod's objections to Detective Jackson's testimony because the district court found that Detective Jackson had the experience and knowledge to present the contested documents, and nothing in his testimony required the district court to make additional findings about the testimony's reliability. See id. At trial, Detective Jackson testified about how he used a Cellebrite device during the course of his investigation to download information from one of the victim's cell phones onto a thumb drive and then testified about the contents of that information. He testified about what Cellebrite does and how he used it in the course of his investigation to extract information from the victim's cell phone. His investigation and Cellebrite use yielded readable text of the downloaded data, a link to images downloaded from the victim's cell phone, and "extraction reports." Detective Jackson also testified that he could select what data to extract from the phone through Cellebrite. In short, Detective Jackson testified about his use and interaction with Cellebrite—and how he extracted data from one of the victim's phones in this case. We have previously allowed testimony similar to Detective Jackson's testimony without requiring that the testimony meet Rule 702's expert testimony requirements. See United States v. Seugasala, 702 F. App'x 572, 575 (9th Cir. 2017) ("The officers who followed the software prompts from Cellebrite and XRY to obtain data from electronic devices did not present testimony that was based on technical or specialized knowledge that would require expert testimony.").

That the Seugasala court reviewed for plain error is a distinction that does not change the fact that our court has previously allowed testimony similar to Detective Jackson's testimony.

Nevertheless, the dissent asserts that Detective Jackson provided expert testimony subject to Rule 702. The dissent believes the district court erred by not making a reliability finding regarding Detective Jackson's testimony and accepting the information obtained through Cellebrite. However, even assuming that the district court erred in admitting Detective Jackson's testimony, the error was harmless. … The record reflects that testimony from one of the victims and from one of the victim's aunts independently supports McLeod's conviction without Detective Jackson's testimony. Therefore, assuming Detective Jackson's testimony was admitted in error, any error was harmless because sufficient evidence in the record supports McLeod's conviction.

In sum, the district court did not abuse its discretion in admitting Detective Jackson's testimony. See Harper v. City of Los Angeles, 533 F.3d 1010, 1030 (9th Cir. 2008) ("We afford broad discretion to a district court's evidentiary rulings."); see also Kumho Tire, 526 U.S. at 149. And even if the district court erred, such error was harmless. See Estate of Barabin, 740 F.3d at 464. …

Accordingly, the district court did not commit any reversible error on any issue on appeal, and we affirm.

AFFIRMED.

MOLLOY, District Judge for the District of Montana, dissenting in part and concurring in part:
I respectfully disagree with the majority reasoning and rulings in this case.

1. The district court abused its discretion when it admitted Detective Jackson's expert testimony without making a reliability finding, and the error prejudiced McLeod.

First, Jackson provided expert testimony. See Fed. R. Evid. 702(a). Jackson testified based on technical knowledge—specifically, technical knowledge about the use of the Cellebrite device.

Jackson testified as to what the Cellebrite device is, what it does, how it works, and what it produces. Jackson also testified he performed a "logical extraction" on the Blackberry, and explained the resulting report to the jurors at length.

The majority cites to United States v. Seugasala for the proposition that testimony by an officer who uses Cellebrite to extract the contents of a cellular device is not expert testimony. 702 F. App'x 572, 575 (9th Cir. 2017). But Seugasala, a non-binding memorandum opinion, involved plain error review, not abuse of discretion. Moreover, simply because the user can follow prompts from the program does not mean that expert testimony is not required or that the underlying technology is reliable. In the context of this case, McLeod presented evidence that the Cellebrite device Jackson used can produce significant errors, including not acquiring files and misreporting data.

Even if Jackson only provided lay testimony, as the majority found, he nevertheless received the court's expert imprimatur in front of the jury. … The district court stated that, based on Jackson's experience, he was "more than qualified" to testify about the Cellebrite report. When McLeod objected to further introduction of the Cellebrite report into evidence, the district court stated Jackson "[had] … the experience and the knowledge to present the[] documents." Later, the district court overruled McLeod's relevance objection as the government questioned Jackson about his continued use of Cellebrite in the years following the extraction at issue in this case. In other words, the district court first endorsed Jackson as an expert, and then permitted the government to bolster Jackson's credentials in front of the jury.

Second, the district court made no reliability finding before admitting Jackson's testimony and the Cellebrite report. "[T]he failure to make an explicit reliability finding [i]s error," even where "the district court's ruling suggests an implicit finding of reliability." United States v. Jawara, 474 F.3d 565, 583 (9th Cir. 2007) (citation omitted).

…

Jackson's testimony and the Cellebrite report were critical to the government's case. The government relied on the report to establish which messages were sent and received, as well as the time of each message. The timing of the messages rebutted McLeod's claim that he did not know the victim was a teenager when the images were produced. The government also used the Cellebrite report to create demonstrative exhibits showing a "more readable" format that was "closer to the way [the messages] would have looked on the [Blackberry] device when the individual was holding it." The government relied on that report in closing.

In sum, the district court erred when it allowed Detective Jackson to provide expert testimony without making a reliability finding and accepted the Cellebrite report he prepared. That error was not harmless because it involved evidence critical to the government's case, prejudicing McLeod. Reversal and a new trial is appropriate for this reason and for the additional reasons set forth below.

2. The district court abused its discretion when it admitted the victim's testimony that McLeod molested him. While that testimony is relevant to McLeod's intent regarding the travel and transportation counts, United States v. Dhingra, 371 F.3d 557, 565 (9th Cir. 2004), it had limited, if any, relevance to the production counts, and its probative value was substantially outweighed by its prejudicial effect, Fed. R. Evid. 403.

Relevant evidence may be excluded "if its probative value is substantially outweighed by a danger of … unfair prejudice." Id. "The probative value of evidence against a defendant is low where the evidence does not go to an element of the charge." United States v. Gonzalez-Flores, 418 F.3d 1093, 1098 (9th Cir. 2005). In this case the proper balance of fairness and "judicial economy" should tip to the defendant's right to a fair trial.

...

On the other hand, the testimony was not probative of any of the elements the government had to prove for the production counts, which were that (1) the victims were minors, (2) McLeod persuaded (or attempted to persuade) them to take part in sexually explicit conduct for the purpose of producing a visual depiction, and (3) the visual depictions were transported in or affecting interstate or foreign commerce by any means, including by computer. 18 U.S.C. § 2251(a). Sexual contact between the defendant and the victim is not an element, as the government conceded in its pretrial brief opposing severance. The district court reasoned that the testimony "would be relevant to rebut a defense that the defendant did not know that the victim was a minor" because "the subsequent acts tend[ed] to rebut the inference raised by such a defense that the defendant would not have committed the acts had he known the minor's age." But the victim testified before McLeod presented his case, which forced McLeod into rebuttal regardless. The same forced hand ruling took place in a pretrial exchange between the court and defense counsel. Further, one of the production counts involved a different victim than the one who testified about the physical contact, meaning not only that the touching testimony was not probative of that count, but was not even relevant.

McLeod suffered unfair prejudice because of the touching testimony. "The term 'unfair prejudice,' as to a criminal defendant, speaks to the capacity of some concededly relevant evidence to lure the factfinder into declaring guilt on a ground different from proof specific to the offense charged." Old Chief, 519 U.S. at 180. "In other words, unfairly prejudicial evidence is that having 'an undue tendency to suggest decision on an improper basis, commonly, though not necessarily, an emotional one.'" Gonzalez-Flores, 418 F.3d at 1098 (quoting Old Chief, 519 U.S. at 180). "Where the evidence is of very slight (if any) probative value, it's an abuse of discretion to admit it if there's even a modest likelihood of unfair prejudice or a small risk of misleading the jury." Id. (quotation and citation omitted). A fair trial cannot be dependent on a "guilty anyway" assessment of the evidence or the joinder of charges.

...

The district court erred by admitting the touching testimony because it was not probative of the production counts and presented an unacceptably high risk of prejudice. Reversal for a new trial on this ground is appropriate.

3. The district court's failure to sever the travel and transportation counts from the production counts after it decided to admit the touching testimony abridged McLeod's right to a fair trial. United States v. Lewis, 787 F.2d 1318, 1321 (9th Cir. 1986), amended on denial of reh'g, 798 F.2d 1250 (9th Cir. 1986).

...

The district court concluded the touching testimony was relevant to the production counts because it rebutted what the court presumed was McLeod's affirmative defense—that he did not know the victims were underage. But the testimony does not go to any element of the production charges, and allowing the government to introduce that evidence in its case in chief put the cart before the horse. Whether McLeod touched the victim after the two exchanged photos and video does not make it more likely that he knew how old the victim was during those exchanges. The problem here was that the testimony was not probative of the production counts but in the jury's mind was more than likely evidence of bad character. See id. at 1322. Such prejudice would have been at its peak regarding the production count concerning the other victim.

...

In my opinion, the district court abused its discretion by refusing to sever the production counts from the travel and transport counts once it decided to admit the touching evidence. Reversal and a new trial with the counts severed is appropriate here.

...

Because the district court erred by admitting Jackson's expert testimony and the Cellebrite report without making a reliability finding, by admitting the touching testimony, and by failing to sever the production counts from the travel and transport counts, I respectfully dissent in part, and I would send this case back for new trials on the severed counts with instructions to the district court to reconsider its evidentiary rulings.

Questions:

1. Why would the defendant argue so vigorously for the application of expert witness requirements in the question of child pornography- claiming that only an expert can prove the correlation of defendant's phone and the child pornography?
2. Does the Court conclude that even if it was error, the proof of the matter could still be proven by the lay investigator and that a failure to distinguish was merely harmless error?
3. Explain the dissenting opinion. Why does the dissenting judge find the error anything but harmless?

Locate the following case using your favorite legal research resource, read the case and answer the questions below.

United States v. Reyes-Garcia, Dec. 18, 2019, No. 18-10144 (11th Cir.)

Questions:

1. How does the court rule the exclusive nature of "code" language drug dealing and transactions? Does such as opinion rely on expert evidence alone?
2. One of the interesting aspects of this case is how "error" aside from the harmless variety already discussed, come in other forms. What are they?
3. Why is the hearsay doctrine treated differently as between in a lay or expert witness?

2.2 QUALIFICATIONS FOR THE EXPERT WITNESS

Whether under the Frye, Daubert or Kumho doctrines, expert witnesses must do more than merely allege expertise but prove specialized capacities that uniquely suit these witnesses to give said testimony. And while it has just been shown that there is often an inevitable crossover between expert and lay witnesses, the array of scientific and specialized expertise call for a series of qualifications to be designated an "expert" witness. The following cases highlight the needed pre-requisites and rules that guide expert witnesses.

The requisites for the expert witness designation generally include education and training, experience and exposure, a legitimate scientific subject matter that is generally accepted by most experts in this field, publications and research, as well as professional reputation. In Anderson, the entire case depends upon a wide assortment of forensic science and the corresponding challenges to the expert's role in the case.

Anderson v. State
220 So. 3d 1133 (Fla. 2017)
March 9, 2017

Charles Anderson appeals an order denying his motion to vacate his conviction of first-degree murder and sentence of death filed under Florida Rule of Criminal Procedure 3.851. Anderson also petitions this Court for a writ of habeas corpus. We have jurisdiction. See art. V, § 3(b)(1), (9), Fla. Const. For the reasons expressed below, we affirm the postconviction court's order as to Anderson's conviction and deny Anderson's habeas petition but vacate the death sentence and remand for a new penalty phase.

...

1. Tire Impression Evidence

At trial, the State presented the testimony of Fred Boyd, a latent print, footwear, and tire-tread analyst. Boyd testified that in the course of his employment with the Broward County Crime Lab, he was asked to compare photos and casts of the tire impressions left at the scene where Keinya Smith's body was found to standards made from the tires on Anderson's car. Boyd concluded that two of the tire impressions left near Keinya's body were of no value, the third "could have been made by" Anderson's right front tire, and the fourth impression "could not have been made by" any of the tires on Anderson's car.

The State laid the foundation for Boyd to be deemed an expert in tire-tread examination by asking Boyd about his training and experience. Boyd testified that he spent the first twenty years of his career as a military policeman in the Army Criminal Investigation Division Crime Lab and retired as chief of the latent print unit in 1988. He was then employed by the Broward County Crime Lab, where he worked in fingerprints, footwear, and tire-tread examination for the next eight and a half years. At the time of Anderson's trial in 1999, Boyd was employed with the Las Vegas Metropolitan Police Department's crime lab as an analyst of fingerprints, footwear, and tires. Boyd testified that the bulk of his training in these disciplines took place when he was in the Army, but he also received continuous training throughout his thirty-one-year career both on the job and through attendance at numerous FBI schools and other seminars. Prior to Anderson's trial, Boyd testified as an expert in tire-tread impression evidence ten times in Broward County. Boyd was tendered to and accepted by the court as an expert in the field of tire examination without objection.

Anderson now claims that counsel was deficient for failing to object to Boyd being deemed an expert because he was not "certified" in tire-tread examination. This claim is without merit. The Florida Evidence Code provides that a witness may be qualified as an expert based on "knowledge, skill, experience, training, or education." § 90.702, Fla. Stat. (1999). There is no requirement that a witness be "certified" in a particular field in order to be deemed an expert and allowed to give opinion testimony. Boyd's specialized knowledge, training, and extensive experience were sufficient for the trial court to qualify him as an expert. An objection on the basis that he was not "certified" in tire examination would have been meritless. "[C]ounsel cannot be deemed deficient for failing to make a meritless objection." Raleigh v. State, 932 So.2d 1054, 1064 (Fla. 2006).

Anderson also claims that trial counsel was deficient for failing to utilize a defense expert to challenge Boyd's testimony. This claim is also without merit because the alleged deficiency is rebutted by the record. During his cross-examination of Boyd, trial counsel brought out the weaknesses in Boyd's testimony and rigorously challenged its significance by pointing out that Boyd could only say that the right front tire on Anderson's car was "pretty consistent" or "relatively consistent, but not a hundred percent consistent" with the type of tire that left one of the impressions near Keinya's body; the impression was not specific to the tire on Anderson's car but only to the type of tire; Boyd did not know how many of that particular type of tire were manufactured, where in the country they were primarily distributed, or which stores carried the tire; while Anderson's right front tire "could have, possibly" made that impression, there was "no question at all" that another one of the impressions was not made by any of the tires on Anderson's car; Boyd's analysis only looked at "the general class design" of the tire; Boyd never went and looked at the actual tires on Anderson's car; and there was no "positive I.D. on the tire."

Anderson did not allege in his motion or on appeal what information could have been obtained from other experts and used to impeach or discredit Boyd's testimony. He made only the general allegation that experts could have impeached Boyd by testifying to "the inadequacy of the science behind the opinion testimony at Mr. Anderson's trial." Such an allegation is insufficient to warrant an evidentiary hearing. See Lebron v. State, 135 So.3d 1040, 1060 (Fla. 2014) (affirming summary denial of ineffective assistance claim based on counsel's failure to properly impeach State witnesses as insufficiently pleaded where defendant did not sufficiently allege what information counsel could have used to impeach the witnesses or otherwise refute their testimony); Reaves v. State, 826 So.2d 932, 939–40 (Fla. 2002) (affirming summary denial of postconviction claim based upon legal insufficiency where defendant merely alleged that counsel failed to prepare for and adequately cross-examine witnesses without indicating what favorable information could have been elicited from those witnesses or how defendant was prejudiced). And the record presents no need for impeachment in light of the admitted inconclusive findings of the State's expert. We therefore affirm the summary denial of this claim.

2. Blood Evidence

Anderson argues that counsel was ineffective for failing to utilize defense experts to challenge the State's evidence regarding the spot of Keinya's blood found on Anderson's car seat as it related to the State's theory that Keinya's body was transported in the car from the location where she was run over to where her body was dumped on the side of the road fifteen miles away. Specifically, Anderson claims that experts could have testified that there was no way to determine when Keinya's blood was deposited on the seat and that there would have been more blood in the car if it had been used to transport her body.

Anderson's claim that trial counsel should have utilized an expert to testify that there was no way to determine when Keinya's blood was left in his car was not raised in his motion before the trial court and is therefore unpreserved and not properly before this Court for review. See Steinhorst v. State, 412 So.2d 332, 338 (Fla. 1982) ("[F]or an argument to be cognizable on appeal, it must be the specific contention asserted as legal ground for the objection, exception, or motion below."). Moreover, even if Anderson raised this claim below, an evidentiary hearing would have been unwarranted because the record demonstrates that counsel elicited testimony during cross-examination of the State's expert that there was no way to date the blood found in Anderson's car. See Reed v. State, 875 So.2d 415, 427 (Fla. 2004) (affirming denial of claim that counsel was ineffective for failing to retain a defense expert where facts necessary for defendant's defense were established through cross-examination of State's experts).Anderson also claims that an expert could have testified that if Anderson ran Keinya over and then transported her "bloody body" fifteen miles in his car to the location where it was found, there would have been much more than one drop of blood in his car. But the fact that a "bloody body" would transfer more than one drop of blood to a car is "within the average person's realm of knowledge." Id. at 423. Trial counsel did not render deficient performance by failing to retain an expert to explain a fact commonly known. See id. Trial counsel drew the jury's attention to the lack of blood in Anderson's car in both his opening statement and closing argument and he argued in closing that a single spot of Keinya's blood on the passenger seat did not prove that her body was transported in the car. It was not necessary to present testimony from a blood expert for counsel to argue a reasonable inference to be drawn from the evidence presented, and Anderson did not specify how trial counsel's challenge to this aspect of the case would have been any different had an expert been utilized.

Because Anderson failed to identify any specific errors or omissions with respect to the blood evidence that show that counsel's performance deviated from the norm such that it fell outside the range of professionally acceptable performance, he was not entitled to an evidentiary hearing on this claim.

3. Grease Pattern Impression Evidence

Anderson also claims that trial counsel performed deficiently in failing to consult experts and challenge the admissibility of the State's expert testimony regarding the grease mark or grease pattern impression evidence. At trial, the State called James Gerhart, an FBI pattern impression expert, to testify regarding his analysis of the grease marks found on the jacket that Keinya was wearing when she was run over. Gerhart compared the grease marks to four coils taken from the underside of Anderson's car. Gerhart testified that two of the coils from Anderson's car were consistent in size and design and could have made the grease impressions on Keinya's jacket, but that the two other coils were not similar in design and size and could not have made the grease impressions on the jacket. He told the jury that he could not say for certain that Anderson's vehicle made the impressions on the jacket.

According to Anderson, grease pattern analysis is a "suspect field" of forensic science and the validity of such analysis is "highly questionable due to varying methods of testing and examination." Anderson claims that trial counsel should have consulted with other experts in this "suspect field" and had those experts available to testify about the shortcomings of grease pattern impression evidence and to explain why Gerhart's testimony was inadmissible under Frye v. United States, 293 F. 1013 (D.C. Cir. 1923).

At the time of Anderson's trial, Florida used the Frye standard to determine the admissibility of expert opinion testimony that relied upon a new or novel scientific principle, theory, or methodology. Under Frye, the principle, theory, or methodology on which the opinion evidence is based must have been scientifically valid, and the procedures followed to apply the technique or process must have been generally accepted in the relevant scientific community. A Frye hearing was "utilized in Florida only when the science at issue [was] new or novel." Overton v. State, 976 So.2d 536, 550 (Fla. 2007) (quoting Branch v. State, 952 So.2d 470, 483 (Fla. 2006)). Pattern impression analysis—which falls under the broader category of trace evidence—was not new or novel at the time of Anderson's trial. See, e.g., Cochran v. State, 222 So.2d 462, 463 (Fla. 2d DCA 1969) (noting the admission of expert testimony that cloth impressions left at crime scene could have been made by defendant's gloves). Thus, Anderson would not have been entitled to a Frye hearing, and counsel therefore did not render deficient performance by failing to request a hearing for which there was no legal basis. Nor is there any reasonable probability that such a meritless request would have yielded a different result at trial.

Moreover, trial counsel was able to expose to the jury the "shortcomings" and limited value of the grease impression evidence without retaining another expert. During his cross-examination of Gerhart, counsel pointed out that Gerhart could not positively identify the grease marks on Keinya's jacket as having been made by the coils on Anderson's car due to a lack of identifying characteristics and the best Gerhart could say is that those coils could have made the impressions on the jacket. Gerhart conceded that the other two coils from Anderson's car definitely did not make the impressions on the jacket. Gerhart also admitted that every car has coils and there was nothing unique about the coils in this case.

"Strickland does not enact Newton's third law for the presentation of evidence, requiring for every prosecution expert an equal and opposite expert from the defense. In many instances cross-examination will be sufficient to expose defects in an expert's presentation." Harrington v. Richter, 562 U.S. 86, 111, 131 S.Ct. 770, 178 L.Ed.2d 624 (2011). Here, the record demonstrates that the cross-examination conducted by Anderson's trial counsel was sufficient to expose the defects in

Gerhart's presentation, and "[t]here is no reasonable probability that re-presenting virtually the same evidence through other witnesses would have altered the outcome in any manner." Atwater v. State, 788 So.2d 223, 234 (Fla. 2001).

Because we conclude that the alleged deficiencies with regard to the grease pattern analysis are rebutted by the record, we find no error in the summary denial of this claim.

4. Fiber Evidence

Trace evidence expert Bruce Ayala testified at Anderson's trial that one of the thirteen fibers found under Anderson's car "could have originated" from the pants Keinya was wearing when she was killed. He concluded that the twelve other fibers taken from the car did not match any of Keinya's clothing. As in the previous claim, Anderson alleges that counsel was ineffective for failing to consult with experts and secure a Frye hearing with regard to the fiber evidence.

We conclude that trial counsel did not render deficient performance because there was no legal basis to support a request for a Frye hearing. Fiber evidence—which also falls under the broader category of trace evidence—was not a new or novel science at the time of Anderson's trial; it has been admitted in Florida courts since at least 1958. See, e.g., Trimble v. State, 102 So.2d 738, 739 (Fla. 3d DCA 1958) (noting the admission of expert testimony that fibers found underneath defendant's car were similar to those taken from victim's blouse); see also Bundy v. State, 471 So.2d 9, 20 (Fla. 1985) (noting that no Frye hearing was held with regard to fiber evidence); Bottoson v. State, 443 So.2d 962, 964 (Fla. 1983) ("There was also expert evidence that clothing fibers similar to those in the victim's clothes … were found inside the car."); cf. Johnson v. State, 660 So.2d 637, 644 (Fla. 1995) ("Gathering any fiber evidence is a common object of any murder investigation … ."). Thus, the fiber evidence would not have been deemed inadmissible under Frye or even subject to a Frye hearing. Counsel cannot be deemed deficient for failing to make a meritless objection. See Valentine, 98 So.3d at 55.

Anderson also asserts that the testing of the fiber evidence was done using "poor analysis and improper methodology" and that counsel performed deficiently in failing to present expert testimony that "could have explained what would have constituted proper methodology." This allegation is conclusory and insufficient to warrant an evidentiary hearing. Anderson also failed to establish that there is a reasonable probability that the outcome of the trial would have been different had counsel utilized an expert to challenge the conclusion of the State's expert that one of the thirteen fibers found under Anderson's car "could have originated" from Keinya's pants.

5. Cumulative Prejudice

Even if we were to conclude that trial counsel rendered deficient performance in failing to consult with experts and present expert testimony in order to more rigorously challenge the tire, blood, grease, and fiber evidence, Anderson would still not be entitled to relief on this claim because he cannot establish prejudice.

The tire, blood, grease, and fiber evidence were equivocal: the tire and grease impressions may or may not have been left by Anderson's car; the drop of blood on the car seat may or may not have been left at the time of the murder; and a single fiber out of the thirteen that were found on the bottom of Anderson's car may or may not have come from Keinya's pants. The other evidence presented at trial established that two eyewitnesses observed a car that they described as similar to Anderson's car running over a person without deviating from its path on the night of the murder in the area where Keinya's hair, scalp, blood, and personal items were found; there was damage to the radiator and splash guard on Anderson's car; an area under the car had been wiped; Anderson's car was seen waiting outside the Publix where Keinya worked just before the murder; Anderson admitted picking Keinya up from work on the night of her murder; Anderson admitted that he had sole control of his car the night Keinya was murdered; within the four days leading up to the murder, Anderson twice used his car to chase after and attack Keinya's co-worker, Patrick, after he drove Keinya home from work; approximately 48 hours before the murder, Anderson hit Keinya and threatened her, and she was so scared that she armed herself with a knife, dialed 911, locked herself in her room, and did not go to work the next day; despite being on probation for eleven counts of attempted capital sexual battery on Keinya and under a condition that he have no contact with Keinya, Anderson continued to have extensive contact with Keinya in violation of his probation and he faced the possibility of life in prison if Keinya exposed the violation.

In light of the other evidence presented, there is no reasonable probability that had trial counsel more rigorously challenged the equivocal tire, blood, grease, and fiber evidence, Anderson would have been acquitted of first-degree murder. Even if we were to assume that counsel was deficient, our confidence in the outcome would not be undermined.

...

IV. CONCLUSION

For the reasons stated above, we affirm the denial of postconviction relief as to Anderson's conviction and deny the petition for a writ of habeas corpus, but we vacate the death sentence and remand for a new penalty phase.

It is so ordered.

Questions:

1. How many areas of forensic science appear in this litigation?
2. What was the defense objection to tire track impression evidence and whether the expert met the qualifications to be designated an expert.
3. How does the Court deal with the challenge to grease pattern analysis?

The question of whether an expert is qualified to give testimony on a particular subject matter is not as specialized as it seems at first glance. In Hall v. Flannery, the court granted broad discretion to an expert on the matter of seizures following surgery while said latitude was not present in matters involving neurological conclusions.

Locate the following case using your favorite legal research resource, read the case and answer the questions below.

Hall v. Flannery, 840 F.3d 922 (7th Cir. 2016).

Questions:

1. Why does Daubert exclusively apply in the Hall decision?
2. When dealing with the expertise of heart trauma and cause of death- did the court insist on the expert witness having specialized training and education to support the testimony?
3. Would an unqualified expert witness be an automatic basis for dismissal?

The admission of expert testimony and forensic evidence to support that testimony comes to depend on selected tests, namely:

> Three requirements must be met before expert testimony can be admitted: (1) The witness qualifies as an expert by reason of his knowledge, skill, experience, training, or education; (2) the subject matter of

the testimony is an appropriate one for expert testimony; and (3) admitting the expert testimony will actually assist the fact-finder in deciding the case.[8]

In Rhomer, the Court deals with all three issues but does so with some twists.

Rhomer v. State
569 S.W.3d 664 (Tex. Crim. App. 2019)
January 30, 2019

A jury convicted Appellant of felony murder for causing the death of another while committing felony driving while intoxicated, and it sentenced him to seventy-five years in prison. The court of appeals affirmed. We granted Appellant's petition for discretionary review to decide whether an accident reconstruction expert can testify about a specific type of accident reconstruction in which he has no formal training and whether accident reconstruction should be governed by the Kelly or Nenno test for evaluating the reliability of the expert testimony. Kelly v. State, 824 S.W.2d 568 (Tex. Crim. App. 1992); Nenno v. State, 970 S.W.2d 549, 560 (Tex. Crim. App. 1998), overruled on other grounds by State v. Terrazas, 4 S.W.3d 720 (Tex. Crim. App. 1999). We affirm the judgment of the court of appeals.

I. Background

A. Relevant Facts

i. The Accident

Appellant left a bar shortly before 3:00 a.m. and drove his car eastbound on Nakoma Drive. The decedent, Gilbert Chavez, was riding his motorcycle in the westbound lane of Nakoma Drive. As both Appellant and Chavez approached the intersection of Nakoma Drive and Colwick Street from opposite directions, the vehicles collided. Mario Negron and Kenneth Ferrer testified that they were driving in the westbound lane of Nakoma Drive around 3:00 a.m. when they came upon the accident. Both testified to driving through the debris of the accident.

Negron and Ferrer saw Chavez was badly injured and lying near his motorcycle, and they called for help. They saw Appellant at the scene stumbling near his car which had crashed into a building. When police officers arrived, Appellant told them that Chavez had driven into his lane of traffic; he later told one of the officers that Chavez had pulled out in front of him as if Chavez had been traveling in the same direction as Appellant and had hit Appellant on the right side.

Chavez was taken to the hospital where he later died from his injuries. Dr. Randy Frost, the Bexar County Chief Medical Examiner, testified the injuries were consistent with an automobile accident, and multiple traumatic blunt force injuries were the cause of death.

ii. Detective Doyle's Testimony

Detective John Doyle was assigned to the San Antonio Police Department's Traffic Investigation Detail. In his twenty-three years as a police officer, he had investigated at least a thousand vehicular crashes and had testified once as an accident reconstruction expert in a case that involved two cars and a pickup truck. To qualify for his position in SAPD's Traffic Investigation Detail, Doyle attended three courses related to vehicle collisions. Doyle learned the basics of crash investigation – including "skid to stop" formulas, diagram drawing and scene measurement – in an intermediate crash investigation course. He later attended an advanced course at Texas A & M University where he learned how to conduct an "energy analysis" to calculate speed. Doyle also took a reconstruction course and courses on accidents involving pedestrians and bicycles. He testified that his accident reconstruction course work totaled 501 hours.

Doyle never took a course specifically related to motorcycle accident reconstruction and testified that he did not know of one offered in Texas. He admitted that there are "different physics, different science, [and] different mathematical principles" when evaluating a crash that involves two cars versus a crash that involves a car and a motorcycle. He testified, however, that the differences typically have to do with speed calculations and that the "basic facts" of an accident are still the same.At the accident scene, Doyle observed Appellant's car crashed into a building and Chavez's motorcycle lying in the adjacent parking lot. Doyle did a visual inspection of the debris, tire marks and vehicular damage. He then used a precision surveying tool, a Sokkia instrument, to map the locations of the vehicles, debris, curb strikes and scrapes, and the dimensions and curvature of the road. Using these measurements, Doyle created a diagram of the accident that showed the spacial relationship between all of the pieces of evidence at the scene of the accident. Doyle could not calculate the speed of the vehicles at the time of the collision due to the huge weight differential between the car and the motorcycle and due to Appellant's car having crashed into a building without displacing it.

Detective Doyle observed a debris field in the westbound lane of Nakoma Drive that included pieces of Chavez's motorcycle. The debris was in front of the area of impact because the momentum of the debris carried it in the direction in which the motorcycle was traveling at the time of impact. There was no debris in Appellant's lane. There was no pre-impact braking by either the motorcycle or the car. There were three curb strikes and two scrape marks between the resting place of the motorcycle and the resting place of the car. Appellant's car sustained damage on its front left corner because the motorcycle struck the car's wheel well. Part of Appellant's front bumper was lodged in the cooling fins on the left side of Chavez's motorcycle.

On impact the motorcycle and Chavez "took two different paths" with the motorcycle pushed up and backwards while Chavez went over and beside the car, flinging bodily tissue onto it from the gaping wound in his leg and leaving a trail of hand prints in the dust on the car.

Doyle formed the opinion that the wreck was "more or less" a head-on collision:

> Basically that the vehicle driven by the defendant straightened out the cur[ve], hit the motorcycle in the – his traffic lane, in the oncoming traffic lane. The motorcyclist was struck by the left front corner of the car. He went over the car and the vehicle was – motorcycle was pushed backwards into the parking lot. The vehicle continued on in its same direction resulting in ultimately the death of the complainant.

Appellant's failure to negotiate the curve was consistent with alcohol impairment. Doyle testified, "That's one of the very, very common factors in alcohol-related crashes is failing to negotiate a curve, basically straightening out a curve, very common." He concluded that Appellant caused the crash due to alcohol intoxication. He dismissed Appellant's alternative, on-scene claims that Chavez came up from behind him and hit him on his right or crossed into Appellant's lane; Doyle found those scenarios to be inconsistent with the physical evidence that he observed.

B. Issues Presented

This Court granted review to resolve three issues raised by Appellant. First, did the court of appeals violate Texas Rule of Evidence 702 in affirming the trial court's decision to admit Doyle's expert testimony though he had no qualifications in motorcycle accident reconstruction? Second, in relying on Nenno instead of Kelly, did the court of appeals apply the wrong standard when deciding that Doyle's testimony was reliable even though he applied no scientific theory or testing from the field of accident reconstruction and had no qualifications in the field of motorcycle accident reconstruction? Third, should the Nenno standard apply when an expert in a technical scientific field chooses not to apply the scientific testing or theory to a particular case? We reject the premises that Doyle had no qualifications and chose to apply no scientific theory or testing. We hold that the trial court did not abuse its discretion in admitting Doyle's expert testimony; the Nenno standard was applicable under the circumstances of this case; and Doyle's testimony was reliable. Consequently, we affirm the judgment of the court of appeals.

C. Standard of Review

An appellate court reviews a trial court's ruling on the admission of evidence for an abuse of discretion. Rodgers v. State, 205 S.W.3d 525, 527 (Tex. Crim. App. 2006); Powell v. State, 63 S.W.3d 435, 438 (Tex. Crim. App. 2001). The trial court abuses its discretion when it acts without reference to any guiding rules and principles or acts arbitrarily or unreasonably. Montgomery v. State, 810 S.W.2d 372, 380 (Tex. Crim. App. 1990).

II. Analysis

A. Expert Witness Testimony

An expert witness may offer an opinion if he is qualified to do so by his knowledge, skill, experience, training or education and if scientific, technical or other specialized knowledge will assist the trier of fact in understanding the evidence or determining a fact in issue. TEX. R. EVID. 702. Witnesses who are not experts may testify about their opinions or inferences when those opinions or inferences are rationally based on the perception of the witnesses and helpful to a clear understanding of the witnesses' testimony or the determination of a fact in issue. TEX. R. EVID. 701. There is no distinct line between lay opinion and expert opinion. Osbourn v. State, 92 S.W.3d 531, 537 (Tex. Crim. App. 2002).

Three requirements must be met before expert testimony can be admitted: "(1) The witness qualifies as an expert by reason of his knowledge, skill, experience, training, or education; (2) the subject matter of the testimony is an appropriate one for expert testimony; and (3) admitting the expert testimony will actually assist the fact-finder in deciding the case." Vela v. State, 209 S.W.3d 128, 131 (Tex. Crim. App. 2006). These conditions are commonly referred to as (1) qualification, (2) reliability, and (3) relevance. The first two are in play with respect to Detective Doyle's testimony.

B. Qualification

The specialized knowledge that qualifies a witness to offer an expert opinion may be derived from specialized education, practical experience, a study of technical works or a combination of these things. Wyatt v. State, 23 S.W.3d 18, 27 (Tex. Crim. App. 2000). "A witness must first have a sufficient background in a particular field, but a trial judge must then determine whether that background 'goes to the very matter on which [the witness] is to give an opinion.' " Vela v. State, 209 S.W.3d 128, 131 (Tex. Crim. App. 2006), quoting Broders v. Heise, 924 S.W.2d 148, 153 (Tex. 1996). "Fit" is a component of qualification, and "the expert's background must be tailored to the specific area of expertise in which the expert desires to testify." Vela, 209 S.W.3d at 133. The party offering expert testimony has the burden to show the witness is qualified on the matter in question. Penry v. State, 903 S.W.2d 715, 762 (Tex. Crim. App. 1995).

To determine whether a trial court has abused its discretion in ruling on an expert's qualifications, an appellate court may consider three questions: (1) Is the field of expertise complex? (2) How conclusive is the expert's opinion? (3) How central is the area of expertise to the resolution of the lawsuit? See Rodgers, 205 S.W.3d at 528. Greater qualifications are required for more complex fields of expertise and for more conclusive and dispositive opinions. Id. The first two

Rodgers' factors – complexity and conclusiveness – weigh in favor of less stringent qualification requirements in this case while the third factor – dispositiveness – weighs in favor of higher requirements.

Accident reconstruction may sometimes be complex, but the reconstruction opinion offered by Doyle was not. He did not calculate pre-impact vehicle speeds; he examined the physical evidence in the context of the accident scene to form an opinion about the area of impact and how the collision occurred. His testimony was more akin to latent print comparison than to DNA profiling because it was relatively simple. See id. Nor was his testimony very conclusive; he testified that the area of impact was an estimation, not a precise point. According to Doyle, the phrase "point of impact" has fallen out of favor because "it's never a precise point, and it's an area of impact." His opinion was dispositive because there was no other evidence about the central issues of location and cause of the accident. Given that two of the three Rodgers factors weigh in favor of upholding the trial court's ruling on Doyle's qualifications, we cannot say the trial court abused its discretion in admitting Doyle's opinions.

Although Doyle had never taken a class that focused on motorcycle-involved crashes, he knew how to analyze an accident scene based on debris, vehicle damage, skid and gouge marks and vehicle resting places, and he had done so hundreds of times. The involvement of a motorcycle in this collision did not interfere with his ability to analyze the physical evidence present at the accident scene, and his background went to the very matter about which he testified.

Appellant argues that Doyle was not qualified to testify as an expert in motorcycle accident reconstruction because he was not trained to conduct speed and energy calculations for accidents involving motorcycles. But Doyle offered no opinions involving such calculations. According to Doyle, such calculations were impossible because of the weight differential between the car and motorcycle and because the car hit a building that was not displaced. His failure to conduct an impossible speed analysis did not make him unqualified to testify about where and how the crash occurred based on an analysis of the physical evidence at the crash site.

Appellant contends that the use of the Sokkia measuring device and the diagraming of the scene were not scientific methods and that Doyle did not apply any scientific theory in this case. We disagree that measuring and diagraming are not scientific methods. But even if they were not were not scientific methods, that would not mean that Doyle was unqualified; an expert does not need to use scientific methods to be qualified. Morris v. State, 361 S.W.3d 649, 654 (Tex. Crim. App. 2011). An expert is qualified by specialized knowledge, training, or experience. Wyatt, 23 S.W.3d at 27. There is no requirement that the expert's specialized knowledge, training or experience be based on scientific principles.

The trial court is supposed to act as a gatekeeper against expert testimony that would not help the trier of fact. This is not the same thing as requiring every expert to be the best possible witness. We agree with the State that the relevant question was not whether Doyle lacked a particular qualification that would have made him the ideal expert witness but whether the qualifications that he did have would have assisted the jury in determining an issue of fact. We hold that Doyle's qualifications would have assisted the jury in determining issues of fact, namely, where and how the collision happened, and we affirm the court of appeals' decision that the trial court did not abuse its discretion in determining Doyle was qualified to offer an expert opinion on these issues.

C. Reliability

The next issues are whether the expert testimony about accident reconstruction in this case should be evaluated for reliability under Kelly or Nenno and whether Doyle's testimony was reliable.

When an expert's testimony is based on a hard science involving precise calculations and the scientific method, the expert must satisfy the test set forth in Kelly, 824 S.W.2d at 573. Morris, 361 S.W.3d at 654. ("When the subject of an expert's testimony is 'scientific knowledge,' then the basis of that testimony must be grounded in the accepted methods and procedures of science."). The Kelly test for reliability of evidence derived from a scientific theory requires that: (1) the underlying scientific theory must be valid, (2) the technique applying the theory must be valid, and (3) the technique must have been properly applied on the occasion in question. Kelly, 824 S.W.2d at 573.

Nenno set forth a framework for evaluating the reliability of expert testimony in fields of study outside the hard sciences. Nenno, 970 S.W.2d at 561. The Nenno test asks whether (1) the field of expertise is a legitimate one, (2) the subject matter of the expert's testimony is within the scope of that field, and (3) the expert's testimony properly relies upon and/or utilizes the principles involved in the field. Id. The distinctions between hard and soft sciences may be blurred, and the reliability inquiry is flexible. Id. at 560-61.

Appellant argues that this case should be analyzed under Kelly because accident reconstruction is a hard science. He also claims that Doyle's testimony would fail the Kelly test and even the less stringent Nenno test because Doyle did not use the scientific methods and principles that were available to him when he failed to calculate the speed of the vehicles before the collision. We disagree. While a speed calculation might fall under the Kelly test, Doyle could not do a speed calculation in this case due to the weight differential between the car and the motorcycle and because the car crashed into a building without displacing it. Under these circumstances, Doyle's failure to conduct a speed calculation was irrelevant to the reliability of his opinions about how and where the collision happened based on the physical evidence he observed at the scene. As the State points out, Doyle used physical evidence to put together a fairly simple jigsaw puzzle.

Judge Walker's concurring opinion cites Wooten v. State, 267 S.W.3d 289 (Tex. App.—Houston [14th Dist.] 2008, pet. ref'd), and Pena v. State, 155 S.W.3d 238 (Tex. App.—El Paso 2004, no pet.), as support for applying the Kelly test here. But those cases are distinguishable from this one because they addressed the admissibility of expert testimony about the pre-impact speed of the defendants' vehicles. Wooten, 267 S.W.3d at 304; Pena, 155 S.W.3d at 246.

We agree with the court of appeals that the Nenno test applied to Doyle's testimony because his opinions were based on his training and experience in evaluating physical evidence at crash scenes more than on a hard scientific inquiry such as calculating a vehicle's pre-impact speed. Rhomer v. State, 522 S.W.3d 13, 17 (Tex. App.—San Antonio 2017). We conclude that (1) the field of accident reconstruction is a legitimate one, (2) the subject matter of Doyle's expert testimony was within the scope of that field, and (3) his testimony properly relied upon and utilized the principles involved in the field, i.e., examining the physical evidence in the context of the crash site to draw conclusions about the location and cause of the crash.

III. Conclusion

Doyle was qualified to testify as an expert in accident reconstruction, and his testimony was reliable. We affirm the judgment of the court of appeals.

Question:

1. Prove this statement True or False: The Court concludes that one can be an expert without any scientific basis or grounding.
2. What was the field of expertise in Rhomer? And does the Court indicate there are both scientific and non-scientific methods involved in accident reconstruction.
3. How does the opposing counsel seek to show the accident expert lacked both the qualifications and the experience to hold himself out as an expert?

Critics of Daubert argue that all of the former standards of Frye have become so watered down that one wonders why any standards exist at all for the admission of expert evidence and testimony. Look closely at how this has taken place in Gissantaner.

United States v. Gissantaner
990 F.3d 457 (6th Cir. 2021)
March 5, 2021

At issue in this case is the reliability of a form of DNA-sorting evidence under Rule 702 of the Federal Rules of Evidence and Daubert v. Merrell Dow Pharmaceuticals, Inc., 509 U.S. 579, 113 S.Ct. 2786, 125 L.Ed.2d 469 (1993).

I.

Daniel Gissantaner became embroiled in an argument with his neighbors. One neighbor called 911, telling the dispatcher that Gissantaner, a convicted felon, had a gun. The police responded to the call and found a pistol inside a chest in Gissantaner's house. The chest belonged to Gissantaner's roommate. When the government charged Gissantaner with possessing a firearm as a felon, it used DNA-sorting evidence, called STRmix, to link Gissantaner to the gun.

Gissantaner moved to exclude the evidence as unreliable under Evidence Rule 702. See Daubert v. Merrell Dow Pharms., Inc., 509 U.S. 579, 113 S.Ct. 2786, 125 L.Ed.2d 469 (1993). Gissantaner and the government retained experts, who took competing positions on the issue. The district court in turn appointed two experts of its own to consider the question. One of these experts said that STRmix evidence is reliable in general and as applied to this case; the other said it is reliable in general but not as applied to this case. The court excluded the STRmix evidence as unreliable. The government filed this interlocutory appeal. See 18 U.S.C. § 3731.

II.

. . .

The Biology DNA Unit of the Michigan State Police laboratory has used STRmix for six years. The laboratory received training on the software starting in March 2015, and it began using the software for cases in February 2016, about three and a half months before receiving the sample in this investigation (Figure 2.3).

In Gissantaner's case, an analyst with the Michigan State Police laboratory took information about the DNA present in the mixture and entered it into STRmix to estimate how much of the DNA came from each person. The resulting DNA profile summary said that one person "contributed" 68% of the DNA in the mixture, a second contributed 25%, and a third contributed 7%. The third contributor supplied 8 or 9 cells (approximately 49 picograms) to the mixture.

STRmix compared the DNA of the suspect—Gissantaner—to this profile with the goal of ascertaining a "likelihood ratio" about his potential contribution to the sample. R.77 at 34. A comparison of Gissantaner's DNA to the profile suggested that he matched the third contributor, generating a likelihood ratio of 49 million to 1. More precisely, if less accessibly, that means the DNA "profile is 49 million times more likely if [Gissantaner] is a donor than if he is not." Id. at 48.

The two "ifs" capture a qualification. The likelihood ratio tells us only that, in the abstract and without considering any other evidence in this case, it would be unusual if this DNA contained no DNA contributed from Gissantaner. The ratio does not on its own tell us how likely it is that Gissantaner illegally possessed a firearm. Determining whether it is likely that Gissantaner, as opposed to someone else, contributed to the mixture requires looking at other facts beyond the scope of DNA evidence. Perhaps other people with similar profiles, say relatives of Gissantaner, were nearby. Perhaps the roommate had locked the chest and lost the key long before Gissantaner moved in and had a chance to touch the gun. Or perhaps the DNA landed on the gun without Gissantaner touching it, say when he and his roommate shook hands

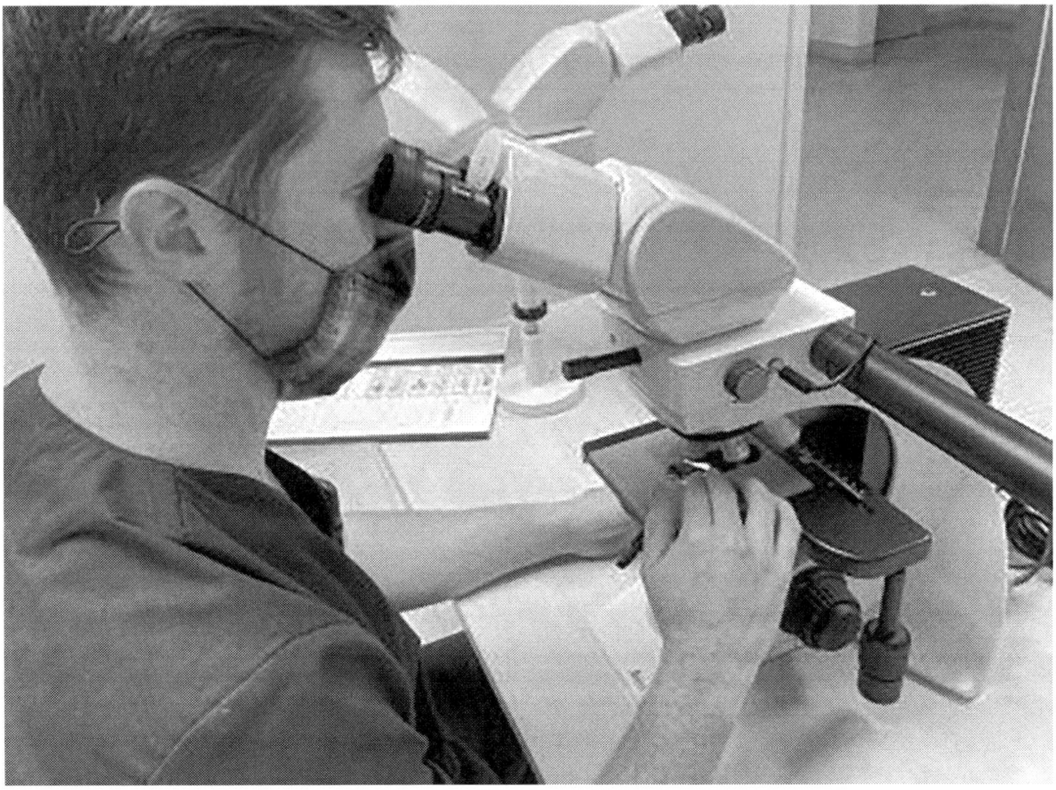

FIGURE 2.3 Forensic pathologist at work. Source Cook County, Illinois Forensic Pathology Fellowship Program.

or when they each touched the same object before his roommate subsequently touched the gun. Even with these qualifications, the evidence remains significant. By accepted measures in the forensic community, a profile match of 1 in 49 million amounts to "very strong support" that Gissantaner somehow contributed DNA to the mixture. R.41-10 at 3. These two bottom-line conclusions bring into view the need to use evidence of this sort carefully. Such conclusions—the 49-million-to-1 ratio and "very strong support"—can be highly probative of guilt or innocence. Yet the mechanisms for obtaining them—the software, the science—are beyond the ken of most jurors and judges. If highly consequential evidence emerges from what looks like an indecipherable computer program to most non-scientists, non-statisticians, and non-programmers, it is imperative that qualified individuals explain how the program works and ensure that it produces reliable information about the case. All of this explains why the courts have developed reliability standards for admitting evidence of this type and why a functioning adversarial system remains critical in handling it.

III.

Rule 702 of the Federal Rules of Evidence sets the framework for determining whether to admit scientific and other technical evidence in federal civil and criminal cases. It permits an expert to testify about scientific knowledge if "(a) the expert's scientific, technical, or other specialized knowledge will help the trier of fact to understand the evidence or to determine a fact in issue; (b) the testimony is based on sufficient facts or data; (c) the testimony is the product of reliable principles and methods; and (d) the expert has reliably applied the principles and methods to the facts of the case." Fed. R. Evid. 702. …

Four inquiries guide the reliability analysis: Is the technique testable? Has it been subjected to peer review? What is the error rate and are there standards for lowering it? Is the technique generally accepted in the relevant scientific community? Daubert, 509 U.S. at 593–94, 113 S.Ct. 2786. Multi-factor tests, especially non-exhaustive tests, run the risk of obscuring the core inquiry. The key handholds of Rule 702 thus bear repeating: To be admissible, any relevant scientific or technical evidence must be the "product of reliable principles and methods" and must have been "reliably applied" in the case. That is what matters most. Otherwise, the central point of Daubert, to establish that Evidence Rule 702 "displaced" the common law Frye test, would be lost and would lead to the replacement of an old common law test with a new (harder to pronounce) common law test. Compare Frye v. United States, 293 F. 1013, 1014 (D.C. Cir. 1923), with Daubert, 509 U.S. at 585–89, 113 S.Ct. 2786. That is not progress.

Through it all, the district court has a "gatekeeping role" in screening expert testimony to ensure that only reliable testimony and evidence go to the jury. Daubert, 509 U.S. at 597, 113 S.Ct. 2786. We give fresh review to a district court's framing of the legal standard, United States v. Pugh, 405 F.3d 390, 397 (6th Cir. 2005), and abuse-of-discretion review to its admissibility decision, Kumho Tire Co. v. Carmichael, 526 U.S. 137, 152, 119 S.Ct. 1167, 143 L.Ed.2d 238 (1999).

IV.

Measured by Evidence Rule 702 and a proper framing of the Daubert factors, the DNA evidence should be admitted on this record.

Testability. An untestable scientific theory is all theory and no science. In the absence of proof that a technology "can be … tested," Daubert, 509 U.S. at 593, 113 S.Ct. 2786, there is no way to show whether it works (its "refutability" or "falsifiability," a scientist would say) and no way to give it "scientific status." Id.; United States v. Bonds, 12 F.3d 540, 559 (6th Cir. 1993). The question on the table is whether a method can be "assessed for reliability," not whether it always gets it right. Fed. R. Evid. 702 advisory committee's note to 2000 amendment; Bonds, 12 F.3d at 558–59. Disputes about the "adequacy of the [theory's] testing" or about the "accuracy of [a theory's] results," generally speaking, provide grist for adversarial examination, not grounds for exclusion. …

STRmix can be tested. Using "lab-created mixtures," in which the actual contributors of the DNA samples are known, scientists have tested STRmix to gauge the reliability of the technology. R.146-14 at 2. Suppose that one person, Aaron, contributed to a lab-created mixture, but another, Britney, did not. Forensic scientists can test STRmix to see whether it suggests that Aaron is a match for the mixture, but Britney is not. If STRmix suggests that Aaron is not a match for the mixture (by outputting a low likelihood ratio), that would be a false negative. If STRmix suggests that Britney is a match for the mixture (by outputting a high likelihood ratio), that would be a false positive. Each possibility shows that STRmix is testable, that lab-created mixtures offer a way to "assess[] [the] reliability" of STRmix. Fed. R. Evid. 702 advisory committee's note to 2000 amendment.

The record from the evidentiary hearings in this case provides a long proof that STRmix is testable and refutable. Almost all of the evidence in the hearings went to these points: How often is it accurate? How often is it not? Similar evidence of DNA testability has sufficed before in our circuit. Bonds, 12 F.3d at 558.

(The astute reader might remind us that Daubert asks whether the scientific theory at issue "can be (and has been) tested." 509 U.S. at 593, 113 S.Ct. 2786. It is not clear what the parenthetical means. The rest of the Daubert considerations—including the next one, peer review—all turn in one way or another on what actual testing of the theory reveals in terms of reliability. At all events, the point makes no difference here. STRmix indeed "(has been) tested" many times before, as the rest of this opinion confirms.)

Peer review. Subjecting a new technology to "peer review and publication" offers another measure of reliability. Id. The "key" is whether "the theory and procedures have been submitted to the scrutiny of the scientific community." Bonds, 12 F.3d at 559. …

Peer review is not student review. It "conditions publication on a bona fide process" of review by other scientists and experts in the field. Daubert v. Merrell Dow Pharms., Inc., 43 F.3d 1311, 1318 n.6 (9th Cir. 1995). The scientific community uses different conventions for publication from most journals published in the legal community. No offense to former, current, and future members of law journals everywhere: But it is one thing to convince lawyers in training to publish a piece; it is quite another to convince peers in a professional community to publish a piece. That is why readership and citation are pivotal when it comes to legal scholarship and why publication itself is noteworthy in scientific scholarship—and ultimately why publication in a peer-reviewed journal alone typically satisfies this Daubert inquiry …

For like reasons, this factor does not demand independent authorship—studies done by individuals unaffiliated with the developers of the technology. Independent studies, to be sure, advance the cause of reliability. Bonds, 12 F.3d at 560. But they are not indispensable. Peer review contains its own independence, as it involves "anonymously reviewing a given experimenter's methods, data, and conclusions on paper." Mitchell, 365 F.3d at 238. If experts "have other scientists review their work" and if the other scientists have the chance to identify any methodological flaws, that usually suffices. Mitchell v. Gencorp Inc., 165 F.3d 778, 784 (10th Cir. 1999). When scientific research is accepted for publication by a reputable journal following the "usual rigors of peer review," that represents "a significant indication that it is taken seriously by other scientists, i.e., that it meets at least the minimal criteria of good science." Daubert, 43 F.3d at 1318.

STRmix clears this bar. At the time of the Daubert hearing in the district court, more than 50 published peer-reviewed articles had addressed STRmix. According to one expert, STRmix is the "most tested and most … peer reviewed" probabilistic genotyping software available. R.77 at 82. At least two of the studies were done by individuals unconnected to the development of the software. This plainly suffices. …

Error rate and standards to lower it. Even ten witnesses to a crime come with risks of error. So too for DNA evidence. This consideration looks to the error rate of the technology and to whether the scientific community has established standards that forensic scientists can use to mitigate the risk of error. Daubert, 509 U.S. at 594, 113 S.Ct. 2786.

Think about Gissantaner's case to see the point. The government would like to use STRmix to match Gissantaner to the DNA on a gun. That is not a good idea—not the "product of reliable principles and methods" under Rule 702—if STRmix has a high error rate, if it has trouble "avoid [ing]" "false positives," and if there are no standards or guidelines to avoid or lessen these risks. Bonds, 12 F.3d at 559 (quotation omitted); Mitchell, 365 F.3d at 241.

How often, then, does STRmix falsely suggest a suspect matches a DNA sample? Not often, the evidence suggests. When examining "false inclusions," one peer-reviewed study concluded, based on an analysis of the DNA of 300,000 people who were known not to be in a mixture, that STRmix had accurately excluded the non-contributors 99.1% of the time. Just 1% of the time, in other words, it gave a likelihood ratio suggesting that someone was included in the mixture who was not actually included in it. Most of these false inclusions, moreover, were associated with low likelihood ratios—meaning that, under STRmix's own estimates, the confidence that the person was included was low. …

General acceptance in the scientific community. One might still be skeptical of the reliability of a relatively new technology like STRmix, and rightly so, if the relevant scientific community has not yet accepted its use. Daubert, 509 U.S. at 594, 113 S.Ct. 2786. The question for debate is "general acceptance," not uniform acceptance within the community. See Bonds, 12 F.3d at 562. Nor must the science be beyond reproach. Id. What matters is whether the relevant scientific community accepts the software. See Daubert, 509 U.S. at 594, 113 S.Ct. 2786; see also Wilden v.

Laury Transp., LLC, 901 F.3d 644, 654–55 (6th Cir. 2018). After that, the long-tested cauldron of cross-examination, not exclusion, is the place to go for accuracy. "[C]onventional [trial] devices, rather than wholesale exclusion under an uncompromising 'general acceptance' test, are the appropriate safeguards where the basis of scientific testimony meets the standards of Rule 702." Daubert, 509 U.S. at 596, 113 S.Ct. 2786. For a technology that is widely used, controversies over its use in a given case usually will be left to the jury. See United States v. Jones, 965 F.3d 149, 160 (2d Cir. 2020).

STRmix satisfies this consideration. It has garnered wide use in forensic laboratories across the country. More than 45 laboratories use it, including the FBI and many state law enforcement agencies. At this point, STRmix is the "market leader in probabilistic genotyping software." R.146-1 at 17.

Consistent with this reality, numerous courts have admitted STRmix over challenges to its general acceptance in the relevant scientific community. See United States v. Lewis, 442 F. Supp. 3d 1122, 1155 (D. Minn. 2020) ("[T]here is no doubt that STRmix has gained general acceptance."); United States v. Washington, No. 8:19CR299, 2020 WL 3265142, at *2 (D. Neb. June 16, 2020) ("Authority and evidence demonstrate that STRmix is generally accepted by the relevant community."); People v. Blash, No. ST-2015-CR-0000156, 2018 WL 4062322, at *6 (V.I. Super. Ct. Aug. 24, 2018); … ("[STRmix's] analyses are based on calculations recognized as reliable in the field."); United States v. Pettway, No. 12-CR-103S (1), (2), 2016 WL 6134493, at *1 (W.D.N.Y. Oct. 21, 2016) (discussing "exhaustive[] research[]" concluding that "the scientific foundations of the STRmix process are based on principles widely accepted in the scientific and forensic science communities"). The Second Circuit determined that the scientific community accepted a different (but similar) DNA-sorting software, Forensic Statistical Tool, even though just one laboratory had used it. Jones, 965 F.3d at 156, 162.

…

All in all, STRmix satisfies Rule 702 and the case law construing it. In the words of Rule 702, it is the "product of reliable principles and methods."

…

Questions:

1. The Court's opinion admits that the science may not be the best of science or the most demonstrable form of science yet still helpful to the Court. True or False
2. Does the court similarly deemphasize peer review as a means of assuring quality in the expert evidence?
3. Gissantaner deals with DNA evidence- a form of forensic that most impute always right and always certain. While the science may be so, the methods and methodologies to extract may be imperfect and even subject to error. What does the Court mean that a method can still be reliable but not necessary right?

> Visit the American Medical Forensic Specialists website and learn about Medical expert witnesses at https://www.amfs.com/resources/what-is-a-medical-expert-witness.

Under the general parameters of expert evidence analysis, trial court's are generally given wide latitude to conduct processes and in the absence of abuse of discretion, are not likely to be overturned by an appellate court. In addition, when errors are discovered, the appeals court generally goes out of its way to declare the error harmless rather than reversible. In United States v. Ray, the opposite result occurs.

United States v. Ray
956 F.3d 1154 (9th Cir. 2020)
April 18, 2020

Patrick Bacon and Daniel Ray were convicted of assault with a deadly weapon with intent to do bodily harm and assault causing serious bodily injury. 18 U.S.C. § 113(a)(3), (6). They were sentenced to 120 months and 100 months, respectively. On appeal Bacon argues that the district court should have allowed his forensic clinical expert psychologist, Dr. Karim, to testify, which would have allowed him to present his insanity defense to the jury. We hold that the district court abused its discretion in excluding Dr. Karim's testimony because the testimony was relevant to Bacon's defense. Because this error was not harmless, and we cannot tell from the record whether the testimony was reliable, we must vacate Bacon's conviction and remand for a new trial.

I.

Bacon and Ray were both incarcerated at Victorville Federal Prison in California. Bacon stabbed inmate Anthony Grecco with a metal shank, fracturing Grecco's sinus cavity and causing stab wounds to his head and chest. Security cameras recorded the attack and events beforehand.

Bacon had entered a housing unit, and when questioned by correctional officers, lied and said he was housed there. He met with Ray, and they both walked to Ray's cell. Ray took a book from the cell. Defendants walked to a table, where Ray put the book in front of Bacon and walked away. Bacon took the book apart. Ray returned and stood next to Bacon, until Bacon left holding something below his waist. Bacon then stabbed Grecco with the shank. Guards responded, broke up the assault, and recovered the shank and book.

A grand jury indicted Bacon and Ray under 18 U.S.C. § 113(a)(3) (assault with a deadly weapon with intent to do bodily harm) and (a)(6) (assault causing serious bodily injury). After a two-day trial, a jury found defendants guilty of both counts.

Prior to trial Bacon gave notice, pursuant to Federal Rule of Criminal Procedure 12.2, that he would assert an insanity defense. The government then moved in limine to preclude Bacon's expert, Dr. Karim. Among other conclusions, Dr. Karim opined: (1) "that a review of Mr. Bacon's psychosocial history confirms that he has suffered from a severe and chronic mental illness (or defect) throughout the course of his adult life" and "he presents with long-standing and chronic mental health disorders"; (2) "there are elements of a downward spiral of isolation, depression, paranoia, and anxiety that resulted in a dissociative state for Mr. Bacon prior to the conduct itself" and (3) as a result "it would be reasonable to conclude with a high degree of clinical certainty that an individual who was suffering from the myriad of severe mental health disorders that Mr. Bacon was facing on October 18, 2016 would have had difficulty understanding the nature and quality of his actions at the time of the offense conduct." Dr. Karim acknowledged that Bacon has "a history of aggression and physical assaults," but concluded that Bacon's psychological deterioration during the months before the assault impacted his ability to "differentiate his actions" at the time of the assault. Dr. Karim further suggested that Bacon's "largely unplanned and unsophisticated criminal history" could be explained by "a diagnosis of an Unspecified Bipolar disorder."

The government moved to preclude Dr. Karim's testimony. The government argued the expert testimony was irrelevant and unreliable under Daubert and Federal Rule of Evidence 702, because

Dr. Karim did not opine that Bacon suffered from "any mental health disorder" on the date of the assault, his opinion about a "dissociative state" was not based on medical literature, he did not explain the results of tests he administered to Bacon, and his opinions appeared to have been based on hearsay.

The district court granted the motion, finding that under Rule 702, Dr. Karim's opinion was not relevant because it would "not help the trier of fact to understand the evidence or determine the issue of sanity." After explaining the Daubert and Rule 702 standards, and summarizing Dr. Karim's opinions, the district court stated: "We start with the question of relevance. In fact, we start and end with the question of relevance." The court found that "Dr. Karim's opinion that an individual who was suffering from a myriad of severe mental health disorders that Mr. Bacon was facing would have had difficulty understanding the nature and quality of his action at the time of the offense conduct is equivocal and will not help the trier of fact to understand the evidence or determine the issue of sanity." The court also stated that Dr. Karim's testimony does "not satisfy the threshold standard of relevance" because "Dr. Karim is unwilling or cannot opine that as a result of Mr. Bacon's mental health issues he was unable, as opposed to [had] difficulty understanding, [or] appreciat[ing] the nature and quality of his acts … ." Thus, according to the district court, Dr. Karim's testimony did not "satisfy the standard to entitle the defendant to [assert] an insanity defense according to the law of this circuit." The court ultimately found "that Dr. Karim's opinions, therefore, are speculative, irrelevant, and unreliable."

In the alternative, the district court found that "Dr. Karim is precluded from testifying as an expert witness because whatever probative value the proffered testimony may have [is] substantially outweighed by undue prejudice, confusion of the issues, and undue waste of time under [Federal Rule of Evidence] 403." Because the district court precluded Dr. Karim from testifying, it barred Bacon's insanity defense, under 18 U.S.C. § 17, the Insanity Defense Reform Act ("IDRA").

We have jurisdiction under 28 U.S.C. § 1291.

II.

We review "the district court's exclusion of expert testimony" for abuse of discretion. United States v. Christian, 749 F.3d 806, 810 (9th Cir. 2014). We first "consider whether the district court identified the correct legal standard for decision of the issue before it" and then we "determine whether the district court's findings of fact, and its application of those findings of fact to the correct legal standard, were illogical, implausible, or without support in inferences that may be drawn from facts in the record." United States v. Hinkson, 585 F.3d 1247, 1251 (9th Cir. 2009) (en banc).

A.

Bacon argues that the district court abused its discretion by refusing to allow Dr. Karim's testimony even though it was relevant and reliable. "The admissibility of expert testimony is generally governed by Federal Rule of Evidence 702, which requires district courts to 'perform a gatekeeping function to ensure that the expert's proffered testimony is both reliable and relevant.'" Christian, 749 F.3d at 810 (quoting United States v. Redlightning, 624 F.3d 1090, 1111 (9th Cir. 2010)). Here the district court focused exclusively on relevance when evaluating Dr. Karim's testimony: "We start with the question of relevance. In fact, we start and end with the question of relevance."

The correct legal standard is for the district court "to determine the relevance of the psychological evaluation the expert conducted and the medical diagnoses he made, not his ultimate legal conclusion regarding the defendant's mental state." Christian, 749 F.3d at 811. Here, the district court instead focused on Dr. Karim's bottom-line opinions, rather than "his proposed expert testimony," id., contrary to our guidance in Christian. There, we emphasized "that a district court deciding whether to admit expert testimony should evaluate whether that testimony 'will assist the trier of fact in drawing its own conclusion as to a fact in issue' and should not limit its consideration to 'the existence or strength of an expert's opinion.'" Id. (quoting United States v. Rahm, 993 F.2d 1405, 1411 (9th Cir. 1993)). We explained this is necessary because the doctor there could not have testified that the defendant "lacked the capacity to form the specific intent to threaten," id. at 812 (citing Fed. R. Evid. 704(b)), and "[i]t would make little sense to require a conclusive opinion in determining admissibility, and then absolutely to forbid expression of the opinion in testimony," id. (quoting Rahm, 993 F.2d at 1411 n.3). So too here. Dr. Karim could not have testified to the jury that Bacon's mental disease and defect prevented him from appreciating the nature and quality or wrongfulness of his acts, because "an expert witness must not state an opinion about whether the defendant did or did not have a mental state or condition that constitutes an element of the crime charged or of a defense." Fed. R. Evid. 704(b). Thus "the absence of an opinion to that effect in his report is not a valid reason to preclude his testimony." Christian, 749 F.3d at 812.

Dr. Karim's report demonstrates that his evaluation of Bacon was relevant to Bacon's insanity defense. For example, Dr. Karim concluded that Bacon "was suffering from a myriad of severe mental health disorders," and that Bacon "would have had difficulty understanding the nature and quality of his actions at the time of the offense conduct." If admissible, testimony about these "severe" mental health disorders and their impact on Bacon's perception at the time of the assault "may well have been helpful to the jury in deciding," Christian, 749 F.3d at 812, whether Bacon was insane at the time. Accordingly, the district court abused its discretion by precluding Dr. Karim's testimony because he did not opine that Bacon was unable to appreciate the nature and quality of his acts at the time of the assault. This was the wrong legal standard. Instead, the district court should have focused on whether Dr. Karim's testimony would have assisted the jury "in drawing its own conclusion as to a 'fact in issue,'" id. at 811—the impact of any serious mental health disease or defect on Bacon's ability to appreciate the nature and quality of his acts.

If otherwise admissible, Dr. Karim's expert testimony "would have been highly probative" of Bacon's mental state and "unlikely to cause significant confusion with the jury if properly constrained by compliance with the rules of evidence." United States v. Cohen, 510 F.3d 1114, 1126–27 (9th Cir. 2007). Thus, even if the district court had explained the Rule 403 exclusion, it likely would have abused its discretion. With no explanation, it clearly did so.

We do not hold that the district court must admit Dr. Karim's testimony on remand, only that the district court abused its discretion in finding the testimony was not relevant to Bacon's insanity defense. On remand, to fulfill its "gatekeeping function" under Rule 702 and Daubert, the district court should consider whether Dr. Karim's testimony is reliable. See Christian, 749 F.3d at 810 (quoting Redlightning, 624 F.3d at 1111). The government, in its Daubert motion, raised a number of very real reliability issues with Dr. Karim's expert testimony, including that Dr. Karim did not explain his reasoning or methodology in arriving at his conclusions and cited no medical literature showing that a "dissociative state" or other mental health disorders suffered by Bacon at the time of the offense are considered mental diseases or defects. We cannot express any view on the admissibility of Dr Karim's testimony under Rule 702 "because the record before us is too sparse to determine whether the expert testimony is … reliable." Barabin v. AstenJohnson, Inc., 740 F.3d 457, 466 (9th Cir. 2014) (en banc).

B.

We must decide whether the exclusion of Dr. Karim's testimony was harmless error. See United States v. Morales, 108 F.3d 1031, 1040 (9th Cir. 1997) (en banc). It was not. If the district court had admitted Dr. Karim's testimony, Bacon's insanity defense would have gone to the jury. Given Bacon's prior mental health diagnoses, an expert witness may have "provided some evidentiary basis for inferring … a link between [Bacon's] obvious mental illness and [his] sole defense." Christian, 749 F.3d at 813. Without this testimony Bacon was unable to present his insanity defense to the jury. Thus, the error was not harmless, and Bacon's "substantial rights were affected by the district court's error." Id.; see also Rahm, 993 F.2d at 1415–16.

III.

We now turn to the proper remedy for the district court's non-harmless error of precluding Bacon's expert testimony: We must vacate the conviction and remand for a new trial. See Christian, 749 F.3d at 814. In Christian, we explained that "Barabin extended a general evidentiary rule requiring a new trial '[w]hen the district court has erroneously admitted or excluded prejudicial evidence' to the admission of expert testimony." Id. (quoting Barabin, 740 F.3d at 466 (alteration in original)). While acknowledging that "Barabin involved the admission of expert testimony in a civil trial," the Christian court held "that Barabin's analysis applies with equal force to" criminal cases in which the district court excluded expert testimony. Id. Absent intervening Supreme Court authority, we are bound by the prior decisions of this Court. See Miller v. Gammie, 335 F.3d 889, 893, 899–900 (9th Cir. 2003) (en banc). Accordingly, we vacate Bacon's conviction and remand for a new trial.

Second, if Bacon again testifies and the government seeks to impeach him with evidence of his prior convictions, the district court should consider the five factors we noted in United States v. Hursh, 217 F.3d 761 (9th Cir. 2000), when "balancing the probative value of evidence of a defendant's prior convictions against that evidence's prejudicial effect," id. at 768. We express no view on the merits of any challenges to that impeachment.

VACATED and REMANDED.

Questions:

1. What area of scientific or technical expertise was at stake in this case?
2. When the defendant sought the admission of an expert psychologist, what was the trial court's conclusion?
3. How does the Appeals Court resolve this dilemma?

Under the Daubert standard, the term "helpfulness" is commonly encountered. For some advocates and judges, literally anything can be construed as helpful when it may really not be. One parameter that gives guidance on the meaning of helpfulness is whether the evidence is "relevant" to the case at hand. In U.S. Williams, that very issue of relevance emerges.

United States v. Williams
865 F.3d 1328 (11th Cir. 2017)
August 1, 2017

In the waning hours of May 9, 2015, the crew of the Rasputin—defendants Vanston Venner Williams, Mario Alin Bent Barker, Carlos Clemente Henry Taylor, Edince Garcia Cardoza, and Hendrick Guillermo Linero Duffis—were traveling away from Colon, Panama when, much like their vessel's namesake, their luck ran out. A Coast Guard cutter approached the Rasputin, which sped away as four of the crew members swiftly threw dozens of packages overboard, none of which were recovered. Although the defendants successfully jettisoned their contraband, they could not elude the Coast Guard, who boarded the Rasputin and arrested its crew. After a trial, a jury concluded that the jettisoned packages contained cocaine, convicting each defendant of conspiracy to distribute at least five kilograms of a substance containing cocaine while on board a covered vessel, in violation of 46 U.S.C. §§ 70503(a)(1) and 70506(b) and 21 U.S.C. § 960(b)(1)(B), and possession with intent to distribute at least five kilograms of a substance containing cocaine, in violation of 46 U.S.C. §§ 70503(a)(1) and 70506(a) and 21 U.S.C. § 960(b)(1)(B). The jury also convicted Williams of failure to heave to, in violation of 18 U.S.C. § 2237(a)(1), and the remaining defendants of aiding and abetting Williams's failure to heave to, in violation of 18 U.S.C. §§ 2237(a)(1) and (2). The defendants appeal their convictions, challenging several of the district court's evidentiary rulings as well as the sufficiency of the evidence supporting their convictions. After oral argument and thorough review of the record, we affirm each defendant's drug convictions and Williams's failure-to-heave-to conviction but reverse for lack of evidence the remaining defendants' aiding and abetting failure-to-heave-to convictions.

I. BACKGROUND

We present only the facts relevant to the issues on appeal, which are the defendants' challenges to the admission of the testimony of government expert Gustavo Tirado, the testimony of several Coast Guard officers that the objects they witnessed being jettisoned resembled cocaine bales seized in previous interdictions, and a document listing the Rasputin's next port of call, as well as the defendants' challenges to the sufficiency of the evidence as to each of their convictions.

The United States Coast Guard cutter "Bear" was patrolling waters off the coast of Colombia and Panama—an area where drug trafficking is known to occur—searching for vessels that might be smuggling drugs. Around midnight, crew member Petty Officer Turner Adair noticed that the Bear's radar was picking up a vessel approximately six to eight nautical miles away. The vessel was heading north-northwest away from Colon, Panama at eight to ten knots. As the Bear approached the vessel, Adair used a forward-looking infrared system (FLIR) to view it. The FLIR system could capture images of body heat miles away. Objects that were body temperature or warmer appeared as black, while objects cooler than body temperature appeared as white. Through the FLIR, Adair could see that the vessel was a fishing boat containing a driver and four other individuals. When the Bear's crew hailed the vessel on the radio, Adair observed that the four individuals who were not driving the vessel began to move about in a "worried manner." Trial Tr. 9/8/2015, Doc. 118 at 189-90, 201. The four individuals began to place objects, each measuring approximately one foot high by three feet long, into a fishing net.

As the four men on the vessel gathered the objects, the vessel increased its speed to 12 to 14 knots and began zig zagging from left to right for no apparent reason. Adair testified that such maneuvers were dangerous given that the waves and swells that night were between six and eight feet. After the four men finished placing at least ten bale-like objects into the fishing net, they tied the net up and threw it into the sea. Noticing that it took all four men to lift the net, Adair estimated (based on bales he had handled) that each bale weighed 65 to 75 pounds. After dumping the first net overboard, the four men loaded five more bales into another fishing net then threw that net overboard as well. The vessel changed direction, then heading south, and the four men dumped a third net full of bales.

...

Fleming discovered that the Rasputin's radio was turned to Channel 16—the channel that the Coast Guard had been using to attempt to communicate with the vessel—and the volume was turned all the way up. Fleming observed two empty fuel drums aboard the vessel and four empty 40-gallon gas containers. He recognized the strong smell of gasoline, noting that the fish hold had more than an inch of gasoline covering the floor. The Rasputin did not run on gasoline, however; it ran on diesel. The presence of gasoline suggested to the officers that it was being used as a masking agent to alter the chemical composition of contraband residue. And despite the fact that the Rasputin was registered in Florida as a commercial fishing vessel, there was no fishing gear, bait, ice, or fish aboard. Fleming and his crew seized three Global Positioning Systems (GPS) from the

Rasputin and one $20 bill each from Williams and Cardoza. The officers also seized from the vessel a "zarpe"—a Colombian document including the names of the defendants and their ports of call. The zarpe listed Colon, Panama as the Rasputin's next port of call. The zarpe was admitted into evidence at trial over the defendants' objection that it was hearsay and unauthenticated.

No quantity of drugs was found aboard the Rasputin, nor were any of the jettisoned packages recovered, as they apparently sank. In an attempt to detect the presence of contraband aboard the vessel, Fleming's boarding team used an IonScan machine. IonScan technology is designed to detect trace amounts of illicit materials—often amounts so small as to be imperceptible to the human eye. Samples, or "swipes," are taken of areas and objects thought to contain contraband. The samples are then run through the IonScan machine, which measures the amount of time it takes for ions from vaporized molecules to drift from one side of a tube into a collector. Because every substance has a unique, predictable drift time, the machine can identify a substance on a sample based on the amount of time it takes for the vaporized molecules to drift into the collector. Fleming collected 34 IonScan swipes aboard the Rasputin.

To protect against contamination and to ensure the boarding team introduced no contraband onto the Rasputin, the officers who boarded had their gear swiped for testing prior to boarding. Aboard the Rasputin, Fleming conducted IonScan swipes by putting on rubber gloves from sealed boxes and using small circular pads (also from sealed boxes) to swipe various surface areas of the Rasputin and its crew. Another member of Fleming's team kept a log of each swipe, assigning a number and description to each swipe. After each swipe, Fleming removed his gloves and placed them into a sealed bag, which was identified with a number corresponding to the number logged for each swipe. The sealed bags containing the IonScan pads were transported back to the Bear for analysis.

Petty Officer Richard Caruso was the IonScan operator aboard the Bear. Caruso had attended a three-day training course where he learned how the IonScan machine worked and how to maintain the machine, calibrate it properly, run swipes through it, and read the results of its analysis. Caruso testified at trial that he had operated the machine more than five times in training and at least eight to ten times since, and that no scientific degree was required to operate it. The Bear's IonScan machine underwent a weekly preventative maintenance check and a daily parts check, and it was located in a secure area with limited access. While the Coast Guard pursued the Rasputin, Caruso turned on the IonScan machine and calibrated it, determining that it was operating correctly. He then swiped himself, the machine, and the area around the machine to ensure there was no contamination.

Caruso received the swipes Fleming had taken aboard the Rasputin. Of the 34 swipes, 13 tested positive for cocaine. After each positive hit, Caruso ran two blank swipes through the machine to clear it, again to prevent contamination. The IonScan analysis detected positive hits for cocaine in the Rasputin's fish hold, marine toilet, and sink, as well as on seat cushions and on a knife. The IonScan also revealed cocaine on the person of each occupant of the Rasputin except defendant Taylor. By contrast, the IonScan tests conducted on the Rasputin's fantail—the area of the ship from which the bales were jettisoned—were negative.

At trial, the government called Senior Chief Petty Officer Gustavo Tirado as its expert witness on IonScan technology. Tirado explained that he had trained as an IonScan machine operator and that he had operated IonScan machines hundreds of times, running thousands of swipes. From 2006 to 2011 and from 2013 on, Tirado was an IonScan machine instructor, teaching others on hundreds of occasions how to use IonScan machines. Tirado was initially trained to operate IonScan machines at a four-day seminar in 1999, where he was taught both how to operate the machine and to interpret its results. In 2006, when Tirado first became an IonScan instructor, he participated in more in-depth training, where he learned in greater detail how to interpret the results of IonScan testing. He attended additional training almost every year from 2006 to 2015. The district court permitted him to testify as an expert on the IonScan process and the meaning of the results of IonScan testing, after a hearing to assess the admissibility of expert testimony under Daubert v. Merrell Dow Pharmaceuticals, 509 U.S. 579, 113 S.Ct. 2786, 125 L.Ed.2d 469 (1993).

Tirado explained at trial the scientific process used by the IonScan machine, including how it measures drift time. He noted that to avoid false positive results, the IonScan machine was programmed to identify the presence of cocaine only where certain other parameters were met. He further explained that when an IonScan machine yields a positive result, it will produce values for three variables—delta, maximum amplitude, and number of segments—that indicate whether the sample had a high or low concentration of a particular substance. Tirado applied those variables to Caruso's IonScan results, explaining that some of the positive hits showed greater concentrations of cocaine than others. He further testified that according to the operator's manual the IonScan machine reported false positives of less than 1%.

Tirado was unable to tell from the IonScan results that any particular quantity of cocaine was ever aboard the Rasputin, however. He was also unable to say whether the positive samples resulted from direct contact with cocaine or indirect contact from traces of cocaine brought aboard the ship by a person or object that had had direct contact with cocaine. Nor did he say how long the cocaine traces detected by the IonScan machine had been aboard the Rasputin.

On cross examination, Tirado testified that Caruso deviated from standard Coast Guard protocol in conducting the IonScan testing in two ways. First, although Caruso properly ran two blanks through the machine in between positive results, he violated protocol by failing to note whether the blanks indicated for an illicit substance. It is unclear, therefore, from Caruso's log alone whether two blanks were run, and if so, whether the blanks indicated the presence of cocaine. Second, Tirado testified that Fleming violated protocol by wearing only one glove on one hand when conducting swipes, when Coast Guard protocol required gloves on both hands and multiple layers of gloves on the swiping hand.

Adair testified at trial that the jettisoned objects he saw on the FLIR resembled cocaine bales he had seen in prior drug interdictions. Other Coast Guard officers gave similar testimony, including Caruso, Fleming, and Petty Officer William Coffey. The defense's objections—and the district court's handling of those objections—were inconsistent. The defense strenuously objected to Fleming's comparison between cocaine bales from prior drug interdictions on the grounds that such testimony was speculative, irrelevant, prejudicial, and unnoticed expert testimony, but let Adair and Caruso give similar testimony largely without objection. The district court sustained an objection to Coffey's testimony that the bales seen on the FLIR looked like bales found during previous cocaine interdictions.

Although several Coast Guard witnesses testified that the objects seen on the FLIR appeared similar in shape and size to cocaine bales from previous interdictions, they also testified that the size and shape were consistent with everyday objects. For example, when Adair was asked whether a "bale" was different from a package, he answered no and explained that a "bale" could be "the same size [and] same weight" as a package. Trial Tr. 9/8/2015, Doc. 118 at 230. Fleming explained that a bale is "slightly larger than a file box" and is "similar to your suitcase or your briefcase." Trial Tr. 9/9/2015, Doc. 119 at 110. Coffey testified that a bale is approximately the size of a burlap sack or a briefcase.

The jury returned a guilty verdict as to all charges against all defendants. Each defendant was sentenced to 10 years' imprisonment on each drug conviction. Williams was sentenced to 5 years' imprisonment, to be served concurrently, on the failure-to-heave-to conviction, as were the remaining defendants for their aiding and abetting failure-to-heave-to convictions. The defendants timely appealed.

…

1. IonScan Evidence

The defendants lodge a number of attacks on Tirado's testimony introducing the results of the IonScan testing conducted on swipes taken from the Rasputin. The defendants challenge Tirado's qualifications to interpret IonScan test results, the relevance of his opinions to the issues in this case, and whether the nature of his testimony was overly prejudicial. We reject these arguments.

Federal Rule of Evidence 702, which governs expert testimony, permits a "witness who is qualified as an expert by knowledge, skill, experience, training, or education" to give an opinion or otherwise testify if "the expert's scientific, technical, or other specialized knowledge will help the trier of fact to understand the evidence or to determine a fact in issue," "the testimony is based on sufficient facts or data ... [and] is the product of reliable principles and methods," and "the expert has reliably applied the principles and methods to the facts of the case." Fed R. Evid. 702.

No one disputes that Tirado delivered expert testimony, the admissibility of which is governed by Rule 702. In Daubert, the Supreme Court defined the contours of a proper Rule 702 inquiry, explaining that Rule 702's requirement that expert testimony "assist the trier of fact to understand the evidence or to determine a fact in issue ... goes primarily to relevance." 509 U.S. at 591, 113 S.Ct. 2786 (internal quotation marks omitted). Indeed, expert testimony unrelated to the issues in the case does not assist the trier of fact. Id. This requirement "has been aptly described ... as one of 'fit.'" Id. In essence, "Rule 702's helpfulness standard requires a valid scientific connection to the pertinent inquiry as a precondition to admissibility." Id. at 591-92, 113 S.Ct. 2786 (internal quotation marks omitted).

Moreover, the proponent of expert testimony bears the burden of demonstrating the expert's qualifications and competence to give his proposed testimony. United States v. Frazier, 387 F.3d 1244, 1260 (11th Cir. 2004) (en banc). Experts may be qualified by scientific training, education, or experience in the relevant field; they need not be formally educated to qualify as experts. Id. at 1260-61. But even if an expert is qualified, and his testimony is relevant and methodologically sound, Federal Rule of Evidence 403 permits the exclusion of all testimony—expert or otherwise—if "its probative value is substantially outweighed by a danger of ... unfair prejudice, confusing the issues, misleading the jury, undue delay, wasting time, or needlessly presenting cumulative evidence." Fed. R. Evid. 403; see also Daubert, 509 U.S. at 595, 113 S.Ct. 2786. Under this framework, we evaluate the district court's admission of Tirado's testimony.

...

ii. Daubert "Fit" and Rule 403

The defendants next argue that Tirado's testimony does not "fit" for Daubert purposes and has insufficient probative value for Rule 403 purposes because it leaves several questions unanswered: for example, Tirado failed to address the quantity of cocaine aboard the Rasputin or how recently cocaine was present, and he was unable to account for a number of purported inconsistencies in the government's theory of the case. We disagree that these gaps required the district court to exclude Tirado's testimony. To be admissible, expert testimony need not answer every question. See id. at 1348 (affirming district court's admission of expert testimony over a Daubert "fit" challenge where the testimony related to one, but not another, of the issues before the court). Indeed, at its core, the Daubert "fit" question is one of relevance, 509 U.S. at 591, 113 S.Ct. 2786, and the positive IonScan tests have some "tendency" to make it "more ... probable" that the jettisoned packages contained cocaine "than it would be without the [IonScan evidence]." Fed. R. Evid. 401 (defining relevant testimony). Tirado's testimony was helpful to the jury because the singular presence of traces of cocaine aboard the Rasputin made it more likely that the jettisoned packages contained cocaine. We cannot say that the testimony wholly lacked relevance—such that the district court abused its discretion in admitting it—simply because the IonScan evidence was imperfect or could not definitively answer every question concerning the presence of cocaine aboard the Rasputin.

Nor can we say that the district court abused its discretion by admitting the IonScan evidence over the defendants' objection that it was unduly prejudicial. Federal Rule of Evidence 403 requires a showing that the prejudicial effect of relevant evidence substantially outweighs its probative value. Expert testimony creates a unique risk of prejudice as it "may be assigned talismanic significance in the eyes of lay jurors." Frazier, 387 F.3d at 1263. Here, though, the district court properly found that the IonScan evidence had probative value. And the defendant's explanation for why Tirado's testimony was prejudicial—it failed to address time period or quantity, and it could have misled the jury into believing that the tests definitively proved that the jettisoned packages contained cocaine—easily could have been mitigated by thorough cross examination. See

Quiet Tech., 326 F.3d at 1341 (noting that "vigorous cross-examination" is a traditional tool for attacking weak, but admissible evidence). Recognizing that expert testimony poses a unique risk of prejudice, we cannot say that the risk of prejudice here so substantially outweighed the IonScan evidence's probative value as to require the district court to exclude it.

...

IV. CONCLUSION

For the foregoing reasons, we affirm the drug convictions of all defendants. We affirm Williams's failure-to-heave-to conviction and reverse the remaining defendants' aiding and abetting failure-to-heave-to convictions. We remand for resentencing of Barker, Cardoza, Duffis, and Taylor.

AFFIRMED IN PART, REVERSED IN PART, AND REMANDED.

Questions:

1. What technology or technical equipment was subject to challenge?
2. What argument was made by defense counsel which allegedly showed irrelevance in the IONScan technology as the Guard scanned for drugs?
3. How does the appellate court handle the drug convictions?

2.3 IMPEACHMENT AND THE EXPERT

Every witness is subject to impeachment and cross-examination even the scientific and technical witnesses with high levels of qualification and experience. Impeachment may deal with the flaws in preparation and fraud in the method; it may deal with character issues such as drug usage, criminal backgrounds and other compromising and conflicts or bias; it may deal with previous discipline or challenge to the expert's reliability and integrity or other matters. What is clear is that while experts may impress, they also can falter.

Taylor v. Shields
United States Court of Appeals for the Third Circuit
April 17, 2018
No. 17-2439

Joshua Taylor challenges the Magistrate Judge's denial of his motion for a new trial in his § 1983 excessive force action against the Defendant, Police Officer Larry Shields. Taylor contends that he is entitled to a new trial on three grounds: (1) the Magistrate Judge failed to bar the Defendant's expert from testifying; (2) the Magistrate Judge unfairly prevented him from offering certain impeachment evidence against the expert; and (3) the Magistrate Judge unfairly precluded him from admitting the Philadelphia Police Department's off-duty policy into evidence.

Because the Magistrate Judge did not abuse his discretion in denying Taylor a new trial on these same grounds, we will affirm.

I. Jurisdiction and Standard of Review

The parties consented to proceed before Magistrate Judge Strawbridge, who had jurisdiction over Taylor's § 1983 action pursuant to 28 U.S.C. §§ 1331 and 636(c). We have jurisdiction over final orders of the district court pursuant to 28 U.S.C. § 1291.

"The decision to grant or deny a new trial is confided almost entirely to the discretion" of the trial court. Blancha v. Raymark Indus., 972 F.2d 507, 512 (3d Cir. 1992). We review the denial of a motion for a new trial for abuse of discretion. McKenna v. City of Phila., 582 F.3d 447, 460 (3d Cir. 2009). To the extent the appeal involves a review of evidentiary rulings, including the decision to admit or exclude expert testimony, we use an abuse of discretion standard. Pineda v. Ford Motor Co., 520 F.3d 237, 243 (3d Cir. 2008). We also review the trial court's case management decisions for abuse of discretion. ZF Meritor, LLC v. Eaton Corp., 696 F.3d 254, 268 (3d Cir. 2012).

II. Facts and Procedural History

Taylor and Shields recount contradictory versions of the incident in question. Both versions of the facts involve a confrontation on April 25, 2011, between Taylor, in and around his residence, and Shields, a Philadelphia police officer who was off duty at the time.

1. Taylor's Version

Taylor alleges that he purchased a new firearm, lawfully, around April 25, 2011. A. 1064-65. At the time, he was living with his family at 4624 Worth Street in Philadelphia. He avers that on April 25, he walked next door to the home of his neighbors, the Thompsons, to show his new gun to his friend, Sean Thompson. While Taylor was on his way, he made eye contact with Shields, who was wearing plain clothes. A. 1068-69. Both parties agree that Taylor's firearm was in plain view on his body. A. 1072-73, 75. When Taylor reached the Thompsons, he states that he learned Sean was not home from Sean's wife, Mary Thompson. A. 1071-72. Taylor claims that he left the Thompsons' house to walk back to his home about 4 to 5 minutes later, when Shields allegedly approached him, did not identify himself as a police officer, drew his firearm, and began to chase Taylor. Br. at 4; A. 1078-80.

Taylor states that he then ran into his home, shut the door, and fell, causing his gun to drop and fall three feet from him. A. 1080-87. He claims that he rolled over onto his right side when Shields opened the door, shot Taylor in the chest with Taylor's children present, and ran out. Id. Taylor survived.

Mary Thompson testified that she observed Shields run into Taylor's home with a gun, so she rushed to Taylor's house and found him lying on the floor, having been shot, attempting to stand up. A. 1737-40. Taylor acknowledged that Shields called 911, but claims that the call was made five minutes after he was shot, rather than immediately.

2. Police Officer Shields' Version

Officer Shields was allegedly helping his sister move on April 25, 2011. A. 897-901. Unfamiliar with the neighborhood, he saw Taylor leave 4624 Worth Street—which he did not know was Taylor's residence—with a handgun in one hand and an ammunition magazine in another, walk into the house next door, and leave only a few seconds later, holding a loaded gun. A. 953-57. Shields alleged that he was concerned that Taylor posed a threat, A. 955-57, and avers that as a police officer, he was required to investigate further. When Taylor emerged from the Thompsons' house, Shields claims to have said, "My man, stop. Police," without displaying a badge. A. 972-73. But, Shields alleges, when Taylor heard Shields say "Police," he ran into 2624 Worth Street with a gun, and Shields followed him. A. 986-87. Shields claims he was concerned that Taylor might pose a threat to someone inside the home. A. 1013-14. When he approached the doorway, Shields states, Taylor was pointing a gun at him. A. 958, 964. Shields admits that he fired a shot at Taylor, and claims it was in fear for his life. A. 958,1013. He then returned to his sister's place, where he and his sister both called 911. A. 944.

III. Procedural History

Taylor sued Shields under § 1983 for one count of using excessive force. As the parties are familiar with the trial timeline, we summarize in brief.

Before the deadline for pre-trial motions, Officer Shields identified Dr. Jonathan L. Arden, a forensic pathologist, as his defense expert. Shields intended to present Dr. Arden to testify on the bullet trajectory from Shields' firearm into Taylor's body. He later served Dr. Arden's expert report on Taylor's counsel. However, Taylor filed a motion in limine to exclude Dr. Arden three months after the deadline, without an explanation for the delay. He argued that Dr. Arden, a forensic medical examiner, was unqualified to opine on ballistics technology. The Magistrate Judge denied the motion and allowed Dr. Arden to testify.

At trial, Taylor introduced multiple witnesses who testified that the bullet path may have been consistent with Taylor's account of the shooting, or otherwise that the evidence did not conclusively demonstrate the bullet path, one way or another. Dr. Arden opined that the evidence was consistent with Shields' version of the events. Taylor also sought to impeach Dr. Arden with evidence of his resignation from a prior job, which the Magistrate Judge did not allow.

The jury returned a verdict for Officer Shields, concluding that his use of force against Taylor was justified. Taylor timely moved for a new trial, with the same arguments as he sets forth in the present appeal: the Magistrate Judge erred in (1) allowing Dr. Arden to testify, (2) excluding impeachment evidence against him, and (3) barring the admission of the Philadelphia Police Department's off-duty policy. After briefing and argument, the Magistrate Judge denied Taylor's motion for a new trial. A.5.

IV. Analysis

1. Dr. Arden

Officer Shields introduced Dr. Arden to opine on Taylor's position when he was shot. Given the relative distance between Taylor and Shields, and given the downward path of the bullet inside Taylor's body, Dr. Arden concluded that Taylor would have either been standing up or sitting up at a 90 degree angle when Shields shot him. A. 29.

Taylor makes two objections to Dr. Arden's expert testimony: (i) he contends that the Magistrate Judge erred in denying his untimely Daubert motion because Dr. Arden was unqualified under Rule 702 of the Federal Rules of Evidence to testify on ballistics and because his opinion was unsound; and (ii) Taylor objects to the Magistrate Judge's ruling that barred him from offering certain impeachment evidence against Dr. Arden. We review the Magistrate Judge's rulings for an abuse of discretion.

First, the trial court was within its discretion under Rule 16(b) of the Federal Rules of Civil Procedure to enforce its scheduling order and therefore deny Taylor's Daubert motion as untimely. Eichorn v. AT&T Corp., 484 F.3d 644, 650-51 (3d Cir. 2007) ("This Court and others have frequently upheld a trial court's exercise of discretion to deny a party's motion to add experts or other fact witnesses after the close of discovery or after a deadline in a scheduling order."). Moreover, Taylor offered no explanation for the tardiness of his motion by three months.

Additionally, the Magistrate Judge initially dismissed Taylor's Daubert motion as untimely, but later, in his opinion denying Taylor a new trial, the Magistrate Judge explained his decision at length. He acknowledged that in spite of enforcing pre-trial deadlines, he did not neglect his underlying gatekeeper duty, namely, to ensure that Dr. Arden's expert testimony met the requirements of Rule 702 of the Federal Rules of Evidence and Daubert v. Merrell Dow Pharmaceuticals, 509 U.S. 579 (1993). Rule 702 provides:

> A witness who is qualified as an expert by knowledge, skill, experience, training, or education may testify in the form of an opinion or otherwise if: (a) the expert's scientific, technical, or other specialized knowledge will help the trier of fact to understand the evidence or to determine a fact in issue; (b) the testimony is based on sufficient facts or data; (c) the testimony is the product of reliable principles and methods; and (d) the expert has reliably applied the principles and methods to the facts of the case.

Fed. R. Evid. 702.

In rejecting Taylor's objections to Dr. Arden's credentials, the Magistrate Judge appropriately explained that Dr. Arden—a forensic pathologist with decades of experience—was eminently qualified to opine on the path between Officer Shields' bullet and Taylor's body. Dr. Arden, a trained anatomic and forensic pathologist, has testified as an expert witness hundreds of times on the very topic of "bullet trajectories within bodies, including the assessment of the relative positioning between a gun and the person shot by the gun." A. 23-24 (quoting Arlen Declaration at 6) (emphasis omitted). The Magistrate Judge also rejected Taylor's objections to Dr. Arden's diagrams that demonstrated the path of the gun and depicted why Dr. Arden believed that Shields' version of the events was consistent with the downward path of the bullet in Taylor's body. The Magistrate Judge deemed Dr. Arden's opinion to be based on facts that were testable, scientifically reliable, and formulated with reliable techniques. Upon reviewing Dr. Arden's credentials, data, and scientific analysis, the Magistrate Judge denied Taylor's objections to Dr. Arden's expert testimony.

Although Taylor may have opposed Dr. Arden's analysis, we agree that Dr. Arden met all of the requirements of Rule 702 and Daubert to opine on the exact topic in question, and thus the Magistrate Judge did not abuse his discretion in allowing Dr. Arden to testify as an expert. Absent an abuse of discretion, we "will not substitute our own judgment for that of the trial court regarding the admission or exclusion of expert testimony." Waldorf v. Shuta, 142 F.3d 601, 627 (3d Cir. 1998).

Moreover, the Magistrate Judge determined that any flaws in Dr. Arden's analysis, such as imprecision in his methodology or assessment, could be addressed on cross-examination and rebutted with other witness testimony. Accordingly, Taylor vigorously cross-examined Dr. Arden and offered multiple witnesses to contradict Dr. Arden's testimony: (i) Mr. Kenneth Katsaris, a police practices expert, opined that the bullet in question could vary in its path within the human body, suggesting that there was no conclusive answer to how Taylor was positioned A. 1487; (ii) Dr. Mohammed Irfan Khan, a trauma surgeon who treated Taylor on the day he was shot, opined that the bullet took a strange, spiral path in Taylor's body A.1235; and (iii) Lieutenant Steven Nolan, an investigator from the Philadelphia Police Department, testified that because there was no exit wound in Taylor's body and because witnesses gave contradicting accounts of what happened, the Lieutenant was unable to conclude with certainty Taylor's position when he was shot. A. 1212-14.

Second, the Magistrate Judge did not abuse his discretion by barring Taylor from offering irrelevant impeachment evidence against Dr. Arden. Taylor sought to introduce the circumstances of Dr. Arden's departure from the Washington Office of the Chief Medical Examiner (WOCME). Dr. Arden was placed on administrative leave following serious accusations of mismanagement and harassment, and resigned. Although Taylor contended that this evidence of workplace safety risk and racial discrimination put Dr. Arden's qualifications and credibility in doubt, the Magistrate Judge disagreed, and also found that this evidence neither met the requirement of Rule 608(b) nor passed the balancing test of Rule 403 of the Federal Rules of Evidence. Rule 608(b) permits courts to allow, on cross-examination, questions about specific instances of a witness's conduct "if they are probative of the character for truthfulness or untruthfulness." Fed. R. Evid. 608(b) (emphasis added). Even so, the court may exclude evidence otherwise permissible under Rule 608(b) if its probative value is substantially outweighed by unfair prejudice under Rule 403.

Here, the Magistrate Judge found that Taylor failed to meet his burden under Rules 608(b) and 403. Evidence of Dr. Arden's departure from WOCME had no bearing on his quality for truthfulness: it was relevant not to his qualifications as a forensic medical expert, implicating Rule 702, but to his administrative responsibilities, the Magistrate Judge found, which was not relevant to his expertise in this case. Moreover, this evidence—far more prejudicial than probative—had a high potential to confuse the jury. A. 17. We agree. Taylor has not shown that the Magistrate Judge abused his discretion in so ruling, and thus we affirm this evidentiary decision.

2. Philadelphia Off-Duty Police Policy

Taylor also sought to introduce the text of Policy 98-1, the Philadelphia Police Department policy that concerns the conduct of off-duty officers. Taylor argues that Shields violated Policy 98-1 by failing to identify himself clearly as a police officer and by delaying calling 911 after he shot Taylor. Taylor sought to use Policy 98-1 to show that Officer Shields used excessive force. While Taylor had referenced 98-1 throughout the trial, the Magistrate Judge later clarified that the policy itself could not be admitted into evidence. Taylor argues that this affected his trial strategy and prejudiced him.

The Magistrate Judge's evidentiary decision was not an abuse of discretion, and Taylor was not prejudiced by this ruling. The Magistrate Judge allowed Taylor to address the substance of the policy at length but did not allow the policy itself to be admitted; the Magistrate Judge noted a concern that if the jurors had the text of the policy, the trial could stray from a question of the use of excessive force to one of violating the off-duty policy. We agree that doing so would be to risk "lead[ing] the jury to equate local policy violations with constitutional violations." McKenna, 582 F.3d at 461. Taylor had the opportunity to cross-examine Officer Shields at length regarding the off-duty policy and practices, notably, Shields' failure to call 911 leading up to the shooting. Taylor also presented Dr. Katsaris as a rebuttal witness to testify about best police practices. In sum, Taylor was not prejudiced by the ruling barring admission of Policy 98-1 into evidence.

V. Conclusion

Taylor has failed to show that the trial court abused its discretion in its evidentiary rulings and in denying Taylor a new trial. For the foregoing reasons, we will affirm the Magistrate Judge's denial of a new trial in its entirety.

Questions:

1. The trial judge rejected a request to allow full impeachment or even removal of the chief witness for the prosecution, Dr. Arden. On what two theories was defense counsel seeking this line of inquiry?
2. What argument did defense counsel specifically make regarding the competency and qualifications of the Medical Examiner?
3. Is it fair to argue that the trial court appears ready, willing and able to live with these flaws or inadequacies in the expert witness?

Aside from impeachment of the witness presenting expert evidence, the challenge may be the forensic science itself. Over the last 40 years, practitioners often hear of defects or flaws in the scientific and technical subject matter, examples being tainted ampoules in intoxilyzer machines, defecting test strips for chemical and blood analysis and in the United States v. Butler decision, a successful challenge to the scientific reliability of microscopic hair comparisons. Hair comparisons were bandied about by the FBI and other labs as a unquestionably reliable technique when the reality is its inherent unreliability.

Locate the following case using your favorite legal research resource, read the case and answer the questions below.

United States v. Butler, 955 F.3d 1052 (D.C. Cir. 2020)

Questions:

1. Was there a disagreement between the expert witness testimony as to the rare error and the reality of error in the actual comparisons?
2. Part of the reason the appellate court was so upset about the use of forensic hair comparison is the willful deceit on the part of the expert but also other parties. Explain.
3. Did the Court construe the presentation as harmless or reversible error?

Impeachment of either the witness or testimony relating to the expertise is an event that can take place at nearly every phase of criminal or civil litigation. During pre-trial Motions in Limine seek advance approval or in the case of challenged science a request for a Daubert hearing tries to clear the air before the case advances. During trial the ebb and flow of expert testimony and cross of same gives clarity and meaning to the forensic application during trial. After trial, challenges to the result from a defendant's perspective frequently employ challenges to the use or misuse of forensic evidence and expert testimony. In Swearingen once encounters a host of post-trial tactics to produce a differing outcome.

In re Swearingen
935 F.3d 415
(5th Cir. 2019)

Larry Ray Swearingen, convicted nineteen years ago of capital murder for strangling nineteen-year-old Melissa Trotter, seeks permission to file a fourth federal habeas corpus petition. He also moves to stay, for the fifth time, his execution date, scheduled for the evening of Wednesday, August 21, 2019. Finding Swearingen's claims fail to meet the strict requirements imposed by 28 U.S.C. § 2244(b) for authorizing a successive petition, we DENY his application and also DENY his motion for stay of execution.

I.

Swearingen was sentenced to death in 2000 after a Texas jury determined that he murdered Trotter by strangulation while committing, or attempting to commit, kidnapping or sexual assault. His conviction was based on a "mountain of inculpatory evidence." .... Over the next two decades, "Swearingen has filed a convoluted tangle of habeas applications, pro se motions, mandamus actions, and amended pleadings," seeking to overturn his conviction and postpone his death sentence. Id. at *6; see also id. at *6–10 (recounting state and federal post-conviction litigation history through 2009); Swearingen, 478 S.W.3d at 719 (recounting history of motions for post-conviction DNA testing). These legal machinations have resulted in Swearingen's execution being put off five times.

...

On March 12, 2019, his sixth execution date was set for Wednesday, August 21, 2019. Seven days before the execution date, Swearingen sought our court's authorization to file a fourth habeas petition based on two claims. First, based on a recent letter from the Texas Department of Public Safety ("DPS"), Swearingen claims the State sponsored "false and misleading" trial testimony regarding blood flecks found under Trotter's fingernails, in violation of Giglio v. United States, 405 U.S. 150, 92 S.Ct. 763, 31 L.Ed.2d 104 (1972). Second, based on another recent DPS letter, Swearingen claims the State withheld evidence that a criminologist had "manufactured" evidence that the torn pantyhose used to strangle Trotter matched the pantyhose found at Swearingen's house, in violation of both Giglio and Brady v. Maryland, 373 U.S. 83, 83 S.Ct. 1194, 10 L.Ed.2d 215 (1963). Based on these claims, Swearingen has also moved for a stay of execution.

II.

A.

Swearingen must receive this court's authorization to file a second or successive petition. 28 U.S.C. § 2244(b)(3)(A); see generally, e.g., In re Raby, 925 F.3d 749, 754 (5th Cir. 2019). We may give that authorization only if we conclude that Swearingen's application makes a prima facie showing that it satisfies the strict requirements in § 2244(b). Id. A prima facie showing is "simply a sufficient showing of possible merit to warrant a fuller exploration by the district court." In re Campbell, 750 F.3d 523, 530 (5th Cir. 2014) (citation omitted). Consequently, if it seems reasonably likely that a successive petition meets the strict requirements provided in the statute, we will grant the motion for a successive petition. Id.

As both parties concede that Swearingen's present claims were not raised in previous federal habeas petitions, Swearingen must therefore make a prima facie showing that he satisfies the requirements of § 2244(b)(2)(B):

> (2) A claim presented in a second or successive habeas corpus application under section 2254 that was not presented in a prior application shall be dismissed unless—...
>
> (B)(i) the factual predicate for the claim could not have been discovered previously through the exercise of due diligence; and
>
> (ii) the facts underlying the claim, if proven and viewed in light of the evidence as a whole, would be sufficient to establish by clear and convincing evidence that, but for constitutional error, no reasonable factfinder would have found the applicant guilty of the underlying offense.

Id. § 2244(b)(2)(B).

This analysis does not address the merits of Swearingen's two claims, but only whether Swearingen "makes a prima facie showing that [he] can meet the [two] requirements of § 2244(b)(2)." Swearingen, 556 F.3d at 347. Within that framework, we consider each of Swearingen's claims in turn.

B.

Swearingen claims that a recent letter from the DPS crime lab director—sent to Swearingen's counsel on August 9, 2019—shows the State sponsored "false and misleading" trial testimony from a DPS analyst, Cassie Carradine, concerning blood flecks found under Trotter's fingernails. Evidence at trial showed that the DNA profile derived from these flecks matched neither Trotter nor Swearingen. Swearingen, 303 S.W.3d at 735. According to Swearingen, Carradine testified this blood came from "contamination" of the sample, leading the jury to discount the possibility that the blood actually pointed to someone else as Trotter's murderer. Now he claims that the 2019 DPS letter shows Carradine's testimony was fraudulently concocted by the State in violation of Giglio. This claim fails to meet both of the § 2244(b)(2)(B) threshold requirements.

At the outset, we observe that Swearingen mischaracterizes what the DPS letter says. He claims that the letter "proves" that Carradine's testimony was "false and misleading" and thus "demonstrates that the State sponsored the false contamination theory for the purpose of misleading the jury." The letter (which we reproduce in full below) says nothing like that. The letter first notes that Carradine "was qualified" to opine about contamination "within the DPS laboratory or based on the packaging and condition of the evidence," but that she lacked "direct knowledge" about how the evidence was "collected or stored prior to its submission to DPS" (emphasis added). Thus, the letter states that a "more appropriate answer" from Carradine "would have been that she could not speak to the possibility of contamination of the samples when they were outside the control of the DPS laboratory." Next, the letter states that Carradine had an "insufficient basis" to opine regarding contamination and that, therefore, she should have testified that "[t]he full range of possibilities include contamination or that it was not contamination and the [DNA] profile did come from the evidence." Contrary to Swearingen's characterization, the letter does not say that Carradine testified "falsely" and says nothing to suggest that the State "sponsored" a "false contamination theory" to "mislead" the jury. Instead, the letter says at most that Carradine lacked a foundation to opine on contamination that may have occurred when the samples were outside DPS custody, and that her answers should have recognized that nuance.

We also note that Swearingen's application neither quotes Carradine's actual trial testimony nor provides transcripts of her testimony. Nonetheless, Swearingen baldly claims that Carradine "definitively told the jury" that the blood flecks were "the result of contamination either at the DPS Crime Lab or at earlier Autopsy [sic]." That is false. The trial record pages Swearingen cites to support that claim tell a different story. After describing how she collected the flakes, Carradine testified only that there was a "possibility" of contamination. And while she briefly referred to the possibility that the flecks may have been deposited "at the time … the sample was being collected," she focused far more on the possibility that the contamination might have occurred in the DPS lab. Furthermore, Swearingen's counsel objected to Carradine's testimony as "speculation," and then cross-examined her extensively on how the fingernail scrapings were packaged and how she handled them. During that cross-examination, Carradine was asked how the samples were transported to DPS and she responded, "I have no way of knowing how they were transported to the office."

Having properly characterized Swearingen's claim regarding the letter, we can now assess whether his claim meets the two prongs of § 2244(b)(2)(B). We conclude it meets neither.

First, the claim's "factual predicate"—that Carradine lacked a foundation for her testimony regarding possible contamination—could have been discovered long before the DPS letter was sent in August 2019. Indeed, the record shows that Swearingen's attorneys were already aware of the issue: His trial attorneys objected to Carradine's testimony on this point as "speculative," they cross-examined her on how she received and handled the fingernail scrapings, and they elicited an admission that she had "no way of knowing" how the samples were handled before arriving at DPS. More broadly, whether the blood flecks came from contamination was vigorously contested at trial. See, e.g., Swearingen, 424 S.W.3d at 39 (observing that "[t]he jury chose to believe that the foreign DNA either was contamination or that it came from outside the context of the crime"); Swearingen, 2009 WL 249778 at *5 n.7 (Cochran, J., concurring) (listing post-conviction court's findings on blood flakes). Any "reasonable attorney would have been put on notice" that they should have probed the basis for Carradine's testimony on this issue—and Swearingen's attorneys did just that at trial. See Blackman v. Davis, 909 F.3d 772, 779 (5th Cir. 2018), cert. denied, — U.S. ——, 139 S. Ct. 1215, 203 L.Ed.2d 236 (2019).

Second, even assuming the facts underlying this claim are true —again, that Carradine lacked a foundation for testifying about possible contamination—we cannot say that no reasonable jury would have found Swearingen guilty "in light of the evidence as a whole." 28 U.S.C. § 2244(b)(2)(B)(ii). As the State points out, Carradine's testimony was cumulative of more detailed expert testimony from Dr. Joye Carter, from which a jury might have readily reached the same conclusion about the blood. And even if we assume the foreign DNA under Trotter's fingernails was not from contamination, that would not clearly and convincingly exonerate Swearingen. We cannot improve on the TCCA's reasoning on this point:

> We are not persuaded that results showing the presence of another DNA donor in the fingernail scrapings would overcome the "mountain of evidence" of the appellee's guilt. Primarily, this is because the victim's having encountered another person would not factually exclude the appellee from having killed her.

Swearingen, 424 S.W.3d at 38–39. Given that "mountain of evidence," jurors could have found the DNA evidence outweighed by the prosecution's case, even without a suggestion that the sample was contaminated. While theoretically possible that DNA evidence from the blood flakes could have swayed the jury, it is not sufficient under § 2244(b) merely to show that evidence "muddies the waters." In re Raby, 925 F.3d at 759. "Clear and convincing" evidence is required. § 2244(b)(2)(B)(ii). Even assuming Carradine's testimony on this point lacked any foundation, Swearingen has not come close to establishing that "no reasonable factfinder" would have found him guilty.

C.

Swearingen next claims that, in violation of Giglio, the State sponsored false trial testimony from a criminologist, Sandy Musialowski, concerning the degree of "match" between the ligature used to strangle Trotter and the torn pantyhose found at Swearingen's home. He also claims that, in violation of Brady, the State withheld exculpatory and impeaching evidence on this issue. His Giglio claim is based on another recent DPS letter which he says "retracts" Musialowski's opinion that the ligature and the pantyhose matched. His Brady claim is based on notes, written by Musialowski during her examination of the torn pantyhose, that he claims the State withheld. We conclude that these interrelated claims fail to satisfy § 2244(b)(2)(B). The Giglio claim turns on a July 19, 2019 letter from DPS that Swearingen again misrepresents. The letter (also reproduced below) does not "retract" Musialowski's testimony—it merely says her testimony about the match between the ligature and the torn pantyhose would today use different terminology. "Today," says DPS, "we would report that the two pieces [of pantyhose] were once joined, but would not include the statement 'to the exclusion of all others.'" At the same time, the letter affirms that Musialowski "did not err in her reporting or testimony regarding [the pantyhose match]." We think it sufficient to dispose of this claim to say this: Even if proven true, we do not think the claim provides clear and convincing evidence that no reasonable jury would find Swearingen guilty. In other words, even if Musialowski's testimony were revised from "the litgature and the pantyhose were a unique match to the exclusion of all others," to "the litgature and the pantyhose were once joined," we see no possibility that this would make any difference to Swearingen's guilt.

Finally, the Brady claim turns on Musialowski's allegedly undisclosed notes. The notes document that, at the initial stages of her examination, Musialowski could not say the ligature and the pantyhose were a "physical match," but as her examination progressed, she was able to determine that they were indeed a "unique physical match." Swearingen claims this is evidence that his attorneys could have used to impeach Musialowski's key conclusion linking him to the murder weapon. We conclude this claim fails both prongs of § 2244(b)(2)(B).

With respect to the due diligence prong, § 2244(b)(2)(B)(i), Swearingen's attorneys were obviously aware at trial of the precise nature of Musialowski's examination of the pantyhose, which is reflected in the allegedly undisclosed notes. For instance, the statement plausibly helpful to Swearingen in the notes is the notation that Musialowski, having initially observed the ligature and pantyhose under glass plates, observed "no physical match between ligature and pantyhose." But this phase of the investigation, and the difficulties Musialowski faced, were discussed in detail at trial. Musialowski testified that she attempted to observe the ligature and pantyhose on glass plates, but "because the ends were rolling, it was difficult for [her] to keep the ends down to compare." So, she used other methods, including pinning the fabric to a foam board, which ultimately led to her conclusion of a match. The notes are consistent with Musialowski's testimony at trial, and thus do not constitute new evidence previously undiscoverable by due diligence. Swearingen's claim thus fails under § 2244(b)(2)(B)(i).

Alternatively, Swearingen's claim about the notes fails the second prong of § 2244(b)(2)(B). As already discussed, Musialowski's notes were basically cumulative of her trial testimony about her developing examination of the pantyhose and ligature. Introducing the notes would simply have established—as her trial testimony did—that Musialowski did not initially see a match when analyzing the items on glass plates, but did see one upon further investigation. These facts were already present in her trial testimony and would have been, at most, marginally enhanced by the notes. Thus, the notes, even if introduced and credited by the jury, would not have led to a trial in which "no reasonable factfinder would have found the applicant guilty of the underlying offense." 28 U.S.C. § 2244(b)(2)(B).

That conclusion is particularly compelling given that § 2244(b) requires us to assess the claim "in light of the evidence as a whole." § 2244(b)(2)(B)(ii). We have before alluded to the "mountain of inculpatory evidence," discussed in detail by numerous courts, state and federal, that seals Swearingen's guilt for Trotter's murder. Swearingen, 303 S.W.3d at 736. We end by quoting a relatively succinct summary of that evidence from a federal district court, which underscores why Swearingen's "new" claims in this latest phase could not possibly have made any difference to the outcome of his trial:

> To reiterate, Swearingen was the last person that Ms. Trotter was seen with alive. Ms. Trotter had been in Swearingen's truck, where he forcibly removed hair follicles. Swearingen's histological evidence does not explain why she was in his house that day, why it was later found to be in disarray, and why he falsely claimed that there had been a burglary there. The evidence itself does not explain why papers belonging to Ms. Trotter were found near the house of Swearingen's parents and her cigarettes were in Swearingen's house. The new information does not explain why Ms. Trotter was found wearing the same clothes as when she disappeared and why she had a note given to her by a friend on December 8 in her back pocket. The new evidence does not show why cell phone records traced Swearingen to a location near where Ms. Trotter was found. Histology does not explain why half of a pair of pantyhose belonging to Swearingen's wife was found in Swearingen's house and the other half around Ms. Trotter's neck. The new evidence does not explain why the same meal Ms. Trotter was last seen eating was found in her stomach. Swearingen lied about his whereabouts, tried to fabricate an alibi, made false police reports, fled from the police, asked friends to lie in his behalf, told others that the police would be after him, and crafted an ultimately inculpatory letter to throw attention away from himself. Swearingen told other inmates, "Fuck, yeah, I did it."

III.

We DENY authorization to file a successive habeas corpus petition. We DENY the motion for stay of execution.

Questions:

1. Defense Counsel poses a series of post-conviction theories involving forensic evidence. While not successful, these are creative examples of defense advocacy and at the same time, these examples illuminate how necessary preparation and proper protocols are. Name some of those forensic challenges.
2. Why did the State err in the matter of pantyhose?
3. Was the State's criminologist able to issue a conclusion regarding the comparative analysis of severed pieces of pantyhose?

NOTES

1 See Michael Stockdale, "Expert Evidence in Criminal Proceedings: Current Challenges and Opportunities," *Journal of Criminal Law* 80, no. 5 (October 2016): 344, https://doi.org/10.1177/0022018316668448.

2 See Douglas G. Smith, "The Increasing Use of Challenges to Expert Evidence Under *Daubert* and Rule 702 In Patent Litigation," *Journal of Intellectual Property Law* 22, no. 2 (Spring 2015): 347, https://digitalcommons.law.uga.edu/jipl/vol22/iss2/4.

3 Fed. R. Evid. 703.

4 Cynthia Lynne Pike, "The Impact of Revised MRE 702 and 703 in Response to *Daubert*," *Wayne Law Review* 52 (2006): 285.

5 Michael Graham, "Evidence and Trial Advocacy Workshop," *Criminal Law Bulletin* (1980): 360, 362. Professor Ronald Carlson, in his outstanding study, Policing the Bases of Modern Expert Testimony, relates in part, "The litmus test of admission or exclusion of an expert's opinion based upon the reports of others is the reasonableness of the testifying expert's reliance on such reports. If the underlying facts or data that help the expert reach conclusions are of a type on which other experts in the field customarily rely, most courts will permit the expert to testify." Ronald Carlson, "Policing the Bases of Modern Expert Testimony," *Vanderbilt Law Review* 39, no. 3 (1986) 577, 578.

6 Dean Weinstein, in his text Evidence, states, "A witness' acquaintance with certain data may have to be shown in qualifying him as an expert pursuant to Rule 702. And the necessity of making the evidence understandable to the jury will often result in the posing of some foundational questions before the opinion is requested." 3 Jack B. Weinstein and Margaret A. Berger, *Weinstein's Evidence* (1981), 705.

7 Brian Simon, "The Basis of Expert Testimony: Ryan v. KDI Sylvan Pools Lets the Expert Have Their Ways," *Rutgers Law Review* 43 (1991): 1235, 1240.

8 Rhomer at 669.

3 Forensic Identity

Proof of identity is central to every criminal action, for to prove the actus reus of a crime, that the crime was committed has little meaning if not tied to a particular agent or individual. Proof of identity is especially evident in the recent, high-profile case of Bryan Christopher Koburger who has been named a suspect in the murder of four University of Idaho students in November of 2022. For six weeks after the crime, the entire justice apparatus has been on a search for identity. The "act" was self-evident, the agency that authored the act, a bit more challenging.

That is what this chapter and its aligned cases address: the matter of identity, from fingerprints to anthropological conclusions about identity. Law enforcement, prosecutors and the trier of fact could not convict without this essential proof.

3.1 FINGERPRINTS AS PROOF OF IDENTITY

The probative value of fingerprints as a source and means of identification is largely not in dispute. From a general sense, the individuality of fingerprints is pretty much a fact lacking debate. In addition, the fingerprint is difficult to alter or change without severe mutilation; and

FIGURE 3.1 Bryan Koburger.

DOI: 10.4324/9781003267126-3

even in those cases, the lower layers of skill still possess similar patterns or ridges and waves. Many of the challenges to prints relate to the processes of extraction and comparison. A historic decision from Illinois from 1911 lays out how fingerprint evidence was welcomed without much reservation.

> Read the story about the Green River Killer DNA evidence at https://www.fbi.gov/history/field-office-histories/seattle.

Locate *People v. Jennings*, 252 Ill. 534 (Dec. 21, 1911) using your favorite internet resource, read and answer the following questions.

Questions:

1. What legal authority did the Court rely upon in terms of cases and statutes?
2. How extensive was the utilization of fingerprints in 1911? And how was the Court satisfied with the background and expertise of those giving testimony on the identification based on the fingerprint?
3. In what sense does the sufficiency of evidence argument make for a good defense strategy?

The Davis decision tests the comparative method known as ACE-V which compares print patterns in three stages, namely: Analysis, Comparison and Verification.

Davis v. State
Court of Appeals Case No. 19A-CR-333 (Ind. App. Dec. 4, 2019)
December 4, 2019

Davis raises three issues on appeal, which we consolidate and restate as whether the trial court properly admitted a fingerprint examiner's opinion that Davis was the source of a latent palm print recovered from the crime scene.

Facts

On September 5, 2015, as cashier Carmen Hernandez-Saucedo worked in Obadiah's Smoke Shop ("the store") in Indianapolis, a tall, African-American man in a blue shirt approached the cash register to make his second purchase of the day. Hernandez-Saucedo went behind the counter to collect payment, and the man repeatedly struck Hernandez-Saucedo's face, took the contents of the cash register, and pointed a gun at Hernandez-Saucedo. When Hernandez-Saucedo attempted to flee, the man shot her in the thigh and fled the store. After the robbery, Hernandez-Saucedo saw the man drive away in a green vehicle with gold hubcaps. Video surveillance footage revealed that the man carried what appeared to be a .380 caliber handgun, wore black Jordan 13 sneakers with red soles, and touched the cash register and the inside surface of the front door of the store. Indianapolis Metropolitan Police Department Detective Jerry Townsend obtained descriptions of the man and the green car from Hernandez-Saucedo;

and still photographs from the surveillance footage yielded a license plate number for the green car. The license plate number was associated with a green 1994 Pontiac Bonneville that was registered to Demetrius Muhammad. Officer David Waterman, an evidence technician for IMPD, lifted four latent prints from the store and delivered them to IMPD latent fingerprint examiner, Rochella O'Neil. Of two palm prints that were recovered from the door of the store, only one was usable. The usable palm print ("the palm print") was comprised mostly of friction ridge skin and was an "excellent quality print" because of its size and "the amount of minutiae that was available." Tr. Vol. III p. 32. O'Neil processed the palm print through the Automated Fingerprint Identification System ("AFIS") database on September 8, 2015. Although Davis' prints were in the AFIS database at the time, Davis was not—to O'Neil's knowledge—among the candidates generated by AFIS as potential sources of the palm print. On September 11, 2015, as Detective Townsend drove in the vicinity of Muhammad's address, Detective Townsend observed a green Pontiac Bonneville with gold hubcaps. Detective Townsend watched as the driver—Davis—exited the car and entered an apartment. Davis wore black Jordan 13 sneakers with red soles. IMPD surveilled the apartment, and Davis eventually emerged and drove away in the green Bonneville. The police initiated a lawful traffic stop when Davis failed to use his turn signal. On or about September 12, 2015, the police obtained a search warrant for the apartment. The search yielded a pair of red-soled black Jordan 13 sneakers, ammunition for a .380 caliber handgun, and a blue shirt that resembled the shirt worn by the robber. On September 16, 2015, the State charged Davis with armed robbery, a Level 3 felony; battery by means of a deadly weapon, a Level 5 felony; and carrying a handgun without a license, a Class A misdemeanor. Three years later, the State filed a motion to fingerprint Davis on October 27, 2018 and notified counsel for Davis that the State intended to analyze the palm print. The State requested that O'Neil compare the palm print to Davis' prints from the AFIS database. O'Neil compared the palm print to Davis' prints and concluded that, "in [her] opinion[,]" the palm print from the store "did belong to Jaquan Davis." Tr. Vol. III p. 36. O'Neil did not compare the palm print to any other offenders' prints because she "[di]d not expect to find th[e] [same] level of corresponding detail in another source[,]" because of the size and detailed minutiae present on the palm print, and because, based on her training and experience, no two people present identical friction ridge detail on their hands. Id. at 37. On October 29, 2018, Davis filed a motion in limine, wherein Davis asserted that "the methodology and techniques used in latent print analysis" are not "based on reliable scientific principles[.]" Conf. App. Vol. II p. 177. In addition to arguing that latent print analysis was unreliable, Davis also argued that any opinion, "even if based on reliable principles," would violate Evidence Rule 403. Id. at 178. On November 21, 2018, the trial court conducted a hearing regarding, among other things, the admissibility of the palm print. Counsel for Davis objected to the palm print evidence on the ground that the State analyzed the palm print and identified it as Davis' palm print three years after the palm print was collected and one week before Davis' jury trial. See Tr. Vol. II p. 13. Defense counsel also advised the trial court that a scheduled deposition of O'Neil was cancelled due to defense counsel's illness. The State countered that "once the State became aware that there was a print that had not been compared to Mr. Davis' prints, [the State] immediately notified Defense counsel, immediately requested that [the palm print] be examined by Ms. O'Neil, and then promptly turned over the identification report[.]" Id. at 14. Counsel for the State added that, although defense counsel was unable to conduct O'Neil's deposition, defense counsel was "able to email Ms. O'Neil questions and receive responses [.]" Id. Over Davis' objection, the trial court found that the palm print was admissible, and that defense counsel had sufficient time to review the evidence. Davis was tried by a jury on November 29, 2018. That morning, O'Neil took an inked palm print, comprised of mostly friction ridge skin, from Davis. During its case in chief, the State called O'Neil to testify. O'Neil testified at length regarding her training and her twenty-two years of experience with IMPD's fingerprint identification unit and fourteen and one-half years of experience with the latent print unit. O'Neil testified further that IMPD's latent print examiners each have a minimum of fourteen years of professional experience and follow the ACE-V methodology—which is endorsed by the International Association for Identification as the accepted standard methodology for performing fingerprint comparisons. O'Neil testified that, in her role as a latent print examiner, she "perform[s] … analysis, comparison, evaluation and verification on any friction ridge detail that [comes] to [IMPD] on latent cards or digital photographs that are recovered from crime scene[s.]" Id. at 18. Regarding the ACE-V methodology, O'Neil testified further:

> Q: Okay. Now, let's talk about the process in comparing that latent print to known prints or exemplars. What is the procedure that you follow?
>
> A: We follow the ACE-V Methodology. ACE-V stands for Analysis, Comparison, Evaluation, and Verification.
>
> Basically during the Analysis part, [] we [] receive the evidence and we examine it to determine whether or not there's any friction ridge detail present on that evidence. We're also looking at the quality to try to make a determination as to whether or not we can make a comparison. If we believe we can make a comparison, then we move to the comparison stage and we place the unknown latent print next to a known exemplar under magnification and we look at the unique identification characteristics as well as the unit relationship between those characteristics.
>
> While we're doing that, we move to the evaluation stage which is where we form an opinion as to whether or not we believe there were dissimilarities which would invoke an exclusion or if we found enough [] correspondence between the minutiae to make an identification.
>
> * * * * *
>
> A. And then we move to the verification stage. Once we form our opinions as to whether or not we believe there was an identification or an exclusion, then we move to the verification stage. And that's where we hand our entire case over to another co-worker or a supervisor who re-works it to make sure they come to the same conclusion.

See Tr. Vol. II pp. 21-22. O'Neil testified further that, in Davis' case, she: (1) received a request from the prosecutor to compare the palm print to Davis' prints; (2) conducted the analysis "with ACE-V"; and (3) concluded that the latent palm print matched Davis' prints because of "an overwhelming amount of minutiae [] that [O'Neil] would not expect to find repeated in another source." Id. at 49. When the State asked O'Neil to opine on the source of the palm print, after O'Neil compared the palm print to Davis' prints, defense counsel objected as follows: "Judge, at this point I'm going to object to this witness testifying to the print matching my client under Rule 702 and [based upon] the prior hearing that we had on this matter." Tr. Vol. II p. 33. The trial court overruled the objection and incorporated its earlier findings. O'Neil then testified, "In my opinion, the latent print that was left on [] the entry door inside was made by Mr. Jaquan Davis, specifically[,] his writer's palm." Id. at 34. At the close of the evidence, the jury convicted Davis on all counts. At Davis' sentencing hearing on January 11, 2019, the trial court sentenced him to eleven years for armed robbery, a Level 3 felony, and three years for battery by means of a deadly weapon, a Level 5 felony, with the sentences to be served concurrently. Davis now appeals.

Analysis

Davis argues that the trial court abused its discretion in admitting O'Neil's identification because O'Neil failed to: (1) "explain[] the basis for her conclusions"; and (2) "show[] if or how the ACE-V method was reliably applied[.]" Davis' Br. p. 27. The State counters that this issue is waived because Davis is making a different argument on appeal than he made at trial. Davis objected below to O'Neil's identification on the ground that the ACE-V methodology is not based on reliable scientific principles. On appeal, Davis implicitly accepts the validity of ACE-V and asserts that O'Neil failed to "show [] if or how the ACE-V method was reliably applied[.]" Davis' Br. p. 27. We agree with the State that this issue is waived. See Pigg v. State, 929 N.E.2d 799, 803 (Ind. Ct. App 2010) (claims are generally waived if raised for the first time on appeal), trans. denied. Waiver notwithstanding, we address the merits of Davis' argument that the trial court abused its discretion when it admitted O'Neil's identification of Davis as the source of the palm print. Absent an abuse of discretion, we will not disturb a trial court's determination that a witness is qualified to testify as an expert and render an expert opinion. Ross v. State, 665 N.E.2d 599 (Ind. Ct. App. 1996). An abuse of discretion occurs if the trial court's decision is clearly against the logic and effect of the facts and circumstances before the court, or the reasonable, probable, and actual deductions to be drawn therefrom. Id. Indiana Evidence Rule 702 provides:

a. A witness who is qualified as an expert by knowledge, skill, experience, training, or education may testify in the form of an opinion or otherwise if the expert's scientific, technical, or other specialized knowledge will help the trier of fact to understand the evidence or to determine a fact in issue.
b. Expert scientific testimony is admissible only if the court is satisfied that the expert testimony rests upon reliable scientific principles.

In Burnett v. State, 815 N.E.2d 201, 209 (Ind. Ct. App. 2004), we held:

> ACE-V methodology is generally accepted by the [International Association for Identification], an international organization in the field. [[] [Fingerprint experts in several different states, including Indiana, Ohio, Illinois, and Michigan, use the ACE-V method for fingerprint identification. Thus, the State established that the ACE-V methodology is generally accepted within [the latent fingerprint examiner-expert]'s relevant field of study and that it is more likely than not that the scientific principles upon which the testimony rests are reliable.

Burnett v. State, 815 N.E.2d 201, 209 (Ind. Ct. App. 2004). Thus, Davis' claim that ACE-V methodology is not based on reliable scientific principles fails. Moreover, O'Neil testified regarding: (1) her extensive professional training and experience; (2) IMPD's policy of training its print examiners to employ the ACE-V methodology; (3) the detailed steps of the ACE-V analytical process; and (4) O'Neil's utilization of the ACE-V methodology when she identified Davis as the source of the palm print. Based on the foregoing, the trial court did not abuse its discretion by admitting O'Neil's expert testimony pursuant to Rule 702. See Bond v. State, 925 N.E.2d 773, 781-82 (Ind. Ct. App. 2010) ("[Examiner] explained the steps involved in the ACE-V methodology and testified that she used ACE-V[.] [] In light of [examiner's] testimony—along with the established reliability of the ACE-V protocol and [examiner's] qualifications as an expert witness—we cannot say that the trial court erred by finding that the ACE-V method was applied and that [examiner's] opinion was admissible under Rule 702."). Moreover, the State presented extensive evidence regarding O'Neil's application of ACE-V methodology and techniques. Counsel for Davis cross-examined O'Neil at length. Indiana does not require a specific "'test' or set of 'prongs' which must be considered in order to satisfy Indiana Evidence Rule 702(b)." See Alcantar v. State, 70 N.E.3d 353, 359 (Ind. Ct. App. 2016) (quoting McGrew v. State, 682 N.E.2d 1289, 1292 (Ind. 1997)). We, thus, find the dispute here goes to the weight attributed to the ACE-V methodology and the conclusions that O'Neil reached in reliance on ACE-V, and not to the admissibility of O'Neil's testimony as to ACE-V. See Alcantar, 70 N.E.3d at 359.

Conclusion

The trial court did not abuse its discretion in admitting O'Neil's identification of Davis as the source of the palm print pursuant to Indiana Evidence Rule 702. We affirm.

Affirmed. Brown, J., and Altice, J., concur.

Questions:

1. Prove this following True or False: Comparison depend upon the sufficiency of ridges and other characteristics that make the comparison possible?
2. How does the Court reconcile the distinction between the substance of proving identity through print comparisons with the process utilized?
3. Does the Court question the validity of the comparison technique nor indicate to admit the results are an abuse of discretion?

State v. Pigott,
325 P.3d 247, 181 Wn. App. 247 (Wash. Ct. App. 2014)

¶ 1 A Frye hearing is not required where there is no evidence involving new methods of proof or new scientific principles. Here, the fingerprint identification method used by the police is generally accepted within the scientific community. The trial court did not err in not conducting a Frye hearing and admitting the evidence. Affirmed.

¶ 2 Michael Pigott was charged and found guilty in juvenile court of two counts of residential burglary. In both homes, detectives discovered latent fingerprints which were identified as belonging to Pigott. Prior to trial, Pigott requested a Frye hearing, arguing that fingerprint identification is no longer generally accepted in the scientific community. The trial court denied the motion. After trial, the court found Pigott guilty and imposed the standard range disposition on each count. Pigott appeals arguing that the trial court improperly denied his motion for a Frye hearing.

¶ 3 We review de novo a trial court's decision whether to hold a Frye hearing. In determining the admissibility of evidence based on novel scientific theories or methods, Washington courts employ the "general acceptance" standard set forth in Frye v. United States. Under Frye, novel scientific evidence is admissible if it is based on a theory or principle which is generally accepted in the relevant scientific community, but not admissible if qualified experts have significant disputes as to its validity. Evidence not involving new methods of proof or new scientific principles is not subject to examination under Frye.

¶ 4 Washington has a long history of admitting fingerprint identification evidence. Here, the fingerprint experts examined the latent prints, compared them to Pigott's known print, and made an evaluation. The results were verified. The State's expert witnesses did not rely on novel or unusual scientific techniques in reaching their opinions. The evidence presented clearly demonstrated that this specific method is commonly used and generally accepted by the fingerprint analysts. Once the scientific community accepts a methodology, application of the methodology to a particular case is a matter of weight and admissibility under ER 702, which allows a qualified expert witness to testify if scientific, technical, or other specialized knowledge will assist the trier of fact.

¶ 5 Pigott argues that new evidence raises questions as to the continued general acceptance of the ACE–V technique used here to analyze fingerprints. ACE–V is an acronym for "analysis, comparison, evaluation, and verification." The ACE–V component was succinctly described by the Third Circuit in United States v. Mitchell:

The basic steps taken by an examiner under this protocol are first to winnow the field of candidate matching prints by using Level 1 detail to classify the latent print. Next, the examiner will analyze the latent print to identify Level 2 detail (i.e., Galton points and their spatial relationship to one another), along with any Level 3 detail that can be gleaned from the print. The examiner then compares this to the Level 2 and Level 3 detail of a candidate full-rolled print (sometimes taken from a database of fingerprints, sometimes taken from a suspect in custody), and evaluates whether there is sufficient similarity to declare a match.

Once a second examiner confirms the conclusion of the first examiner, it is verified. The expert witnesses used generally accepted methods in analyzing Pigott's fingerprints. Thus, the methods were not "novel," and did not require a Frye hearing. Because such methodology is generally accepted in the scientific community, the evidence was properly admitted.

¶ 6 Pigott relies on a high profile misidentification of Brandon Mayfield made by the Federal Bureau of Investigations in its investigation into the terrorist bombing of a train in Madrid, Spain. As a result of this misidentification, the Office of the Inspector General (OIG) initiated an investigation of the Mayfield case. Citing to the OIG report, Lyn Haber, Ph.D. and Ralph Haber, Ph.D. argue that fingerprinting is not an exact science. In support of their theory, the Habers cite to a 2009 report by the National Research Council of the National Academy of Sciences, which recommended additional testing to determine the reliability of latent fingerprint analysis generally and of the ACE–V methodology in particular.

¶ 7 However, the reliability of fingerprint identification has been tested in our adversarial system for over a century and routinely subjected to peer review. The trial court considered all of this in reaching its conclusion that a Frye hearing was not needed. The trial court expressly found that the identification analysis is accepted in the relevant scientific community. Once the evidence is accepted as scientifically acceptable, the question of admissibility turns on whether the witnesses qualify as experts and whether the proffered testimony would be helpful to the trier of fact. Under the circumstances here, the trial court did not err in admitting the evidence under ER 702. Any objections that Pigott had to the fingerprint analysis addressed the weight of the testimony, not its admissibility. The trial court did not err.

¶ 8 Affirmed.

Questions:

1. The Pigott decision unequivocally supports the science of fingerprints. This being so, how could the Court support the request for a Frye hearing?
2. What second phase, once a Frye hearing is rejected, could defense counsel launch a challenge?

The sufficiency of comparative points and markers remains an important consideration for those advancing any fingerprint evidence at trial. At a minimum, even though there is no universal requirement of comparison points, also known as "Galton" markers, most states have minimum expectations.

United States v. Saunders
826 F.3d 363 (7th Cir. 2016)

Christopher Saunders and Rashid Bounds sold heroin on the west side of Chicago. They were indicted, and went to trial. A number of their co-conspirators testified against them, and they were convicted of conspiring to distribute at least 100 grams but less than one kilogram of heroin. At sentencing, the district court held them responsible for between three and ten kilograms of heroin and sentenced each of them to 216 months' imprisonment. On appeal, the defendants contend that the court erroneously denied their motion to exclude the government expert's fingerprint testimony because the government's pretrial disclosures did not sufficiently disclose the basis of the expert's opinion. While we agree that the government's disclosure failed to meet the requirements of Federal Rule of Criminal Procedure 16, we find the error to be harmless because there was overwhelming evidence of the defendants' guilt. Saunders and Bounds also contend that the court erroneously admitted a stipulation regarding a traffic stop of two alleged co-conspirators who drove away from the police while tossing packets that resembled heroin from their car. But we find that the stipulation was relevant to the government's drug conspiracy case and its prejudicial effect did not outweigh its probative value. Finally, the defendants appeal their sentences, arguing that the jury specifically found that less than one kilogram of heroin was involved and the district court erred by reexamining that finding. However, we find that the special verdict form, properly interpreted, does not contain such a finding from the jury. So the district court did not err in finding that more than one kilogram was involved. In addition, defendants contend that the district court erred in failing to articulate a methodology for arriving at its drug quantity finding. We do not agree. The district court properly identified and articulated a reliable basis for its calculation of the drug quantity. Therefore, we affirm the defendants' convictions and sentences.

I. BACKGROUND

In July 2012, defendants Christopher Saunders and Rashid Bounds, along with Joenathan Penson and Terrence Penson, were charged with conspiring to possess with intent to distribute and to distributing heroin in violation of 21 U.S.C. §§ 841 and 846. The indictment alleged that they conspired with David Price and others to sell heroin in Chicago. Saunders and Bounds were tried jointly before a jury, and were found guilty of conspiring to possess and distribute at least 100 grams, but less than one kilogram, of heroin.

At trial the government called as witnesses Mokece Lee, Joenathan Penson, and James Brown, who each pled guilty. They testified that the defendants were their co-conspirators in the heroin trade. That conspiracy began no later than November 2007 and lasted until at least March 2008, and Price, another co-conspirator, supplied heroin for distribution to the defendants and others on the west side of Chicago. The defendants managed and supervised specific blocks where the heroin was sold to lower-level workers. The resulting profits were split with Price and sometimes other co-conspirators.

Throughout the conspiracy, the base of operations was an apartment under Price's control known as "Up Top." While no one lived at Up Top, the defendants had keys to the apartment. Access to Up Top was, with a few exceptions, limited to those involved in the conspiracy. The defendants and their co-conspirators mixed Price's raw heroin with Dormin, a sleeping pill, to create larger quantities of product. They packaged the heroin mixtures, took turns making deliveries, and collected money from the purchasers.

As part of a task force investigation by the Chicago Police Department, surveillance photographs and videos were taken at Up Top. The surveillance revealed that the defendants and other co-conspirators were coming and going to Up Top during the conspiracy period. The investigation also produced evidence from six weekly trash pulls from outside Up Top. Evidence of heroin mixing and packaging was found, including Dormin bottles, cardboard Dormin containers, red plastic capsules, aluminum foil pieces, plastic baggies and their empty cardboard containers, and spools of tape. From this evidence, a forensic chemist also found trace amounts of heroin and diphenhydramine, the active ingredient in Dormin, and a fingerprint specialist, Joseph Ambrozich, found latent prints that matched the defendants' fingerprints.

A. The Rule 16 Disclosure

Because Ambrozich was to testify for the government as an expert witness, the government filed a Rule 16 disclosure. See Fed. R. Crim. P. 16(a). The disclosure explained the Analysis, Comparison, Evaluation and Verification (ACE–V) method of fingerprint examination, the method Ambrozich used to find a match. The Rule 16 disclosure explained the four steps of the ACE–V method, noting that in the second step, the fingerprint expert compares how many matching points of verification are found in the prints. These points of verification, sometimes called Galton points, are what justify a conclusion of a positive fingerprint identification. But the government did not disclose how many matching Galton points Ambrozich found during the examination or what the points were.

Because the number of Galton points was missing from the Rule 16 disclosure, the defendants moved to strike the expert opinion. That motion was denied, and the district court allowed the expert testimony, without requiring pre-trial disclosure of the number of points or offering any other remedy to the defendants. During trial, the government questioned Ambrozich about the number of Galton points he used to match the fingerprints, and he responded that there is no set number of points required to make a positive identification in the United States. Instead, he testified as to his personal preference of finding at least ten or twelve points of identification, which he said made him a conservative fingerprint examiner as compared to other examiners. He stated that in this case he found between twelve and twenty shared points of identification between the latent prints from the trash pulls and the defendants' fingerprints. During cross-examination, Ambrozich acknowledged that because there is no set standard in the United States, experts might differ not only in the number of points required, but in what qualifies as a point of identification as well.

The jury found the defendants guilty, and they moved for a new trial, partly on the basis of the government's failure to provide an adequate Rule 16 disclosure for the fingerprint expert. The district court denied the motion, concluding that the defendants had received a fair trial and had not been unfairly prejudiced by the government's incomplete disclosure.

...

II. ANALYSIS

The defendants raise four arguments on appeal. Two are evidentiary and two are related to their sentencing. We will discuss each in turn.

A. Evidentiary Issues

On appeal, the defendants challenge two evidentiary decisions. First, they argue that the district court abused its discretion in admitting the government's fingerprint expert testimony without requiring the government to disclose the bases and reasons for the expert's opinions. Second, they claim that the district court erroneously admitted the stipulation regarding the traffic stop of Carr and Price.

1. Expert Disclosure Insufficient, But Harmless Error

Specifically, the defendants contend that the district court erred in admitting testimony of the government's fingerprint expert, Ambrozich, without requiring the government to disclose before trial the number of Galton points Ambrozich used to make a positive fingerprint identification. We review the district court's denial of the defendants' motion to exclude for abuse of discretion. United States v. Stevens, 380 F.3d 1021, 1025 (7th Cir. 2004). If the district court abused its discretion in erroneously admitting evidence, we would reverse and remand for a new trial only if the error was not harmless. United States v. Thornton, 642 F.3d 599, 604 (7th Cir. 2011) ; United States v. McGee, 408 F.3d 966, 981 (7th Cir. 2005) ; Fed. R. Crim. P. 52(a).

The first step in our harmless error analysis is to determine whether there was actual error, which in this case would be a Rule 16 violation. Federal Rule of Evidence 702 requires expert testimony to be based on sufficient facts or data, and to be the product of reliable scientific or other expert methods that have also been reliably applied. And for expert testimony to be admissible, the proponent of the evidence must establish that the expert's testimony is reliable (and relevant) by a preponderance of the evidence. Daubert v. Merrell Dow Pharm., Inc., 509 U.S. 579, 589, 592–93, 113 S.Ct. 2786, 125 L.Ed.2d 469 (1993). In criminal cases when the government seeks to call an expert witness, the government must at the defendant's request provide a summary of any expert testimony the government intends to use, including "the witness's opinions, the bases and reasons for those opinions, and the witness's qualifications." Fed. R. Crim. P. 16(a)(1)(G) (emphasis added). We agree with the defendants that the government's Rule 16 disclosure was insufficient because it failed to provide the number of points of identification that were the basis for Ambrozich's opinion that the fingerprints were a match. See United States v. Robinson, 44 F.Supp.2d 1345, 1347–48 (N.D. Ga. 1997) (excluding fingerprint testimony for failure to comply with defendant's request for disclosure of documentation of the points of comparison relied on by expert to make identifications).

The government makes much of the fact that its disclosure discussed the ACE–V method, which the expert used to conclude that the defendants' fingerprints matched the trash pull evidence. The ACE–V method is "the standard method for determining whether two fingerprints are from the same person." United States v. Herrera, 704 F.3d 480, 484 (7th Cir. 2013). But the ACE–V method is a process. It is not the basis or reason for an expert's conclusion any more than the scientific method is the basis or reason for other scientific conclusions. We have found the ACE–V method to be sufficiently reliable under Daubert. Id. at 484–87. But this case is about disclosure of the basis and reasons for the expert opinion, not the reliability requirement of Daubert for admission of expert testimony.

The ACE–V method begins with a three-level analysis of whether the latent print is of value. The examiner starts with level 1 characteristics, which are the familiar pattern of loops, arches, and whorls that are generally visible to the naked eye. United States v. Mitchell, 365 F.3d 215, 221 (3d Cir. 2004). At level 2, the examiner analyzes the more detailed characteristics of the fingerprint, including "ridge characteristics" which are the patterns of islands, dots, and forks formed by the ridges. Id. The points where those ridges terminate or bifurcate are often referred to as Galton points. Id. "The typical human fingerprint has somewhere between 75 and 175 such ridge characteristics." Id. Finally, at level 3, the examiner looks at the microscopic variations in the ridges, which are the most vulnerable to distortion. Id.

After the analysis of the fingerprint, the examiner proceeds to the comparison step, comparing what he found on the latent fingerprint to what is found on a known fingerprint. Id. at 222. An evaluation is made as to whether there is sufficient similarity to declare a match. Id. The last step of the process involves a verification where another examiner repeats the observations to see if he comes to the same conclusion. Id. ; Herrera, 704 F.3d at 484. So the ACE–V method boils down to recognizing and categorizing distinctive features in prints and "it is the distinctiveness of these features, rather than the ACE–V method itself, that enables fingerprint examiners to match fingerprints." Id. at 485 (emphasis added). The distinctiveness of these features—whether referred to as Galton points, points of comparison, points of verification, or ridge characteristics—are what Ambrozich used to determine that the fingerprints in this case were a match.

The government contends that "the absence in the expert's report of a set number of Galton points on which he relied for his conclusions … did not mean that his opinion had no basis." But in this very statement the government concedes that the number of Galton points is the basis of the expert's opinion because it is what the expert "relied" on for his conclusion. The absence of the number of points in the report does not mean that his opinion has no basis, but it does mean that he has failed to disclose his basis.

The government argues that because there is no governing standard for how many Galton points are required to justify a fingerprint match, its Rule 16 disclosure was adequate. But the lack of a nationally accepted number of Galton points to make a fingerprint match does not undermine the importance of the number of Galton points here. For some fingerprint experts, and seemingly, for Ambrozich, the number of points is the basis for the determination that there is a match. See Mitchell, 365 F.3d at 222 ("An n -point match refers to a match between an unknown latent print and a known full print in which the examiner has identified n corresponding Galton points in the correct geometry relative to one another. A number of jurisdictions both outside the United States and within seem to rely on a system where a minimum number of corresponding points must be found before a match may be declared, irrespective of Level 3 detail."). Some experts use an alternative approach that combines quantity and quality whereby "[i]f ridge characteristics are abundant, then the quality of Level 3 detail is unimportant; but a paucity of Galton points can be compensated for by high-quality Level 3 detail." Id. We make no statement today about the preferred method of fingerprint identification or what methods are sufficient under Daubert. Instead, we apply the plain language of Rule 16. Whatever standard a fingerprint expert chooses to use to determine a match—whether it be a set number of Galton points or a combination of points and quality—the basis for the opinion that two prints are a match must be disclosed, upon the defendants' request.

In fact, being such a crucial, yet subjective, variable in the process arguably makes the number of points and what those points are even more important. Without knowing the Galton points used by the government's expert in advance, a defendant's ability to prepare an attack on the validity of the identification may be hindered. See Robinson, 44 F.Supp.2d at 1346 (finding that the defendant's expert was unable to review the basis of the opinion rendered by the government's fingerprint expert where the government did not properly disclose the location of each point of comparison).

Having concluded it was error not to disclose the Galton points we must still determine whether the error was harmless. In determining whether an evidentiary error is harmless, we consider whether the prosecution's case would have been significantly less persuasive in the mind of the average juror if the erroneously admitted evidence had been excluded. Thornton, 642 F.3d at 605. Here, there was an abundance of other evidence supporting the government's case against the defendants. The surveillance photographs and videos of the defendants coming and going from Up Top, the physical evidence from the trash pulls, and the testimony of the co-conspirators are all pieces of evidence that make the government's case a strong one. So without the fingerprint testimony, there was plenty of strong evidence for the jury to determine the defendants' guilt, and the government's case would not have been significantly less persuasive. While the district court should have required the government to supplement its disclosure, admission of the evidence was harmless.

...

III. CONCLUSION

For the foregoing reasons, we Affirm the defendants' convictions and sentences.

Questions:

1. Why is the following paragraph central to the defense claim that fingerprint evidence was improperly admitted?
   Whatever standard a fingerprintexpert chose to use to determine a match—whether it was a set number of Galton points or a combination of points and quality—the basis for the opinion that two prints were a match had to be disclosed, upon the defendants' request.
2. Prove True or False: The Prosecution failed to notify defendant that the fingerprint analysis did not meet the minimum comparative points?
3. Is the disclosure of comparison methodology on fingerprints used by experts that should be done automatically.

Finding fingerprints is not as easy as most people conclude. Prints can smear and destroy the comparison quality. Prints can deteriorate or be environmentally impacted or intentionally cleaned up. At the other end of the spectrum, is it proper for experts to testify concerning a surface being a poor repository for prints?

> In the Black Dahlia case, fingerprint analysis resulted in identification. Read the story at https://archives.fbi.gov/archives/news/stories/2006/october/dahlia_102006.

United States v. Tavares
843 F.3d 1 (1st Cir. 2016)

Convicted of being a felon in possession of a firearm, Verissimo Tavares appeals both the conviction and his sentence. He claims that the district court erred in admitting improper and prejudicial expert testimony concerning the absence of fingerprints on the gun that the jury found Tavares to have possessed. He also claims that the district court erred in its guideline sentencing calculations by treating his convictions for resisting arrest and assault and battery with a dangerous weapon as "crime[s] of violence." For the following reasons, we affirm the conviction but remand for reconsideration of the sentence.

I. BACKGROUND

Before dawn on August 4, 2013, two Boston police officers responded to a dispatch about a disturbance in a Boston neighborhood. As the officers dispersed the crowd, they heard gunshots emanating from the next street; when they went to investigate the source, they caught sight of a figure, subsequently revealed to be Tavares, riding away on a motor scooter. A chase ensued.

One of the pursuing police officers testified that he saw an object in Tavares's hands. Another testified that he saw Tavares throw an object into the yard of a nearby dwelling at 71 Clarkson Street. The chase ended when the police took Tavares into custody. Searching for the hurled object, one of the officers discovered a silver handgun in the yard of 71 Clarkson Street. Detectives were summoned to examine the firearm. They determined that the handgun was a semiautomatic firearm, loaded with five rounds of ammunition.

A federal grand jury charged Tavares as a felon in possession of a firearm. See18 U.S.C. § 922(g)(1). He maintained his innocence, and the case went to trial. During the government's case in chief, several police officers testified about the chase, the arrest, and the retrieval of the firearm. The government also presented the testimony of Richard Auclair, a fingerprint expert who held the position of Criminalist II in the Latent Print Unit (the Unit) at the Boston Police Department. The defense rested without presenting any evidence. The jury found Tavares guilty.

In anticipation of sentencing, the probation department prepared a presentence investigation report (the PSR). The Report recommended that the district court apply a four-level career offender enhancement under the sentencing guidelines, seeU.S.S.G. § 2K2.1(a)(2), based on a conclusion that Tavares's prior Massachusetts convictions for resisting arrest, seeMass. Gen. Laws ch. 268, § 32B(a), and for assault and battery with a dangerous weapon (ABDW), see id. ch. 265, § 15A(b), were both "crime[s] of violence." Tavares objected to the classification of his putative predicate offenses as crimes of violence under the residual clause of the sentencing guidelines' career offender provision. SeeU.S.S.G. § 4B1.2(a)(2) (Nov. 2014 ed.); id. at § 2K2.1(a)(2), cmt. n.1. Relying on our decisions in United States v. Glover, 558 F.3d 71, 80–81 (1st Cir. 2009), and United States v. Almenas, 553 F.3d 27, 33–34 (1st Cir. 2009), the district court overruled this objection. In so doing, the district court deemed both prior convictions to be for crimes of violence under the residual clause. The career offender enhancement, coupled with other adjustments not now in issue, yielded an advisory guideline range of 120–150 months, necessarily capped at 120 months by the ten year maximum applicable to the statute of conviction. See18 U.S.C. § 924(a)(2). Using the advisory guideline range as a "place to start" and mulling the factors enumerated in 18 U.S.C. § 3553(a), the court imposed an 84-month prison term. This timely appeal followed.

II. ANALYSIS

Challenging his conviction, Tavares argues that the district court erred, to his prejudice, by admitting over his objection a portion of Auclair's expert testimony. Alternatively, he argues that the district court erred in calculating his guideline sentencing range by counting his prior convictions as convictions for crimes of violence. We address each argument in turn.

A. Expert Testimony

The principal issue at trial was whether Tavares had possessed the gun that the police found in the yard at 71 Clarkson Street. Mainly through cross-examination and argument, Tavares sought to show that the government had not proved his possession of the weapon beyond a reasonable doubt. To bolster this claim, he suggested (among other things) that the police officers' testimony about his involvement with the weapon was inconsistent; that the weapon, when found, did not bear his fingerprints and, thus, had not been in his hands; and that the police had rushed to judgment. The government countered, in part, by presenting Auclair's testimony.

After being duly qualified as a criminalist, Auclair testified about the significance of the fact that the examination of the gun by the police laboratory revealed only a very partial print that was itself insufficient to implicate or exclude Tavares. Auclair delineated the factors that affect recovery of usable prints (including the quality of ridge skin, the texture of the surface involved, the nature of print deposition, the treatment of the surface after print deposition, and environmental conditions). After explaining the procedures used by the Unit to preserve prints, Auclair testified that the firearm removed from the yard at 71 Clarkson Street did not reveal any usable prints.So far, so good. During direct examination, however, the prosecutor asked Auclair about the percentage of cases in which usable prints were recovered from examined firearms, that is, what percentage of examined guns were found to contain fingerprints with sufficient ridge detail to allow the authorities to make an identification. Over the defendant's objection, the court allowed Auclair to opine, based primarily on the Unit's experience over a period of nearly nine years, that usable prints had been recovered from approximately 16% of firearms examined. Under cross-examination, Auclair explained that his opinion derived in part from a compilation of the Unit's fingerprint analyses completed by an intern: we say "completed" because the Unit regularly kept such data on a series of spreadsheets, and the intern had simply updated those data and tabulated them. Auclair could not say, however, either what procedures were used in the process of compilation or what oversight of the intern was provided by Unit staff. He could opine, however, that the 16% figure was generally consistent with his own personal experience in examining hundreds of guns. After cross-examination, Tavares moved to strike Auclair's opinion. The court denied his motion.

In the context of fingerprint examination, "deposition" is the act of depositing something (such as sand, snow, or mud) on a surface, especially over a period of time.

On appeal, Tavares challenges the court's rulings admitting and refusing to strike this portion of Auclair's testimony. Specifically, Tavares argues, first, that the testimony lacked a proper foundation; and second, that the testimony was both not relevant and unfairly prejudicial.

We review a trial court's decision to admit or exclude evidence for abuse of discretion. SeeUnited States v. Pires, 642 F.3d 1, 10 (1st Cir. 2011) ; United States v. Stierhoff, 549 F.3d 19, 27 (1st Cir. 2008). In carrying out this task, we afford "broad deference to the determination made by the district court as to the reliability and relevance of expert testimony." Beaudette v. Louisville Ladder, Inc., 462 F.3d 22, 25 (1st Cir. 2006). Absent a material error of law, we will not upset such a determination unless it appears that the district court "committed a meaningful error in judgment." Ruiz–Troche v. Pepsi Cola of P.R. Bottling Co., 161 F.3d 77, 83 (1st Cir. 1998) (quoting Anderson v. Cryovac, Inc., 862 F.2d 910, 923 (1st Cir. 1988)).

Against this backdrop, we turn first to Tavares's argument that the challenged testimony lacked a sufficient foundation. The touchstone for the admission of expert testimony in federal court litigation is Federal Rule of Evidence 702. The rule provides in relevant part that, as a precursor to giving expert testimony, an expert must be "qualified ... by knowledge, skill, experience, training, or education" and must possess specialized knowledge that "will help the trier of fact to understand the evidence or to determine a fact in issue." Fed. R. Evid. 702. The rule further demands that such opinion testimony rest on "sufficient facts or data." Id.

These requirements obligate a trial court to act as a gatekeeper in order to ensure, as a condition of admissibility, that proffered expert testimony rests on a sufficiently trustworthy foundation. SeeDaubert v. Merrell Dow Pharms., Inc., 509 U.S. 579, 597, 113 S.Ct. 2786, 125 L.Ed.2d 469 (1993). Where, as here, the factual basis of an expert's testimony is called into question, the district court must determine whether the testimony has "a reliable basis" in light of the knowledge and experience of the relevant discipline. Kumho Tire Co. v. Carmichael, 526 U.S. 137, 148, 119 S.Ct. 1167, 143 L.Ed.2d 238 (1999) (quoting Daubert, 509 U.S. at 592, 113 S.Ct. 2786). We, in turn, review such determinations for abuse of discretion. Seeid. at 142, 119 S.Ct. 1167.Tavares does not question Auclair's credentials as a fingerprint expert. Nor could he: Auclair earned a master's degree in forensics, underwent additional training upon joining the Unit, worked in the field for several years, and passed a series of annual proficiency tests.

Expert testimony, however, can for the most part be no better than the information provided to the expert. That principle is summed up in the familiar phrase "garbage in, garbage out." Tavares says that, whatever Auclair's qualifications, he lacked a sufficiently reliable basis for offering an opinion about the rate at which usable fingerprints appear on examined firearms.

This aspect of Tavares's challenge focuses on the alleged inadequacies of the compilation of data upon which Auclair drew in reaching his opinion about the 16% rate of usable prints recovered from examined firearms. The last step in that compilation was taken by an intern (not working under Auclair's supervision), and Auclair did not profess to know what procedures the intern had followed in compiling and tabulating the data.

Although Auclair was not aware of the specific procedures used to compile and tabulate the data that went into the intern's report, he made clear that the report was neither an ad hoc nor an informal production. Rather, the report was the latest iteration in ordinary course of a type of statistical compilation that the Unit had periodically produced on earlier occasions. These past reports were kept by the Unit in the ordinary course of its operations and were based on data that the Unit had collected and maintained in spreadsheets over a number of years. These spreadsheets had been prepared by Unit staff (including individuals with responsibility for technical information within the Unit). The district court did not abuse its discretion in finding that Auclair, given his position and expertise, was entitled to rely on these spreadsheets. SeeUnited States v. Corey, 207 F.3d 84, 89 (1st Cir. 2000) (approving expert's reliance on "materials maintained at ATF 'research libraries,' which contained information on approximately five thousand different firearms"); cf.United States v. Smith, 566 F.3d 410, 412 (4th Cir. 2009) (upholding admission of ATF agent's testimony that relied on "an ATF computerized database that had been compiled 'over many, many years as agents have done this practice' " in response to a challenge under the best evidence rule); Clausen v. M/V New Carissa, 339 F.3d 1049, 1059–61 (9th Cir. 2003) (upholding expert testimony that relied on history and reports created by others, even where there was no supported peer-reviewed literature). The intern's report merely updated these spreadsheets and tabulated the results, and it was within the court's discretion to find that Auclair could reasonably rely on that report as well.

What is more, Auclair testified that the percentage identified in the report for the Unit as a whole was "generally consistent" with his own long experience and that the scientific publications he had consulted did not affect this assessment. This testimony provided a basis for concluding that the department's report was materially reliable. It also provided an independent basis for the basic point being made: it was by no means unusual to find no usable prints on a gun. On that point, it could have hardly made any difference whether the percentage of guns found to have usable prints was exactly 16% or "generally" 16%. All in all, we think that it was within the district court's discretion to accept Auclair's determination that the updated Unit statistics were reliable.

We likewise reject the defendant's related argument that Auclair's testimony lacked a proper foundation because he had not performed or supervised the work that produced the compilation. An expert may rely on information not itself admitted into evidence when forming an opinion. SeeJones ex rel. United States v. Mass. Gen. Hosp., 780 F.3d 479, 494 n.8 (1st Cir. 2015). So, too, an expert may rely on information that is not independently admissible. SeeCorey, 207 F.3d at 89. Nor is there any requirement that the information relied on by an expert must have been compiled by him or under his supervision. SeeCrowe v. Marchand, 506 F.3d 13, 18 (1st Cir. 2007). And though "the entirety of [an expert's] testimony cannot be the mere repetition of 'the out-of-court statements of others,'" United States v. Luna, 649 F.3d 91, 105 (1st Cir. 2011) (quoting United States v. Cormier, 468 F.3d 63, 73 (1st Cir. 2006)), that was not the case here: Auclair's reliance on the compilation represented only a small fraction of his testimony on the subject of unusable prints and was corroborated by his familiarity with past calculations by the Unit and his own experience.

Tavares's embrace of our decision in United States v. Giambro, 544 F.3d 26 (1st Cir. 2008), does not advance his cause. There, the trial court found that the basis for the expert's testimony was "purely anecdotal," and was otherwise unreliable. Id. at 33. Our affirmance of that finding as not an abuse of discretion simply does not mean that it was an abuse of discretion to admit Auclair's testimony that rested in its material force on several independent, non-anecdotal grounds.

To say more on this point would be to paint the lily. In the circumstances here, we think that any question about the factual underpinnings of Auclair's opinion goes to its weight, not to its admissibility. SeeMilward v. Acuity Specialty Prods. Grp., Inc., 639 F.3d 11, 22 (1st Cir. 2011). So, too, deciding whether the data were of a type that Auclair could reasonably rely upon under Federal Rule of Civil Procedure 703 was well within the trial court's discretion. SeeCorey, 207 F.3d at 92.

The defendant's challenge to the relevance of Auclair's testimony is equally unavailing. "Evidence is relevant if: (a) it has any tendency to make a fact more or less probable than it would be without the evidence; and (b) the fact is of consequence in determining the action." Fed. R. Evid. 401. In this instance, we think that evidence reflecting the overall rate at which usable fingerprints are recovered from firearms was plainly relevant and likely helpful to the jury in determining what significance, if any, should be accorded to the absence of fingerprints on the firearm found at 71 Clarkson Street. Surely, such evidence had a tendency to make a fact of consequence more probable: it suggested that the absence of usable prints did not mean, ipso facto, that the weapon was never in the defendant's hands, or that the police work was shoddy. See, e.g., United States v. Burdeau, 168 F.3d 352, 356–57 (9th Cir. 1999). Seen in this light, the evidence "assisted the jury in understanding that … certain objects are not particularly conducive to finding prints." United States v. Glover, 479 F.3d 511, 518 (7th Cir. 2007). Absent Auclair's testimony, "the jury may not have understood how [the defendant] could have possessed the weapon without leaving prints." Id.

In an effort to deflect the force of this reasoning, Tavares points out that the 16% figure did not distinguish between firearms that were subjected to the so-called "fuming" process before they were sent to the lab (like the firearm in this case) and those that were not. This omission, Tavares submits, rendered the testimony too general to be relevant.

As explained by Auclair, "fuming" is the process in which a firearm is placed in a chamber filled with a heated glue substance. The glue then adheres to the moisture in the fingerprint, hardens the moisture, and turns the fingerprint white, rendering the print visible and less likely to be rubbed away.

This argument is futile. There is simply no requirement that statistics must in all instances separately account for every potentially significant variable in order even to be relevant. SeeMorgan v. United Parcel Serv. of Am., Inc., 380 F.3d 459, 468–69 (8th Cir. 2004). That is true of the "fuming" variable here. The defendant had the right—which he exercised—to cross-examine Auclair about the chances that the recovery percentage might differ materially in cases in which fuming was performed earlier. No more was exigible: after all, district courts have "broad latitude … with respect to the determination of the admissibility of expert testimony," Crowe, 506 F.3d at 18—and the limits of that broad discretion were not exceeded here.

Tavares has a fallback position. He contends that the challenged testimony, even if relevant, ought to have been excluded under Federal Rule of Evidence 403. Rule 403 provides, in pertinent part, that "[t]he court may exclude relevant evidence if its probative value is substantially outweighed by a danger of … unfair prejudice, confusing the issues, misleading the jury," or the like. Fed. R. Evid. 403. The rule is addressed to the district court's informed discretion and its due administration recognizes that "[t]his balancing is best performed by the trial judge, who has an intimate familiarity with the ebb and flow of the case and with its nuances." United States v. Raymond, 697 F.3d 32, 38 (1st Cir. 2012). "[O]nly rarely—and in extraordinarily compelling circumstances—will we, from the vista of a cold appellate record, reverse a district court's on-the-spot judgment concerning the relative weighing of probative value and unfair effect." Id.(quoting Freeman v. Package Mach. Co., 865 F.2d 1331, 1340 (1st Cir. 1988)).

Under this generous prescription, the district court's admission of the challenged testimony cannot be faulted. As we have already explained, the evidence was plainly relevant. Tavares, in turn, points to no prejudice that was so substantial as to compel exclusion. To be sure, Auclair's opinion was prejudicial in the sense that it aided the government's theory of the case and diminished the force of Tavares's theory of the case. But Rule 403 guards only against unfair prejudice, seeUnited States v. Benedetti, 433 F.3d 111, 118 (1st Cir. 2005), and the probative value of this evidence, though modest, was not substantially outweighed by any unfairly prejudicial effect. Hence, there was no abuse of discretion in admitting that evidence.

…

CONCLUSION

We affirm Tavares's conviction, but remand for reconsideration of his sentence consistent with this opinion. If the district court concludes that either the Resisting Arrest conviction or the ABDW conviction did not qualify under the career offender guideline, it should vacate and resentence. Otherwise, it should vacate and then re-enter the present sentence.

Questions:

1. What surface is alleged to be a tough place to find a print?
2. Which rule guides the determination on the admissibility of expert testimony as to the difficulty in leaving prints- Daubert or Frye?
3. Why is the defendant so adamant about excluding the 'absence of fingerprint" from a particular surface of a weapon.

3.1.1 Palmprints

The road to fingerprint admission was not all that challenging given the basic science, but did Palmprints face significant opposition?

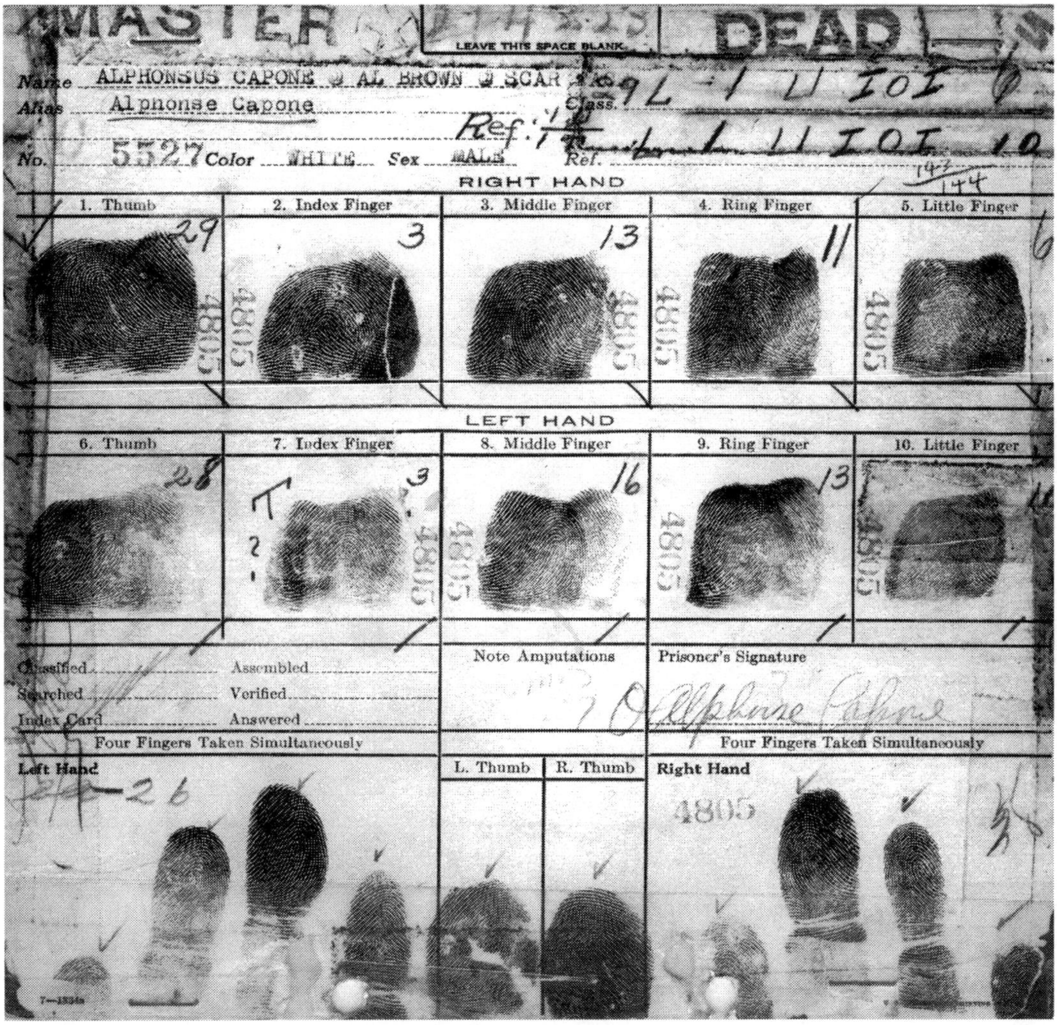

FIGURE 3.2 Al Capone's fingerprint card

State v Kuhl
Supreme Court of Nevada
July 1918 decided
No 2327

. . .

The appellant, Kuhl, with his co-defendants, were jointly informed against by the district attorney of Elko County for the crime of murder. They were specifically charged with the killing of one Fred M. Searcey, a United States mail-stage driver, at or near Jarbidge, in Elko County, Nevada. The testimony was wholly circumstantial. One of the elements in the case was an envelope, secured from one of the rifled mail sacks, on which was a bloody print or impression of a portion of the palm of a human hand. The trial of the defendant Kuhl resulted in a verdict of murder in the first degree, by reason of which the death penalty was imposed. From the judgment, and from the order denying a new trial, this appeal ensues.

It is the contention of appellant here that the trial court erred in admitting the testimony of the witnesses Stone and Botorff, offered in behalf of the state, as experts on palm-print identification. From the record it is disclosed that the impression found upon the envelope taken from the rifled mail sack was made by that portion of the palm which is immediately below the base of the little finger of the left hand. In offering the testimony of the experts, photographic enlargements and projectoscope views were used and presented to the jury. Objections were interposed to these methods of presenting the evidence, and with such we will deal during the course of the opinion.

\The first question which we propose to discuss is a novel one, inasmuch as our research has failed to disclose an expression from any court from which we might gain aid or guidance. After the arrest of the appellant, Kuhl, and while he was confined in the jail at Elko, an impression was taken of the palm of his left hand, and particularly that portion of the palm below the base of the little finger. The fact that the witnesses Stone and Botorff testified that the two impressions were made by the same hand gives rise to that phase of the appeal most strongly contended for by appellant.

Before testifying to their opinion as to the identity of the defendant's palm print with the impression found upon the bloody envelope taken from the mail sack, each of the witnesses fully explained his qualifications. Mr. Stone related in detail as to his study on the subject of finger-print identification and classification. It is disclosed that his investigation and research in this line had taken up his time almost continuously from the year 1908 or 1909 to the time of the trial; that during that time he had been engaged by at least two recognized identification bureaus, one under the state police department of the State of Nevada, the other under the police department of the city of Fresno in California. He testified to having visited numerous identification bureaus and to having attended conventions held by those engaged in this science in the United States. The witness Botorff related an experience entailing research and investigation in the line of finger-print identification and classification continuing from the year 1903 up to the time of the trial. Each of the witnesses was, as the record discloses, exhaustively and skilfully cross-examined on every phase of the subject that would bring forth to the jury their ability or lack of ability to give a correct or worthy conclusion as to the identity of finger-print impressions.

Were we dealing here with a finger-print impression, or the question of the comparison or identity of finger-print impressions, our course would be easy, for the courts of this county, and of England as well, have paved the way for the recognition of this science as an evidentiary element in criminal prosecutions. The main contention here is that the experts who testified were not qualified to give an opinion as to the identity of palm-print impressions; and, as we understand the contention of appellant, it is that science has not yet developed this question sufficiently to bear out the conclusion of an expert on the subject. Will the same rule which has led the courts to recognize experts on finger-print identification permit such experts to testify as to their conclusion upon palm-print identification? This is the one vital question here.

The origin of finger-print identification may be traced back to a period a hundred years before the birth of Christ. Scientific American, April 1, 1916, p. 356. By a Japanese scholar, Mr. Kumagusu Minakata, in an article entitled "The Antiquity of the Finger-Print Method," we are told that the discovery of this phenomenon of identity, as it may be termed, was made by the Chinese. In a most interesting article, entitled "History of the Finger-Print System" (Annual Report of the Board of Regents of the Smithsonian Institution for the year ending June 30, 1912, p. 631), Mr. Berthold Laufer traces the subject back to an era before the birth of Christ. He refers to the writings of Kai Kung-Yen, an author who wrote about the year 650 A. D., and who makes allusion to the employment of finger-print impressions in his time, and earlier, for the purposes of identification.

. . .

All of the writers upon the subject, to whose lines we have had access, agree that the palmar surface of the hands and the soles of the feet in men and monkeys are covered with minute ridges that bear a superficial resemblance to those made on the sand by wind or flowing water. Galton first gave expression to this fact; and Sir E. R. Henry, commissioner of police of the metropolis of London, corroborates with the statement that the inner part of the hand and the sole of the foot are traversed in all directions by lines of varying length. He says that the most conspicuous are the creases caused by the folding of the skin, and the least conspicuous but much more numerous lines are the papillary ridges which exist over the whole palmar surface, giving it an appearance that may be likened to that of a newly plowed field with its ridges and furrows, or to sand which the water, in receding from, has left ribbed. In Mr. Frederick A. Brayley's book, entitled "Finger-Prints Identification," we find the following significant language:

"'God's finger-print language,' the voiceless speech, and the indelible writing imprinted on the fingers, hand palms, and foot soles of humanity by the All-Wise Creator for some good and useful purpose in the structure, regulation, and well-being of the human body, has been utilized for ages before the civilization of Europe as a means of identification by the Chinese, and who shall say is not a part of the plan of the Creator for the ultimate elimination of crime by means of surrounding the evilly disposed by safeguards of prevention, and for the unquestionable evidence of identity in all cases where such is necessary, whether it be in wills, deeds, insurance, or commercial mediums of finance, as well as in the discovering and identification of lawbreakers."

Mr. Tighe Hopkins, in his work, "Wards of the State," makes extended reference to the papillary lines as covering the palms of the human hands and the soles of the human feet. In a work entitled "Criminal Investigation," translated by John and J. Collyer Adam from the work entitled "System der Kirminalistik," by Dr. Hans Gross, extended reference is made to the general subject. In a pamphlet published by Sir Wm. J. Herschel, dealing with the subject of finger-print identification, we find a most interesting history of experiments made by the learned author while acting in the British service in India. He interestingly relates of his experimentation with his own whole hand and with his right foot, which he says after an interval of fifty-seven years remained irresistibly unchanged. "The Origin of Finger Printing," by Sir Wm. J. Herschel, Humphrey Milford, Oxford University Press, June, 1916, p. 11. In his work "Guide to Finger-Print Identification," by Henry Faulds, late surgeon superintendent of Tsukiji Hospital, Tokyo, Japan, reference is made to the papillary ridges found covering the face or palmar surface of the hands and feet of the human being. In a terse and graphic little work entitled "Hints on Finger Prints," written by Rai Sahib Hem Chandra Bose, finger-print expert of Bengal, India, and a pupil of Sir Edward Henry, we find that, after dwelling on the possibility of error in finger-print comparison, the author makes this most significant assertion:

> "In fact, the indications on the inner surface of the hand are so numerous that, if half a square inch of any part of it were all that remained, that would be enough in that it would prove identity by comparison."

In his work entitled "The Finger-Print Instructor," Mr. Frederick Kuhne, of the bureau of criminal investigation of the police department of the city of New York, after dwelling at length on the basis of finger-print identification and the methods of classifying finger-print impressions, and especially upon the extent and usefulness of such identification, tells us that in some European cities impressions of the palms of the hands are utilized as an additional means of identification, especially because numerous patterns and characteristics appear in the palms as well as in the fingers, and in his work (page 96) he sets forth an illustration vividly portraying the truth of his assertion.

The lines on the palms of the human hand and the soles of the feet, which form the basis of individual identification, are the papillary ridges. They serve the office of raising the mouths of the ducts, so as to facilitate the discharge of the sweat, and perhaps perform the additional functions of aiding the sense of touch and of giving elasticity to the skin of the hand, and, having a vacuumistic tendency, they assist in preventing against slipping. These papillary ridges form figures, patterns, or designs, which research, study, and science have divided into classes named after their particular form, to wit, arches, loops, whorls, and composites. These patterns, as they have been established and named by those who have become devotees to the science of finger-print identification, while they have been discussed principally in connection with finger impressions, are not confined to the human finger alone, but are found with equal importance and equal persistency in the human palm and the sole of the human foot.

Mr. Harold J. Shepstone, in an article entitled "The Finger-Print System of Identification," appearing in the Scientific American of date October 1, 1910, at page 256, after dwelling upon the wonderful lineations in the form of ridges and patterns which adorn the palmar surface of the human hand, says:

> "One of the most interesting facts about this system is that every member of the human race, irrespective of age or sex, carries in person certain delicate markings by which identity can be readily established."

The learned author illustrates his article by a whole-hand impression showing the systems and the identifying markings. In the issue of the Scientific American of date August 19, 1911, there appears an article entitled "No Two Finger Prints Alike," and there reference is made to a communication addressed to the French Academy of Science by Mr. M. V. Balthazard, a student of the finger-print science. This learned authority declares his findings to the effect that, if any finger print be divided into a hundred squares, each square will contain some distinctive mark. He says that two finger prints will differ from each other, either in the arrangement of the marks in the different squares or in the character of the marks in a particular square. He says the total number of combinations of the two kinds of marks (branching or termination of ridges) in the 100 squares in the 100th power of 4. "This," says the author, "is a number that no one can possibly imagine. It is equal approximately to a number that would be represented by the figure 1 followed by 60 zeros. This means that there are possible just so many different kinds of prints, and that no particular combination will occur more frequently than others. The chances of any particular combination of marks occurring may be represented by a fraction with 1 as the numerator and a denominator represented by 1 followed by 60 zeros, a very tiny fraction of a chance indeed." To the suggestion as to how many points must agree in two finger prints to make sure of identity, this author bases his reply on mathematical grounds, to the effect that, when two finger prints agree in 17 out of the 100 squares, it is practically certain that they were made by the same finger. His reasoning in establishing this basis is most interesting and instructive.

In the issue of Law Notes of February, 1917, there appears an article entitled "Finger-Print Evidence," in which the subject is treated at some length. In an issue of the same publication of date January, 1918, and under the same caption, attention is directed to the learned discussion of the subject by Judge Wadhams, of the Court of General Sessions of the Peace of New York, where, in the case of People v. Sallow, 100 Misc. Rep. 447, 165 N. Y. Supp. 915, the matter is historically dealt with in the consideration of the validity of an act requiring the taking of the finger prints of persons arrested for crime.

In a most exhaustive work that has just come to our attention, in the writing of which Mr. Harris Hawthorne Wildes, Ph.D., and professor of zoology in Smith College, and Mr. Bert Wentworth, former police commissioner of Dover, N. H., collaborate, a system somewhat similar to that established by Galton is made the basis of sectional investigation by which palm-print identification may be carried out by means of the human hand. On page 138 of the work, illustrations are set forth in which the several sections of the hand are portrayed. By this work our attention is called to a term applied to the skin of the palmar surface of the hand and the sole of the foot which we think most appropriate, to wit, "friction skin." After dwelling upon the nature and character of this, the authors reassert the statement found in the works of all the other writers to whom we have referred, to the effect that this friction skin is covered by breaking, forking, splitting ridges, which ridges form patterns and designs most irregular and individual, and, say the authors:

> "As these features remain absolutely constant throughout the entire life, and are far too complicated to make a duplication of even a single ridge probable, it naturally follows that a small area of friction skin, no matter where taken, is sufficient for an absolute and positive identification, provided only that a record of it, in the form of a 'print,' or some other form of accurate reproduction, has been previously made and is available for comparison."

On page 126 of the work the authors refer to an instance where a small square area was cut out from the same place in the hand prints of two individuals, the place selected being one which has never occasioned any special interest among investigators, and where the ridges run monotonously in straight or slightly curved parallels. The area from which this patch of friction skin was taken lies above the proximal end of the metacarpal bone of the thumb, which would be approximately at the base of the first system of the hand under Galton's plan of subdivision. The authors refer to the fact that this region, or area, from which the experimental patch was taken, is the most featureless and monotonous of any of the parts of the human hand. Yet they say a careful scrutiny of the prints, especially when aided by a slight magnification, shows such marked differences that even a beginner in the work of identification would have no trouble in distinguishing them at once.

In concluding a most interesting chapter on the subject of structure and development of friction ridges on the palmar surface of the human hand and the sole of the human foot, dealing with details of their course and arrangement, the authors conclude that these surfaces furnish a basis upon which to found a system of identification positive and absolute. Here we find the unequivocal declaration that the patterns of the friction skin are individual, and, taken together, impossible to duplicate in another individual. Further, they declare that the separate ridges, too, show numerous details, which are in themselves so individual that a small area of friction skin, taken even in the most featureless portion, cannot be matched by any other piece. "Personal Identification," by Wildes and Wentworth.

All of the learned authors, experts, and scientists on the subject of finger-print identification, and each of those to whom we have heretofore referred, agree that these patterns, formed by the papillary ridges on the inner surface of the human hand and the sole of the foot, are persistent, continuous, and unchanging, from a period in the existence of the individual extending from some months before birth until disintegration after death. While most of the experts on finger-print identification deal most extensively with impressions of the human fingers, we find that some, of whom Mr. Galton is first and foremost, have divided the palmar surface of the human hand into what they term well-marked systems of ridges. Mr. Galton in his work fixes these systems thus:

> First, that which runs over the ball of the thumb and adjacent parts of the palm, bounded by a line which starts from the middle of the palm, close to the wrist, and sweeps around the ball of the thumb to the edge of the palm on the side of the thumb, which it reaches about half an inch, more or less, below the base of the forefinger. The second system is bounded toward the thumb by the base line of the first system and toward the little finger by a line starting from about the middle of the little-finger side of the palm and ending on the opposite side just below the forefinger. The third system is bounded thumbwards by the base line of the second until that line arrives at a point immediately below the axis of the forefinger. There the boundary of the third system leaves this base line and skirts the base of the forefinger, until it reaches the interval which separates the fore and middle fingers. The upper boundary of the third system consists of a line which leaves the middle-finger side of the palm a small distance below the base of the little finger and terminates between the forefinger and middle fingers. Galton on Finger Prints, Macmillan & Co., London and New York, 1892, p. 53.

...

2, 3. It is contended by appellant here that the court erred in admitting certain photographic enlargements and in permitting the witnesses Stone and Botorff to illustrate their testimony by the use of a projectoscope by means of which an enlarged photograph of the impressions was displayed to the jury. It might suffice to say that no question is raised as to the accuracy of the photographic exhibits, nor is any question raised as to the method of identifying the photographs or as to the manner in which the palm-print impression of the defendant was taken or as to the correctness of the enlargements. The appliances used and the methods resorted to, so far as we are able to determine, were those appliances and methods recognized by science. By these appliances, the jury was afforded an opportunity to follow the testimony of the experts in their direct and cross-examination. By this means they were better able to judge of the correctness of the testimony as it was being given and to estimate its weight and significance. This method of presenting proof has received the sanction of the highest authority. Wharton on Criminal Evidence (8th Ed.), sec. 544; Wigmore on Evidence, vol. 1, sec. 795; Rogers on Expert Testimony (2d Ed.), sec. 140; Dederichs v. Salt Lake C. R. Co., 14 Utah, 137, 46 Pac. 656, 35 L. R. A. 802, and note; State v. Connors, 87 N. J. Laws, 419, 94 Atl. 812. That instruments may be photographed for the purpose of so enlarging as to make the proportions plainer, and such photographs, when already in evidence, may be projected to illustrate the testimony of witnesses, is a rule that has found general sanction. First National Bank v. Wisdom, 111 Ky. 135, 63 S. W. 461; United States v. Ortiz, 176 U. S. 422, 20 Sup. Ct. 466, 44 L. Ed. 529; Howard v. Illinois Trust Co., 189 Ill. 568, 59 N. E. 1106; Marcy v. Barnes, 16 Gray (Mass.) 161, 77 Am. Dec. 405.

4. Appellant complains of the act of the trial court in admitting photographs of the palm impressions, when upon such photographs there were certain lines and markings, placed there by the witnesses Stone and Botorff before their testimony was given. These lines, as appears from what record there is before us, were placed on the photographs by the experts for the purpose of more clearly illustrating their testimony. They indicated the points of similarity and identity to which the experts testified. Their existence and significance were fully explained by the witnesses. These markings in no wise affected the photographs, and we are at a loss to discern any prejudice or injury that could have thus accrued to the appellant.

...

The judgment is affirmed. The court below is directed to fix a time and make all necessary orders for having its sentence carried into effect by the warden of the state penitentiary.

Let the order be entered accordingly.

Questions:

1. In assessing the reliability and admissibility of palmprints, what language in the Court's opinion displays a confidence in both the science, the uniqueness and permanency of palmprints?
2. How did the Court reconcile the usage of photographic enlargements of the palmprints?
3. How does the Court reconcile the argument that fingerprints and palmprints are not compatible in method or science?

3.2 ANTHROPOLOGY AND IDENTITY

The tools, techniques and science behind forensic anthropology aid the investigative effort to identify human remains, gender and age, and a whole host of other applications that target identity. While not as technically precise, anthropological determination is a process of elimination or a sort of winnowing out. Consider the 1850 case which draws said inference from a footprint in the snow.

FIGURE 3.3 An FBI Evidence Response Team member examines a recovered skull at the Anthropology Research Center in Knoxville, Tennessee, during a Human Remains Recovery course in March 2018. The outdoor site—popularly known as the "Body Farm"—consists of 2.5 shady acres where human donors are buried, partially covered, or left out in the elements so forensic anthropologists can study how bodies break down and decompose under different conditions. See https://www.fbi.gov/news/stories/body-farm-20th-anniversary-032019 for more details.

Locate the following case using your favorite legal research resource, read the case and answer the questions.

Commonwealth v. Webster, 59 Mass. 295, 5 Cush. 295 (1850) Supreme Court of Massachusetts, Suffolk and Nantucket

Questions:

1. How does the Court describe the anthropological reference as acceptable circumstantial evidence that leads to a conclusion on human agency.
2. When compared to fingerprints, is it fair to say that a good portion of anthropology may rule some things as in or out. In other words, a fingerprint leads to identity based on points or references that match.

An interesting slant on the use of anthropology identity can be gleaned from Parker v. State- a 1949 Indiana decision. The actual body of a murder victim could not be found whole, but only parts. The defendant fully confessed to the murder but under common law principles, the lack of a body- a corpus delicti gives rise to reasonable doubt.

Parker v. State
88 N.E.2d 556 (Ind. 1949)

The appellant was charged by indictment in four counts with first degree murder. He was tried before a jury, which returned a verdict of murder in the second degree and he was sentenced to serve a life term in the Indiana State Prison. He urges on appeal to this court, among other grounds for reversal, that the evidence was not sufficient to sustain the verdict and that there was error in admitting a written confession which he contends was obtained by physical mistreatment and persistent questioning over a period of several days.

On April 7, 1943, Lillian Johnson disappeared from the home which she owned in Terre Haute, Indiana, where she and appellant had been living together for about two years. Accepting the testimony most favorable to the State, as we must, it may be said that she has never been seen again, although there were witnesses who testified they had seen her in Indianapolis with appellant after she is said to have disappeared from Terre Haute, and appellant's daughters testified they visited her for two weeks at her home in Terre Haute in June, 1943. She and appellant were together on the date of her disappearance and subsequently appellant removed household goods from her home and wrote a letter in her name to a daughter of Mrs. Johnson in Terre Haute. There was also a telegram to Mrs. Johnson's employer, which was received on April 11, four days after Mrs. Johnson's disappearance, which read, "Have decided to go to Laramie, Wyoming. Will send for clothes later. Sorry Lillian." The author of this telegram was not identified other than by its signature, and the Western Union records had been destroyed.

Appellant likewise disappeared from Terre Haute on or about April 7, 1943, and for a time thereafter worked in Indianapolis, and later served a term in Federal prison and then reappeared in Terre Haute in September, 1947. On September 18, 1947, he was arrested for alleged assault upon a girl and was taken to the police station, where he was questioned. The charge upon which he was arrested does not seem to have materialized, but the police continued to hold him and question him about his relations with Mrs. Johnson and her disappearance. As a result of this questioning, he signed a written confession on September 20, 1947, in which he admitted the killing of Mrs. Johnson. The police checked the details of this confession and reached the conclusion that he had not told the truth and proceeded to question him further and, on September 22, 1947, he signed another confession in which he told in detail how he killed Mrs. Johnson and that he had dismembered her body and disposed of parts in three different places. The police found a skull and the upper part of a skeleton near the place designated by appellant. The flesh and tendons had entirely disappeared and only bare bones were found at the places where appellant said he had disposed of the other parts of the body. There was absolutely no evidence of any probative value to identify the bones found as part of the skeleton of Mrs. Johnson, except as stated by appellant in his confession, and there was no direct evidence of Mrs. Johnson's death, except as stated in the confessions. Appellant repudiated his confessions and claimed he had signed them as a result of threats and violence and prolonged questioning by the police without rest. The police denied all threats and violence, but admitted frequent questioning, and that they had taken appellant from the jail to the different places appellant had said he had buried parts of the body of Mrs. Johnson. After hearing evidence of the circumstances under which the confessions were obtained, the court admitted them and this is urged as error.

Appellant's counsel contend that there was insufficient evidence to sustain the verdict because, among other things, there was no evidence of the corpus delicti, 1. except as contained in the confession, and in this connection they urge also that without independent proof of the corpus delicti a confession is not admissible and is not sufficient to sustain a verdict. As we view the case these are the only questions we need consider. Proof of the "corpus delicti" means proof that the specific crime charged has actually been committed by someone. Hawkins v. State (1941),219 Ind. 116, 129, 37 N.E.2d 79; Hunt v. State (1939),216 Ind. 171, 178, 23 N.E.2d 681; 1 Wharton's Criminal Law (12th Ed.), § 347, p. 450; 14 Am. Jur., Criminal Law, § 6, p. 58; 23 C.J.S., Criminal Law, § 916, p. 181.

The rule seems to be well established generally that an extra-judicial confession will not be admitted in evidence and a conviction will not be upheld until and unless the corpus 2. delicti has been established by clear proof independent of the confession. Gaines v. State (1921), 191 Ind. 262, 268, 269, 132 N.E. 580; Hunt v. State, supra; Messel v.State (1911), 176 Ind. 214, 219, 95 N.E. 565; Griffiths v.State (1904), 163 Ind. 555, 558, 559, 72 N.E. 563; 1 Wharton'sCriminal Law (12th Ed.), §§ 359, 360; 20 Am. Jur., Evidence, § 484; 22 C.J.S., Criminal Law, § 839, pp. 1471-2; 23 C.J.S., Criminal Law, § 916b, p. 183; Underhill's Criminal Evidence (4th Ed.), § 36; Anno., 127 A.L.R., 1130, 1134.

In Wharton's Criminal Law, supra, it is said:

"Where a person is charged with the commission of a particular crime, before he can be found guilty thereof, it is essential that the existence of the corpus delicti be established, which cannot be done by mere extrajudicial confession of the accused; it must be done by direct and positive proof aliunde, and beyond a reasonable doubt. Facts ascertained by reason of the confession may be used for the purpose of establishing the corpus delicti; but this will not dispose of the rule requiring that the corpus delicti must be proved independently of the confession, and beyond a reasonable doubt, before evidence of the confession is admissible." (§ 359.)

In Gaines v. State, supra, the force of a confession not made in open court before the corpus delicti had been proved was under discussion, and this court said:

"A naked confession is one which is not corroborated by independent proof of the corpus delicti. Upon such a confession made in open court, as, for example, by a plea of guilty, a conviction of any crime and sentence may be had. But in the case of all extrajudicial confessions it is the rule that the corpus delicti must be proved by additional evidence before a conviction upon the naked confession alone will be upheld. Underhill, Criminal Evidence (2d ed.) § 147.

"In Gillet, Indirect and Collateral Evidence, § 117, it is stated that: 'In the United States the doctrine is thoroughly established that an extrajudicial confession will not be received as plenary evidence, and further, that there must be also proof in such cases of the corpus delicti.' (Citing many cases.)"

In reversing the case, this court said:

"… If the 'confession' of the appellant is eliminated from the evidence there is no evidence at all tending to show that any burglary was committed. The verdict is therefore not supported by the evidence, in one of the essential elements of the offense and is not according to law."

In Indiana the independent evidence alone need not be sufficient to establish the corpus delicti beyond a reasonable doubt, but there must be some evidence of probative value 3, 4. aside from the confession to prove that the crime charged was committed. When there is some independent evidence tending to prove that the crime charged has been committed by someone the confession may be considered with the independent corroborating facts in determining whether the corpus delicti has been established beyond a reasonable doubt. Griffiths v.State, supra; Messel v. State, supra; Hunt v. State, supra.

In Griffiths v. State, supra, this court said at p. 558:

"… We are mindful of the rule that the extrajudicial confession of a defendant is not alone sufficient to make out the corpus delicti, and that as applied to a prosecution for larceny it is required that there must be proof of the commission of the particular larceny charged. We deny, however, that such is the operation of the rule that the confession of the defendant can not in any case be used to accelerate the force or inferences concerning the fact of ownership, or that his confession can not be considered, along with the proper corroborative evidence, in determining whether the fact of the commission of the crime charged has been made out.

In Messel v. State, supra, this court said, at p. 219:

"The extra-judicial confession of the defendant alone is not sufficient to prove the corpus delicti; but such confession may be considered with independent corroborative facts, not of themselves sufficient to prove the corpus delicti beyond a reasonable doubt, to prove that the offense was committed… . (citing authorities.)"

However, it seems established over the years that the corroboration required is not of incidental facts stated in the confession but that the offense charged had been committed. 5. In Underhill's Criminal Evidence (4th Ed.), § 36, p. 43, this subject is covered in the following language:

"… The 'corroboration' of a confession or admission which is required in order to prove the corpus delicti refers not merely to facts proving the confession but to facts concerning the corpus delicti, or evidence independent of the confession. The corroboration of a confession does not necessarily prove the corpus delicti."

This view of the matter is confirmed in the quotation from Gaines v. State, supra, which we have set out earlier in this opinion. There Gillett on Indirect and Collateral Evidence is quoted with approval to the effect that in this country it is thoroughly established that an extrajudicial confession will not be received as plenary evidence, but that there must also be independent proof in such cases of the corpus delicti. In the Gaines case, this court indicated that if it eliminated the confession there was no evidence of the corpus delicti and the verdict, therefore, was not supported by evidence of one of the essential elements of the offense and was contrary to law. We think this is sound and that, in the light of many cases and many texts, it means that it is not sufficient that a fact stated in the confession is corroborated, such as the statement in the confession in this case that he buried the body of Mrs. Johnson at a certain place, but it is necessary to show independently of the confession that Mrs. Johnson was murdered.

In People v. Ranney (1908), 153 Mich. 293, 116 N.W. 999, 1,000, corroboration of the corpus delicti was involved and the following is quoted from the syllabus in that case:

"The word 'corroborate' in the rule that the confession of the accused, establishing the corpus delicti, must be corroborated, refers to facts which concern the corpus delicti, and does not mean merely tending to produce confidence in the truth of the confession."

In People v. Cuozzo (1944), 292 N.Y. 85, 54 N.E.2d 20, the defendant had made a confession that he killed a girl by knocking her unconscious and then placing her body upon the railroad tracks. Her mangled body was found alongside the tracks described by defendant in his confession, and there were other facts in his confession which were corroborated. The court held, however, that this was not sufficient corroboration to establish the corpus delicti independently of the confession. It was pointed out that there was no independent evidence that the girl had not been struck accidentally by the train when walking along the track. Other evidence corroborating incidental facts contained in the confession also were pointed out. It was said, however, that finding the body was not proof of a homicidal killing and that corroboration of the other facts enumerated in the confession, would go only to the weight to be given the confession.

In the case before us, the finding of a part of the skeleton, even though not at the exact place described in the confession, might go to the weight to be given the confession, if and when it was offered in evidence, but it would not amount to substantial evidence that Mrs. Johnson had been murdered. The only evidence that Mrs. Johnson had been killed, outside the confession, was that bones and a skull were found and that she had disappeared. There was absolutely no evidence of any probative value that the bones or skull were hers. The coroner did no more than testify that the bones were those of a female and a professor of physical anthropology of Indiana University testified as a witness for the State that there was no conclusive evidence of the cause of the death of the person whose bones were found, but that in his opinion they were the bones of a female between 47 and 65 years of age, and that they had been buried for more than two years but not more than 25 or 30 years. The State says that the facts that there was evidence that the appellant took property out of Mrs. Johnson's home after she disappeared and that he wrote a letter to the daughter of Mrs. Johnson after Mrs. Johnson had disappeared, and signed Mrs. Johnson's name to it, were corroborative of his confession that he killed Mrs. Johnson. This evidence might serve to corroborate facts found in appellant's admissions and confessions and go to its weight but it fails to prove independently of the admissions and confessions that the skeleton and skull were those of Mrs. Johnson.

We hold that there was no evidence of the corpus delicti in this case of any probative force, except the confessions and admissions of the accused, and, therefore, the evidence was 6. insufficient and the case must be reversed.

In view of our conclusion, it is not necessary to determine whether there was error in admitting the confession, or in other respects urged by appellant.

The judgment is reversed, with instructions to sustain appellant's motion for a new trial.

Questions:

1. In the defendant's confession, what information was provided authorities on where he disposed of the body.
2. What did law enforcement discover after this confession?
3. Given these facts, would the same issue arise today? What other forensic forms, combined with anthropology would have generated a dependable identity conclusion?

Forensic Anthropology is quite effective in ruling out categories, such as sex/gender, age and development and other bodily variables that are edified by this intense form of examination. At the same time, the field can determine, given the pathology or unique markings on the human body, how a person dies or whether or not it is an accident or intentional by design.

Yepez v. State
No. 04-17-00358-CR (Tex. App. Jul. 3, 2019)

A jury found appellant Cristian Yepez guilty of murder and tampering with evidence and assessed punishment at thirty years' and fifteen years' confinement, respectively. On appeal, Yepez argues: (1) the trial court erred in admitting expert testimony; (2) the trial court abused its discretion in denying Yepez's motion to suppress; (3) the trial court erred in denying Yepez's request to include a concurrent cause instruction in the jury charge; and (4) the evidence is insufficient to corroborate accomplice witness testimony. We affirm.

BACKGROUND

An unidentified caller contacted 9-1-1 and reported a domestic disturbance between a male and a female at an apartment complex in Houston, Texas. Fifteen minutes later, another 9-1-1 operator received an anonymous call regarding the same domestic disturbance. The second caller stated a woman, later identified as Janette Pantoja, came outside of the apartment and told the caller she had been kidnapped and the kidnapper had killed her three-year-old daughter Jasleen. Officer Daniel Lunceford of the Houston Police Department responded to the disturbance and spoke to Pantoja. Officer Lunceford described Pantoja as crying hysterically. Pantoja stated her boyfriend, Yepez, had hit her with a closed fist, and Officer Lunceford observed bruising above her eyebrow. Pantoja also told the officer she was from Chicago and Yepez was keeping her against her will after killing her daughter, Jasleen, when they were in Laredo. Officer Lunceford took Pantoja to the police station, where Detective Robert A. Klementich questioned her.

Detective Klementich was already familiar with Pantoja. Over the past couple of weeks, he had been working on a missing child case with the Chicago Police Department, which reported the missing child's mother and the mother's boyfriend - Pantoja and Yepez - were staying with Yepez's relatives in Houston. Pantoja's relatives had contacted the Chicago Police Department about the missing child after Pantoja and Yepez returned to Chicago from Laredo without the child. According to Pantoja's relatives, Pantoja and Yepez told her family the child had been kidnapped. The Chicago police gave Detective Klementich two Houston addresses, and Detective Klementich sent officers to both locations. At the second location, one of the investigating officers encountered a group of individuals, which included a woman who gave the officers a false name and acted "squirrely." Later, Detective Klementich and the investigating officers received photographs of Pantoja and Yepez from the Chicago police and realized the woman they encountered was Pantoja. However, when the investigating officers returned to the location, Pantoja was no longer there. When Detective Klementich subsequently met Pantoja at the police station, he knew Pantoja was from Chicago and connected to the missing child case. He questioned Pantoja concerning Jasleen's whereabouts, and she stated Yepez had assaulted her at the Houston apartment and killed Jasleen when they were in Laredo.

Detective Klementich dispatched two officers to bring Yepez to the station for questioning. He believed he had probable cause to arrest Yepez for assault family violence. He also believed he had enough information to detain Yepez to investigate Jasleen's whereabouts. After the dispatched detectives arrested Yepez for assault family violence and unlawful restraint, they brought him to the police station. Detective Klementich informed Yepez of his Miranda rights. Although Yepez initially waived his Miranda rights, during the interview he indicated he wanted to speak to a lawyer. Detective Klementich ended the interview and informed Yepez he was going to be charged with unlawful restraint. Detective Klementich then left, leaving Yepez handcuffed in the interview room.

Approximately an hour later, Yepez asked to speak to the detectives and told them Jasleen had been kidnapped. Yepez then changed his story and told the detectives that on the day Jasleen died, she had been swimming at the motel pool when they were in Laredo. After swimming, she started throwing up and was highly irritable and crying. Yepez stated that when he, Pantoja, and Jasleen returned to the motel room, he put Jasleen in "timeout" and placed her face to the wall to calm her down. She continued to cry and throw up until she collapsed and lost consciousness. Yepez said he contacted a man who helped him bury the body.

Detective Klementich contacted the Laredo Police Department and shared Yepez's statements about burying Jasleen's body. Laredo Police Detective Greg Cantu drove to a specified location - an empty lot near a gas station - where he found a child's skeletal remains. Detective Cantu secured the scene, contacted his supervisor, and waited for a crime scene investigator to arrive. Chief Medical Examiner Dr. Corinne Stern arrived at the scene and determined the empty lot had recently been cleared, exposing the remains. Dr. Stern collected the remains and sent them to Dr. Oscar Harrell Gill-King, a forensic anthropologist who analyzed the remains to determine the cause of death. Dr. Gill-King identified the remains as Jasleen. In his report, Dr. Gill-King indicated the skull displayed signs of "perimortem vertical loading," and concluded the child's death was caused by her skull striking a static object.

A grand jury indicted Yepez for felony murder, serious bodily injury to a child, and tampering with evidence in connection with Jasleen's death. At trial, the State argued Yepez murdered Jasleen by striking her head with blunt force. The defense argued she died due to dry drowning, after which Yepez and Pantoja panicked and buried the body. The jury heard testimony from several witnesses including Pantoja, the detectives and officers working on the case, and Dr. Gill-King. Pantoja testified Yepez grabbed Jasleen by her ankles, held her upside down, struck her head against the bathroom floor and later against the bathroom sink. Pantoja testified Jasleen reacted by "just sitting" on the countertop. Pantoja testified Yepez then took Jasleen to another room and slammed the bathroom door, leaving Pantoja in the bathroom. Thereafter, she heard another bang and Jasleen was unconscious when she opened the door. She testified she unsuccessfully attempted CPR, but Jasleen never regained consciousness. Ultimately, Pantoja explained Yepez decided to bury the body.

In addition, Dr. Gill-King testified he determined the child's cause of death was homicide caused by vertical loading. The forensic anthropologist also testified the injuries to Jasleen's skull were consistent with someone holding her by the ankles with her head hanging down and then hitting her head against a hard surface. On cross examination, Dr. Gill-King further opined that earth moving equipment used to clear the lot where Detective Cantu found the remains did not cause the injuries to Jasleen's body. The jury also saw photographs of Jasleen's remains and watched a videotape recording of Yepez's statements made to Houston police.

At the charge conference, Yepez requested a jury instruction on concurrent cause, arguing the evidence showed the child died from dry drowning. The trial court denied the request. The jury found Yepez guilty of felony murder and tampering with evidence. Yepez appealed.

ANALYSIS

Expert Testimony

In his first issue, Yepez argues the trial court abused its discretion by admitting Dr. Gill-King's expert testimony regarding the child's cause of death. Yepez claims Dr. Gill-King's testimony was unreliable because the scientific theory and technique he used were not valid and there was a large analytical gap between the evidence and his proffered opinion.

Standard of Review

We review a trial court's decision on the admission of expert testimony for an abuse of discretion. Rhomer v. State, No. PD-0448-17, 2019 WL 408186, at *3 (Tex. Crim. App. Jan. 30, 2019). "[A] trial court abuses its discretion when it acts without reference to any guiding rules or principles or acts arbitrarily or unreasonably." Id. We will uphold the trial court's ruling on the admission of evidence unless it lies outside the zone of reasonable disagreement. Layton v. State, 280 S.W.3d 235, 240 (Tex. Crim. App. 2009).

Applicable Law

A witness who qualifies as an expert by knowledge, skill, experience, training, or education may testify if scientific, technical, or other specialized knowledge will assist the trier of fact in understanding the evidence or determining a fact issue. TEX. R. EVID. 702. The proponent of the testimony must establish by clear and convincing proof that the proffered testimony is sufficiently relevant and reliable to assist the factfinder in reaching an accurate result. Vela v. State, 209 S.W.3d 128, 134 (Tex. Crim. App. 2006). To assess the reliability of an expert's opinion based on a hard science, the Court of Criminal Appeals has set forth the three-prong Kelly test, requiring that "(1) the underlying scientific theory be valid, (2) the technique applying the theory must be valid, and (3) the technique must have been properly applied on the occasion in question." Rhomer, 2019 WL 408186, at *4 (citing Kelly v. State, 824 S.W.2d 568, 573 (Tex. Crim. App. 1992)). In addition, courts should consider:

1. the extent to which the underlying scientific theory and technique are accepted as valid by the relevant scientific community, if such a community can be ascertained;
2. the qualifications of the experts testifying;
3. the existence of literature supporting or rejecting the underlying scientific theory and technique;
4. the potential rate of error of the technique;
5. the availability of other experts to test and evaluate the technique;
6. the clarity with which the underlying scientific theory and technique can be explained to the court; and
7. the experience and skill of the person(s) who applied the technique on the occasion in question.

Kelly, 824 S.W.2d at 573.

When expert testimony concerns a field of study outside the hard sciences, the Court of Criminal Appeals instructs us to apply the Nenno test to evaluate the reliability of an expert's testimony. Rhomer, 2019 WL 408186, at *5 (citing Nenno v. State, 970 S.W.2d 549, 550 (Tex. Crim. App. 1998), overruled on other grounds by State v. Terrazas, 4 S.W.3d 720 (Tex. Crim. App. 1999)); Weatherred v. State, 15 S.W.3d 540, 542 (Tex. Crim. App. 2000). "The Nenno test asks whether (1) the field of expertise is a legitimate one, (2) the subject matter of the expert's testimony is within the scope of that field, and (3) the expert's testimony properly relies upon and/or utilizes the principles involved in the field." Rhomer, 2019 WL 408186, at *5; Weatherred, 15 S.W.3d at 542. In applying this framework, we must refrain from developing rigid distinctions between "hard" and "soft" sciences, because the distinction between various types of testimony can be blurred. Morris v. State, 361 S.W.3d 649, 654 (Tex. Crim. App. 2011). Under either the Kelly or Nenno test, reliability should be evaluated by referencing the standards applicable to the particular professional field in question. Coble v. State, 330 S.W.3d 253, 274 (Tex. Crim. App. 2010).

Application

Dr. Gill-King identified himself as a forensic anthropologist who analyzes skeletal remains to identify people and determine their cause of death. He testified forensic anthropologists are generally contacted by authorities to examine remains. When the remains are brought to him, he inventories the material, cleans it, and analyzes it. He explained that forensic anthropologists use various cleaning methods depending on the situation. He added that sometimes authorities provide background information regarding the scene where the remains are found, and he prefers to have limited information regarding the facts of the case to avoid bias. After analyzing the material, he creates a biological profile and writes a report with his interpretations. He explained that forensic anthropologists make objective and subjective observations, and their reports are not subject to a formal peer review process. He further added that no standards of practice have been established to guide forensic anthropologists on how to make interpretations, but official standards of practice are being developed.

Turning to the instant case, Dr. Gill-King testified he received a "nearly complete skeletonized remains of a human infant" to examine along with information that the Laredo police discovered the remains in a shallow grave. He explained the remains were in an advanced state of decomposition because they were recovered in a shallow grave and were exposed to a high level of bacteria, which accelerated the decomposition process.

Dr. Gill-King also testified the remains displayed no sign of antemortem trauma, eliminating any indication that the child had old injuries or was abused over an extended period. He opined the remains displayed perimortem trauma - trauma that occurred at the time of death. Such trauma was significant to Dr. Gill-King because it could explain the cause of death. In this case, he observed "the skull was broken in several pieces." He also noted it exhibited a diastatic fracture - breakage at a suture - at the top of the head and fractures on both sides. According to Dr. Gill-King, these characteristics - multiple fractures, diastasis, and fractures on the sides of the skull - constituted three of the five key criteria forensic anthropologists use to indicate signs of "cranial death" - death caused by a head injury or injuries.

Dr. Gill-King further testified the fractures in Jasleen's skull were consistent with fractures caused by vertical loading, an injury that occurs when the skull strikes a static object, transferring energy from the top of the skull, down through its base, and along the spinal axis. For support, he pointed to the diastatic fracture, an area where a small piece of bone was pressed in when Jasleen's skull struck a static object. He explained diastasis is a classic sign of head trauma in infant deaths. He further explained the fractures at the base of Jasleen's skull and radiating out from her spine indicated that her head had been struck against a static object. When her skull was struck against the static object, the shock travelled through her body and transmitted force through the base of the skull, resulting in the fractures. He ultimately concluded the injuries were static as opposed to dynamic, highlighting that the terms "static" and "dynamic" were widely used in his field.

Yepez contends Dr. Gill-King's testimony was unreliable because the scientific theory and technique he used were not valid. Specifically, Yepez argues Dr. Gill-King's report was not subject to peer review and another expert disagreed with the report's conclusions. However, the Court of Criminal Appeals has stressed that "the absence of peer review does not necessarily undercut the reliability of the testimony." Nenno, 970 S.W.2d at 562. Rather, any doubt created by the absence of peer review goes to the weight of the evidence rather than its admissibility. Id. Similarly, disagreement between experts also goes to the weight of the evidence, and the jury can believe one expert over another. See Sharp v. State, 707 S.W.2d 611, 614 (Tex. Crim. App. 1986).

Yepez also argues Dr. Gill-King's testimony was unreliable because a large analytical gap existed between his opinion and the underlying facts. According to Yepez, Dr. Gill-King's opinion did not coincide with the State's theory that Yepez killed Jasleen by striking her head against a static object. We disagree. After Dr. Gill-King opined the child's fractures were perimortem and caused by static force, he specifically testified the injuries would be "consistent with" someone holding a child upside down and striking the child's head against a hard surface, like the floor. He testified that based on his experience with examining remains of injured children, he believed Jasleen died after her head was struck against a static object.

The record reflects Dr. Gill-King based his opinion on a methodology generally used by forensic anthropologists when analyzing skeletal remains and on his years of experience analyzing skeletal remains and injuries in children. See Coble, 330 S.W.3d at 274. And although Dr. Gill-King stated his report was not subject to peer review, the record is clear that he based his opinion on principles relevant in his field of study. See Nenno, 970 S.W.2d at 562; Coble, 330 S.W.3d at 274. Accordingly, we hold the trial court did not abuse its discretion in admitting Dr. Gill-King's expert testimony regarding Jasleen's cause of death. We therefore overrule Yepez's first argument.

...

CONCLUSION

We affirm the trial court's judgment.

Questions:

1. Does the Court accept the expert's view that leads to a finding of cranial death in the child's body?
2. Can the forensic anthropologists properly reach other conclusions as to how and in what manner these head or cranial injuries were sustained?
3. Prove True or False: The Court describes forensic anthropology as a mix of objective science and subjective interpretation.

3.3 TRACE EVIDENCE AND IDENTITY

The concept of trace evidence includes a series of evidentiary forms such as hair, fibers, clothing threads and other microscopic substances that are generally tied to the human agent or actor. Put another way, trace forms are tied to a particular defendant. So in a case hair analysis, whether its be from the head, the eyebrows or other facial feature, may lead to a positive id although this form of evidence is from perfect. This is evident in our first case from 1882.

Knoll v. State
Supreme Court of Wisconsin
April 14, 1882, Argued; May 10, 1882, Decided

The defendant, and plaintiff in error, was charged and tried for the crime of having Avilfully and felon-iously killed, with malice aforethought, one Charles Rohde on the 4th of March, 1879. The evidence relied upon by the state to prove the accused guilty of the crime charged against him was mainly circumstantial. It appeared that the deceased had becu in the employ of the defendant for a short time, and the weight of testimony tends to show that he was last seen alive in the early part of the evening of the 3d of March, at the defendant's saloon. His body was found on the 25th of the same month in an unfrequented place in a swamp, about forty rods south of a railway track, and about one-half mile south-west of the defendant's residence. There were four cuts or stabs on his left breast, and two on the front part of *252each leg between the knee and thigh. Medical witnesses, who mhde %, postmortem examination- of the body, testified that there was a fracture of the skull on the right side of the forehead extending backward and downward to about the margin of the ear; that his bowels had been severely injured from blows; and that death was probably caused by the injury to the head and bowels combined. The theory of the prosecution was, that the defendant had killed Rohde on the night of March 4th in his barn, and had then taken the body to the place where it was found for the purpose of concealing the crime.

Among other evidence given to sustain this theory the state produced as a witness one John Timmens, who lived very near the railroad, and who testified that about midnight, March 3d, he saw from the window of his house a man come along the track with a wheelbarrow and something in it. The witness described the wheelbarrow and the :man. There was a wheelbarrow in the possession of the defendant answering this description, and the defendant in size was about such a man as witness saw. There was some blood and hair found upon the wheelbarrow which the defendant had. One Ur. Piper-was sworn as a medical expert. He had made a microscopical examination of blood found on the barrow and on pieces of wood taken from the barn; had examined pieces of cloth and hair,— hair -taken from the skull of the deceased, and hair found on the wheelbarrow. This witness was permitted to state, against defendant's objection, that he had made a comparision of the hair found on the wheelbarrow and that taken from the skull, such comparison being founded on his experience, he having made, as he said, a very careful study of hair. He was asked to state, and did state under objection, the result of that comparison. He said that the hair was precisely the same in every respect, in length, magnitude, color, and in every other respect, so that any person could have told it as *253well as himself, and he added: " As the result of that comparison, I can say that it was from the head of the same person."

"We have detailed enough of the facts to show the very important bearing of this inculpatory testimony. One exception relied on for a reversal of- the judgment is the admission of this testimony against the defendant's objection; and the question presented for decision is, whether. it was competent and proper testimony under the circumstances. The objection to its admission is, that the witness was permitted to state or give his opinion upon a vital fact in the case, which it was the province of the jury to determine from the evidence givén. The witness said that the hair which he had examined found on the wheelbarrow and that taken from the skull of the deceased were from the head of the same person. The witness reached this conclusion, as we understand his testimony, not from any scientific tests, or peculiarities in the structure of the hair which an examination by a microscope would disclose, but from the length, magnitude and color, or those obvious marks and resemblances which one person of good vision would observe as readily as another. The comparison made required no peculiar skill nor scientific knowledge. It was no more in the province of an expert than of an ordinary person to make it. It related to a matter of common observation. The jury were as competent to make the comparison from the description given of the hair, and draw the conclusion whether it came from the head of the same person, as was the witness. The opinion of the witness as to the fact that the hair came from the head of the same person was not admissible on the ground that the inquiry related to a scientific subject—one which required peculiar knowledge or previous study and experience to give information about. JBut it related to a matter within the ■observation, judgment and knowledge of any ordinary man; for the resemblances relied upon in making the comparison, *254as the length, magnitude and color of the hair, were as open to the observation of the jury, or the jury could draw their inference from these resemblances as well as any one. The witness, then, could not testify to his opinion on' the ground that the subject matter of the inquiry related to a scientific subject, and was expert testimony.

Is there any other principle upon which the testimony would be admissible? At first we had some doubt whether it should not be received on the ground that the witness was merely stating his opinion as to the identity of the hair, and that it was admissible upon the same principle as an opinion in respect to the value of., property, or damage done to it, or the identity of a chattel or person, or facts of that nature. In regard to this class of facts a witness can only testify by using language which amounts to little more than giving his opinion about them. But this kind of evidence is admitted in that class of cases from necessity, because it is impossible, by any mere words of description, to give the jury a proper understanding of the facts. But, of course, the general rule is that a witness cannot testify as to his opinion, but is limited to stating facts. Respectable authorities may be found which go nearly, if not quite, the length of sustaining the admission of the testimony which we are considering. See Comm. v. Dorsey, 103 Mass., 412; Comm. v. Sturtivant, 117 Mass., 122. But it seems to us such evidence is of a most dangerous character, especially when a witness is allowed to testify, as Dr. Piper did, that in his opinion the hair found on the wheelbarrow and that removed from the skull of the deceased were from the head of the same person. The witness had stated without objection that the hair found on the wheelbarrow was human hair. Possibly this might be said to involve a question of special knowledge, learning or experience. But the witness then described the hair, and said that from comparison of its length, magnitude and color it must, in his opinion, all have come from the same head. *255That conclusion was the precise fact which the jury were called upon-to determine. It is not entirely clear from the record whether the hair taken from the skull and that found on the wheelbarrow were 'before the court and jury, though we infer such to be the case. If so, it is obvious the jury could make the comparison for themselves, for the resemblance or marks of similarity were obvious. But, if we are mistaken in this supposition, the hair in both instances had been so fully described—the points of resemblance or identity had been so fully given—that the jury could draw their own conclusion as to whether it came from the head of the same person or not.

In a number of cases which will be found in our reports the rule has been laid down as to when and upon what questions a witness may testify to his opinion as a conclusion of fact. Luning v. State, 2 Pin., 284; Burnham, v. Mitchell, 34 Wis., 133; Oleson v. Tolford, 37 Wis., 337; Benedict v. Fond du Lac, 44 Wis., 495; Mellor v. Town of Utica, 48 Wis., 457; Yanke v. State, 51 Wis., 464; Noonan v. State, post, p. 258. But none of these cases go the length of sanctioning the admission of such testimony as that given by Dr. Piper. We think it was clearly incompetent, and must work a reversal of the judgment.

While this result disposes of the case, it may be proper to make a few further remarks on one or two other points which were much discussed by counsel. Dr. Piper also testified as an expert in regard to examinations made by him with a microscope of certain blood stains found upon pieces of cloth and wood. He gave it as his opinion, founded upon such examination, that some of these stains were caused by human blood corpuscles. Eor the purpose of discrediting the witness it was proposed on the part of the defense to read opinions" stated in certain medical works on this subject. The court would not permit this to be done, holding, in effect, that as Dr. Piper had not'referred to any medical work, and did not rely upon the authority of medical writers to support his views, but testified from his own knowledge and experience, it was not proper to read from medical works to contradict him. 0There can be no doubt of the correctness of this decision, which is sustained by the authorities referred to by Mr. Justice Cassoday in Stilling v. Town of Thorp, 54 Wis., 528.

A number of objections are taken to the charge of the court given on the trial. It is perhaps sufficient to say, in ■answer to all df them, that the record does not present a single exception to the charge, nor is it pretended that any was ever taken to it. But, notwithstanding this, the learned counsel for the defendant insisted that it was the duty of ■this court to review the charge, even though it was not excepted to, and if found incorrect in the propositions of law laid down for the guidance of the jury to reverse on that ground. That would be contrary to the uniform practice of .this court since its organization. In no case, civil or criminal, has this court reviewed the charge of the trial court where no exception was taken to it. In the absence of all ■legislation on the subject, we do not conceive it to be our .duty to change a rule of practice so long established in this and other courts. In this case the defendant was assisted •bj7 able and intelligent counsel, abundantly competent to protect his rights and secure for him a fair trial. If they sdid not see anything in the charge of the court which they .deemed unfavorable to their client, or objectionable in law, why should this court be called upon to review it? Counsel rsays that it is a humane principle recognized in criminal .law, especially in a case of murder, that the accused stands upon all his rights, waiving nothing which can possibly .prejudice him. In the sense in which this principle is -sought to be applied, it is not strictly correct. For example: .an objection to the admission of improper testimony must ■be taken in time to be available; so must an exception to *257the charge. There are many things in the conduct of the case which the accused loses the right to object to unless he takes the objection at the proper time. It is needless to mention them, for they will readily occur to every intelligent lawyer. The claim, therefore, that the accused waives nothing, or that he can always insist upon and take advantage of any error in the proceedings, cannot be maintained. Often a plea of guilty is entered by the accused and acted upon by the court. In that case the party waives his right to a trial. We have said this much in reply to the argument, which was so seriously pressed upon us, that we could review the charge though no exception was taken to it. But no inference must be made from this that we consider the charge erroneous in any material point. We simply decline to review it.

To guard against any possible misapprehension, we deem it proper to say a word on a remark in the charge. The circuit court, among other things, told the jury that their verdict must be simply murder in the first degree or not guilty, according as they should find the fact. This same charge was given in the case of Dickerson v. State, 48 Wis., 288, and held by this court not to be error. Of course what is there said in the opinion had reference to the facts and circumstances of that particular case. There it was plain, from all the evidence, that, if the accused was guilty of the commission of any offense whatever, it was of the crime of murder in the first degree. The charge was considered with reference to that state of facts. But whether the charge would have been sustained had there been evidence in the case which would support a conviction of a lesser grade of homicide, is a question not decided in the Dickerson Case. This observation, it must be understood, has no reference to the facts or circumstances of the case before us. It would be vevy improper for us to indicate any impression as to the guilt of the defendant, which we may have derived from an *258examination of the evidence, and we do not do so. The Dickerson Case was commented on in the argument at the ibar, and hence this explanatory remark as to what was intended to be decided in that case.

By the Gourt.— The judgment of the circuit court is reversed, and the cause remanded for a new trial. The warden of the state prison will surrender the plaintiff in error to the sheriff of Outagamie county, who will hold him in custody until he shall be discharged or his custody changed by due course of law.

Questions:

1. Does the Court's opinion seem to reject or at least partially hesitant about the reliability of hair comparisons? And in addition, if there is no expert science nor expert witness, would not or could not a lay witness reach a similarity judgment?
2. How does the court describe the admission of hair comparison for identity or criminal agency purposes?
3. Is the following statement True or False: In one way hair comparisons can be assessed and analyzed to demonstrate similarities of color, strand and the like. In another way, matching a hair pulled from a skull, is a far different evidentiary burden because it requires deeper forensic analysis.

Read the statement regarding DNA evidence in JonBenet Ramsey case at https://bouldercolorado.gov/news/news-release-jonbenet-ramsey-homicide-update.

Look up the following case using your favorite legal research resource, read and answer the questions.

Brown v. Booker, Civil Action No. 7:16-cv-00576

United States District Court for the Western District of Virginia, Roanoke Division

March 31, 2021, Decided

Questions:

1. Does the Court object to hair comparison as a general proposition.
2. Are there arguments made concerning the integrity of the chain of custody? Are these arguments successful? Why or Why not?
3. Even if a claim of identity without reservation was made during the expert's testimony on the exact comparison of hairs, the Court still indicates, the other evidence leading to defendant's conviction is simply too substantial to not find guilt. How does the Court deal with this issue?

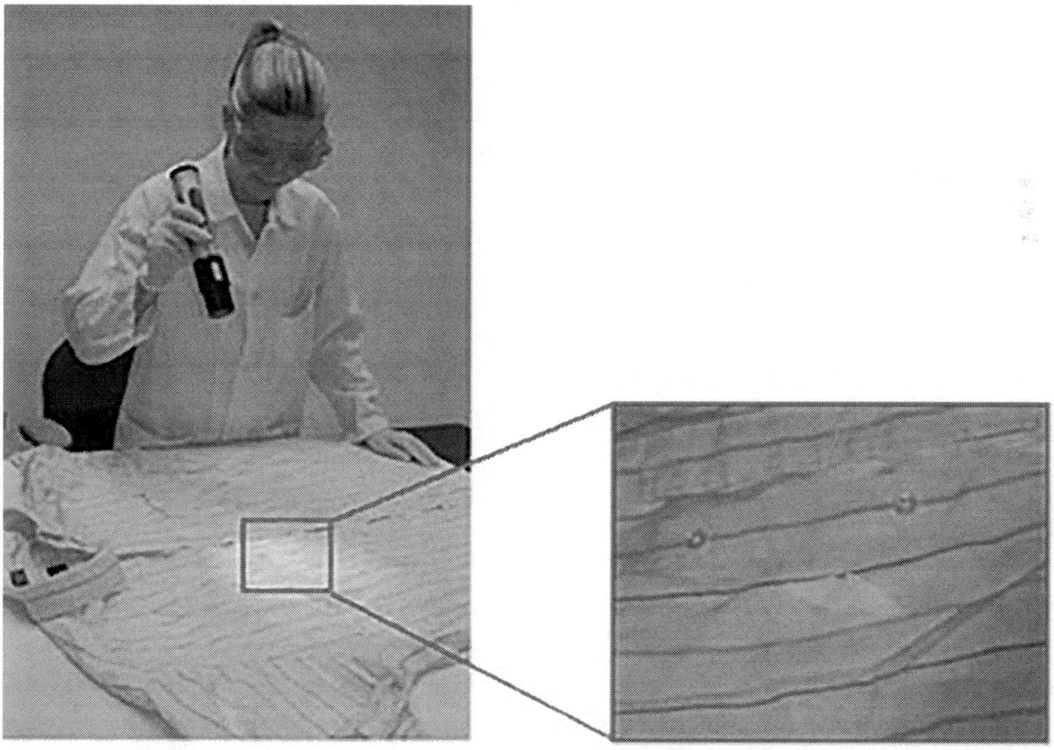

FIGURE 3.4 Examining evidence with an Alternative Light Source (ALS).

Source: New Jersey State Police.

In re Hill
United States Court of Appeals for the Eleventh Circuit
June 13, 2019, Decided
No. 19-12045-G

Pursuant to 28 U.S.C. § 2244(b)(3)(A), George Hill, Sr. has filed an application seeking an order authorizing the district court to consider a second or successive petition for a writ of habeas corpus. Such authorization may be granted only if:

> (A) the applicant shows that the claim relies on a new rule of constitutional law, made retroactive to cases on collateral review by the Supreme Court, that was previously unavailable; or
>
> (B)(i) the factual predicate for the claim could not have been discovered previously through the exercise of due diligence; and
>
> (ii) the facts underlying the claim, if proven and viewed in light of the evidence as a whole, would be sufficient to establish by clear and convincing evidence that, but for constitutional error, no reasonable factfinder would have found the applicant guilty of the underlying offense.

28 U.S.C. § 2244(b)(2). "The court of appeals may authorize [*2] the filing of a second or successive application only if it determines that the application makes a prima facie showing that the application satisfies the requirements of this subsection." Id. § 2244(b)(3)(C); see Jordan v. Sec'y, Dep't of Corr., 485 F.3d 1351, 1357-58 (11th Cir. 2007) (explaining that this Court's determination that an applicant has made a prima facie showing that the statutory criteria have been met is simply a threshold determination).

In his application, Hill indicates that he wishes to raise one claim in a successive § 2254 petition. Hill contends that DNA testing has revealed that pubic and head hairs found on or near the murder victim do not match his DNA. He asserts that DNA testing was unavailable at the time of his trial, but that he moved to have the hairs tested in 2012, and a laboratory report revealing the results was made public in July 2016. Hill argues that the hair evidence was important at his trial, as a crime laboratory analyst testified that the hair fibers found on the victim possessed the same characteristics as Hill's hair fibers and, thus, "were similar to Petitioner's hairs." He contends that the DNA test results would provide the reasonable doubt necessary to acquit him in a new trial. Hill attached a copy of the laboratory report to his petition.

Hill's proposed claim relies on newly discovered evidence in so much as he alleges that DNA testing and the resulting laboratory report were not available when he was initially tried in the late 1980's. Applying the first prong of § 2244(b)(2)(B), it is unclear whether Hill exercised due diligence in pursuing the DNA test results, which he sought in 2012, because he does not allege any facts regarding when DNA testing became available to him. He also does not explain the nearly three-year delay between when the laboratory report became available to him and when he filed this application. However, the relevant inquiry is whether the evidence could have been discovered before Hill's initial § 2254 proceedings, and it is at least arguable that DNA testing was not widely considered a reliable form of evidence when Hill's prior § 2254 petitions were filed in 1994 and 1996. See In re Boshears, 110 F.3d 1538, 1540 (11th Cir. 1997).

As for the second prong of § 2244(b)(2)(B), Hill states that because the State's theory of prosecution was that he attempted to sexually batter the victim and killed her during the ensuing struggle, the DNA test results lead to the conclusion that he did not commit either crime. However, the only inculpatory trial evidence that Hill identified was evidence that hairs found on the victim were "similar" to Hill's hair. Hill did not discuss any other trial evidence, much less place the hair evidence in context of other trial evidence.1 Thus, it is impossible to determine whether, in light of the evidence as a whole, Hill could show by clear and convincing evidence that no reasonable juror would have convicted him. See 28 U.S.C. § 2244(b)(2)(B)(ii).

But even if Hill satisfied all the other statutory criteria under § 2244(b)(2)(B), his claim fails because he has not shown, or even alleged, that the DNA test results establish that a constitutional violation occurred. See 28 U.S.C. § 2244(b)(2)(B)(ii). Under § 2244(b)(2)(B)(ii), a prisoner must allege "both clear and convincing evidence of his actual innocence and another constitutional violation." In re Bolin, 811 F.3d 403, 409 (11th Cir. 2016). Hill has not done so.

Accordingly, because Hill has failed to make a prima facie showing of the existence of either of the grounds set forth in § 2244(b)(2), his application for leave to file a second or successive petition is DENIED.

Questions:

1. Is there a distinction or difference in the quality of microscopic comparison of hair and the DNA analysis.
2. How does the defendant argue that he was entitled to his petition based on a scientific method that did not exist at his time of trial and adjudication?
3. Does the use of microscopy in hair comparisons violate any constitutional rights?

Trace evidence like hair from the head, a beard or the pubic area may never yield a perfect comparison in search of identity. However, trace evidence often provides a basis for probable cause determination for subsequent arrests and searches. The Supreme Court of Delaware provides a lucid example of this in the following case.

Locate the following case using your favorite legal research resource, read the case and answer the questions below.

Thompson v. State, 539 A.2d 1052 (Del. 1988)

Questions:

1. Prove this statement True or False. "Investigators used in the case advised Delaware authorities that the comparison conclusion could not completely certain nor can it ever be."
2. Why does the court feel comfortable about hair comparison in matters of probable cause and at the same time, less confident in hair analysis as a perfect, objective science?
3. In testimony at trial, hair comparisons, while avoiding claims of identicality, are described in other ways, which may lead to an inference of absolute similarity. How so?

The impact of overplaying the power of any forensic science, or mischaracterizing its capacity and power, can result in a conviction being overturned. In U.S. v. Ausby, the exaggerated character of hair evidence resulted in an overturned conviction.

United States v. Ausby
United States Court of Appeals for the District of Columbia Circuit
November 5, 2018, Argued; March 1, 2019, Decided
No. 17-3077

In 1972, the government charged John Milton Ausby with the murder and rape of Deborah Noel. The prosecution introduced several pieces of evidence at trial connecting Ausby to the crime, including the testimony of a forensic expert who claimed that hairs found at the crime scene were microscopically identical to Ausby's hair. The jury subsequently convicted Ausby of the rape and murder and Ausby received a life sentence.

The government now concedes that the testimony of the forensic expert was false and misleading and that the government knew or should have known so at the time of Ausby's trial. Ausby moved to vacate his conviction under 28 U.S.C. § 2255 and the holding in Napue v. Illinois, 360 U.S. 264 79 S.Ct. 1173, 3 L.Ed.2d 1217 (1959). The district court denied Ausby's motion because it determined that the forensic expert's testimony was not material to the jury's guilty verdict. United States v. Ausby, 275 F.Supp.3d 7, 32 (D.D.C. 2017). We, however, conclude that Ausby has demonstrated a "reasonable likelihood" that the forensic expert's admittedly false testimony "could … have affected the judgment of the jury." Napue, 360 U.S. at 271, 79 S.Ct. 1173. We therefore reverse.

I. BACKGROUND

Deborah Noel returned to her apartment on December 14 for the first time in two weeks. Shortly after her arrival, she was raped and murdered in her bedroom. A grand jury indicted Ausby on six counts related to Noel's death: felony murder, premeditated murder, rape while armed, rape, burglary while armed and burglary.

At trial, the prosecution introduced substantial evidence connecting Ausby to Noel's rape and murder. First, the prosecution called to testify two individuals who encountered a black male in garb resembling Ausby's in Noel's apartment building in the days leading up to her murder. One of the two identified Ausby himself from a photo array and in the courtroom during trial. Second, the prosecution presented evidence that Ausby had left a thumbprint inside Noel's apartment sometime within ten days of the murder. Third, the prosecution presented evidence that vials of scented oil left inside Noel's apartment and beneath her windows belonged to Ausby. Fourth, the prosecution presented evidence that the bullet that killed Noel potentially, but not definitively, matched the handgun Ausby was carrying when he was arrested.

At issue here, the government also elicited testimony from FBI Special Agent Robert Neill, a microscopic hair analysis specialist, regarding hair found at the scene of the crime. According to Agent Neill's testimony, microscopic analysis of hair involved categorizing hair on the basis of between fifteen and twenty-five characteristics "which tend to be more or less unique to a particular individual." To compare two hairs, Agent Neill placed them side-by-side under a special microscope and then compared them using the hairs' observable characteristics.

Agent Neill testified that hairs taken from inside Noel's apartment and on her body were "microscopically identical" to Ausby's hairs. Although Agent Neill acknowledged that "microscopic hair comparisons do not constitute a basis of positive personal identification," he opined that "the questioned hairs … either originated from the head of Mr. Ausby or from some other person … whose head hairs or pubic hairs are microscopically identical."

During its closing statement, the prosecution reviewed the testimony from those who encountered Ausby in the apartment building and the evidence connecting him to the oils found in Noel's bedroom and under her windows. The prosecutor then asked the jury: "Now, could I not have rested my case right there? Could you not have said, Why are you boring me with anything further in this case? Is that not enough to convict this defendant?" But continuing on, the prosecutor revisited Agent Neill's hair-comparison testimony as well as the evidence regarding Ausby's thumbprint and the potential match between Ausby's handgun and the bullet that killed Noel.

In his closing, defense counsel admitted that Ausby had entered Noel's apartment but argued that he did so during Noel's two-week absence, not on the day of her rape and murder. He challenged the reliability of Agent Neill's purported identification of Ausby's hairs on Noel's body, particularly given that Agent Neill had conceded that microscopic hair comparison analysis cannot produce a positive identification. In response, the prosecutor asserted during his rebuttal that microscopic hair comparison analysis "is not a positive means of identification but it amounts to a positive means here."

The jury convicted Ausby of felony murder and rape while armed. The court then sentenced Ausby to life imprisonment for Noel's murder and ten to thirty years' imprisonment for her rape. This Court affirmed Ausby's conviction and sentence. United States v. Ausby, 489 F.2d 1273 (Table) (D.C. Cir. 1974) (per curiam). Ausby has fully served his rape sentence, leaving his life sentence for murder.

In 2012, the FBI and the Department of Justice began reviewing cases in which the government had introduced testimony regarding microscopic hair comparison analysis to assess whether the government's forensic expert gave false or misleading testimony that exceeded the limits of science. After reviewing Ausby's case, the FBI determined that Agent Neill misled the jury by implying that he could positively identify the hairs taken from the crime scene as belonging to Ausby. In light of Agent Neill's admittedly misleading testimony, the United States conceded error and waived any statute of limitations and procedural-default defenses in the event Ausby sought relief under 28 U.S.C. § 2255. The government, however, took "no position regarding the materiality of the error" at that time.

In 2012, the FBI and the Department of Justice began reviewing cases in which the government had introduced testimony regarding microscopic hair comparison analysis to assess whether the government's forensic expert gave false or misleading testimony that exceeded the limits of science. After reviewing Ausby's case, the FBI determined that Agent Neill misled the jury by implying that he could positively identify the hairs taken from the crime scene as belonging to Ausby. In light of Agent Neill's admittedly misleading testimony, the United States conceded error and waived any statute of limitations and procedural-default defenses in the event Ausby sought relief under 28 U.S.C. § 2255. The government, however, took "no position regarding the materiality of the error" at that time.

II. ANALYSIS

A federal prisoner may move to have his sentence vacated under 28 U.S.C. § 2255 if "the sentence was imposed in violation of the Constitution or laws of the United States." 28 U.S.C. § 2255(a). Under Napue v. Illinois and its progeny, "the prosecution's introduction of false testimony" deprives a defendant of a fair trial as required by the Fifth and Sixth Amendments. United States v. Straker, 800 F.3d 570, 602 (D.C. Cir. 2015) (per curiam). The government commits a Napue violation "when the government introduces false or misleading testimony or allows it to go uncorrected, even though the government knew or should have known that the testimony was false." Id. at 603 (citations omitted). If the government introduces false or misleading testimony, however, "the grant of a new trial is not automatic," Vega, 826 F.3d at 529; the false evidence must also be material to justify a new trial, Giglio v. United States, 405 U.S. 150, 154, 92 S.Ct. 763, 31 L.Ed.2d 104 (1972). We review a Napue claim de novo. United States v. Sitzmann, 893 F.3d 811, 821 (D.C. Cir. 2018) (per curiam).

The government has conceded that Agent Neill's testimony regarding microscopic hair comparison analysis was false and that the government knew or should have known it was false at the time of Ausby's trial. The parties dispute only whether Agent Neill's false statements were material to Ausby's conviction.

As first formulated in Napue and repeated in the decades of cases that have since applied its holding, the government's introduction of false testimony is material if the evidence "could … in any reasonable likelihood have affected the judgment of the jury." Giglio, 405 U.S. at 154, 92 S.Ct. 763 (alteration in original) (quoting Napue, 360 U.S. at 271, 79 S.Ct. 1173); see also Sitzmann, 893 F.3d at 828; Straker, 800 F.3d at 603. The "reasonable likelihood" standard does not require the defendant to show "that he `more likely than not' would have been acquitted" absent the false statements. Wearry v. Cain, ___ U.S. ___, 136 S.Ct. 1002, 1006, 194 L.Ed.2d 78 (2016) (per curiam) (quoting Smith v. Cain, 565 U.S. 73, 75, 132 S.Ct. 627, 181 L.Ed.2d 571 (2012)).1 Rather, the defendant need show only that the false testimony "'undermine[s] confidence' in the verdict." Id. (quoting Smith, 565 U.S. at 76, 132 S.Ct. 627). Thus, even if the false testimony "may not have affected the jury's verdict," it is material if the evidence reasonably could have affected the verdict. Id. at 1006 n.6 (emphasis added). As the Supreme Court's explication makes clear, the "reasonable likelihood" standard is outcome-driven—that is, the relevant question is whether the false testimony "could have altered the outcome of the case." Vega, 826 F.3d at 531. This standard "is quite easily satisfied." United States v. Williams, 233 F.3d 592, 594 (D.C. Cir. 2000).2 This Court's Napue decisions have also held that the burden of proving materiality lies with the defendant. See Sitzmann, 893 F.3d at 829; Straker, 800 F.3d at 605; United States v. Burch, 156 F.3d 1315, 1328-29 (D.C. Cir. 1998). Ultimately, then, the Napue materiality standard requires Ausby to establish that Agent Neill's false testimony reasonably could have altered the outcome of his case, thereby undermining confidence in the jury's guilty verdict.

Despite the consistency with which Napue decisions have articulated the materiality standard, Ausby asks us to strike out "reasonable likelihood" and insert the phrase "reasonable possibility." While Ausby posits that the two phrases are substantively "synonymous," he also maintains that the semantic difference "is meaningful" for clarifying the materiality standard. Even taking Ausby's position as internally consistent, we reject it. The only support Ausby offers in favor of his new semantic gloss are the opinions of two Supreme Court Justices, both of which failed to garner majority support. See United States v. Bagley, 473 U.S. 667, 679-80, 679 n.9, 105 S.Ct. 3375, 87 L.Ed.2d 481 (1985) (opinion of Blackmun, J.) (joined in relevant part by Justice O'Connor); Strickler v. Greene, 527 U.S. 263, 299, 119 S.Ct. 1936, 144 L.Ed.2d 286 (1999) (Souter, J., concurring in part and dissenting in part) (writing for himself in relevant part). These two non-precedential opinions fail to outweigh the nigh sixty years of Supreme Court precedent consistently employing the phrase "reasonable likelihood." We therefore decline Ausby's invitation to reformulate the Napue materiality standard.

In addition to quibbling over wording, Ausby argues that the "reasonable likelihood" standard is substantively equivalent to the harmless-error standard of Chapman v. California, 386 U.S. 18, 87 S.Ct. 824, 17 L.Ed.2d 705 (1967). Under what he christens the "Napue/Chapman standard," Ausby argues that he is entitled to a new trial unless the government can prove that Agent Neill's false testimony was harmless beyond a reasonable doubt. For support, Ausby first relies on the opinions of Justice Blackmun in United States v. Bagley and Justice Souter in Strickler v. Greene. But just as we decline to rely on these non-precedential opinions as definitive articulations of the "reasonable likelihood" standard, we also decline to rely on them as reliable explanations of the standard's substantive meaning.

Next, Ausby relies on this Court's decision in United States v. Vega. Vega stated in dicta that the Napue materiality standard is equivalent to the harmless-error standard of Chapman v. California. Vega, 826 F.3d at 529 n.4. In its holding, however, Vega applied and rested its decision exclusively on the "reasonable likelihood" standard laid out in Napue without reference to Chapman. See id. at 531. That is this Circuit's standard and the one we apply here as well. To the extent Ausby tries to wring out of Vega's footnote a rule imposing the burden of proving immateriality on the government or otherwise changing the "reasonable likelihood" standard, he is flatly incorrect. Consistent with longstanding precedent, Vega left on the defendant the burden of establishing that the false testimony "could in any reasonable likelihood have affected the judgment of the jury." Vega, 826 F.3d at 529 (quoting Burch, 156 F.3d at 1329); see id. at 530-31; Straker, 800 F.3d at 605; Burch, 156 F.3d at 1328-29.

The Supreme Court's and this Court's Napue decisions fall at one end of the materiality spectrum or the other. In some cases, the false testimony concealed facts that would have undermined the credibility of the government's key witness. See Giglio, 405 U.S. at 151, 154-55, 92 S.Ct. 763 ("key" cooperating witness falsely testified that he received no promise that he would not be prosecuted if he testified); Napue, 360 U.S. at 265, 269-71, 79 S.Ct. 1173 ("principal" cooperating witness falsely testified that he was not promised a reduced sentence if he testified); Alcorta v. Texas, 355 U.S. 28, 29, 31-32, 78 S.Ct.

103, 2 L.Ed.2d 9 (1957) (prosecution's only eyewitness falsely testified that he was not having an affair with the accused's wife at the time of the crime); United States v. Iverson, 637 F.2d 799, 801-05 (D.C. Cir. 1980) ("key" cooperating witness falsely testified that she had already been sentenced, intimating that she had nothing to gain by testifying). Because this evidence plainly could have affected the jury's verdict, the false statements were material. In other cases, the false statements involved facts that were either redundant or clearly irrelevant to the jury's decision. See Sitzmann, 893 F.3d at 829 (single purportedly false statement in a five week trial was not material given "abundant evidence" demonstrating defendant's participation in the charged conspiracy); Vega, 826 F.3d at 531 (witness's false testimony regarding photoarray identification was not material because several other witnesses identified the defendant); Straker, 800 F.3d at 604-08 (false testimony either played almost no role in the trial or was significantly out-weighed by the defendants' confessions); Burch, 156 F.3d at 1328-29 ("allegedly" false statements were not material because they could not have affected the credibility of the government's evidence and witnesses); United States v. Anderson, 509 F.2d 312, 327 & n.110 (D.C. Cir. 1974) (one allegedly false statement regarding a witness's identification did not bear "such significance to [the defendant's] bribery trial as a whole that it 'could … in any reasonable likelihood have affected'" the jury's judgment (alteration in original) (quoting Giglio, 405 U.S. at 154, 92 S.Ct. 763)). The false statements in these cases, therefore, were obviously not material to the jury's verdict.

We believe Agent Neill's false hair-comparison testimony during Ausby's trial falls somewhere between the two ends. Agent Neill's testimony was neither the sole piece of evidence on which the prosecution hung its case nor redundant or irrelevant. We ultimately conclude, however, that Agent Neill's testimony falls on the material side of the spectrum. Agent Neill's testimony was the primary evidence that directly contradicted Ausby's defense theory—that Ausby had been in Noel's apartment during her two-week absence but not on the day of her rape and murder. Ausby's defense theory plausibly explained the remaining evidence. First, the sightings of Ausby near Noel's apartment occurred four or five days before the murder and thus were consistent with Ausby's presence in Noel's apartment on an earlier date. Second, Ausby could have left the thumbprint and vials of oil in Noel's apartment while in her apartment sometime before the date of her rape and murder. And third, the ballistic evidence could not yield a positive identification. That Agent Neill's testimony played a key role in debunking Ausby's defense is borne out by the prosecution's emphasis in its closing rebuttal that Agent Neill's microscopic hair-comparison analysis "is not a positive means of identification but it amounts to a positive means here." Thus, without Agent Neill's hair-comparison testimony, there is a reasonable likelihood that the jury could have accepted Ausby's defense theory. Ausby has therefore carried his burden of demonstrating that Agent Neill's testimony "could … in any reasonable likelihood have affected the judgment of the jury." Giglio, 405 U.S. at 154, 92 S.Ct. 763 (alteration in original) (quoting Napue, 360 U.S. at 271, 79 S.Ct. 1173). Because the government has conceded that Agent Neill's testimony was false and that the government knew or should have known so at the time of Ausby's trial, Ausby has presented a valid claim under Napue that he was convicted in violation of the Fifth and Sixth Amendments. Accordingly, the district court should have granted Ausby's § 2255 motion to vacate his conviction.

For the foregoing reasons, the judgment of the district court is reversed and the case is remanded for proceedings consistent with this opinion.

So ordered.

Questions:

1. What is the primary basis for the Court to overturn the conviction?
2. Does the Court declare erroneous testimony on hair evidence can and did potentially effect the outcome?
3. Can you envision and outcome different than this one? Provide how a conviction might be upheld there was error regarding hair identity comparisons?

3.4 IDENTITY AND ODONTOLOGY

Identity can also be discerned from dental structure, records of care and unique characteristics. And when dental repairs occur and a comparison analysis of those records with actual teeth can be shown, the power of odontology to identify is quite substantial. It is also possible to make an identity claim based on bitemarks, but that conclusion has been under challenge in appellate cases, and very successfully so.

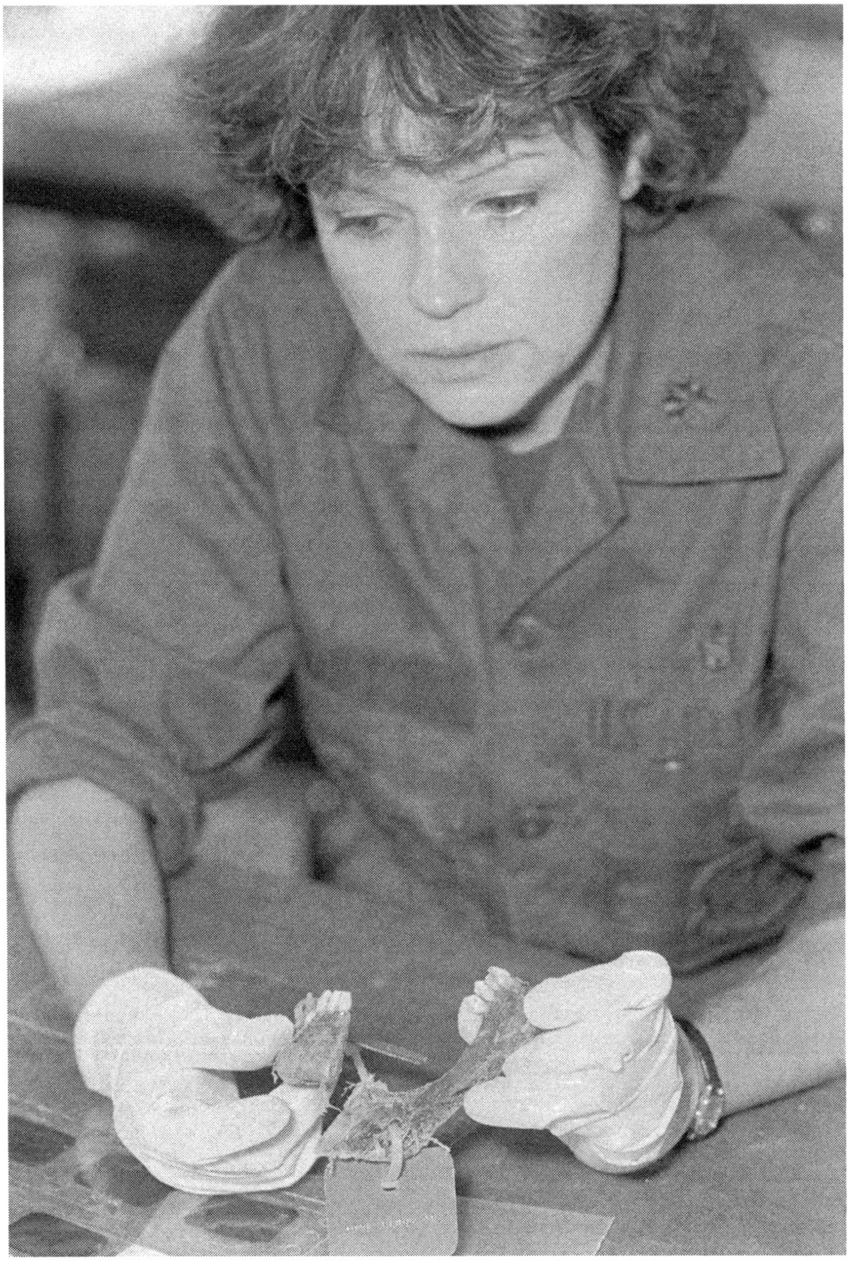

FIGURE 3.5 Major (MAJ) (Dr.) Bert Orch, forensic dentistry officer, inspects remains during the Forensic Dentistry Indentification Laboratory at the Thomas Recreation Center.

Source: National Archive ID 6426154.

Locate the following case using your favorite legal research resource, read the case and answer the questions below.

Ex parte Chaney, 563 S.W.3d 239 (Tex. Crim. App. 2018)

Questions:

1. During trial, does the expert testimony on "bitemarks" concludes that the marks can be individualized as to identity. At the time, the science associated with bitemarks appears to allow this conclusion. Where in the Court's opinion can this finding be supported?
2. During trial, is it fair to say that the prosecutor elevates bitemark evidence to a level of certainty that no longer exists?
3. Does the defense strategy counter the conclusion on individuation of bitemarks and at the same time indicate the science is less than settled?

The stress on bitemark evidence continues to the very present. While the aim of the identification is usually in good faith and on its face seems sensible, the claim of identicality of person and the bitemark impression left behind can no longer be made.

Locate the following case using your favorite legal research resource, read the case and answer the questions below.

Howard v. State, 300 So. 3d 1011 (Miss. 2020)

Questions:

1. Where were bitemarks evident on the victim of the rape?
2. Precisely, how does the expert's actual testimony cause the appeals court to order a new trial?
3. Were the qualifications, background in education and experience satisfactory to be declared an expert in his subject matter?

What if an expert witness later recants his or her testimony because what the expert thought ironclad and scientifically rock-solid yet over time becomes quite faulty and without the support of reputable science or methodology? Is this change of opinion a basis for a post-conviction relief like a new trial, release from custody under habeas corpus principles or simply harmless given that no conviction can be fully supported with odontology alone?

Locate the following case using your favorite legal research resource, read the case and answer the following questions.

In re Richards, 63 Cal.4th 291, 202 Cal. Rptr. 3d 678, 371 P.3d 195 (Cal. 2016)

Questions:

1. Does the Supreme Court of California admit that bitemark or dental identification as conducted in the late 1990's is no longer as reliable as once held?

2. What does the Court mean by the term “false evidence?” Would this include or exclude expert witness testimony that was once valid but over time becomes invalid?
3. Two remaining issues emanate from this decision. First, was Dr. Sperber qualified? Second, does this or could this decision have an impact on highly publicized cases? Any cases that stand out for consideration if the defendants were still alive?

Visit the American Board of Forensic Odontology at https://abfo.org.

Locate the following case using your favorite legal research resource, read the case and answer the following questions.

People v. Prante, 180 N.E.3d 875 (Ill. App. Ct. 2021)

Questions:

1. Even though Prante is an Illinois case, the majority reviewed and critiqued the Texas Forensic Science Commission report on bitemark analysis as to identity. What does the commission say about the method when compared to a fingerprint? Secondly, as in other rulings, here appears to a subjective component to this sort of forensics. Why is this a negative?
2. Why does the Court conclude the evidence can no longer withstand Frye review and meet its basic standard?
3. Even despite these inadequacies and changes to the once embraced science, the Court does not declare the petitioner “innocent” for it is not enough to argue false science. The Court requires that the error or flawed science be so significant that it automatically manifests innocence.

3.5 IDENTITY AND DNA

The certitude attributed to DNA evidence is not without justification. While some DNA is a bit more remote, as in the genealogical variety where distant relatives with similar matches can be traced back to a particular person as was just done in the case of Byran Christopher Kohburger, there are more direct lines of identity. Matching the DNA in hair, fiber, fluid, blood or other bodily excretion, are perfect matchs without rival.

Locate the following case using your favorite legal research resource, read the case and answer the questions below.

People v. Williams, 35 N.Y.3d 24, 147 N.E.3d 1131, (N.Y. 2020)

Questions:

1. How does the Court of Appeals characterize the nature and importance of DNA in our justice system?
2. Does the Court reach the conclusion that DNA methodology was incorrectly applied in this case? Secondly, what is the net effect on the Court’s final decision?

3. Whether LCN or HCN- the testing methodology for DNA is largely the same although the amount of facilities that can conduct LCN testing is smaller.

The power of DNA to target a clear-cut identity is tough to challenge. However, proof of identity may or may not be the right identity, or there might be other or alternative explanations. Solid investigative work is not enslaved to the world of DNA but rather it relies upon not only DNA but also other forms of evidence to prove a case in chief.

People v. Clark
214 P.3d 531 (Colo. App. 2009)

Defendant, Rosten Lee Clark, appeals the judgment of conviction entered on a jury verdict finding him guilty of one count of sexual assault. ... Defendant's most substantial contention is that the principal evidence against him, a DNA match of semen stains on articles of the victim's clothing she was not wearing at the time of the assault, is insufficient to sustain his conviction. We conclude that the evidence is sufficient and reject his other challenges to the conviction. ...

...

I. Sufficiency of the Evidence

Defendant contends that the DNA match is the principal evidence linking him to the crimes, and it is insufficient to prove beyond a reasonable doubt that he was the perpetrator. We disagree.

A. Standard of Review

When assessing the sufficiency of the evidence supporting a guilty verdict, we must determine whether any rational trier of fact might accept the evidence, taken as a whole and in the light most favorable to the prosecution, as sufficient to support a finding of guilt beyond a reasonable doubt. People v. Sprouse, 983 P.2d 771, 777 (Colo. 1999). We give the prosecution the benefit of every reasonable inference that might be drawn from the evidence. People v. McIntier, 134 P.3d 467, 471 (Colo.App. 2005). Additionally, the determination of the credibility of witnesses is solely within the province of the fact finder, and we will not act as a thirteenth juror and set aside a verdict because we might have reached a different conclusion. Id. at 471-72.

B. Evidentiary Standards

DNA evidence has obvious value to the criminal justice system. Countless criminals have been removed from the nation's streets and dozens of wrongfully convicted persons have been freed from its jails since the dawn of forensic DNA technology. See ABA Standards for Criminal Justice: DNA Evidence intro., at 18-19. Colorado courts have previously addressed the accuracy and admissibility of DNA evidence, see, e.g., People v. Shreck, 22 P.3d 68 (Colo. 2001), but those issues are not argued here because defendant never contended that the DNA was not his.

Although DNA evidence has provided circumstantial evidence of identity for many years, its use as an exclusive means for proving the identity of the perpetrator in criminal cases is relatively rare. Several jurisdictions have held that DNA evidence alone may support a criminal conviction without further corroborative proof. In most of those cases, however, the crime scene DNA sample was taken from a location that effectively eliminated most innocent explanations for its presence. See, e.g., State v. Freeman, 269 S.W.3d 422, 425-26 (Mo. 2008) (match between defendant and DNA collected from instrument of strangulation found around victim's neck sufficient to convict); People v. Rush, 165 Misc.2d 821, 630 N.Y.S.2d 631 (Sup.Ct. 1995) (match between defendant and DNA collected from victim's anal and vaginal swabs sufficient to convict), aff'd, 242 A.D.2d 108,

672 N.Y.S.2d 362 (1998); State v. Toomes, 191 S.W.3d 122, 128-32 (Tenn.Crim. App. 2005) (match between defendant and DNA collected from victim's anal swabs sufficient to convict).

We have found only one sexual assault case in which DNA obtained from a crime scene—rather than directly from a victim's person—was the primary identification evidence the prosecution proffered. See People v. Soto, 21 Cal.4th 512, 88 Cal.Rptr.2d 34, 981 P.2d 958, 960-62 (1999). However, in that case, the trial court also admitted evidence of "spontaneous utterances" by the victim, indicating her belief that the defendant was her assailant. Id. at 962. Additionally, it does not appear that the defendant there raised a sufficiency of the evidence argument on appeal.

We have found no Colorado authority on point; thus, whether crime scene DNA evidence, without further significant corroborative proof, is sufficient to identify the perpetrator and therefore sustain a criminal conviction is an issue of first impression in this state.

The legal standards associated with the use of fingerprint identification provide a useful analytical model. See Toomes, 191 S.W.3d at 130 (finding fingerprint identification analogous and employing its framework to determine sufficiency). Our courts have long accepted that latent fingerprints are a corporeal signature of their maker, capable of conclusively identifying the individual who impressed them. See Medina v. People, 168 Colo. 255, 258, 450 P.2d 662, 664 (1969) (fingerprint "testimony" alone sufficient to support burglary conviction); People v. Jennings, 252 111. 534, 96 N.E. 1077 (1911) (one of the first United States cases admitting fingerprint evidence). But establishing that a fingerprint belongs to a particular person does not, by itself, prove that the person committed a particular crime. Courts have therefore established standards to ensure that fingerprint-based proof of identity is not conflated with proof of culpability. See, e.g., People v. Ray, 626 P.2d 167, 170-71 (Colo. 1981).

In Ray, the supreme court articulated the applicable analysis for assessing the sufficiency of evidence where an accused is tied to a crime by fingerprints alone:

> It has been generally recognized that where, as here, the only evidence of guilt of accused persons consists of their fingerprints found at the scene of the crime, the evidence, to be legally sufficient to sustain a conviction, must be coupled with evidence of other circumstances tending to reasonably exclude the hypothesis that the prints were impressed at a time other than that of the crime. (Citations omitted.) Such "other circumstances" need not be those completely independent of the fingerprint and may properly include circumstances such as the location of the print, the character of the place or premises where it was found and the accessibility of the general public to the object on which the print was impressed.

Ray, 626 P.2d at 170-71 (quoting with approval Edmonds v. State, 5 Md.App. 132, 141, 245 A.2d 618, 623 (1968)).

Thus, a defendant's fingerprints found inside a burglarized home, when coupled with testimony from the owners of the home that the defendant had never been an invited guest, were found sufficient to convict for second degree burglary and theft in People v. Angel, 701 P.2d 149, 151 (Colo.App. 1985). However, a defendant's fingerprints found on the milk chute of a burglarized home, where the prosecution produced no evidence to prove when the prints were impressed, were deemed insufficient to support conviction in Ray, 626 P.2d at 171. See also Solis v. People, 175 Colo. 127, 128-29, 485 P.2d 903, 904 (1971) (defendant's fingerprint found on window glass broken during the burglary of a building, where the prosecution could not establish when it was left or whether it had come from the inside or outside of the window, deemed insufficient).

The lesson of both the Ray standard and the cases applying it is that to support a criminal conviction, physical evidence of fingerprint identification must be corroborated by additional evidence that the identified person was indeed the perpetrator of the crime. This so-called fingerprint evidence rule does not call for a special jury instruction and does not alter a jury's ability to weigh evidence. See People v. Mandez, 997 P.2d 1254, 1271 (Colo.App. 1999). It does, however, mark the minimum requirements of evidentiary sufficiency for a case either to go to a jury or to survive appellate review. Id. We conclude that this requirement is as applicable to DNA evidence as it is to fingerprints.

However, the Ray standard cannot be applied to DNA evidence without modification. The raw material by which DNA profiles are constructed exists within every human cell. Oxford Illustrated Companion to Medicine 397 (Stephen Lock et al. eds., 3d ed. 2001). Humans naturally and continuously shed hair and tissue cells that—as surely as fingerprints—serve as a signature of their identity. Gray's Anatomy: The Anatomical Basis of Clinical Practice 152 (Susan Standring ed., 40th ed. 2008) (hair shedding is part of a recurrent cycle); DNA Initiative: Identifying DNA Evidence, http://dna.gov/basics/evidence_collection/identifying/(last visited Feb. 11, 2009) (listing multiple sources of DNA evidence). Unlike fingerprints, however, these microscopic signatures can be left upon objects that have not been handled directly, and can be transferred by innumerable mechanisms into areas where their authors have scarcely traveled. See D.H. Kaye Michael E. Smith, DNA Identification Databases: Legality, Legitimacy, and the Case for Population-Wide Coverage, 2003 Wis. L. Rev. 413, 435-36 (2003) (regarding shed DNA).

Human biology and common sense thus dictate that the evidentiary value of DNA obtained from a hair recovered in a public place is far different from that of DNA extracted from semen found in a private residence. Hence, where connection to a crime is established by DNA evidence alone, it is necessary to consider the type of biological material from which the crime scene DNA sample was obtained and its susceptibility to transfer by innocent means.

Guided by the Ray standard, we conclude that where, as here, the only direct evidence connecting an accused person to the crime is the presence of DNA at the scene of a crime, the evidence, to be legally sufficient to sustain a conviction, must be coupled with evidence of other circumstances tending to reasonably exclude the hypothesis that the DNA was deposited at a time other than that of the crime. Such other circumstances may include the source material of the DNA and its susceptibility to transfer, the location of the DNA, the character of the place or premises where it was found, the accessibility of that place or premises to the general public, and the object upon which the DNA was found. See Ray, 626 P.2d at 170-71.

C. The DNA Evidence

Here, defendant's semen was on a sweatshirt and headband recovered from the scene of the assault. The victim testified that she purchased the sweatshirt several days before the attack. If believed, her testimony thus provided the jury with a narrow window of time in which defendant's semen was likely deposited.

The sufficiency of the evidence here depends upon the nature of the biological material recovered, and the fact that it was found on an article of personal possession. Semen is neither deposited unintentionally nor transferred as readily as hair or tissue cells. Thus, there are only three possible explanations for the presence of defendant's semen on the victim's sweatshirt: (1) the semen was deposited before the victim acquired the sweatshirt, an argument defendant does not make here; (2) it was deposited during consensual sexual activity; or (3) it was deposited during nonconsensual sexual activity.

Defendant testified that his semen was deposited during consensual oral sex with the victim's mother before the sexual assault occurred. However, he had denied having sexual relations with the mother when he was initially interviewed by authorities. In addition, the mother denied during her testimony that she had ever had sexual contact with defendant. Defendant's explanation thus called upon the jury to decide whether to believe him or the mother. The jury obviously did not credit defendant's testimony, and we cannot substitute our assessment of witness credibility for that of the fact finder. See McIntier, 134 P.3d at 471.

Once "the jury did disbelieve [defendant], it was further entitled to consider whatever it concluded to be perjured testimony as affirmative evidence of guilt." Wright v. West, 505 U.S. 277, 296, 112 S.Ct. 2482, 120 L.Ed.2d 225 (1992) (opinion of Thomas, J., joined by Rehnquist, C.J., and Scalia, J.) (citing Wilson v. United States, 162 U.S. 613, 620-21, 16 S.Ct. 895, 40 L.Ed. 1090 (1896)); accord id.

at 297, 112 S.Ct. 2482 (White, J., concurring in this aspect of Justice Thomas's opinion); id. at 315, 112 S.Ct. 2482 (O'Connor, J., joined by Blackmun and Stevens, JJ., concurring that evidence was sufficient) ("[Defendant] took the stand and gave an explanation that the jury rejected, thereby implying a finding that the explanation was false."); see also Reeves v. Sanderson Plumbing Products, Inc., 530 U.S. 133, 147, 120 S.Ct. 2097, 147 L.Ed.2d 105 (2000) (applying in civil context "the general principle of evidence law that the factfinder is entitled to consider a party's dishonesty about a material fact as `affirmative evidence of guilt'" (quoting Wright v. West, 505 U.S. at 296, 112 S.Ct. 2482, and citing Wilson, 162 U.S. at 620-21, 16 S.Ct. 895)).

Taken in the light most favorable to the prosecution, the narrow window of time during which the victim had owned the sweatshirt, combined with the mother's denial and defendant's initial denial of any consensual sexual encounter, suffice to reasonably exclude the hypothesis that the DNA was deposited at a time other than that of the crime. Cf. Angel, 701 P.2d at 151 (defendant's fingerprint inside of burglarized home, combined with testimony from homeowners that defendant was never an invited guest, sufficient to convict). Thus, we conclude that the evidence was sufficient to convict. See McIntier, 134 P.3d at 471.

Defendant presented extensive evidence at trial suggesting that a neighbor was the perpetrator. But that evidence did not explain the presence of defendant's DNA. And in any event, it was the jury's function to weigh that evidence and resolve conflicts, testimonial inconsistencies, and disputes. Id. We cannot act as a thirteenth juror and overrule its assessment of the evidence. Id.

...

III. Shifting the Burden of Proof

Defendant next contends that his conviction must be reversed because the trial court allowed the prosecution to shift the burden of proof to him. Defendant argues that this occurred when the prosecutor asked an expert witness whether defendant had conducted a particular scientific test on a piece of evidence. Because we conclude that any error was harmless beyond a reasonable doubt, we reject defendant's contention that reversal is required.

A. Standard of Review

As noted previously, we review a trial court's decision regarding the admissibility of evidence for abuse of discretion. Ibarra, 849 P.2d at 38. Even if an abuse of discretion has occurred, however, we will not reverse a trial court's determination if the error was harmless. See Chapman v. California, 386 U.S. 18, 21-22, 87 S.Ct. 824, 17 L.Ed.2d 705 (1967). When errors are of constitutional dimension, reversal is required unless the error was harmless beyond a reasonable doubt. See id. at 24, 87 S.Ct. 824; Golob, 180 P.3d at 1013.

B. The Burden of Proof

The Due Process Clause protects an accused against conviction "except upon proof beyond a reasonable doubt of every fact necessary to constitute the crime with which he is charged." In re Winship, 397 U.S. 358, 364, 90 S.Ct. 1068, 25 L.Ed.2d 368 (1970). This burden of proof cannot be shifted from the prosecution to a defendant during a criminal trial, and a defendant may not be called upon to affirmatively prove that he or she did not commit a crime. See Leonard v. People, 149 Colo. 360, 372, 369 P.2d 54, 61 (1962). Accordingly, the prosecution cannot place upon a criminal defendant the burden of proving innocence through the testing of evidence. See id.; People v. Beasley, 384 Ill.App.3d 1039, 323 Ill.Dec. 558, 893 N.E.2d 1032, 1040 (2008) (though defendants may be able to submit evidence for fingerprint analysis, they have no burden to do so); Barbara E. Bergman Nancy Hollander, Wharton's Criminal Evidence § 2:2 (15th ed. 1997).

In some circumstances, improper questioning of an expert witness could imply to jurors that a defendant carries this burden. See People v. Lehmkuhl, 117 P.3d 98, 105 (Colo. App. 2004), (discussing Hayes v. State, 660 So.2d 257, 265 (Fla. 1995)); Beasley, 323 111. Dec. 558, 893 N.E.2d at 1032. However, such circumstances are not present in this case.

C. Application

Here, the prosecution called an expert in the fields of serology, forensic DNA, and blood stain analysis. Defendant cross-examined this witness regarding a sample of stained carpet taken from the crime scene that had tested presumptively positive for the presence of semen. His examination of the expert revealed that a test that could have conclusively confirmed the presence of semen was available but was not performed. The expert stated that the test was not performed because doing so would have consumed the stain.

On redirect examination the prosecutor asked why the stain could not be consumed. The witness replied: "We are mandated to set aside half of every stain unless we have permission to consume. And that's for future retesting typically by the Defense." The prosecutor then asked—over defendant's objection—whether defendant had done any testing of the stain.

We perceive that the prosecution's second question, regarding whether defendant conducted his own testing of the carpet stain, was improper. Such questions may imply to juries that defendants have some obligation to prove their own innocence, which clearly they do not. Thus, the trial court should have sustained defendant's objection to the question.

D. Harmlessness

The trial court's error was, however, harmless beyond a reasonable doubt. Defendant hoped to use the stain to show that someone else's semen was found near the scene of the crime. Indeed, defendant's cross-examination of the expert revealed that a substance that was either semen or some other bodily fluid had been found and could not be attributed to him. Thus, the prosecution's question did not detract from the stain's evidentiary value to defendant.

Additionally, the expert did not know the answer to the prosecutor's question. The jury thus gleaned little from this expert that could have led it to believe that defendant bore, yet failed to carry, some burden to independently test evidence.

The remainder of the trial proceedings served to deemphasize the question. Defendant stated in closing argument that he had no obligation to test the carpet stain. The prosecution's closing argument did not reference whether defendant tested the stain, nor did the prosecutor argue that defendant had any obligation to do so. Additionally, the trial court provided the jury with instructions that clearly placed the burden of proof on the prosecution. We presume that the jury understood and heeded these instructions. People v. Dunlap, 975 P.2d 723, 744 (Colo. 1999).

Under these circumstances, the prosecution's single improper question in no way contributed to defendant's conviction. See Golob, 180 P.3d at 1013. We thus conclude that the error was harmless beyond a reasonable doubt.

...

The judgment is affirmed. Defendant's sentence is vacated, and the case is remanded to the trial court for resentencing.

Questions:

1. In a criminal prosecution for rape or aggravated sexual assault, would DNA findings from semen automatically target the person identified?
2. While the judgment was affirmed, what issue does the Court send back to the lower court to remedy?
3. Part of the Court's opinion deals with the prosecution effort to shift the burden of forensic testing like DNA to the onus of the defendant. How does the court react to this tactic?

Post-conviction remedies frequently tackle the DNA question and in diverse ways. First, after DNA testing, which had not been done during the original offense, now needs to be done since it exonerates the defendant. Or, defendant may seek to challenge the validity of the conviction based on errant DNA analysis or method. In the following New Hampshire case, the defendant tries to stretch the definition of DNA into a Non-DNA framework.

State v. Breest,
155 A.3d 541 (2017)

In 1973, the defendant, Robert Breest, was convicted of murdering Susan Randall. Although we affirmed his conviction on direct appeal, see State v. Breest, 116 N.H. 734, 367 A.2d 1320 (1976), the defendant has maintained his innocence since his conviction and has instituted numerous collateral proceedings in an effort to secure his freedom. Since 2000, the defendant has succeeded in obtaining multiple rounds of DNA testing of fingernail clippings taken from Randall's body. All of this testing, including the latest round conducted in 2012, showed that the defendant could not be excluded as a contributor of DNA material found on the clippings. However, the most recent testing revealed for the first time that the clippings also contained DNA material from another unidentified male contributor. Based upon the latest test results, the defendant moved for a new trial.

At the hearing on his new trial motion, the State sought to exclude non-DNA evidence that the defendant had proffered, but that was not presented at his original trial. The Superior Court (Smukler, J.) granted the State's motion to exclude the non-DNA evidence and, following a hearing, denied the motion for a new trial. The defendant appeals both rulings. We affirm.

...

II

On appeal, the defendant contends that the trial court's decision denying him a new trial should be vacated because the trial court erred by: (1) concluding that, to obtain a new trial under RSA 651-D:2, the defendant was required to show that a new trial would result in an acquittal rather than merely a hung jury; (2) not finding a hung jury to be a "different result" under the Cossette standard, which the trial court applied in its RSA 651-D:2 analysis; (3) excluding new non-DNA evidence that, the defendant asserts, would alter the outcome in a retrial; and (4) failing to properly weigh the evidence. We address these arguments in turn.

A

The defendant first contends that, to obtain a new trial, he must show only that there would be "at least a hung jury" rather than an acquittal, were he to be retried. This standard, he asserts, is "consistent with the text of RSA 651-D:2," the statute governing post-conviction DNA testing, because a hung jury would be "favorable" under the terms of the statute.

RSA 651-D:2 states, in relevant part:

> In addition to any other substantive or procedural remedies provided by applicable law, if the results of DNA testing conducted under this section are favorable to the petitioner, the court shall order a hearing and shall enter any order that serves the interests of justice, including an order vacating and setting aside the judgment, discharging the petitioner if the petitioner is in custody, resentencing the petitioner, or granting a new trial.

RSA 651-D:2, VI(b).

Resolution of this issue requires us to engage in statutory interpretation. "The interpretation of a statute is a question of law, which we review de novo." State v. Costella, 166 N.H. 705, 709, 103 A.3d 1155 (2014) (quotation omitted). "In matters of statutory interpretation, we are the final arbiters of the intent of the legislature as expressed in the words of a statute considered as a whole." Id. (quotation omitted). "When examining the language of the statute, we ascribe the plain and ordinary meaning to the words used." Id. (quotation omitted). "We interpret legislative intent from the statute as written and will not consider what the legislature might have said or add language that the legislature did not see fit to include." Id. (quotation omitted).

The defendant argues that an acquittal standard is inconsistent with the text of RSA 651-D:2, VI (b) because the text contemplates relief when the new DNA test results are "favorable," and a hung jury would be a "favorable" outcome.

A plain reading of the statute shows that the word "favorable" refers to the results of the DNA testing, not the ultimate disposition of the court case for which the testing is conducted. See RSA 651-D:2, VI(b) (stating that a hearing will occur "if the results of DNA testing … are favorable to the petitioner"). The results of a DNA test are favorable insofar as they are exculpatory and unfavorable insofar as they are inculpatory. A hung jury is not an outcome, favorable or otherwise, of a DNA test. To be sure, a hung jury is an outcome that may be obtained if a new trial is ordered. The statute, however, requires only a hearing if the DNA results are favorable, not a trial. See id. After the hearing, a court may order a trial if the evidence produced at the hearing demonstrates that such relief is warranted, but the court is not required to grant a new trial (or any other relief) merely because the DNA test results were "favorable." See id. Rather, the statute requires the court to "enter any order that serves the interests of justice," which may, in certain circumstances, warrant no relief other than the hearing itself. See id. Thus, in RSA 651-D:2, VI(b), "favorable" defines the prerequisite condition, i.e., favorable DNA test results, that can give rise to every possible outcome that may follow from a hearing. See id.

…

C

The defendant next argues that the trial court erred by excluding from the hearing non-DNA evidence that he asserts would be available at a new trial and would undermine the State's case against him. The specific evidence that the court excluded was the affidavits of Kelleher, Neuschatz, and Morris. The defendant argues that the exclusion of this evidence was inconsistent with the language of the new trial standard, the policy and legislative history of RSA 651-D:2, and the applicable case law from other jurisdictions.

"We review challenges to a trial court's evidentiary rulings under our unsustainable exercise of discretion standard and reverse only if the rulings are clearly untenable or unreasonable to the prejudice of a party's case." State v. Noucas, 165 N.H. 146, 158, 70 A.3d 476 (2013) (quotation omitted). "In determining whether a ruling is a proper exercise of judicial discretion, we consider whether the record establishes an objective basis sufficient to sustain the discretionary decision made." Id.

The defendant's assertion that the new trial standard generally considers the import of new evidence in terms of its bearing on the likely outcome of another trial is correct. Yet, as applied to this case, his argument overlooks a critical step: the procedure outlined by RSA 651-D:2. Because RSA 651-D:2, VI(b) creates a mechanism for a defendant to obtain relief based upon new DNA evidence in circumstances in which a new trial would not otherwise be available, see RSA chapter 526 (2007), the conditions and limitations imposed by the statute as prerequisite to such relief must be strictly observed.

Here, the trial court's exclusion of the non-DNA evidence is in accord with the terms of RSA 651-D:2. Under the statutory scheme, the petitioner must first request new DNA testing. See RSA 651-D:2, I. If the results of that testing are favorable, then a hearing is granted. See id. at VI(b). At the hearing, the court shall "enter any order that serves the interests of justice," including an order setting aside the conviction or granting a new trial. Id. The hearing functions as a preliminary step that precedes a full repetition of the original trial, and is occasioned not by the discovery of any type of new evidence, but by the discovery of "favorable" new DNA evidence. See id. At no point in RSA chapter 651-D is non-DNA evidence mentioned. See RSA ch. 651-D. Thus, we conclude that the legislature did not intend to expand the scope of the hearing required by RSA 651-D:2, VI (b) to consider the potential impact of non-DNA evidence.

We are concerned by the potential for abuse that could arise under the defendant's interpretation of the statute. RSA 651-D:2 creates a unique mechanism for overturning a criminal conviction based upon DNA evidence. Absent that mechanism, an individual's motion for a new trial is barred by the statute of limitations after three years. See RSA 526:4 (2007) (providing that "[a] new trial shall not be granted unless the petition is filed within three years after the rendition of the judgment complained of, or the failure of the suit"). An expansive reading of RSA 651-D:2 could undermine this purpose by permitting even marginally favorable DNA evidence to serve as the catalyst for the grant of a new trial based upon the untimely production of non-DNA evidence. Accordingly, we conclude that the trial court did not err in excluding the non-DNA evidence from the hearing.

D

Finally, the defendant argues that the trial court's ruling that the new DNA test results did not warrant a new trial was against the weight of the evidence. Specifically, the defendant contends that the presence of a second male DNA profile inevitably leads to the conclusion that Randall was attacked by a second assailant, which, in turn, upends the State's theory of the case at the original trial that the defendant acted alone in murdering Randall. To the extent that the trial court considered other explanations for the second male DNA profile, like contamination and casual contact, the defendant claims the court "credited speculation." We disagree.

"Whether newly discovered evidence requires a new trial is a question of fact for the trial court." Cossette, 151 N.H. at 361, 856 A.2d 732 (quotation omitted). We will sustain the trial court's decision unless its conclusion is clearly unreasonable. Id. (quotation omitted). Contrary to the defendant's assertion, the evidence presented at the hearing does not point definitively to the conclusion that the second DNA profile arose from another assailant. There is ample support in the record for the trial court's finding that the other unknown male DNA found on the victim's fingernail clippings could have resulted from either casual contact or contamination. The record makes it clear that these alternatives to the defendant's "second assailant" theory were not mere speculation.

The State's expert, Dr. Word, opined that casual contact, which included handshakes, eating at a restaurant, and even "passing people," could have deposited foreign DNA under Randall's fingernails. Both of the experts who testified conceded that the specter of contamination was a significant concern. The defendant's expert, Huma Nasir, stated that it was possible for a medical examiner conducting an autopsy in the 1970s to use the same nail clippers on multiple bodies. According to Word, this practice would have led to significant contamination. The defendant points to the conclusions in a 2014 book authored by Professor Peter Gill, a "leading authority" in the field of DNA evidence. The book summarized several studies, each asserting that the chances of observing a foreign DNA profile under an individual's fingernails absent "intimate contact" are low. Word, however, testified that there are "many studies" contradicting those findings.

To support his argument that the trial court credited "speculation" when considering the possibility of contamination, the defendant points to Nasir's testimony regarding the measures she took in preparing the DNA samples for the 2012 test. These measures included using isopropanol and a Bunsen burner to sterilize the lab equipment. Word, however, noted that she knew of no studies indicating that Nasir's methods were effective in removing casual-contact-based or contamination-based DNA.10 With regard to the isopropanol used to remove the DNA, Word observed that it "[would] do nothing to the DNA." On the contrary, Word noted that if there were already contaminating DNA on the instruments, "the isopropanol [would] help preserve it quite nicely." Other complications arising from DNA testing were mentioned as well. These complications include stochastic effects, which distort the observed relative concentrations of multiple DNA profiles, and DNA degradation.

More importantly, all four of the DNA tests conducted since 2000 have failed to exclude the defendant as the source of some of the male DNA under Randall's fingernails, thus further contradicting his position at trial that he had never interacted with Randall and was not in the State when she was killed. This is especially consequential in light of the relatively low prevalence of the alleles (the variational forms of a gene appearing on a chromosome) tested on the fingernail clippings, according to evidence presented to the trial court. The alleles, which are consistent with the defendant's alleles, statistically appear in the Caucasian population only "roughly one time out of 140,000 individuals[,] in the African American population roughly once in 26,000 individuals and in the Hispanic population approximately once in 32,000 individuals." Viewed together with the evidence presented at the defendant's original trial, we agree with the trial court that, even considering the 2012 test results, the DNA evidence is "more inculpatory than exculpatory."

The defendant relies upon what he asserts is an analogous case, an unreported Illinois trial court order, People v. Andersen, 80 C 01405, at 1-3 (Ill. Cir. Ct. Cook Cty. July 20, 2015). There, the petitioner was convicted of first degree murder in 1980 based on his confession that he had killed the victim with a knife that had been found by police near the crime scene. Id. at 2, 8. The police tested blood found on the knife and discovered that it was Type A blood, which matched the victim's blood type. Id. at 8. In 2013, the petitioner successfully moved for DNA testing to be conducted on the knife and on fingernail clippings from the victim. Id. at 4-5, 8-9. The petitioner was definitively excluded from both samples. Id. at 4. In addition, the victim was excluded as a source of DNA for the blood found on the knife. Id. at 4. At the hearing on the petitioner's motion for a new trial, the State "question[ed] the reliability" of the test outcomes, asserting contamination as the source of the aberrant results. Id. at 5. Yet the State did not present any experts or scholarly articles to support its claim, nor did it contest the academic works that the petitioner had submitted. Id. at 7.

We find that Andersen is inapposite here. First, the results obtained from both the knife and the fingernail clippings in Andersen definitively excluded the petitioner, whereas the latest DNA results here only exclude the defendant as the contributor of one of the DNA profiles. Id. at 4. Second, there was only one test in Andersen, whereas here there were three previous DNA tests on the victim's fingernails, each of which failed to exclude the defendant as the DNA source. Id. at 4-5, 8-9. Third, the State in Andersen presented no evidence to support its contamination theory, nor did it rebut the academic articles that the petitioner had submitted. Id. at 5, 7. Here, the State called an expert in forensic science who stated that contamination was a definite possibility, and rebutted the academic studies cited by the defendant.

Furthermore, the new DNA evidence must be considered in light of the original trial evidence. See Cossette, 151 N.H. at 361, 856 A.2d 732 (requiring a defendant seeking a new trial to prove that the new evidence "is of such a character that a different result will probably be reached upon another trial"). At trial the defendant claimed that he had no contact with Randall at any time. He further claimed that he was not in New Hampshire at the time she was murdered.

However, the State presented evidence that the defendant left Haggett's Manchester residence around 12:10 a.m. heading in the direction of Granite Square. Multiple witnesses testified to seeing a woman matching Randall's description hitchhiking near Granite Square shortly after midnight. Other witnesses testified that the woman got into a car that matched the appearance of the defendant's car. Furthermore, one witness gave a description of the driver that was consistent with the defendant. The same day that Randall's body was found, the defendant went to the police station and stated that he heard a body had been found and thought the police would want to speak with him.

Doyon interviewed the defendant two weeks after Randall's body was found and noticed scratches between the defendant's knuckles. At trial, Doyon testified that those scratches were consistent with "defensive" wounds received in a fight. During the trial, the State presented forensic evidence linking the defendant's car to Randall's clothing, as well as testimony from Carita, who claimed that the defendant had confessed to murdering Randall.

We cannot conclude that a fact finder, having heard: (1) all of the evidence offered against the defendant at his original trial; (2) evidence unavailable at that trial that the defendant could not be excluded as the contributor of at least some of the DNA found under Randall's fingernails; and (3) evidence unavailable at that trial of the possible origins of the other unknown male DNA contributor, would probably return a verdict different from that rendered by the original jury. Consequently, we hold that the trial court's denial of the defendant's motion for a new trial was not clearly unreasonable or contrary to the weight of the evidence.

Affirmed.

Questions:

1. What is non-DNA versus DNA? Does the Court provide a suitable explanation?
2. What type of DNA determination qualifies for post-conviction relief?
3. How does the Court dispose of this post -conviction petition?

State v. Tucker
920 N.W.2d 680 (2018)

Carlos A. Tucker appeals his convictions and sentences for one count of first degree sexual assault of a child and two counts of incest, related to an incident with his girlfriend's children. Evidence at trial showed that Tucker engaged in sex acts with M.T., age 11, and that M.T. and her two brothers, E.T., age 12, and R.T., age 10, engaged in sex acts upon Tucker's instructions. The main issue presented by this appeal is whether the district court abused its discretion in admitting "Y-STR" DNA evidence over Tucker's objections. We conclude that it did not. We further reject Tucker's contentions that the evidence was insufficient to support his convictions and that the district court imposed excessive sentences. Finding no error, we affirm.

BACKGROUND

Charges Against Tucker

The State charged Tucker with one count of first degree sexual assault of a child in violation of Neb. Rev. Stat. § 28-319.01(2) (Reissue 2016) and two counts of incest with a person under 18 years of age in violation of Neb. Rev. Stat. § 28-703 (Reissue 2008). The charges arose out of allegations by M.T., E.T., and R.T. that Tucker, their mother's live-in boyfriend, had engaged in sex acts with M.T. and that M.T. had engaged in sex acts with E.T. and R.T. after being instructed to do so by Tucker.

Pretrial Proceedings

Prior to trial, Tucker filed a motion in limine seeking to exclude all evidence of Y-STR DNA testing pursuant to Daubert v. Merrell Dow Pharmaceuticals, Inc., 509 U.S. 579, 113 S.Ct. 2786, 125 L.Ed.2d 469 (1993), and Schafersman v. Agland Coop, 262 Neb. 215, 631 N.W.2d 862 (2001) (Daubert/ Schafersman). He also alleged that such evidence would confuse the jury and that its prejudicial effect would outweigh its probative value.

At a pretrial hearing on the motion, the district court heard expert testimony by Shannen Bishop, a DNA analyst at the University of Nebraska Medical Center (UNMC). Bishop testified concerning the Y-STR DNA analysis she conducted on DNA found on the interior of the shorts M.T. wore on the day of the alleged assault. Bishop also explained the origins, mechanics, and limitations of Y-STR DNA testing, as well as the extent to which it is accepted in the scientific community. Bishop testified that Y-STR DNA testing looks only at the male chromosome portion of DNA, while autosomal DNA testing looks at all 23 chromosomes inherited by each person. She dismissed as irrelevant several journal articles submitted by Tucker purporting to discredit the application of Y-STR DNA testing to small ethnic populations. Bishop explained that the articles examined very small, specific sample sizes and that Y-STR DNA science has improved since the articles were published in the early 2000's. She further pointed out that the U.S. Y-STR database, which she used in her analysis, was not even established when most of the articles were written.

Following the hearing, the district court denied Tucker's motion in limine. It applied the Daubert/Schafersman analytical framework and determined the reasoning and methodology behind Bishop's opinions and Y-STR DNA testing to be valid and reliable. The district court further rejected Tucker's argument that the prejudicial effect of Y-STR DNA evidence outweighed its probative value.

Evidence at Trial

At the jury trial, M.T., E.T., and R.T. testified that on the day at issue, their mother was at work and Tucker was home with them. The children testified that Tucker, then age 31, invited them to play a series of games in which he would give the winner candy....A stain found on the living room rug tested positive for semen. DNA testing showed that the semen had the same genetic profile as Tucker and that the probability of randomly selecting an unrelated individual with a DNA profile matching the sample was 1 in 752.4 quintillion. The State also introduced Y-STR DNA evidence found on the shorts M.T. wore on the day of the alleged offenses. This evidence is discussed in more detail in the section below.

Tucker testified in his own behalf. He denied engaging in any sex acts with the children.

Y-STR DNA Evidence

The State called Bishop to testify regarding Y-STR DNA testing. Tucker's counsel made a continuing objection to the Y-STR DNA testing and any opinions derived from such testing based on Daubert/Schafersman, because such evidence was "inherently unreliable and unfairly prejudicial and otherwise not relevant." The district court overruled the continuing objection.

Bishop testified that she is a forensic science DNA analyst at UNMC and that she had performed Y-STR DNA testing on extractions from the interior of M.T.'s shorts. Bishop stated that Y-STR DNA testing was a method of looking only at the male Y-chromosome. Bishop explained Y-STR DNA testing is often used in sexual assault cases because it can identify a male's contribution to a sample, such as a vaginal swab, that may have many more cells from a female contributor than a male contributor. Bishop testified that UNMC has performed Y-STR DNA analysis since the early 2000's and that UNMC has used the particular Y-STR DNA kit used in this case since 2012. Bishop stated that Y-STR DNA analysis is an accepted forensic tool in the forensic analysis community that has been available since the early 2000's.

Bishop explained that the Y-STR DNA testing process is essentially the same as autosomal DNA testing, but admitted that there is a great deal of difference between the discriminatory power of autosomal DNA testing versus Y-STR DNA testing. She testified that autosomal DNA testing can produce results showing that a particular profile is extremely rare in the population, but because the same Y-STR DNA profile is passed to all males in the same lineage and additionally may be present in unrelated members of the general population, Y-STR DNA testing results will not show that a particular profile is as rare. She acknowledged that a coincidental random match might occur one in several quintillion times with autosomal DNA testing but one in a few thousand times with Y-STR DNA testing.

Bishop testified that Y-STR DNA testing identified Y-STR DNA present on the interior of M.T.'s shorts and that it consisted of a mixture of at least two male individuals. Bishop testified that she could not determine what type of cell contributed the Y-STR DNA profile to M.T.'s shorts or how it was deposited there. However, she stated that the major Y-STR DNA profile she found on the shorts matched Tucker at all of the loci obtained and that, consequently, Tucker was not excluded as a potential major source of the DNA tested.

Bishop testified that to calculate the frequency of Tucker's Y-STR DNA profile within the population, it was necessary to consult a database. Bishop testified and her report reflected that according to the U.S. Y-STR database, the probability of randomly selecting a second individual with the same Y-STR DNA profile, given that Tucker expresses such a profile, was 1 in 1,842 for African Americans. Tucker does not dispute the he is African American.

Bishop's report reflected that the U.S. Y-STR database was maintained by a national institute in the forensic science field. Bishop admitted that the U.S. Y-STR database had changed since her analysis because it is always gaining contributors. She further admitted that it was possible that the more individuals that contribute, the better the database will be at discerning the likelihood of a particular profile appearing in the general population.

A forensic scientist with the Nebraska State Patrol Crime Laboratory's biology unit also explained the statistical limitations arising from the patrilineal recurrence of Y-STR DNA and the possible occurrence in the general population, consistent with Bishop's testimony.

...

ANALYSIS

Admissibility of Y-STR DNA Evidence

Tucker challenges the admission of the Y-STR DNA testing and any opinions derived from such testing. His brief contends that the probative value of this evidence was outweighed by its prejudicial effect because the evidence is unreliable. We understand Tucker to be making two basic arguments against the admissibility of the Y-STR DNA evidence. The primary argument is a contention that the inherent nature of Y-STR DNA evidence makes its admission unfairly prejudicial in any case.

The second challenges the reliability of the conclusions regarding the Y-STR DNA evidence reached in this case. We take up each of these contentions in turn.

We first address Tucker's arguments that the Y-STR DNA evidence was unfairly prejudicial. Evidence is relevant if it tends in any degree to alter the probability of a material fact. State v. Grant, 293 Neb. 163, 876 N.W.2d 639 (2016). Under Neb. Evid. R. 403, Neb. Rev. Stat. § 27-403 (Reissue 2016), relevant evidence may be excluded if its probative value is substantially outweighed by the danger of unfair prejudice. Grant, supra. Most, if not all, evidence offered by a party is calculated to be prejudicial to the opposing party. State v. Chauncey, 295 Neb. 453, 890 N.W.2d 453 (2017). Unfair prejudice means an undue tendency to suggest a decision based on an improper basis. Id. It speaks to the capacity of some concededly relevant evidence to lure the fact finder into declaring guilt on a ground different from proof specific to the offense charged, commonly on an emotional basis. Id.

Tucker's primary argument against the admission of the Y-STR DNA evidence is that juries will inevitably base their decision on an improper basis by incorrectly perceiving the Y-STR DNA "match" testimony as conclusively connecting the defendant to the sample. Tucker asserts that jurors are inclined to think of any evidence of a DNA "match" of having an extremely small probability of being the result of coincidence. But, as he correctly points out, the Y-STR DNA testing in this case led to a 1-in-1,842 chance of a coincidental match. According to Tucker, this relatively greater chance of a random match renders the evidence "completely unreliable." Brief for appellant at 37.

We have previously recognized a risk that a jury might give undue weight to DNA evidence if it is introduced without proper context. See State v. Johnson, 290 Neb. 862, 862 N.W.2d 757 (2015).

In Johnson, we observed that "'[b]ecause the potential precision of DNA testing is so well known, a jury might assume that any DNA profile match is extremely unlikely and therefore extremely probative'—even when this is not true." 290 Neb. at 883, 862 N.W.2d at 774, quoting Peters v. State, 18 P.3d 1224 (Alaska App. 2001). We have not, however, concluded that the appropriate measure to prevent a jury from making such an assumption is the wholesale exclusion of DNA evidence falling below a certain threshold of precision. Rather, we have emphasized the need for evidence of DNA testing to be accompanied by evidence of the statistical significance of the findings if it is to be admitted. Johnson, supra.

In this case, Tucker cannot contend that the State sought to introduce the Y-STR DNA evidence without the necessary statistical context. As we have noted, the State introduced much evidence regarding the statistical context for the Y-STR DNA evidence. The jury heard explanations of the relative probative value of Y-STR DNA testing and autosomal DNA testing. Both forensic scientists testified that males share the same Y-STR DNA profile with other males in the same paternal lineage, as well as others in the general population. Their testimony clearly illustrated that due to this recurrence, Y-STR DNA testing results are not as probative as autosomal DNA testing results. They both explained that Y-STR DNA statistical analysis would not render coincidental random match probabilities such as one in several hundred quintillions, showing a particular profile is extremely rare in the population. Instead, Bishop testified, Y-STR DNA testing might produce a random match probability of one in a few thousand, as it did in this case, where Y-STR DNA statistical analysis revealed a 1-in-1,842 probability for African-American contributors that a random Y-STR DNA profile unrelated to this case would match the profile found on M.T.'s shorts. Because the Y-STR DNA testing results were accompanied by the required statistical context, its admission was consistent with the principles we set forth in Johnson.

Moreover, we do not believe that Y-STR DNA evidence is so unique that something other than the principles of Johnson should govern its admissibility. The probabilities of a coincidental match may be exponentially greater with Y-STR DNA evidence than with autosomal DNA evidence, but, if those differences are explained to the jury, we see no reason why jurors would be incapable of grasping the difference. We have previously rejected claims that jurors would not be capable of assigning appropriate weight to "the statistical analysis that accompanies DNA evidence," State v. Bauldwin, 283 Neb. 678, 703, 811 N.W.2d 267, 288 (2012), and reject any suggestion that would be the case with Y-STR DNA evidence.

Y-STR DNA evidence may be less probative than other DNA evidence, but if we were to find it inherently prejudicial, as Tucker urges, we would be treating such evidence differently from other types of evidence that have similar probative value and that are introduced for the same purpose. As the district court and other courts have observed, Y-STR DNA evidence can be used in much the same manner as shoe imprint evidence. Shoe imprint evidence is routinely admitted to show that an imprint at a crime scene matches shoes owned by a defendant even though any number of individuals may own shoes identical to the defendant's.

> The coincidence that [a Y-STR DNA] profile matches that of defendant is probative of his guilt in the same manner as if he had owned shoes that matched a foot imprint found at the crime scene. It is up to the jury to weigh the probative value of that evidence in light of the fact that a significant number of other individuals may possess the same profile.

State v. Calleia, 414 N.J.Super. 125, 152, 997 A.2d 1051, 1066 (2010), reversed on other grounds 206 N.J. 274, 20 A.3d 402 (2011). See, also, People v. Stevey, 209 Cal.App.4th 1400, 148 Cal.Rptr.3d 1 (2012).

\Finally, we note that our conclusion that the Y-STR DNA evidence introduced in this case was not unfairly prejudicial is not a novel conclusion. Courts in a number of other states have reached the same conclusion. See, e.g., State v. Escalante-Orozco, 241 Ariz. 254, 386 P.3d 798 (2017), abrogated on other grounds, State v. Escalante, 245 Ariz. 135, 425 P.3d 1078 (2018) (Y-STR profile evidence is not misleading, nor is its probative value outweighed by risk of unfair prejudice; jury is capable of understanding limited probative value of this evidence and giving it whatever weight it deserves); State v. Jones, 345 P.3d 1195 (Utah 2015) (Y-STR DNA evidence properly explained to jury such that risk of unfair prejudice through confusing or misleading jury did not substantially outweigh probative value of evidence); People v. Wood, 307 Mich.App. 485, 862 N.W.2d 7 (2014), vacated in part on other grounds 498 Mich. 914, 871 N.W.2d 154 (2015) (limitations of Y-STR DNA testing were presented to jury such that there was no danger of confusion or other unfair prejudice that would substantially outweigh probative value).

Having rejected Tucker's argument that Y-STR DNA evidence is inherently unfairly prejudicial, we move to Tucker's assertion that the testimony offered regarding the Y-STR DNA evidence in this case was unreliable. Here, Tucker claims that the database used by Bishop to arrive at her statistical conclusions may not be representative of the population in a given area and that therefore, there is a risk that Tucker's Y-STR DNA is even more common than Bishop acknowledged and thus there is an even greater chance of a coincidental match. This seems to be an attempt at a Daubert/Schafersman challenge, even though Tucker's brief does not cite to Daubert/Schafersman or address its framework. But even if we liberally construed Tucker's brief as having framed his argument as a Daubert/Schafersman issue, the argument fails.

Tucker's argument is premised in part on contentions about the U.S. Y-STR database made in articles published in 2003. Bishop, however, testified at the Daubert/Schafersman hearing that she was familiar with the articles Tucker relies upon. She explained that she found these articles to be irrelevant, because they were about very small sample sizes and Y-STR DNA science has improved since their publication. In fact, she testified that the U.S. Y-STR database upon which she relied was not even established at the time the articles upon which Tucker bases his challenge were written. The remaining article Tucker cites to support this argument was not presented to the district court and is not in the record before us for consideration. See State v. Patton, 287 Neb. 899, 845 N.W.2d 572 (2014) (party's brief may not expand evidentiary record). Given Bishop's testimony concerning the articles Tucker offered to the district court, we see no abuse of discretion in the admission of the Y-STR DNA evidence.

In sum, we conclude that the Y-STR DNA evidence in this case did not suggest a decision based on an improper basis and that thus, its prejudicial effect did not outweigh its probative value. Moreover, Tucker's Daubert/Schafersman arguments also fail. Therefore, the district court did not abuse its discretion in admitting such evidence at trial.

…

CONCLUSION

We conclude that the district court did not abuse its discretion in admitting Y-STR DNA evidence at trial. We further determine that the evidence presented at trial was sufficient to support Tucker's convictions and that the district court did not abuse its discretion in sentencing him. Consequently, we affirm.

AFFIRMED.

Questions:

1. While defendants have little chance of objecting to DNA as a substantive field or endeavor, process challenges still can effective especially the process is flawed or defective. In State v. Tucker, what was the basis for the appeal?
2. When evaluating Y-STR testing, the Court makes an interesting comparison to another evidence form to prove the inherent value, though different approach of Y-STR. What evidence form?

3. How does the Court deal with issue of prejudice alleged in the use of Y-STR DNA testing?

United States v. Morgan
United States Court of Appeals for the Second Circuit
January 12, 2017, Decided
15-2696-cr

Defendant-appellant Johnny Morgan appeals an August 18, 2015 judgment of the district court, convicting him, after a jury trial, of illegally possessing a firearm in violation of 18 U.S.C. § 922(g)(1). The district court sentenced Morgan principally to eighty-four months' imprisonment and three years' supervised release. On appeal, Morgan argues that the district court abused its discretion in admitting: (1) reports and expert testimony concerning a Low Copy Number DNA analysis ("LCN") that showed the presence of DNA consistent with Morgan's genetic profile on a .40 caliber pistol recovered by police in the vicinity of his arrest; and (2) testimony about a 911 call reporting a "man with a gun" near the same area. We assume the parties' familiarity with the underlying facts, procedural history, and issues on appeal.

I. DNA Evidence

Morgan argues that the district court erred in admitting the government's expert testimony and other evidence concerning LCN testing performed by the New York Office of the Chief Medical Examiner ("OCME") because LCN testing is unreliable in all cases, and, in any event, OCME's testing methods cannot be reliably applied to the sample obtained from the pistol in this case.

We review a district court's decision to admit expert testimony for abuse of discretion and will vacate the judgment only if any error was not harmless. United States v. Litvak, 808 F.3d 160, 179 (2d Cir. 2015); United States v. Rea, 958 F.2d 1206, 1219 (2d Cir. 1992).

Although "the proponent of expert testimony has the burden of establishing by a preponderance of the evidence that the admissibility requirements of Rule 702 are satisfied, the district court is the ultimate 'gatekeeper.'" United States v. Williams, 506 F.3d 151, 160 (2d Cir. 2007). At bottom, Federal Rule of Evidence 702 requires the district court to ensure "that an expert's testimony both rests on a reliable foundation and is relevant to the task at hand." Daubert v. Merrell Dow Pharms., Inc., 509 U.S. 579, 597, 113 S. Ct. 2786, 125 L. Ed. 2d 469 (1993).

In assessing expert testimony and evidence, "the district court should consider the indicia of reliability identified in Rule 702, namely, (1) that the testimony is grounded on sufficient facts or data; (2) that the testimony 'is the product of reliable principles and methods'; and (3) that 'the witness has applied the principles and methods reliably to the facts of the case.'" Amorgianos v. AMTRAK, 303 F.3d 256, 265 (2d Cir. 2002) (quoting Fed. R. Evid. 702). Moreover, the district court may consider other factors, including whether a theory or technique has been or can be tested, "whether the theory or technique has been subjected to peer review and publication," the technique's "known or potential rate of error," "the existence and maintenance of standards controlling the technique's operation," and whether the technique has gained general acceptance in the relevant scientific community. See Daubert, 509 U.S. at 593-94.

We have held that to "warrant admissibility ... it is critical that an expert's analysis be reliable at every step." Amorgianos, 303 F.3d at 267. Indeed, "any step that renders the [expert's] analysis unreliable under the Daubert factors renders the expert's testimony inadmissible" as a whole. Id. (quoting In re Paoli R.R. Yard PCB Litig., 35 F.3d 717, 745 (3d Cir. 1994)). Still, "a slight modification of an otherwise reliable method will not render an expert's opinion per se inadmissible" -- a "judge should only exclude the evidence if the flaw is large enough that the expert lacks 'good grounds' for his or her conclusions." Id. (internal quotations marks omitted). Instead, "[v]igorous crossexamination, presentation of contrary evidence, and careful instruction on the burden of proof are the traditional and appropriate means of attacking shaky but admissible evidence." Daubert, 509 U.S. at 596.

Here, the district court admitted the challenged evidence only after reviewing the parties' extensive briefing and conducting a hearing in which it received documentary evidence and testimony concerning LCN testing and OCME's methods from both the government's and Morgan's expert witnesses. See United States v. Morgan, 53 F. Supp. 3d 732, 734-39 (S.D.N.Y. 2014). The government presented evidence that OCME conducted validation studies of its testing methods, the results of which were published in a peer-reviewed journal, and that the DNA Subcommittee of the New York Commission on Forensic Science approved OCME's use of LCN testing in criminal casework. Id. at 737-39. Morgan, for his part, challenged the accuracy of OCME's testing, emphasizing, among other indicia of unreliability, the limitations of OCME's validation studies, the increased risk of error in LCN analysis as compared to traditional DNA analysis, that no other publicly funded lab in the country performs forensic LCN analysis, and that the FBI does not accept the results of such analysis in its national DNA database, the Combined DNA Index System. Id. at 741-46.

Upon review of the record, we conclude that, although LCN analysis is supported by significantly weaker evidence of reliability than traditional DNA analysis, the district court did not abuse its discretion in this case in holding that the proffered expert evidence met the reliability standards of Rule 702: We cannot say that its ruling here was "manifestly erroneous." Amorgianos, 303 F.3d at 265. We express no opinion on the propriety of admitting the results of LCN testing in other cases and note that OCME is discontinuing its use of LCN testing in favor of newer technology that produces reliable results in most of the sensitivity range for which it previously employed LCN testing. See Timothy D. Kupferschmid, NYC Office of the Chief Medical Examiner, Department of Forensic Biology is Implementing New Technologies -- a New STR Kit, a New STR Analysis Software and a New Probabilistic Genotyping Software 2 (Sept. 19, 2016).

2. 911 Call

Morgan also argues that testimony from two police officers concerning a 911 call reporting a "man with a gun" in the vicinity where he was arrested was unduly prejudicial and should have been excluded under Federal Rule of Evidence 403.

Like the district court's Rule 702 determination, we review its other evidentiary rulings for abuse of discretion and will vacate the conviction only for error that is not harmless. United States v. Curley, 639 F.3d 50, 56-58 (2d Cir. 2011); United States v. Mercado, 573 F.3d 138, 141 (2d Cir. 2009). "[S]o long as the district court has conscientiously balanced the proffered evidence's probative value with the risk for prejudice," a determination under Federal Rule of Evidence 403 "will be disturbed only if it is arbitrary or irrational." United States v. Awadallah, 436 F.3d 125, 131 (2d Cir. 2006).

The district court did not abuse its discretion here. It allowed the officers to testify that they were responding to a 911 call about "a person with a gun," but it gave the jury the following limiting instruction to reduce any risk of unfair prejudice:

> [Y]ou heard some testimony about officers' receipt of a 911 dispatch call for [a] man with a gun. That information was introduced solely for the purpose of you understanding why it was the officer reported to the scene. You should not take that evidence as indicating that there was in fact a man with a gun at the scene. You should not take that information for the truth of the matter. The reason why is that information came in strictly to understand why it was that the officer reported to the scene, and you should not use it for any other purpose in your deliberations.

App. at 1622.

"Generally, we presume that juries follow limiting instructions" unless "there is an overwhelming probability that the jury will be unable to follow the court's instructions and the evidence is devastating to the defense." United States v. Becker, 502 F.3d 122, 130 (2d Cir. 2007) (internal quotation marks omitted). Morgan has not overcome this presumption. Although the jury asked to rehear the testimony of the "person who stated [sic] the second 911 call" during its deliberations and the district court granted that request, the district court also issued the limiting instruction excerpted above a second time and provided the written transcript of the instruction to the jury along with the requested testimony. App. at 2296-97. Morgan has not shown that there is an overwhelming probability that the jury was unable to follow an instruction that the district court gave, in effect, three times. Therefore, we reject his challenge to the district court's evidentiary ruling on this matter.

We have reviewed Morgan's remaining arguments and conclude they are without merit. Accordingly, we AFFIRM the judgment of the district court.

Questions:

1. Prove the following True or False: Courts generally reject DNA if the better test not be used or the better technology.
2. Why does the Daubert decision show more tolerance for error than Frye in this case.
3. Why is the "abuse of discretion" so central to the court's ruling on the DNA methodology?

3.5.1 DNA Fingerprinting

While technically not traditional fingerprinting, the use of DNA derived from bodily fluids, hair and other bodily matter, has been an acceptable method of identity proof for the last 30 years. The method has been strikingly apparent in the University of Idaho murders of four students, where fingerprints in the traditional sense were hard to come by while DNA fingerprints were readily discoverable.

People v. Flores, a respected New York decision tackles its usage in its earliest days.

Locate the following case using your favorite legal research resource, read the case and answer the questions below.

People v. Flores, 61 Misc. 3d 1219, 2018 N.Y. Slip Op. 51611, 111 N.Y.S.3d 517 (N.Y. Crim. Ct. 2018).

Questions:

1. How does the Court reconcile the Frye doctrine with the "newness" of the DNA technology?
2. And while the court does not exclude under the Frye principle, it holds that a past usage history in a particular non-criminal case give the science the longevity it needs to be recognized. What practice is referenced by the Court?
3. What else does the Court look to establish a consensus mentality about the new technology?
4. Explain the distinction on DNA fingerprint usage where it is ok to use and not ok in other cases?

4 Forensic Biology

The biological foundations of forensic science are the most commonly witnessed field of expertise, for in biology, one examines the functions and measures of the human body and anatomy: the composition of excretions and secretions, from saliva to seminal fluid, from sweat and bacteria and infectious wounds. On top of this, biology gauges and assesses blood through serological analysis, the contagion and transmission of disease and other pathology and generally most evidence that relates to how the body our being operates. Forensic Biology is a large, large world, without equal in the field of forensic science.

> Visit the NIST's Forensic Science Program page at https://www.nist.gov/spo/forensic-science-program.

Forensic Biology involves crime scene analysis of biological fluids and tissues or extractions from physical evidence for the purpose of identification and individualization. A forensic biologist employs a wide range of laboratory methods and instruments for body fluid identification, cellular composition and the emerging area of DNA analysis.

4.1 BLOOD ANALYSIS

The value and importance of forensic testing on blood and blood spatters cannot be fully appreciated. In murder, assault and rape cases, to name just a few, the collection, identification and interpretation of blood is always a critical function.

The first case dealing with blood and blood spatters manifests the importance.

People v. Long
Court of Appeal of California, Fourth Appellate District, Division Two May 3, 2018, Opinion Filed
E066388

In December 2005, a jury found defendant and appellant Kimberly Louise Long, guilty of second degree murder. (Pen. Code, § 187, subd. (a).) The trial court sentenced defendant to prison for a term of 15 years to life. Defendant petitioned the California Supreme Court for a writ of habeas corpus based, in part, upon ineffective assistance of trial counsel and defendant's actual innocence. In 2015, the Supreme Court issued an order to show cause, before the trial court, as to why defendant's trial counsel was not ineffective and why defendant is not actually innocent of the murder. The trial court granted the writ of habeas corpus on the ground of ineffective assistance of trial counsel. The trial court found defendant failed to meet her burden of proving actual innocence.

DOI: 10.4324/9781003267126-4

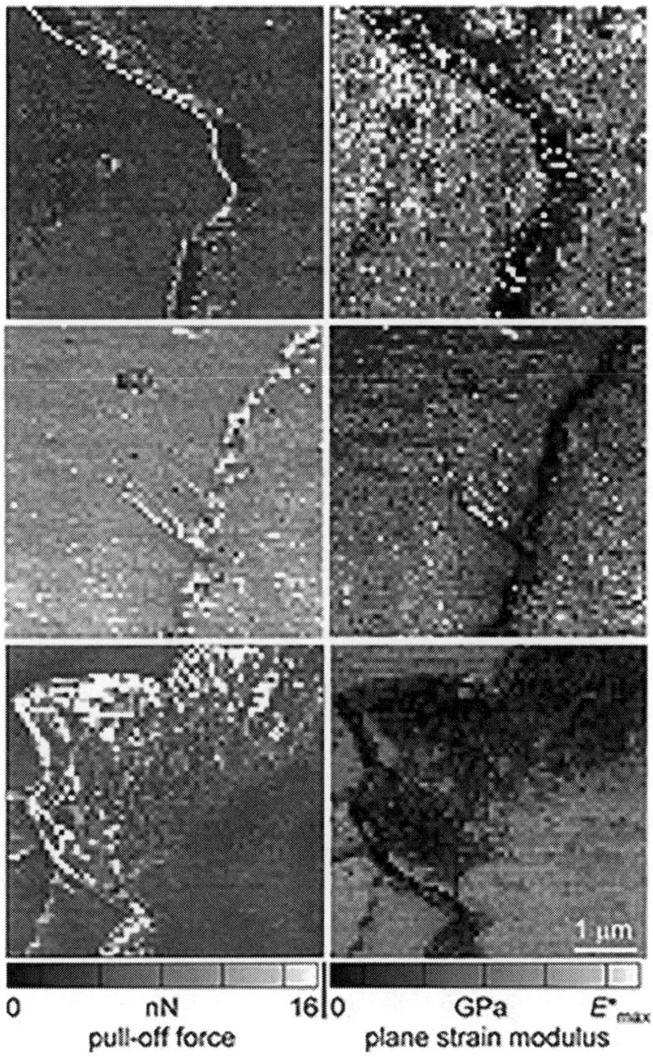

FIGURE 4.1 NIST researchers used atomic force microscopy (AFM) to measure the stickiness (pull-off force) and stiffness (plane strain modulus) of hair that was untreated (top), bleached (middle) and conditioned (bottom). The clear differences in the measured results for various hair treatments suggest that AFM might be used to analyze forensic evidence. The measured forces are very small, in units of nanonewtons or nN (1 nanonewton is about the weight of a pollen grain). Modulus is in units of gigapascals or GPa and ranges from 1–5 for the hair samples (for comparison, the plastic lens material polycarbonate measures about 3).

Source: DelRio/NIST.

The People contend the trial court erred by granting the writ. In particular, the People assert the trial court erred by finding defendant's trial counsel was ineffective for (1) failing to present a time of death defense; and (2) failing to provide further evidence that defendant was wearing the same clothes on October 5, 2003, that she was wearing when police arrived at her house on October 6. We reverse the order.

FACTUAL AND PROCEDURAL HISTORY

...

Detectives' Search

When Corona Police Detective Robert Newman went to defendant's house on October 6, he was trying to locate the weapon that injured the victim. Newman did not find any blunt objects that were marked with blood or hair, or blunt objects that appeared to have been recently washed. Newman noticed two baseball bats by the front door. Newman surmised that the bats were by the door when the victim was struck because there was blood spatter on the bats. Newman did not see any golf clubs inside the residence. No potential weapons were seized from the House or vehicles. On October 9, police officers searched the yards of the homes on defendant's street and the immediately surrounding streets, and asked if neighbors noticed anything discarded in their yards, but nothing was found.

...

Forensic Technician

Corona Police Forensic Technician Daniel Verdugo looked for fingerprints at defendant's house, but no usable fingerprints were found. No footprints were located in or around the House. The only bloodstains in the House were in the garage and the living room. The victim's body was seated, fully dressed, on the loveseat in the living room; he was "slouched down." Drops of blood were located 360-degrees around the victim. The floor and every wall in the living room were stained with spots of blood, but there was no blood spatter on the ceiling. Blood spatter located on the back of the front door indicated the door was closed when the attack occurred. Blood spatter located on the door between the living room and the garage indicated that door was partially open when the attack occurred. Blood spatter in the garage was located on the washer/dryer set and a comforter.

Because there was blood spatter 360 degrees around the victim, it indicated that the weapon was swung horizontally toward the victim. It also indicated that the victim was attacked in the location where he was found. A long weapon, such as a golf club, which would allow for "a big swing" could cause blood to be splattered close to 360-degrees.

In the bloodstain patterns, there were no indications of a struggle or of the victim trying to get assistance; the only disruption in the blood stains appeared to have been caused by the paramedics. In the area of the front door, where two baseball bats were located, there were no gaps in the blood spatter. In other words, it appeared the bats were in place by the door during the attack.

Various items were on the kitchen floor: some coins, pieces of glass from a broken shot glass, and a portable telephone handset. A champagne bottle was in the trash. Various items were on the floor: defendant's helmet, two novelty hats, defendant's sandals, and defendant's jacket. Defendant's sandals and helmet were on the floor behind the loveseat, in the entryway. One of defendant's novelty hats was in the entryway, beyond the back of the loveseat, in the space where one could turn into the living room. Defendant's jacket and the second novelty hat were in the living room, near the wall opposite the front door. No blood was found on defendant's helmet, hats, and jacket. The helmet was tested for blood in a laboratory, but the shoes and novelty hats were not—they were visually inspected at the House but not collected by police.

Verdugo observed that the hot tub in the backyard of the House was running, i.e., the cover was off, the water was bubbling, and the water felt warm. While Verdugo was photographing the House, defendant's dog was "creating a ruckus," being "mean," barking, and howling. Detective Newman gave Verdugo female clothing for lab testing. Verdugo found no bloodstains on the clothing.

Forensic Pathologist

Dr. Joseph Cohen was the Chief Forensic Pathologist for Riverside County. Dr. Joseph Pestaner conducted the autopsy on the victim. Pestaner worked out of state at the time of trial, so Cohen testified. The victim suffered several blunt force injuries on the right side of his face and head, in the form of abrasions, contusions, and lacerations. The victim had a group of injuries on and behind his right ear, including a complex laceration, where a large portion of the tissue on his upper right ear was torn away. The victim had a large fracture at the base of his skull, which ran from one side of his skull to the other side. The victim also had a fracture on the roof of his mouth, on the right side. The victim suffered bleeding on the surface of his brain, i.e., a subarachnoid hemorrhage.

The cause of the victim's death was blunt force injuries to his head. The primary injury causing death was a three- and three-quarter inch laceration behind his ear, which led to the fracture at the base of his skull, which then led to the subarachnoid hemorrhage. Without treatment, a person with the victim's injuries would die within one to 20 minutes, most likely within 10 or 15 minutes. The injuries would lead to bleeding into the lungs, choking on one's own blood, and then death within minutes.

"[A] lot of force" would be required to inflict the victim's injuries, such as that created by a bat, brick, or golf club, rather than a fist. It appeared the weapon used against the victim was long because several of the injuries were long and narrow. Any healthy adult—male or female—would have been capable of inflicting the victim's injuries. It appeared the victim was struck three to eight times. The victim's injuries would have caused him to lose consciousness "nearly immediately."

The victim had no injuries on his hands. There were no defensive injuries on the victim's body. It was not possible to decipher from the injuries where the attacker was located in relation to the victim at the time the injuries were inflicted.

...

Keen

Keen, defendant's trial counsel, began working at the Riverside County Public Defender's Office in 1997. Keen worked on defendant's case from 2003 to 2006. Keen thought it was "very likely" that the person who killed the victim would have blood on his or her person as a result of the killing. At trial, in order to establish defendant did not change her clothing, Keen presented defendant's testimony and Officer Welde's testimony—the officer who sat with defendant on October 6.

Prior to the preliminary hearing, Keen had a copy of Dills's statement to police describing defendant's October 5 clothes. At the preliminary hearing, Keen did not question Dills about defendant's October 5 clothing. Keen had no tactical reason for not questioning Dills about the clothing. Keen did not think defendant's clothing was a critical issue at the preliminary hearing. When asked why he did not think the issue was critical, Keen responded, "It was a preliminary hearing. I didn't—I didn't ask those questions at [the] preliminary hearing. I don't know why not. I should have."

At trial, Keen did not move the court to admit Dills's statement describing defendant's October 5 clothing. Keen thought he had sufficient evidence concerning defendant's clothing with defendant's testimony and Officer Welde's testimony. The direct evidence at trial concerning defendant's clothes was presented only by the defense and it reflected defendant did not change her clothes.

...

i. Rogers

Michelle Rogers was defendant's appellate counsel for her direct appeal. Rogers spoke to Keen while working on defendant's appeal. Rogers asked Keen why Dills's statement describing defendant's clothes was not admitted at trial. Keen said, "I didn't think there was any way I could have gotten that in." Rogers agreed there was no direct legal authority for admitting Dills's hearsay statement. Rogers opined that "[i]t would be difficult" to have Dills's statement admitted, but Keen could have sought to have Dills's hearsay statement admitted under a due process exception to the hearsay rule.4

...

A. CONTENTION

The People contend the trial court erred by finding Keen rendered ineffective assistance.

B. LAW

"Under both the Sixth Amendment to the United States Constitution and article I, section 15 of the California Constitution, a criminal defendant has the right to the effective assistance of counsel. [Citations.] Specifically [s]he is entitled to the reasonably competent assistance of an attorney acting as [her] diligent and conscientious advocate. [Citation.] This means that before counsel undertakes to act, or not to act, counsel must make a rational and informed decision on strategy and tactics founded upon adequate investigation and preparation.

"There are two components to a claim by a defendant that [her] counsel's assistance was so defective as to require reversal of a conviction 'First, the defendant must show that counsel's performance was deficient.' [Citation.] This requires a showing that 'counsel's representation fell below an objective standard of reasonableness.' [Citation.] In evaluating a defendant's showing of incompetence, we accord great deference to the tactical decisions of trial counsel. 'A fair assessment of attorney performance requires that every effort be made to eliminate the distorting effects of hindsight, to reconstruct the circumstances of counsel's challenged conduct, and to evaluate the conduct from counsel's perspective at the time.'

"The second component requires that the defendant show prejudice resulting from counsel's alleged deficiencies. 'It is not enough for the defendant to show that the errors had some conceivable effect on the outcome of the proceeding. [¶] The defendant must show that there is a reasonable probability that, but for counsel's unprofessional errors, the result of the proceeding would have been different. A reasonable probability is a probability sufficient to undermine confidence in the outcome.'" (In re Marquez (1992) 1 Cal.4th 584, 602-603, 3 Cal. Rptr. 2d 727, 822 P.2d 435.)

...

C. CLOTHING

1. FACTS

On the morning of October 6, during a police interview, a female officer took defendant's clothes from her for evidentiary purposes. No blood was found on defendant's clothes. Defendant's clothing included (1) a sleeveless T-shirt, with small silver metallic rings along the seams and a design that included a skull, flames, playing cards, and the words "punk lives!"; (2) jeans; and (3) a black belt with silver metallic studs. In a police interview of Dills, the following exchange took place:

...

During the prosecutor's closing argument at the retrial he said, "The blood evidence. [¶] Well, yeah, there's no blood on her. [¶] Officer Welde told you there's no blood, because, hey, she's not going to let her hug her and climb all over her in her street—personal clothes if she's got blood on her.5 [¶] Who can tell us that those were the clothes that she was wearing that day? You didn't hear any evidence other than from her. You've got to rely upon her again that those were the clothes that she was wearing that day. [¶] And isn't it interesting to you, ladies and gentlemen, that this nurse who works in an emergency room, who renders medical aid on a daily basis doesn't have any blood on her? Any blood on her. Not a smudge. Not a trace of blood".

"Not only does she not have blood on her, but every item of clothing that she strewn—throws, how's that. Every item of clothing that she throws about the house, that doesn't have blood on it either. With all the blood in that scene, she is somehow miraculously lucky enough not to put anything down in her hurried state in an area where there would be blood."

During Keen's closing argument, he said, "Now, Mr. Verdugo. He's the evidence guy. And he draws this diagram labeled Defense A. [¶] On this diagram he pinpoints everywhere he found blood. He found blood on every wall in the living room. He found blood on the door—the front door, the blinds. That table had a fine mist of blood all over it. He found blood over on the wall where the fireplace is. Found blood on the window. The blinds there. On the TV, which is in that entertainment center. He found blood over that little table that had some pictures on it. Blood above that. And then he found blood on the wall behind [the victim]. What he didn't find blood on was any of [defendant's] clothes. Not on her shoes that are right behind [the victim]. Not on her helmet, which is right next to blood spatter. Not on those funny little [novelty] hats. And not on her jacket that she said she was wearing. There was no blood on it."

"If the theory is it was there when it happened, blood went over that way, but there's no blood on her jacket. If their theory is, well, she had it on, then no blood got on her. No blood on her hair, her face. She hasn't cleaned herself. Officer Welde doesn't smell any fresh shampoo, doesn't smell any cleaner on her. Her clothes aren't fresh and laundered. You didn't hear any evidence about a shower having been wet or moist or sinks had been used or water or anything like that. There's no blood on her jacket. There's no blood on her. But there is blood all over the room. [¶] This isn't a gruesome scene, like somebody dumped a bucket of blood. There's blood spatter all over, but not on her."

Keen did not seek to have Dill's statement about defendant's clothes admitted into evidence at defendant's trial because he "didn't think there was any way [he] could have gotten that in." Keen also believed defendant's testimony regarding not having changed her clothes and Officer Welde's testimony about not recalling any signs of washing or cleaning on defendant was sufficient to prove the clothing issue. Michelle Rogers was defendant's appellate attorney for her direct appeal. Rogers said there was not "legal authority directly allowing the admission of hearsay," referring to Dills's statement about defendant's clothing.

…

2. ANALYSIS

a) Preliminary Hearing

We examine whether Keen was ineffective at the preliminary hearing for not questioning Dills about defendant's clothing.

…

Because a preliminary hearing is not a trial, "the cross- examination of a witness at a defendant's preliminary hearing is seldom adequate. The cross-examiner does not have the same motivation to cross-examine a witness at a preliminary hearing when the prosecution's only burden is to establish probable cause, as he later has at the defendant's trial when the People must prove the defendant guilty beyond a reasonable doubt. Thus, the cross-examination is seldom a searching exploration into the witness'[s] motives or other facets of his testimony to test his credibility. In fact, experienced trial lawyers often defer cross-examination entirely until the time of trial when it is most effective." (People v. Vinson (1969) 268 Cal.App.2d 672, 676, 74 Cal. Rptr. 340, fn. omitted.)

…

Analysis

The trial court concluded that the clothing issue was significant. However, the trial court fails to explain how the issue is significant, when the prosecution presented no evidence on the issue, …

Nevertheless, assuming the clothing issue is significant, the trial court fails to explain why it is unreasonable for an attorney to decide not to seek admittance of hearsay. The evidence reflects there was no direct legal authority for admitting Dills's out-of-court statement. It is objectively reasonable for an attorney not to move the trial court to admit hearsay evidence for which there is no exception for admissibility. Accordingly, we disagree with the trial court's analysis.

...

Defendant's argument places too great an expectation on the shoulders of a reasonable attorney. A reasonable attorney can be expected to make arguments that are within the law, but we do not expect a reasonable attorney to necessarily advance the law; again, we are looking at reasonable attorneys, not the best attorneys. Accordingly, we find defendant's argument to be unpersuasive.

Keen's performance being deficient, we find such reasoning to be unpersuasive because a reasonable attorney could choose not to argue a motion based upon Federal Rules of Evidence rule 807 in state court.

...

DISPOSITION

The order is reversed. The matter is remanded to the trial court to address the issue of custody because defendant was granted bail pending the outcome of this appeal.

Questions:

1. What one fact causes extraordinary levels of doubt concerning the presence of blood literally everywhere except what essential place?
2. How does the claim for ineffective counsel relate to blood analysis?
3. If there was a bleeding point on the victim, it was discoverable upon pathological examination. Explain where and caused by what?

The value of blood evidence is elevated due to its importance in DNA testing protocols.

Edwards v. State
453 Md. 174, 160 A.3d 642 (Md. 2017)

This is a direct appeal pursuant to the DNA testing provisions of the DNA Evidence–Post Conviction Review. Md. Code (2001, 2008 Repl. Vol., 2016 Supp.), § 8–201 of the Criminal Procedure Article (Crim. Pro.). Appellant Richard A. Edwards seeks our review of the denial, by the Circuit Court for Saint Mary's County, of his Petition for Post–Conviction DNA Testing. For the reasons set forth below, we shall vacate the judgment of the Circuit Court and remand this case to that court with directions to order the DNA testing of the cigarette lighter.

BACKGROUND

...

A. The Post-Conviction Petition

In his post-conviction Petition, Appellant asserted that there was a reasonable probability that DNA testing of the requested items has the scientific potential to produce exculpatory evidence relevant to

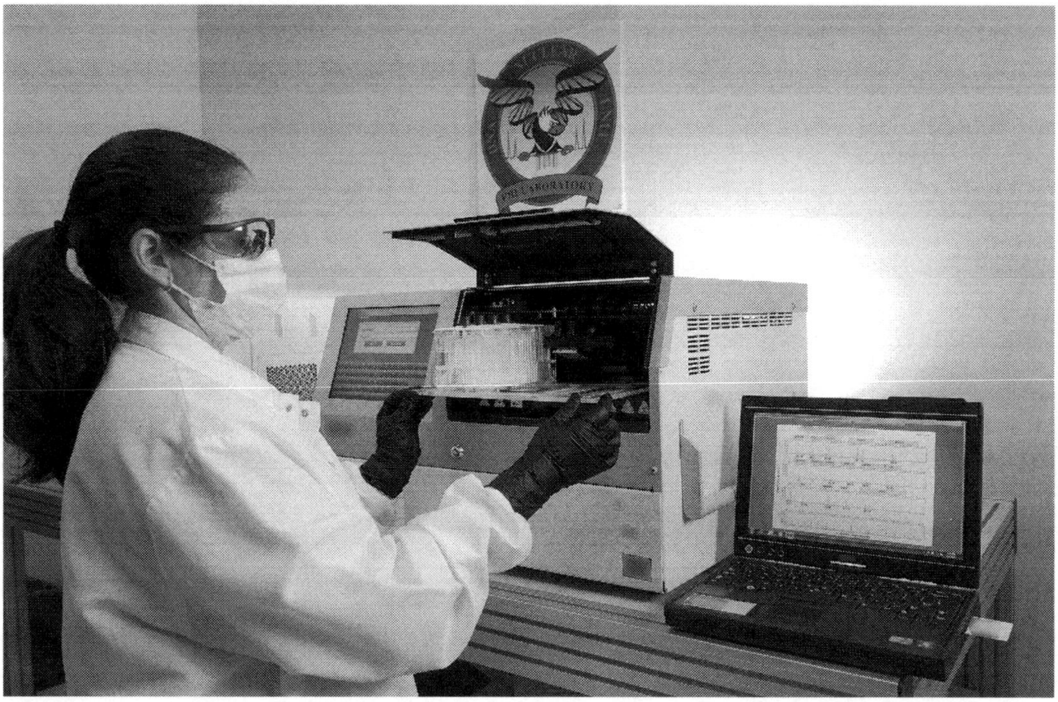

FIGURE 4.2 An FBI Laboratory professional performs nuclear DNA analysis. FBI Media at https://multimedia.fbi.gov/item?type=image&id=3348.

his claim of wrongful conviction. Appellant noted that the victim testified that the perpetrator used her lighter and two witnesses testified that the victim told them the perpetrator asked her for a cigarette. Appellant maintained that it is likely that the perpetrator transferred epithelial cells to the lighter when he used it and that the perpetrator could also have touched the Forever 21 bag and the cigarette pack given his proximity to those items when he sat in the passenger seat.

Appellant contended that "even with minute or degraded quantities of DNA, it is 'frequently possible to obtain successful DNA results from cellular material transferred from the skin of an individual who has simply touched an object.'" (quoting Ray A. Wickenheiser, Trace DNA: A Review, Discussion of Theory, and Application of the Transfer of Trace Quantities of DNA Through Skin Contact, 3 J. FORENSIC SCI. 442 (2002)). The petition noted Wickenheiser's observation that cigarette lighters specifically are an "unusual exhibit material yielding successfully DNA profiles using polymerase chain reaction (PCR) and short tandem repeat (STR) typing."

Appellant further asserted that the testing could show that his DNA was absent on all of the tested items but that the DNA profile of another individual may instead be present on the items. Appellant argued that this would not only support his testimony that he was falsely identified as the perpetrator but it also has the potential to identify the actual perpetrator. Finally, the petition asserted that the DNA testing requested, including polymerase chain reaction (PCR) and short tandem repeat (STR) testing, are generally accepted as reliable in the relevant scientific community.

As noted, a hearing on the petition was held on June 14, 2016 in the Circuit Court for St. Mary's County. At the hearing, the State did not challenge whether the DNA testing requested by Appellant is employed by a method generally accepted within the relevant scientific community. Rather, the focus of the hearing was whether there was a reasonable probability that the testing requested has the scientific potential to produce exculpatory evidence relevant to Appellant's claim that he was not the individual who assaulted Ms. K.On July 6, 2016, the Post-conviction Court issued an order denying Appellant's petition for DNA testing. The order was accompanied by a written opinion, which stated in pertinent part:

> As previously stated, the standard in CP § 8–201(d) [(1)](i) is that there must be a "reasonable probability ... that the DNA testing has the scientific potential to produce exculpatory or mitigating evidence." Because of the highly speculative nature of what petitioner hopes to gain from DNA testing of these items, this [c]ourt can see no possibility that a DNA test performed on the items requested would exonerate Petitioner. Even if petitioner's DNA is not found on the items and other persons' DNA is found, that does not prove or even suggest that any of those other people committed this crime, even if one of them turns out to be a convicted sex offender.
>
> Here, there is no instrument that was used to commit the crime, and therefore, no relevance to finding the DNA of random people on the items Petitioner seeks to have tested. The [c]ourt need not "conduct a fishing expedition to indulge every permutation" that Petitioner might imagine.

The Post-conviction Court continued:

> In this case, it is noteworthy and even dispositive that, at trial, the victim positively identified Petitioner as the perpetrator. She was able to give the police a description of him and what he was wearing. She testified that she got a good look at him because "[h]e was in my face quite a bit ... I had seen him earlier in the bar ... I said that guy looks like a creeper because he kept looking at me as we were dancing." In addition, the victim identified Petitioner for the jury and testified to identifying Petitioner in a photo array. Trial counsel cross-examined the victim and she did not waiver in her identification of Petitioner.

The judge distinguished cases where relief had been granted:

> This is clearly different than Gregg where it was highly probative that the killer had handled a gun which was the murder weapon and the DNA on the gun would have a reasonable probability of producing exculpatory evidence that would corroborate defendant's claim that he was not the shooter. This was especially true given the ambivalent evidence which did convict the Petitioner. It is also different than Brown where the defendant was challenging whether the crime had actually happened and wanted the items tested to disprove it had. Here, there is no such probability that the testing would produce exculpatory or mitigating evidence for the Petitioner.

(citations omitted).

On July 21, 2016, Appellant noted a direct appeal to this Court pursuant to § 8–201(k)(6), and presents the following question on appeal:

> Did the Circuit Court err when it denied Mr. Edwards' Petition for Post–Conviction DNA Testing, where a reasonable probability exists that the requested testing has the scientific potential to produce exculpatory evidence relevant to Mr. Edwards' claim of wrongful conviction?

For the following reasons, we shall answer the Appellant's question in the affirmative and remand the case to the Post-conviction Court for to order testing.

DISCUSSION

...

Parties' Contentions

On appeal, Appellant maintains that a reasonable probability exists that the testing has scientific potential to produce exculpatory evidence relevant to Appellant's claim of wrongful conviction. Ms. K. told the jury that her attacker used her lighter and moreover, photographs of the crime scene show a Forever 21 bag and a pack of cigarettes on the edge of the passenger seat where the attacker sat before the assault. Appellant maintains that it is likely that the perpetrator of the crime transferred epithelial cells onto the lighter when he touched it and that it is also likely the perpetrator could have touched the cigarette pack or Forever 21 bag given his proximity to those items.

Further, Appellant urges that the hearing judge applied the wrong standard in denying the petition when the judge ruled that "the [c]ourt can see no possibility that a DNA test performed on the items requested would exonerate the Petitioner." Appellant insists that the proper standard is not whether DNA testing would "exonerate" him or that such testing would "prove" someone else committed the crime. Instead, Appellant contends that the proper inquiry is whether a reasonable probability exists that the testing has the scientific potential to produce exculpatory or mitigating evidence relevant to the claim of wrongful conviction or sentencing. Appellant posits that this case turns upon the meaning of exculpatory, which, he argues, sweeps more broadly than the term exonerating. Appellant also argues that the inclusion of the terms "or mitigating" and "or sentencing" in the statute show that the standard is one of exculpatory evidence and not exonerating evidence. Finally, Appellant avers that the absence of his DNA on an item where one would expect the perpetrator's DNA, although not necessarily exonerating, is certainly exculpatory to the extent that it would tend to establish that Appellant was not the person that assaulted Ms. K. In other words, negative DNA results would satisfy the broader, less stringent threshold for entitlement to DNA testing than the standard advocated by the State and applied by the post-conviction court. Urging that we affirm, the State avers that Appellant's hoped-for outcome of DNA testing would not be exculpatory. The State first notes that there is no evidence that the assailant touched the Forever 21 bag and the cigarette pack, thus the absence of his DNA on these items would in no way tend to exonerate him. The State explains that the testing would reveal one of two things—either Appellant's DNA is on the lighter or it is not—and argues that neither is exculpatory. The State contends that the best result for Appellant is that Appellant's DNA is not on the lighter and notes that the jury was already told that there was no DNA evidence in this case inculpating Appellant.

Crim. Proc. § 8–201

We agree with Appellant that the standard to be employed by a post-conviction court in the analysis of a petition for DNA testing does not require the proponent to establish that the results must, in all instances, exonerate the defendant. We also conclude that our decision in Brown is, on this record, inapposite.

Maryland's post-conviction DNA testing statute, which was enacted by the General Assembly in 2001, is codified at § 8–201 of the Criminal Procedure Article. " Section 8–201entitles persons convicted of certain serious crimes to pursue DNA testing of physical evidence that is in the possession of the State and might produce exculpatory or mitigating evidence relevant to a claim of wrongful conviction or sentencing." Simms v. State, 409 Md. 722, 727, 976 A.2d 1012, 1015–16 (2009). See also Md. Rule 4–703(2)(A).

The statute was enacted "in line with a nationwide trend to adopt post-conviction DNA testing statutes designed to provide an avenue for the exoneration of the actually innocent." Blake v. State, 395 Md. 213, 219, 909 A.2d 1020, 1023 (2006). We have stated that the purpose of Crim. Proc. § 8–201 is to "facilitate the establishment of claims of actual innocence for serious crimes." Thompson v. State, 395 Md. 240, 252, 909 A.2d 1035, 1042 (2006).

Under Crim. Proc. § 8–201, persons convicted of certain crimes of violence may file a petition requesting "DNA testing of scientific identification evidence that the State possesses that is related to the judgment of conviction." Crim. Proc. § 8–201(b)(1). "Scientific identification evidence" is defined in Crim. Proc. § 8–201(a)(5) as evidence that

i. is related to an investigation or prosecution that resulted in a judgment of conviction;
ii. is in the actual or constructive possession of a law enforcement agency or agent of a law enforcement agency; and

iii. contains biological evidence from which DNA may be recovered that may produce exculpatory or mitigating evidence relevant to a claim of a convicted person of wrongful conviction or sentencing if subject to DNA testing.

Crim. Proc. § 8–201(a)(5). "Biological evidence" is defined as evidence that "includes, but is not limited to, any blood, hair, saliva, semen, epithelial cells, buccal cells, or other bodily substances from which genetic marker groupings may be obtained." Crim. Proc. § 8–201(a)(2). The statute mandates that a court grant a petition for DNA testing if the court finds:

i. a reasonable probability exists that the DNA testing has the scientific potential to produce exculpatory or mitigating evidence relevant to a claim of wrongful conviction or sentencing; and
ii. the requested DNA test employs a method of testing generally accepted with in the relevant scientific community.

Crim. Proc. § 8–201(d)(1).

Here, the parties do not disagree that the requested DNA test employs a method that is generally accepted within the relevant scientific community. Hence, there is no dispute that Crim. Proc. § 8–201(d)(1)(ii) is satisfied. At issue in this case is whether a reasonable probability exists that the requested testing has the scientific potential to produce exculpatory evidence relevant to the Appellant's claim of wrongful conviction.

Again, in denying the petition, the post-conviction judge reasoned:

> Because of the highly speculative nature of what Petitioner hopes to gain from DNA testing of these items, this [c]ourt can see no possibility that a DNA test performed on the items requested would exonerate Petitioner. Even if Petitioner's DNA is not found on the items and other persons' DNA is found, that does not prove or even suggest that any of those other people committed this crime, even if one of them turns out to be a convicted sex offender.

(Emphasis added). The theory of the court's denial, that the proposed DNA analysis would not exonerate Appellant, leaves no room for the broader reach of the statute as indicated by the term "exculpatory." The State argues that the hearing judge applied the appropriate standard. On this point, the parties are in disagreement, specifically over the meaning and application of the term "exculpatory."

The term "exculpatory" as employed in the statute embraces a far broader scope of relief than the "zero sum" standard characterized by the term "exonerate" and advanced by the State and applied by the Post-conviction Court. The sought after testing of objects described by Ms. K., most notably the cigarette lighter, have the "scientific potential to produce exculpatory or mitigating evidence." Nothing in the plain language of the statute suggests that the testing results must "exonerate" a petitioner or "prove" that someone else committed the crime.

Although the General Assembly did not define the word "exculpatory" in the statute, we are able to discern the term "exculpatory" with its plain and ordinary meaning. See Ali v. CIT Tech. Fin. Servs., Inc ., 416 Md. 249, 262, 6 A.3d 890, 897–98 (2010) ("When the Court can ascertain the Legislature's intent from the plain meaning of the verbiage, the Court need not delve deeper. ... In seeking to apply the plain-meaning rule, it is proper to consult a dictionary or dictionaries for a term's ordinary and popular meaning."). Black's Law Dictionary defines exculpatory as "evidence tending to establish a criminal defendant's innocence." BLACK'S LAW DICTIONARY at 577 (7th ed. 1999). This is consistent with the policy of the statute—to facilitate claims of actual innocence.

...

The absence of Appellant's DNA has the potential to exculpate Appellant to the extent that it would tend to prove that he either did or did not use the lighter that Ms. K. testified was used by the man who assaulted her. Where criminal agency is an issue, such as in this case, evidence tending to prove or disprove that the accused's DNA is present on items that the perpetrator touched or may have come into contact with has a great potential to exculpate. Thus, the State's attempt to distinguish Gregg on the basis that the item to be tested in that case was the instrument of the crime, is not dispositive with respect to all cases of DNA testing. Although the absence of Appellant's DNA on the lighter would not conclusively prove that Appellant did not assault Ms. K., as the attacker may not have transferred any trace DNA to the lighter, absolute certainty is not the standard.

In assessing whether there is a reasonable probability that DNA testing may produce exculpatory or mitigating evidence, where the State has possession of an item that a perpetrator allegedly touched, a court may take into account factors such as the nature of the item (e.g., whether it is an instrumentality of the crime), the physical proximity between where the item was located and where the crime occurred, and the temporal proximity between when the perpetrator touched the item and when the crime occurred. Applying these factors to the instant case, we conclude that Appellant has established that DNA testing is warranted as to the cigarette lighter. Although the cigarette lighter is not an instrumentality of the crime, the physical proximity between where the lighter was located and where the crime occurred and the temporal proximity between when the perpetrator touched the lighter and when the crime occurred are factors that weigh heavily in favor of granting Appellant's request. The perpetrator gained access to Ms. K.'s vehicle by asking to borrow her lighter, and the crime occurred immediately after the perpetrator touched the lighter. In sum, as discussed above, the standard is simply whether there is a reasonable probability, or fair likelihood, that the testing has the potential to produce exculpatory evidence, which in turn, is evidence that tends to disprove or negate a petitioner's guilt.

CONCLUSION

The hearing judge incorrectly applied a more stringent standard that would require Appellant to show that the DNA testing he seeks would exonerate him. Given the extant record, we conclude that Appellant has established that DNA testing is warranted in light of the proper standard set forth above. Accordingly, we shall vacate the order of the hearing judge and remand the case to the Circuit Court to enter an order directing DNA testing on the cigarette lighter. Crim. Proc. § 8–201(d). See Gregg v. State, 409 Md. at 721, 976 A.2d at 1012 (remanding a post-conviction DNA case for circuit court to direct DNA testing). Cf. Simms v. State, 409 Md. at 733–34, 976 A.2d at 1019–20 (holding petition facially sufficient to warrant DNA testing; remanding for Circuit Court to direct State to respond to petition).

JUDGMENT OF THE CIRCUIT COURT FOR ST. MARY'S COUNTY VACATED. CASE REMANDED TO THAT COURT WITH DIRECTION TO ISSUE AN ORDER FOR DNA TESTING CONSISTENT WITH THIS OPINION. STATE TO PAY THE COSTS.

Questions:

1. How has the State of Maryland codified blood and DNA testing? Under what category of evidence is blood listed?
2. Part of the state's argument regarding blood being a form of evidence qualifying for DNA testing during post-conviction relief, aside from its biological nature, is its capacity to be:
3. How does the Court deal with the State's claim that blood DNA testing must lead to the defendant's exoneration.

Patterns of blood placement, whether blood spatter or smearing from falling victims are also crucial forensic issues. As in the first case, where the victim lacked any blood remnants of any sort, the absence of blood on a shooter raises a host of questions.

FIGURE 4.3 Floyd Yuselew killed his friend in this house on the Zuni Reservation in New Mexico during a confrontation after a night of drinking in 2009. The victim was sitting in a chair when he was struck in the forehead from behind with an ax. Investigators determined the murderer and the victim were friends who had been drinking well into the morning of the attack. See https://www.fbi.gov/news/stories/journey-through-indian-country-part-3 for details. FBI Media at https://multimedia.fbi.gov/item?type=image&id=5214.

People v. Tree
Court of Appeal of California, First Appellate District, Division Five
August 3, 2017, Opinion Filed
A143390

Rhett August was shot and injured one evening when he answered the doorbell at his Eureka home. A few nights later, Alan "Sunshine" Marcet and his girlfriend Christina Schwarz were shot and fatally wounded in their makeshift bed in the living room of a home where they were staying in Arcata. A jury determined that appellant Bodhi Tree was the shooter on both occasions and convicted him of two counts of second degree murder and one count of premeditated attempted murder, along with findings that as to each offense, he had personally discharged a firearm causing great bodily injury or death. (Pen. Code, §§ 187, subd. (a), 664, 12022.53.) The court imposed consecutive terms of 15 years to life for the second degree murder counts, a consecutive life term for the attempted premeditated murder count, and consecutive firearm enhancements of 25 years to life for each of those counts, for a total prison term of 105 years to life plus life.

Appellant contends the judgment must be reversed because: (1) the prosecution violated its duty to disclose exculpatory evidence under Brady v. Maryland (1963) 373 U.S. 83, 83 S. Ct. 1194, 10 L. Ed. 2d 215 (Brady) by withholding information that a criminalist called as a People's witness had received negative workplace performance reviews; (2) the testimony of three jailhouse informants describing incriminating statements by appellant should have been excluded under

Evidence Code section 352; and (3) the court should have excluded certain DNA evidence as a sanction for the prosecution's late disclosure of that evidence and its failure to preserve a DNA sample for testing by the defense. We affirm.

FACTS

A. Prosecution Evidence

On the evening of May 13, 2013, the appellant attended a party at Rhett August's house on J Street in Eureka. …

August was at home that night along with Upton and Boone. The doorbell rang at 10:00 p.m., but no one was at the door when August answered. August stepped outside, and someone crouching near a neighboring building fired a gun at him several times, hitting him once. According to Butler-Smith, appellant "popped out from around the corner and fired four times." Boone, who had a black Mercedes, drove August to the hospital, where August remained for 19 days receiving treatment for his injuries. The bullet that struck August had passed through his body and was found on the steps; another damaged bullet was discovered lodged in the siding of a building just south of August's home. The absence of shell casings at the scene suggested the use of a revolver.

After the shooting, appellant ran back to the Land Rover and said "go" or "drive." He seemed very excited and said, "I got that motherfucker. I dropped him for sure." …

At 2:00 a.m., residents of the Eye Street house awoke to gunshots and discovered that Schwarz and Marcet had been shot. Schwarz had been shot three times, including once in the head, and was dead by the time paramedics arrived. Marcet was shot four times and died at the hospital; some of his injuries were consistent with the bullet having first passed through Schwarz's body. Appellant was not in the house.

Police responded shortly after the shooting. They found blood on one wall near the victims located approximately 32 inches from the ground and blood on another wall located between 24 and 70 inches from the ground. A 32-ounce Coors beer bottle was sitting on the bookcase near the victims and a 40-ounce Mickey's Malt Liquor bottle was found on a breakfast bar in the kitchen/dining room area. DNA matching appellant's was found on the Mickey's bottle. The absence of expended shell casings indicated the shots had been fired from a revolver.

On the night of the shooting, Butler-Smith was staying at the home of his cousin Shane Goodrich in Arcata. Appellant, who was friends with Goodrich, arrived at the house at 2:30 a.m., said "I got two more," and placed a snub-nose .38 revolver on the table. According to Butler-Smith, appellant said he was at the party and "some dude and his girlfriend were talking shit to him," so he "drank his beer" and shot them before leaving.

…

Deborah Stonebarger, a criminalist with the California Department of Justice (DOJ), examined the clothing appellant was wearing when he was arrested. She did not see any blood on his sweatshirt, and blood found on his pants was determined to be his own. Given the physical dynamics of the shooting, with the victims lying on the floor, Stonebarger would not have expected there to be blood spatter on the shooter's clothing because blood from the victims would be projected away from the shooter by the force of the bullets. She reviewed photographs of the crime scene at the prosecutor's request and believed that "four drops of possible blood" on a wall near the victims was not blood back spatter and was more likely "castoff" blood from a person at the scene, such as an emergency medical responder, handling something saturated with blood. Stonebarger had not attempted to collect gunshot residue (GSR) from appellant's clothing because she was not asked to do so. Stonebarger understood such evidence to be of limited value: "Those particles can be on the presence of a hand of someone who did not fire a gun. Likewise, the particles can be absent on someone who has fired a gun. So in the end, you do a lot of work to say, 'Maybe he was there and maybe not.' 'Maybe he fired the gun and maybe not.'"

Dale Cloutier, also a criminalist with the DOJ, testified as the primary forensics expert for the prosecution and explained that back spatter of blood onto a shooter was not inevitable, meaning the lack of blood on a suspect's clothing would not exonerate the suspect. However, a shooter within a few feet of his victim was within range of back spatter if it is present. Cloutier also examined the bullets recovered from the May 15 shooting of Rhett August and the May 18 shooting of Christina Schwarz and Allen Marcet. He concluded it was possible the same gun was used on both occasions, but was unable to say so definitively. As to the failure to perform GSR testing in this case, Cloutier explained that such testing "is very limited and has very little use."

B. Defense Case

The defense strategy at trial was to discredit the prosecution's lay witnesses, particularly Sean Butler-Smith and the jailhouse informants, based on their criminal records, drug addiction, and/or their motive to lie.4 The defense also attempted to show that the absence of the murder victims' blood on appellant's clothing meant he was not the shooter in Arcata, and that Rhett August was shot by one Melvin Matthews, who had attempted to rob Steve Upton (August's guest on the night of the shooting) several days earlier and had been seen by a detective driving by August's house after the shooting.

Daniel Schumaker, an expert in crime scene reconstruction, analyzed the scene of the Arcata shooting and generated a 3-D model based on the relevant reports. He opined that the shooter had been standing at the victims' feet, and that blood droplets found on the wall behind the shooter were consistent with back spatter from Christina Schwarz moving her head following an injury from a previous shot.

Jacobus Swanepoel, a senior forensic scientist at Forensic Analytical Sciences and an expert in crime scene reconstruction, testified that back spatter may occur when the blood moves in the opposite direction of a bullet that perforates the head. The blood will move in the shape of a cone backwards from the point where the bullet contacts the blood, though back spatter does not always occur. Blood found on the wall of the Eye Street house near the victims was in a pattern that might have been a full cone, or might have been a partial cone where an intervening object, such as the shooter, blocked the wall from the rest of the cone. In Swanepoel's opinion, based on the trajectory of the victims' wounds, the shooter had to be standing near their feet in the three to four-and-a-half foot area in front of the wall where the blood droplets were found. Given the crime scene reconstruction evidence in this case, he would expect to find blood on the shooter from back spatter. Swanepoel testified that spatter from a gunshot wound will travel upward against gravity, including onto the shooter, though gravity may be a factor in how the blood moves.

...

Questions:

1. How does the Court reconcile the feasibility of a shooter/defendant lacking any blood evidence?
2. Depending upon the angle of the shooter to a victim, accompanied by specified victim movement, can a "cone" of blood spatter or other pattern develop? Does the Court accept this conclusion?
3. What turns a full cone into a partial cone of blood pattern?

> Read the press release regarding blood splatter evidence in dog fighting case at https://www.justice.gov/opa/pr/fifth-defendant-multi-state-dog-fighting-prosecution-sentenced-24-months-prison.

Blood spatter analysis is a complex and challenging form of forensic science. While serological determinations yield consistent and dependable findings, the interpretative side of blood spatter has its critics on two fronts: a lack of universal standards and a lack of qualifications that verify expertise.

State v. Langston
Court of Criminal Appeals of Tennessee, At Jackson
January 5, 2017, Session; May 12, 2017, Filed
No. W2015-02359-CCA-R3-CD

...

Special Agent Cervinia Braswell of the Tennessee Bureau of Investigation, who was accepted as an expert in the field of firearms identification, analyzed the gun involved in this case. She confirmed that the cartridge case and the bullet found at the scene had been fired by this gun. After conducting several test fires of the gun, Agent Braswell concluded that it was not malfunctioning. She noted that the handgun's floor plate, which held the magazine in place, was cracked so that it would not remain on the gun and that because the floor plate was plastic, it could be broken if the gun was thrown. She further observed that the gun's two safety mechanisms, a manual safety on the side of the gun and a quarter-cock function that kept the hammer from resting on the firing pin, were both functioning and would have prevented the weapon from being fired unintentionally. Agent Braswell stated that the gun would not fire if it was dropped or if something hit the back of the hammer and would only fire if someone pulled the trigger. She also observed that the handgun did not have a hair-trigger because the trigger pull was in the medium range at 6.5 pounds of pressure. She said that if the gun had been wrapped in a shirt when it was fired, then the shirt could have impeded the rearward travel of the gun's slide, which could have resulted in a cartridge casing not ejecting properly. Finally, Agent Braswell said that the gun's trigger guard prevented an individual from accidentally pulling the trigger.

Dr. Erica Curry, a medical examiner, was accepted as an expert in the field of forensic pathology. She stated that she had reviewed the victim's autopsy report generated in her office and noted that the victim sustained a fatal gunshot wound to the head, with the bullet entering at the victim's left eyebrow and transecting the brain stem and cerebellum before exiting the back of the skull. Dr. Curry said the areas of the brain damaged by the bullet controlled respiration, heartbeat, motor function, and consciousness, which made the victim's injuries fatal. She also noted that the bullet's trajectory was slightly downward from the front to the back because the entrance wound was approximately one inch higher than the exit wound. Dr. Curry stated that the gunpowder stippling near the entrance wound indicated that the gun was fired at the victim from an intermediate range of between six inches and three feet.

Lieutenant Anthony Mullins, who was accepted as an expert in the area of blood spatter analysis, stated that he reviewed the crime scene photographs, the crime scene sketches, Langston's statement, the photographs of Langston's person, the victim's autopsy report, and the officers' reports and supplements before writing a detailed report regarding his opinion in this case. He opined that it was "more likely that [the victim] was sitting down than she was standing at her full height" at the time she was shot. While acknowledging that the victim "could be kneeling" or "could be standing and stooped over," Lieutenant Mullins said he did not believe either of these options were likely given the final position of the victim's body. Instead, he thought "it was more likely . . . that she's lower to the ground and probably seated on the floor." Lieutenant Mullins said that if the victim had been standing up, he would have expected the blood spatter from the exit wound to have hit the wall around the victim's height of five feet and four inches and below.

Instead, the blood spatter on the wall was at the height of four feet and below. Moreover, if the victim had been standing upright, he would have expected to see blood spatter from the entrance wound on the top of her feet, but there was no blood present there. He also concluded that the blood on Langston's shirt indicated that he was fairly close to the victim when he fired the shot. He added that if the victim had been standing when shot, he would have expected the blood to be higher on Langston's shirt.

Although Lieutenant Mullins opined that the victim was most likely seated when she was shot, he could not rule out the possibility that she was standing or standing while stooped over at the time of the shooting, which was consistent with her lunging to stab someone. He acknowledged it was difficult for him to form an opinion about the victim's position at the time of the shooting because he was given limited photographs and sketches of the scene. He acknowledged that he was not present at the crime scene and did not have control over what photographs were taken at the scene, which meant that inconsistent blood stains could have existed that were not depicted in the available photographs. When asked about whether the trajectory rod used at the scene affected his opinion, Lieutenant Mullins stated that he could not "say with any scientific certainty that [the victim] wasn't standing or kneeling or, you know, stooped" but asserted that he did not "think that the trajectory [rod] would completely exclude [the victim being] seated in some position there." Lieutenant Mullins observed that emergency medical personnel likely moved the victim's body at least slightly and that this movement could have affected the accuracy of his opinion. When asked by the defense if it would be fair to say that his opinions, at best, were highly speculative, Lieutenant Mullins answered:

> What I would say is this. Not being on the scene, not having the photographs that I would like to have, not having the measurements that I would like to have, not having seen it with my own eyes, I have to speculate about some things for example the knife and the cell phone based on where they are in the photographs. Some things I can tell you definitively and some things I can't say and exclude all other possibilities. I wouldn't say it's speculation but it's the best that I can do with the information that I'm given.

Ross Gardner, who was accepted an expert in the field of crime scene reconstruction and blood stain pattern analysis, testified for the defense. Gardner opined that the victim had been in a generally upright position at the time she was shot. He explained that he owned a forensic education and consulting group, had a military and law enforcement background, had written many articles on crime scene investigation, had completed over 3000 hours in formal training in criminal investigation, including 1000 hours in crime scene investigation, and was a member of several international professional associations for blood pattern analysis, identification, and crime scene reconstruction. He said that he had been accepted as an expert for the prosecution and the defense in the fields of crime scene investigation and blood pattern analysis throughout the United States. He also said his report in this case had been peer reviewed.

Gardner said he reviewed the crime scene photographs, the autopsy report and photographs, the crime scene sketch, the report from Lieutenant Mullins, and the police reports before forming an opinion as to the victim's position at the time of the shooting. He asserted that the ballistics evidence, when he combined with his reconstruction software, established that the victim was generally in an upright position at the time she was shot. Given the location of the bullet hole in the wall, the estimated distance between the victim and the wall, and the trajectory for the bullet, Gardner opined to a reasonable degree of scientific certainty that the victim was not seated at the moment she was shot. While he believed that the victim's "wound correlate[d] to the trajectory if she's in an upright position," Gardner acknowledged that the victim could have been slightly crouched at the time she was shot. He also opined that the knife was "dynamically involved in the incident" and had not been staged after the fact.

Gardner recognized some difficulties regarding the nature of the evidence. He acknowledged that because there were not many crime scene measurements, he was required to estimate a substantial amount of data, including the measurements of a nearby crib mattress from which he deduced other measurements. He also acknowledged that he used the word "approximation" thirteen times in his report and that he did not have all the photographs and measurements he would have liked when making his three-dimensional reconstruction of the crime scene.

. . .

Following the conclusion of the proof, the jury convicted Langston of second degree murder, and the trial court imposed a mid-range twenty-year sentence. A judgment of conviction was entered, and Langston filed a motion for new trial on November 24, 2015, which was denied. Langston then filed a timely notice of appeal.

. . .

I. Acceptance of Officer as Expert. Langston also contends that the trial court abused its discretion in allowing Lieutenant Anthony Mullins to testify as an expert in the area of blood spatter analysis and to opine that the victim was seated at the moment she was shot. Langston claims Lieutenant Mullins had insufficient formal education in blood spatter analysis, was unfamiliar with the current terminology and methodology in this field, and reached conclusions based on incomplete data. He adds that Lieutenant Mullins's opinions were not peer-reviewed, that he was unfamiliar with the potential rate of error for the methodology he employed, and that he failed to provide his opinion to a reasonable degree of scientific certainty. Finally, Langston asserts that because Lieutenant Mullins was not qualified to offer an expert opinion, his testimony that the victim was seated at the time she was shot, which was uncorroborated and negated by his own expert, was "highly speculative," "extremely prejudicial," and "inflammatory."

During voir dire, Lieutenant Anthony Mullins stated that he had been an officer with the Memphis Police Department for twenty seven years and had been employed in the homicide bureau for nearly ten years, where he had worked on 500 to 800 homicide cases. He said he attended a basic crime scene investigation course lasting fifty-six hours and a basic blood stain pattern analysis course in 2003 lasting forty hours that was taught by expert Paulette Sutton. Lieutenant Mullins said he had worked some crime scenes with Paulette Sutton and had gained experience in crime scene investigation and blood pattern analysis from the many cases he worked as a homicide detective. He also said he had testified as an expert in the areas of blood pattern analysis and crime scene investigation in six or seven of the criminal courts in Shelby County.

Lieutenant Mullins acknowledged that he had not taken any advanced classes in the areas of blood pattern analysis or the effect of blood stains on fabrics and that he had not completed any math or physics courses on blood stain analysis. He also acknowledged that he had not undergone a formal mentorship program in the area of blood pattern analysis and had not completed a shooting reconstruction course. When asked by the defense if he considered himself a forensic analyst, Lieutenant Mullins responded:

> What I consider myself is a Homicide detective, okay. I don't consider myself the best person I know for blood stain pattern analysis or crime scene investigation. My expert tag, if you will, my expert moniker is only in the courtroom. My view of it is because I've had training more than the average officer, more than the average Homicide investigator and I've used it more than the average [officer or investigator]. But like I said, I'm not the best that I know and I'm definitely not the best out there.

Lieutenant Mullins admitted that he had not obtained blood pattern analysis certifications, did not belong to any international associations in that field, and did not read current journals or magazines on blood pattern analysis. He said he had not subjected his opinion in this case to peer review because he did not have the opportunity to do so and that he had not done an investigation into the potential rate of error for his methodology. Lieutenant Mullins stated that he had used the skills he learned in blood pattern analysis when asked to look at sketches and photographs of crimes scenes he did not personally investigate and in his current job of supervising the investigation of officer-involved shootings. When asked if his report was flawed because he used principles of blood pattern analysis that were no longer valid because of the passage of time, Lieutenant Mullins replied: "Like I said, I don't consider myself the best that I even know. As far as the basic principles and some of the terms that I use, some of those terms may be no longer used, but some of those principles are still the same, despite the term that might be used."

During a bench conference following voir dire, defense counsel argued, citing McDaniel v. CSX Transp., Inc., 955 S.W.2d 257 (Tenn. 1997), that Lieutenant Mullins should not be qualified as an expert in the field of blood pattern analysis because he was unfamiliar with the current terminology and methodology in this field, because his opinion had not been peer-reviewed, and because he had received only "limited training" in this "highly technical specialty." Despite the defense's objection, the trial court ruled that "based on [Lieutenant Mullins's] experience, training, and education that he is going to be qualified as an expert and allowed to give his opinion." The court reminded defense counsel that he would be allowed to cross-examine Lieutenant Mullins. When the bench conference concluded, the trial court accepted Lieutenant Mullins as an expert in the field of blood spatter analysis and instructed the jury as follows:

> And I'll caution you and I'll tell you again as I've told you for every expert that testifies. Merely because an expert witness has expressed an opinion does not mean that you are bound to accept the opinion. The same as with any other witness, it's up to you to decide whether you believe the testimony and choose to rely on it.
>
> Part of that decision will depend on your judgment about whether the witness' background or training or experience is sufficient to give the expert opinion that you heard.
>
> So again, it will be as always, you all will be the triers and the finders of facts and so I give you the guidance but it will be up to you.

Determinations regarding the qualifications, admissibility, relevance, and competence of expert testimony fall within the broad discretion of the trial court and will be overturned only for an abuse of that discretion. State v. Davidson, 509 S.W.3d 156, 208 (Tenn. 2016) (citing McDaniel, 955 S.W.2d at 263-64; State v. Scott, 275 S.W.3d 395, 404 (Tenn. 2009)). "A trial court abuses its discretion when it applies incorrect legal standards, reaches an illogical conclusion, bases its decision on a clearly erroneous assessment of the evidence, or employs reasoning that causes an injustice to the complaining party." Scott, 275 S.W. 3d at 404-05 (citing Konvalinka v. Chattanooga Hamilton Cnty. Hosp. Auth., 249 S.W.3d 346, 358 (Tenn. 2008)).

Questions:

1. Part of defense appeal and petition for post-conviction relief relates to the lack of expert status in the subject matter. In addition, did not meet the general standards or criteria for an expert either under Frye or Daubert.
2. Even though traditional qualifications as to education, professional associations, peer review and the like were missing, does the expert witness have another avenue to meet the criteria for a qualified expert?
3. What position was victim in based on blood spatter analysis? Is this conclusion defensible or might here be another explanation.

4.2 SEMINAL FLUID AND SPERMATOZOA

Sexual offenses frequently deal with the collection of seminal fluid to demonstrate two issues: first, penetration or other violation sexually; and second, DNA identity testing that matches the offender to the offense. Standard rape kits used by investigators in the field provided for this collection or for later review by medical personnel.

Bembo v. Fla. Dep't of Corr.
United States Court of Appeals for the Eleventh Circuit
March 30, 2017, Decided
No. 16-16571-C

Kevin Bembo is a Florida prisoner serving a total life sentence after a jury convicted him of one count of burglary of a structure with assault or battery and three counts of sexual battery while wearing a mask. He seeks a certificate of appealability ("COA") and leave to proceed in forma pauperis ("IFP"), in order to appeal the dismissal of his pro se 28 U.S.C. § 2254 federal habeas corpus petition as untimely.

Where the district court denied a habeas petition on procedural grounds, the petitioner must show that jurists of reason would find debatable (1) whether the petition states a valid claim of the denial of a constitutional right, and (2) whether the district court was correct in its procedural ruling. Slack v. McDaniel, 529 U.S. 473, 484, 120 S. Ct. 1595, 146 L. Ed. 2d 542 (2000). If the petitioner fails to satisfy either prong of this two- part test, this Court should deny a COA. Id. This Court "review[s] de novo a district court's denial of a habeas petition as untimely." Chavers v. Sec'y, Fla. Dep't of Corr., 468 F.3d 1273, 1274 (11th Cir. 2006).

The AEDPA imposes a one-year statute of limitations for filing a § 2254 federal habeas petition that begins to run from the latest of four possible events:

A. the date on which the judgment became final by the conclusion of direct review or the expiration of the time for seeking such review;
B. the date on which the impediment to filing an application created by State action in violation of the Constitution or laws of the United States is removed, if the applicant was prevented from filing by such State action;
C. the date on which the constitutional right asserted was initially recognized by the Supreme Court, if the right has been newly recognized by the Supreme Court and made retroactively applicable to cases on collateral review; or
D. the date on which the factual predicate of the claim or claims presented could have been discovered through the exercise of due diligence.

28 U.S.C. § 2244(d)(1).

A state prisoner's conviction generally becomes final when the Supreme Court denies certiorari or issues a decision on the merits, or when the 90-day period in which to file a certiorari petition expires. Nix v. Sec'y for Dep't of Corr., 393 F.3d 1235, 1236–37 (11th Cir. 2004). The 90-day period for seeking certiorari review of a conviction runs from the date of entry of the judgment sought to be reviewed, not from the issuance of the mandate. Chavers, 468 F.3d at 1275.

Additionally, the one-year federal limitations period is statutorily tolled during times in which a properly filed application for state post-conviction relief is pending. 28 U.S.C. § 2244(d)(2). Further, the AEDPA limitations period may be equitably tolled, but the petitioner must show "(1) that he has been pursuing his rights diligently, and (2) that some extraordinary circumstance stood in his way and prevented timely filing." Holland v. Florida, 560 U.S. 631, 649, 130 S. Ct. 2549, 177 L. Ed. 2d 130 (2010) (quotation omitted). Equitable tolling is "an extraordinary remedy [that] is typically applied sparingly." Arthur v. Allen, 452 F.3d 1234, 1252 (11th Cir. 2006) (quotation omitted).

Moreover, under the miscarriage-of-justice exception, "actual innocence, if proved, serves as a gateway through which a [§ 2254] petitioner may pass whether the impediment is a procedural bar … or … expiration of the statute of limitations." McQuiggin v. Perkins, 569 U.S. 383, 133 S. Ct. 1924, 1928, 185 L. Ed. 2d 1019 (2013). To make a credible showing of actual innocence, a petitioner must present new reliable evidence that was not presented at trial, and "must show that it is more likely than not that no reasonable juror would have convicted him in the light of the new evidence." Id. at 1935. For purposes of this miscarriage-of-justice exception, "actual innocence" means factual innocence, not mere legal insufficiency. McKay v. United States, 657 F.3d 1190, 1197 (11th Cir. 2011).

Reasonable jurists would not debate the district court's determination that Bembo's § 2254 petition was untimely. Specifically, the record reveals that the state appellate court affirmed Bembo's convictions and sentences on December 7, 2012. His judgment became final 90 days later, on March 7, 2013, when the time for filing a petition for writ of certiorari expired. Contrary to Bembo's claim, the district court did not err in calculating the 90-day window for seeking certiorari from the date the state appellate court affirmed his convictions and sentences, instead of the date on which the mandate issued or the date on which the time for filing a motion for rehearing expired. See Chavers, 468 F.3d at 1275 (providing that the 90-day period for seeking certiorari review of a conviction runs from the date of entry of the judgment sought to be reviewed, not from the issuance of the mandate). Accordingly, as the district court concluded, the AEDPA limitations period began to run on March 8, 2013, and, absent any properly filed tolling motions, Bembo had until March 8, 2014, to file a timely § 2254 petition.

Although Bembo filed a motion for post-conviction DNA testing on July 25, 2013, pursuant to Fla. R. Crim. P. 3.853, it did not toll the federal limitations period. See Brown v. Sec'y Dep't of Corr., 530 F.3d 1335, 1337-38 (11th Cir. 2008) (holding that a Rule 3.853 motion for DNA testing is not "an application for post-conviction or other collateral review with respect to the pertinent judgment, for purposes of the AEDPA tolling statute" (quotation omitted)).

Thus, the federal limitations period ran untolled from March 8, 2013 until October 29, 2013, when Bembo filed a motion for post-conviction relief, pursuant to Fla. R. Crim. P. 3.850. Bembo's Rule 3.850 motion tolled the federal limitations period until the state appellate court issued its mandate affirming the trial court's decision on December 9, 2014. Nyland v. Moore, 216 F.3d 1264, 1267 (11th Cir. 2000) (explaining that, where a petitioner appeals the denial of his state post-conviction motion, his post-conviction motion remains pending until issuance of mandate by the state appellate court). The federal limitations period resumed on the following day, December 10, 2014.

While Bembo filed a second Rule 3.850 motion on January 23, 2015, it did not serve to toll the federal limitations period because it was dismissed as untimely. See Gorby v. McNeil, 530 F.3d 1363, 1366 (11th Cir. 2008) ("An application that is untimely under state law is not 'properly filed' for purposes of tolling AEDPA's limitations period."). Accordingly, the federal limitations period expired on April 20, 2015, and, therefore, reasonable jurists would not debate the district court's determination that Bembo's § 2254 petition, which was filed on April 27, 2015, was untimely.

Further, reasonable jurists would not debate the district court's determination that Bembo was not entitled to equitable tolling. Bembo did not allege any extraordinary circumstances that allegedly prevented him from filing a timely § 2254 petition. Rather, he argued that his petition should be considered on the merits in order to prevent a manifest miscarriage of justice because he is actually innocent. Specifically, Bembo argued that, had the jury heard a variety of "new evidence," no juror would have found him guilty beyond a reasonable doubt, namely, evidence that (1) the medical staff did not find any trace of spermatozoa on the vaginal and cervical swabs collected from the victim; (2) the sexual assault kit was tampered with based on a nine-day delay in the Florida Department of Law Enforcement ("FDLE") receiving it; (3) pubic hair collected from the victim and hair and toilet tissue collected at the crime scene were not submitted for DNA analysis; (4) the victim and others provided local authorities with information about potential suspects other than Bembo; (5) there was police misconduct because the police notified the victim that sperm had been discovered on the swabs collected as part of the sexual assault kit before the FDLE had completed the DNA testing; (6) the DNA profile obtained from the sexual-assault kit did not match any of the DNA profiles in the Combined DNA Index System ("CODIS"), but Bembo's DNA had been on file with law enforcement since 1991; (7) the buccal swabs collected from Bembo were red in color, but the buccal swabs submitted at trial were not red in color; and (8) Bembo's girlfriend would have testified that they were together at the time the crime occurred.

However, none of this "new evidence" establishes that Bembo is actually, factually innocent and that it is "more likely than not that no reasonable juror would have convicted Bembo in the light of the new evidence." McQuiggin, 133 S. Ct. at 1935; McKay, 657 F.3d at 1197.

...

For these reasons, reasonable jurists would not debate the district court's determination that Bembo failed to satisfy the miscarriage-of-justice exception.

Accordingly, Bembo's motion for a COA is DENIED. His IFP motion is DENIED AS MOOT.

Questions:

1. The Bembo decision is a mix of process and substance where I appears that process trumps the arguments regarding the lack of detection of sperm on the collection swabs. What other procedural problems caused by the defendant hurt his overall case.
2. What else does the defendant allege were not processed- and this be true, how powerful are these allegations in light of a sexual offense prosecution?
3. Why is the concept of "new" evidence so critical to the defendant's argument?

Cowels v. Fed. Bureau of Investigation
936 F.3d 62 (1st Cir. 2019)

Appellants Michael Cowels and Michael Mims were convicted of murder in state court and spent twenty years behind bars serving life sentences. After new testing of trial evidence cast doubt on the verdict, they were granted a new trial. Subsequent DNA testing of a swab taken from the inside of a condom recovered in the vicinity of the victim during the initial investigation revealed an unknown male DNA profile. Cowels and Mims obtained a state court order requiring Massachusetts to upload the DNA profile into a state database of DNA records for comparison purposes. No matches were found. The FBI, however, refused to upload the profile into the national DNA database after determining that it was ineligible for upload. Cowels and Mims went to federal court to compel the FBI to upload the profile, but the district court dismissed their suit based on its conclusion that the FBI's eligibility determination is unreviewable.

Without suggesting that the district court erred in its analysis, we assume that the FBI's eligibility determination is reviewable. Having done so, we conclude that the determination was not arbitrary and capricious. We therefore affirm the dismissal of appellants' suit.

I.

A. Legal Background

The DNA Identification Act of 1994 authorized the FBI Director to establish a DNA index, including DNA identification records of persons charged or convicted of crimes and "analyses of

DNA samples recovered from crime scenes," 34 U.S.C. § 12592(a)(1)-(2), "to facilitate law enforcement exchange of DNA identification information," id. § 12592 (title). ... The FBI's NDIS Operational Procedures Manual ("the Manual") provides additional guidelines for determining whether DNA records are eligible for inclusion in the NDIS. Of relevance to this appeal, pursuant to the Manual, a DNA record that "originate[s] from and/or [is] associated with a crime scene" is eligible for upload if it is "believed to be attributable to the putative perpetrator."

B. Factual Background

In 1994, a Massachusetts jury convicted Cowels and Mims of murdering Belinda Miscioscia, who was found brutally stabbed to death behind a woodworking shop in a yard known as a location for sexual trysts. ... At trial, a state forensic scientist also testified about collecting "an older, wrinkled condom ... covered with dirt and debris as well as sawdust" from the vicinity of the body. The forensic scientist testified that she tested the condom for hair and fibers and swabbed the inside of the condom, confirming the presence of seminal fluid residue. The condom was not tested for DNA.

Twenty years into serving their life sentences, Cowels and Mims were granted a new trial by the Massachusetts Supreme Judicial Court based on new DNA testing of the previously tested towel. See Commonwealth v. Cowels, 470 Mass. 607, 24 N.E.3d 1034, 1037 (2015). The new testing confirmed that the blood did not come from either man or from the victim. In preparation for a new trial, other items collected during the initial investigation were also DNA-tested. Testing by a state forensic scientist of the swab taken from inside the condom indicated sperm and non-sperm male DNA from more than one contributor. Only one of the DNA profiles was suitable for comparison but it did not match either Cowels or Mims. However, the forensic scientist concluded that this DNA profile was ineligible for upload to CODIS.

Cowels and Mims filed a motion in Massachusetts Superior Court to compel the Commonwealth to submit the condom DNA profile to the SDIS and to share the results. They contend that uploading the profile may lead to apprehension of the true killer, who they speculate could be any one of a number of violent and jealous men the victim was involved with in the months before her death. The Commonwealth opposed the motion, arguing that the DNA profile did not qualify for submission to CODIS pursuant to FBI standards. Recognizing that uploading the profile "risks implicating a person entirely innocent of this murder, who merely happened to be having sex in the same area, unrelated to th[e] victim or to the time of her death," the Superior Court nonetheless ordered the Commonwealth to submit the DNA profile to the SDIS. Sup. Ct. Order Dec. 4, 2017, at 2-3. The Commonwealth entered the DNA profile into the state database, but there was no match.

Dorothea Collins, the CODIS administrator for the Massachusetts State Police Crime Laboratory, emailed the FBI, informing the agency about the court order and reiterating the Commonwealth's view that the DNA profile is ineligible for upload to either the SDIS or the NDIS, but requesting that the FBI review for itself whether the DNA profile is eligible for upload to the national database. Collins attached to her email Cowels's and Mims's motion in the Superior Court, the court order, and her affidavit in the Superior Court matter, in which she stated (1) that the victim's body was found "clothed, outside on a platform," and that the condom "was found on the ground between the platform and a tank, covered in sawdust, dirt and vegetation"; and (2) that the victim's DNA was not part of the mixture taken from the inside of the condom. Collins thus concluded in her affidavit that, "[a]lthough the condom was collected from the crime scene during the course of the investigation, its connection to Ms. Miscioscia is not established to support a CODIS upload." Paula Wulff, Unit Chief of the FBI Office of the General Counsel's Forensic Science Law Unit, responded to Collins's email that the FBI had reviewed the attached materials and determined that the DNA profile was not eligible for upload to the NDIS because the condom was not sufficiently linked to the victim.

An assistant district attorney followed up with a request that the FBI consider performing a manual keyboard search -- which is a method of comparing a DNA record to other records in

CODIS without uploading the record -- even if they would not upload the profile to the NDIS. The assistant district attorney explained, "[w]hile we have shared the FBI's view of the relevance of this evidence in our murder case, … I am respectfully requesting that this search be done as a courtesy to me and my office, out of respect for the [c]ourt's prior order, and my desire to avoid the eventual trial judge, and possibly even the jury, misunderstanding the failure to upload the DNA profile in question to NDIS as 'the government being obstructionist.'" Wulff also rebuffed this request and shared a more in-depth written response explaining the FBI's determination that the DNA profile was not eligible for upload to the NDIS or for a manual keyboard search.

The response explained that the FBI had reviewed case materials, including the Superior Court order, Collins's affidavit, and case notes from the 1993 murder investigation. The response stated that the case notes described the condom as being discovered "under sawdust and debris -- covered with sawdust, dirt, dried vegetation etc. breaking apart." (Emphasis by the FBI.) The response also stated, "[t]here is no indication in the material provided that the condom was forensically connected to the victim." Citing the NDIS Operational Procedures Manual, the response concluded, "[f]rom the information that has been provided to the FBI, nothing forensically demonstrates a link between the victim and the subject condom to consider the obtained profile as coming from a putative perpetrator."

…

Cowels and Mims contend that the FBI's eligibility determination was arbitrary and capricious because the reasons the agency provided to support its determination that the DNA profile was not "attributable to the putative perpetrator" -- the standard from the Manual -- do not withstand scrutiny. We disagree. In the explanation of its eligibility determination, the FBI focused on (1) the condition of the condom, and (2) the lack of any forensic connection between the condom and the victim. Both of these reasons support the FBI's eligibility determination. The condition of the condom when found -- "covered with sawdust, dirt, dried vegetation etc. breaking apart" -- supports a conclusion that the condom was not temporally related to the murder. And the lack of DNA evidence tying the condom to the victim supports a conclusion that, given the condition of the condom, uploading the DNA profile could implicate an innocent person rather than reveal the perpetrator.

… But the record before us does not support the contention that the Commonwealth necessarily viewed the condom as connected to the putative perpetrator. As noted, the forensic scientist who collected the condom testified at trial that it was "an older, wrinkled condom … covered with dirt and debris as well as sawdust" (emphasis added), and it does not appear that the condom featured in the prosecution's case at trial, see Cowels, 24 N.E.3d at 1038–1040. Similarly, we cannot glean from the mere fact that the Commonwealth DNA-tested multiple items, including the condom swab, after the men were granted a new trial, that the Commonwealth saw the condom as related to the perpetrator until it was determined to be potentially exculpatory. Most importantly, we do not see how any shift in the Commonwealth's position undermines the FBI's ability to make its own independent judgment, on the basis of undisputed facts, about the condition of the condom and its temporal connection to the murder.

Regarding the lack of a forensic connection between the condom and the victim, appellants contend that the FBI "ignore[d] the fact that the Swab DNA Profile was not taken from the Condom but from the Swabs of the inside of the Condom, and that the Condom itself" -- that is, the outside of the condom, where the victim's DNA would more likely be found -- "was never tested for DNA." They therefore suggest that the FBI's eligibility determination was based in part on "the arbitrary and capricious view that the inside of the Condom could somehow include [the victim]'s DNA." Some of the language in the FBI's written explanation of its eligibility determination does suggest that the agency was not fully attuned to the distinction between the DNA testing of the swab from the inside of the condom and testing of the outside of the condom, which was never performed. See, e.g., FBI Response to NDIS Upload Request, at 1 (noting the absence of the victim's DNA "as would be expected had the condom come in contact with the victim").

However, the FBI's entire explanation makes reasonably clear that the agency's focus was on the absence of any DNA connection in the record between the condom and the victim -- a lack of connection that appellants do not contest -- rather than on a misguided understanding that testing had definitively established that the condom had never come into contact with the victim. In other words, the FBI correctly noted that the record before it did not establish any DNA connection between the condom and the victim. And we readily conclude that this absence of evidence was a rational reason, along with the condom's condition, for the FBI's determination that the swab DNA profile could not be "attribut[ed] to the putative perpetrator," and therefore was ineligible for upload to the NDIS.

III.

For the foregoing reasons, we affirm the dismissal of appellants' claims.

Questions:

1. During an investigation for rape and murder, the forensic units found a condom lying in dirt and sawdust in the vicinity of the victim. What evidence could be discoverable in a routine swab of the condom?
2. Why did the released defendants seek to have the DNA results uploaded into the FBI's national database? Under what Act is this a legitimate request?
3. Why did the FBI reject the request to upload the DNA semen analysis into CODIS? And why did the Court find the rejection reasonable and not arbitrary?

Part of the challenge in many sexual offenses, is the existences of multiple contributors when identifying seminal fluid cases. Conducting a rape kit protocol will in select cases pick up the DNA of multiple parties. See how Finch tries to reconcile this dilemma.

In re Finch
United States Court of Appeals for the Eleventh Circuit October 9, 2020, Filed
No. 20-13591-A

Pursuant to 28 U.S.C. § 2244(b)(3)(A), Joseph Finch has filed an application seeking an order authorizing the district court to consider a second or successive petition for a writ of habeas corpus. Such authorization may be granted only if:

(A) the applicant shows that the claim relies on a new rule of constitutional law, made retroactive to cases on collateral review by the Supreme Court, that was previously unavailable; or

(B)(i) the factual predicate for the claim could not have been discovered previously through the exercise of due diligence; and

(ii) the facts underlying the claim, if proven and viewed in light of the evidence as a whole, would be sufficient to establish by clear and convincing evidence that, but for constitutional error, no reasonable factfinder would have found the applicant guilty of the underlying offense.

28 U.S.C. § 2244(b)(2). "The court of appeals may authorize the filing of a second or successive application only if it determines that the application makes a prima facie showing that the application satisfies the requirements of this subsection." Id. § 2244(b)(3)(C); see also Jordan v. Sec'y, Dep't of Corr., 485 F.3d 1351, 1357-58 (11th Cir. 2007) (explaining that this Court's

determination that an applicant has made a prima facie showing that the statutory criteria have been met is simply a threshold determination).

Background Facts

Finch is a Florida prisoner serving two consecutive life sentences for armed sexual battery on a victim 12 years of age or older and kidnapping. …

The crime was committed in May 2000 and Finch was charged in May 2005. The trial took place in April 2006.

Finch's Application

In his application, Finch indicates he wishes to raise four claims in a successive § 2254 petition. First, he claims that he received ineffective assistance of counsel because his counsel failed to have a private forensic analyst check the DNA evidence obtained from the victim. He argues that this claim rests upon newly discovered evidence. He explains that at trial, the State's experts testified that: (1) the test of the victim's vaginal swab showed the presence of three people's DNA—Finch's, the victim's, and an unknown person's; 2) Finch was the major contributor and the unknown person was the minor contributor; and (3) their opinion was that the minor contributor's DNA was transferred to the victim by Finch from a previous sexual partner of Finch's (since the victim had stated she was raped by one person and had not had sex before that for about a month). He states that in January 2017, he received a letter from the state attorney's office informing him that the BSO's experts had used improper methods for analyzing mixed-contributor DNA samples, and as a result he submitted a public records request about the State experts‘ DNA analysis. In the State's reply, he obtained a November 2013 Florida Department of Law Enforcement DNA match report that showed that the DNA of a man named Charles Windsor, which was collected when he went to prison on unrelated charges, matched DNA found in the vaginal swab of the victim in Finch's case. Finch states that for years he had persistently filed public records requests with the BSO and the State, but had never received this 2013 report that identified Windsor as the unknown contributor until 2017, so he could not have discovered it earlier through the exercise of due diligence. He further states that Windsor matches the victim's physical description of the rapist (which she gave in pretrial depositions, though she testified at trial she could not identify him because he was masked) better than Finch, and that the 2013 DNA match report matched Windsor's DNA with a 12 STR loci match, whereas the reports presented at the time of Finch's trial showed that Finch matched the sample with a 9 STR loci match, and the unknown sample matched with a 3 STR loci match.

As to this claim, he argues that counsel should have hired a private forensic analyst to test the DNA, which would have shown that the unknown sample was from a male. He asserts that his counsel could have argued a theory that the unknown person was the rapist who carried Finch's DNA through a shared sexual partner. Finch contends that his counsel could have argued that the unknown person was the major contributor, and that Windsor matched the victim's description better than Finch did, which the newly discovered evidence showed to be true. Finch also argues that his counsel could have impeached the testimony of Baird, Noppinger, and Tracey with the testimony of a private analyst. Finch asserts that, but for his counsel's failure to analyze the DNA evidence, there is a reasonable probability that the trial would have resulted in an acquittal, and the DNA test result is evidence that would prevent any reasonable factfinder from finding Finch guilty beyond a reasonable doubt.

Second, he claims that Florida failed to disclose exculpatory evidence in the form of a 2005 BSO DNA test report. He explains that this report, which the State generated before his trial but did not disclose to him until 2017, "analyzed the DNA profile of the unknown … contributor" from the vaginal swab. He states that at his trial, Noppinger testified that the unknown was a minor contributor, the unknown person's DNA was a 6 STR loci match, and Finch's DNA was a 9 STR loci match; whereas Tracey inconsistently testified that the unknown person's DNA was a 3 STR loci match. He further states that the undisclosed 2005 report "reflects the fact that the unknown [person's DNA] was a striking and significant 15 STR loci match." He argues that this was newly discovered evidence not previously discoverable through due diligence because he had persistently filed public records requests, but did not receive anything until 2017. He asserts that he could have used this evidence to impeach the State's experts by showing that the unknown person was the major DNA contributor who actually committed the sexual battery, which would have been consistent with the fact that he did not match the victim's physical description of the rapist. By suppressing the 2005 report, he argues, the State "was able to support a specious theory arguing that the unknown was just a minor, stray strand of DNA transferred to the victim by Finch." He argues that in light of this evidence, and without its suppression, no reasonable factfinder would have found him guilty beyond a reasonable doubt.

...

Analysis

Finch does not propose a claim that meets the statutory criteria. Finch's first and third claims are actually one claim, of ineffective assistance of counsel for failure to retain and present testimony from a DNA expert. This claim must be dismissed because Finch presented the same claim in his initial § 2254 petition. See 28 U.S.C. § 2244(b)(1) (stating that a § 2254 claim presented in a prior § 2254 petition "shall be dismissed"). Finch argued in Ground 13 of his original § 2254 petition that he received ineffective assistance of trial counsel because his trial counsel refused to hire an independent expert to examine the DNA sample, and that claim was denied on the merits. His present application asserts the same claim in proposed claims one and three. Even though Finch asserts he has newly discovered evidence in support of the claim, it still must be dismissed. See In re Hill, 715 F.3d 284, 294-95 (11th Cir. 2013) (holding that "new evidence supporting the claim" does not make a prior claim a new claim for the purpose of § 2244(b)(1)). Further, this claim does not actually even rest on new evidence, as Finch's proffered evidence—DNA reports from 2008 and 2013, years after his 2006 trial—does not bear on whether his trial counsel acted deficiently by not retaining an expert. Nor does the post- trial evidence relate to the prejudice prong, in terms of revealing what a retained defense expert might have found. Windsor's sample does not appear to have been added to the state DNA database until 2013, so no defense expert could have found it before trial. And the 2008 DNA report does not indicate, as Finch asserts, that Finch's DNA does not match. The target sample in the 2008 report appears to be Finch, as it is a 2004 sample for a convicted offender, and the STR numbers match what Noppinger testified were obtained from Finch's sample. And Finch's target sample is a match for the candidate sample, which is labeled "1.5," and thus appears to be the vaginal swab sample.

As to Finch's second and fourth proposed claims, his newly discovered evidence does not establish by clear and convincing evidence that no reasonable factfinder would have found Finch guilty of the underlying offense. See 28 U.S.C. § 2244(b)(2)(B)(ii); see also In re Boshears, 110 F.3d 1538, 1541 (11th Cir. 1997) (holding that the application must be denied if "any rational trier of fact could have found the essential elements of the crime beyond a reasonable doubt"). Finch's proposed second and fourth claims both rest upon the 2005 DNA report. The 2005 DNA report shows a 15 STR loci match between the minor contributor's sample and an unknown target sample. That can be deduced from the fact that the candidate sample in the 2005 report is labeled "MINOR" and has the same values at the same loci as the 2013 report has for the candidate sample clearly labeled as "MINOR" and "MINOR MIX." The target sample in the 2005 report is not clearly identified, so its source is unclear, but it too appears to be labeled "MINOR." The 2005 results do not match what Noppinger testified at trial were both Finch's and the major contributor's.

Finch asserts that the 2005 report contradicts Noppinger's testimony that the unknown person was a minor contributor, because Noppinger testified that Finch's DNA was a 9 STR loci match and the unknown person's DNA matched at fewer loci, whereas the 2005 report "reflects the fact that the unknown [person's DNA] was a striking and significant 15 STR loci match." This assertion is incorrect. Whether one is a major or minor contributor is based on how much DNA from each person was found in the swab, not the number of loci that are tested or match. Therefore, no match results would change which contributor is major and which is minor. Baird testified that the major and minor contributors were "very distinct and easily discernable," and the 2005 report is a test on the minor contributor's profile. That the minor contributor was a 15 STR loci match means little, since it is unclear what sample it is matched to, and the unknown target sample it matches appears also to be labeled "MINOR." Furthermore, the fact that the match is at 15 loci rather than 9 is consistent with the trial testimony that STR tests used to test at 9 loci but had been changed to test at 15 loci.

Finch does appear to be correct that the 2005 report represents evidence that the State had reason to know before trial that the minor contributor was male. That is because the 2005 report was created before the 2006 trial, and it shows that the minor contributor profile had "XY," indicating a male. It therefore seems to contradict Baird's testimony that it was impossible to tell whether the minor contributor was male or female. But this does not show by clear and convincing evidence that no reasonable factfinder could have found Finch guilty beyond a reasonable doubt. The 2005 report neither contradicts nor undercuts the key testimony at trial—that Finch's DNA matched that of the major contributor in the victim's vaginal swab. The jury already knew that there was an unknown minor contributor whose DNA was also present, and that no one knew for sure why a second person's DNA was present. All the 2005 report changes is that the minor contributor was definitely male, instead of possibly male, and therefore the State's speculation that the minor contributor's DNA was present because the rapist had sex with another woman shortly before raping the victim must be incorrect. This latter fact arguably makes it more likely that the second person's DNA arose through contamination. But Baird's unrebutted trial testimony was that contamination might explain the presence of the minor contributor's DNA, but not the major contributor's DNA. And the major contributor was Finch. Consequently, Finch's new evidence, accepted as true, does not demonstrate clearly and convincingly that he is innocent. See 28 U.S.C. § 2244(b)(2)(B)(ii); Boshears, 110 F.3d at 1541.

Accordingly, because Finch has failed to make a prima facie showing of the existence of either of the grounds set forth in § 2244(b)(2), his application for leave to file a second or successive petition is hereby DENIED IN PART AND DISMISSED IN PART.

Questions:

1. This case involves successive petitions under a theory of habeas corpus whereby a prisoner alleges that newly discovered evidence, based on the analysis of bodily fluids like seminal remains, indicate a wrongful conviction. What federal law permits or allows this argument?
2. The Act recognizes 3 rationales for this extraordinarily relief of a defendant petitioner. What are those?
3. How does the Court resolve the request?

Lewis v. State
306 Ga. 455, 831 S.E.2d 771 (Ga. 2019)

Freddie Lewis was convicted of malice murder, rape, and burglary in connection with the death of Evelyn Wise. On appeal, Lewis argues that the trial court erred in (1) denying his motion for a directed verdict, (2) denying his motion in limine to exclude DNA evidence based on inadequate chain of custody, … The trial court committed no error, because the evidence was sufficient to support convictions on the crimes charged, and the DNA evidence, … were properly admitted … . We affirm.

Viewed in the light most favorable to the verdicts, the evidence shows that Mary Wise and her daughter, Evelyn, lived in a townhouse-style apartment near the Vine City MARTA station; Mary slept downstairs, and Evelyn had a room upstairs. On the morning of March 3, 1991, an Atlanta Police Department (APD) officer went to the apartment in response to a 911 call regarding a robbery with an injured person. Mary reported that an unidentified male tried to strangle her in the middle of the night. She said that she had gone to bed sometime around 11:30 p.m. after talking to Evelyn, and was awoken around 1:00 a.m. by noises coming from the upstairs area. After hearing the noises, Mary saw a man run down the stairs. The man jumped on her, wrapped a string around her neck, and began to pull on it. Mary fought off the man, but he then choked her with his hands and smothered her with a pillow, causing her to lose consciousness. When Mary regained consciousness, she noticed that some of her money had been taken. Mary did not get a good look at the assailant because it was dark inside the apartment at the time of the attack. The responding officer observed that Mary had a small bruise on her cheek, dried blood on one side of her face, and some slight discoloration around her neck.

After talking to Mary, the responding officer went upstairs. The officer found Evelyn's dead body lying face-up on the floor in her bedroom. Evelyn had a shoestring wrapped around her neck, her pants had been pulled down, her legs were spread so that her genitals were exposed, and she had one shoe on. The officer also saw that a flower pot had been placed on an air conditioning unit outside so that someone could access a ledge leading to the open bedroom window. All other entries to the apartment were locked.

... Because the medical examiner also suspected that Evelyn was sexually assaulted, he swabbed Evelyn's breasts, vagina, and rectum, and this evidence was sent to the GBI.

A GBI examination revealed the presence of intact and partially intact sperm from the vaginal swabs and partially intact sperm from the rectal swabs. A GBI DNA analyst testified that the presence of intact sperm indicated that the semen was deposited 24 to 36 hours before the swabs were collected, and, based on this evidence, the APD detective concluded that the sexual contact occurred at or near the time of Evelyn's death.

The GBI sent the swabs to the FBI, which had been investigating the murders of four elderly women in the Vine City area. An FBI analyst analyzed the sperm cells from the vaginal swabs of Evelyn, determined that the sperm came from only one person, and concluded that the DNA profile was not a match for any known individuals and was not linked to the crimes being investigated by the FBI. After completing his exams, the FBI analyst returned the swabs to the GBI in September 1991. Because there were no leads to pursue, the APD investigation into Evelyn's murder went into an inactive status.

In 2004, the APD created a Cold Case Squad to investigate homicides having a sexual component by reexamining the rape kits in those cases. The Cold Case Squad made repeated requests to the APD's property section to obtain Evelyn's rape kit, but the kit was not located until January 2008. A detective then sent the swabs to a private lab for DNA analysis, leading to the creation of a DNA profile. The GBI reviewed the private lab's work, uploaded the profile into a national database of DNA profiles known as CODIS, and issued a report in April 2008 concluding that the DNA profile was a match for Lewis.

Police began investigating Lewis and learned that he had lived in the same apartment complex as Evelyn at the time of her death; Lewis's building was located about 75 yards from Evelyn's. An APD detective interviewed Lewis, who after being read his Miranda rights and waiving them, admitted that he knew the victim but did not say that he was in a relationship with her or that he ever had sexual intercourse with her. When the detective asked Lewis for a DNA sample, Lewis grew visibly shaken, became extremely nervous, and began stumbling over his words. The detective took two buccal swabs from Lewis and sent them to the GBI for testing. The DNA from Lewis's swabs matched the DNA from the vaginal swab taken from Evelyn in 1991, thus confirming the CODIS match. During his investigation, the detective found no evidence that Lewis had a consensual sexual relationship with Evelyn. The evidence showed that Mary never let any men inside the apartment and was very protective of Evelyn because she had an intellectual disability.

1. Lewis argues that the trial court erred in denying his motion for a directed verdict because the evidence was insufficient to support his convictions. In particular, he argues that there is no direct evidence that he committed the crimes, the evidence showing the presence of his sperm inside the victim's vagina did not rule out the possibility that he and the victim had consensual sex and someone else committed the crimes, and the DNA identification evidence was suspect because the rape kit went missing for a few years and items within it were not clearly identified. We conclude that there was sufficient evidence to authorize his convictions.

The standard of review for the denial of a motion for a directed verdict of acquittal is the same as for determining the sufficiency of the evidence to support a conviction under Jackson v. Virginia, 443 U. S. 307, 319, 99 S.Ct. 2781, 61 L.Ed.2d 560 (1979). See Green v. State, 304 Ga. 385, 387 (1), 818 S.E.2d 535 (2018). Under this standard, we review whether a rational trier of fact could have found the defendant guilty beyond a reasonable doubt, but we do not "reweigh evidence or resolve conflicts in testimony; instead, evidence is reviewed in a light most favorable to the verdict, with deference to the jury's assessment of the weight and credibility of the evidence." Thomas v. State, 300 Ga. 433, 436 (1), 796 S.E.2d 242 (2017) (citation and punctuation omitted).

To sustain a conviction based on circumstantial evidence, former OCGA § 24-4-6, in effect at the time of Lewis's trial, provided that "the proved facts shall not only be consistent with the hypothesis of guilt, but shall exclude every other reasonable hypothesis save that of the guilt of the accused." The evidence need not exclude every hypothesis save that of guilt, only those that are reasonable, and it is for the jury to determine whether an alternative hypothesis is reasonable. See Brown v. State, 304 Ga. 435, 437 (1), 819 S.E.2d 14 (2018). We will not disturb the jury's finding in this respect unless the evidence is insupportable as a matter of law. Id.

(a) The evidence was sufficient to sustain Lewis's convictions. The state in which Evelyn's body was discovered (partially undressed with her genitals exposed), her defensive wounds, and her manner of death supported a finding that Evelyn succumbed to strangulation after attempting to fight off an assailant who was raping and strangling her. The evidence also shows that sperm collected from Evelyn came from only one male, Lewis, and the condition of the sperm showed that her death occurred at or near the time of the sexual contact.

Notwithstanding this evidence, Lewis argues that there was a reasonable hypothesis that he had consensual sex with Evelyn and left the apartment before an unknown assailant entered the victim's apartment and committed the crimes charged. But the jury could – and did – reject this hypothesis as unreasonable. There was no evidence that Evelyn had a relationship with any man, much less Lewis, as Mary was very protective of Evelyn. And despite admitting to police that he knew Evelyn, Lewis never said that he had a relationship with Evelyn or had consensual sex with her. The jury was therefore authorized to reject Lewis's hypothesis. See Daniels v. State, 298 Ga. 120, 123 (1), 779 S.E.2d 640 (2015) (jury authorized to reject theoretical possibility that the defendant had consensual sexual intercourse with the victim within 72 hours of her murder and sometime later an unknown assailant also had sexual intercourse with the victim and murdered her without leaving a trace of DNA evidence recoverable through sexual assault evidence kit).

(b) To the extent Lewis challenges the integrity of the DNA identification evidence based on an insufficient chain of custody of the rape kit, this claim fails. The medical examiner who took the swabs from the victim testified that he sealed the items before sending them to the GBI Crime Lab. Witnesses from the GBI, the FBI, and the private lab that tested materials from the rape kit over the years all testified that the materials were sealed and showed no signs of tampering when they were received, and the witnesses resealed the materials after performing DNA tests. Although the APD did not locate the items for several years after the APD Cold Case Squad first requested them, there were no signs of tampering when the rape kit was finally located in 2008, and there was no evidence the rape kit was removed from the APD's property room during that time. Given this evidence, the State established a reasonable certainty that the original victim swabs were the same evidence that was tested. See, e.g., Johnson v. State, 271 Ga. 375, 382 (13), 519 S.E.2d 221 (1999) ... "Absent affirmative evidence of tampering, mere speculative doubt as to the handling of [the] evidence" is a matter for the jury to resolve. Id. (citation and punctuation omitted). The jury clearly resolved this issue adversely to Lewis, and it was authorized to do so by the evidence presented. See Bates v. State, 293 Ga. 855, 855 (1), 750 S.E.2d 323 (2013) ...); Whitaker v. State, 291 Ga. 139, 140 (1), 728 S.E.2d 209 (2012) (the resolution of evidentiary conflicts "adversely to the defendant does not render the evidence insufficient" (citation and punctuation omitted)).

2. Lewis raises two arguments on appeal regarding the admissibility of the DNA evidence. Neither of his arguments has merit.

(a) Challenging the sufficiency of the State's chain of custody over the swabs taken from Evelyn's body, Lewis argues that the trial court erred in denying his motion in limine to exclude the DNA evidence obtained from those materials. We disagree.

On the morning of the first day of Lewis's trial, as the jury was about to be selected, Lewis informed the trial court that he had a pending motion in limine, which he had filed a few days prior. Lewis argued that the State's DNA evidence should be excluded because the State would be unable to establish an adequate chain of custody because it could not show the location of the evidence from September 1991 (when the GBI received the swabs from the FBI) until January 2008 (when the APD located Evelyn's rape kit and sent it to the Cold Case Squad). The trial court asked Lewis whether he had any evidence that someone had tampered with the rape kit, and Lewis conceded that he did not, noting that there was no evidence about the location of the rape kit. The trial court remarked that the motion should have been filed sooner and that it would have been helpful to have some testimony on the issue in order to make a preliminary determination. After hearing the parties' arguments, the trial court denied Lewis's motion in limine but noted that the State would have to lay a proper foundation at trial.

Lewis complains about the trial court's pretrial ruling, arguing that the court should have "made a finding that would have uncovered the problem with the chain of custody and would have exposed the tests, or lack thereof, of the techniques and procedures that were valid and were capable of producing reliable results." It is not clear what "finding" Lewis wanted the court to make. In any case, he has not shown that the trial court's ruling was erroneous.

A party's motion in limine to exclude evidence as inadmissible "should be granted only if there is no circumstance under which the evidence is likely to be admissible at trial." Grant v. State, 305 Ga. 170, 172 (3), 824 S.E.2d 255 (2019) (citation and punctuation omitted). Given this framework, "the grant of a motion in limine excluding evidence is a judicial power which must be exercised with great care." Andrews v. Wilbanks, 265 Ga. 555, 556, 458 S.E.2d 817 (1995). We review a trial court's denial of a motion in limine for an abuse of discretion. Grant, 305 Ga. at 172 (3), 824 S.E.2d 255.

Here, the DNA evidence was certainly relevant to the issues in the case. Indeed, as described above in Division 1, it was the critical piece of evidence in this case. Lewis's primary challenge regarding chain of custody is that APD could not locate the rape kit for several years when the Cold Case Squad wanted to retest it. Lewis raised the chain of custody issue on the eve of trial and presented no evidence that anyone had tampered with the rape kit. The State argued that it would present testimony at trial establishing that the swabs in the rape kit were the same ones that were collected from the victim. Given this assurance (which was later fulfilled), Lewis could not establish a circumstance in which the evidence would be ruled inadmissible. His argument raised only a bare speculation of tampering, which was an insufficient basis to exclude the evidence. See Hurst v. State, 285 Ga. 294, 296 (2), 676 S.E.2d 165 (2009) ("When there is only a bare speculation of tampering, it is proper to admit the evidence and let what doubt remains go to the weight." (citation and punctuation omitted)). Therefore, the trial court did not abuse its discretion in denying his motion in limine.

...

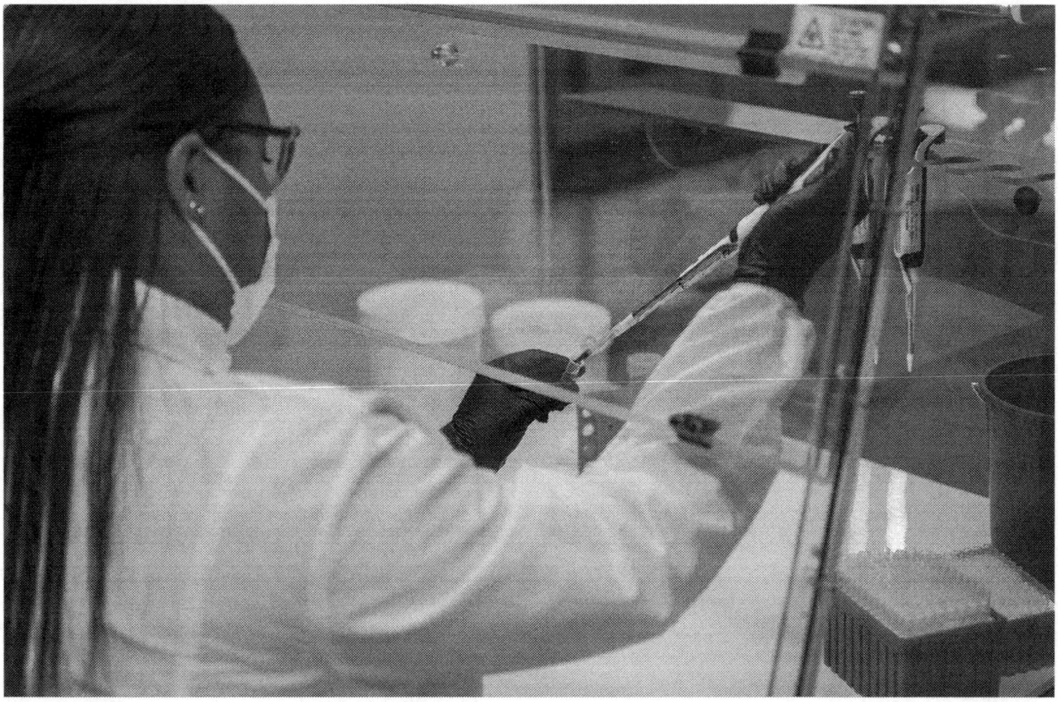

FIGURE 4.4 A biologist performs DNA testing at the Terrorist Explosive Device Analytical Center (TEDAC) in Huntsville, Alabama. TEDAC was formally established in 2003 to serve as the single interagency organization to receive, fully analyze and exploit all terrorist improvised explosive devices, or IEDs, of interest to the United States. On February 16, 2016, TEDAC celebrated the completion of its new laboratory facilities at Redstone Arsenal in Huntsville. FBI Media at https://multimedia.fbi.gov/item?type=image&id=4359.

Judgment affirmed.

Questions:

1. Can a chain of custody argument be an effective strategy as posed by defense counsel?
2. The defendant engages in a sort of alternative explanation for how or why his sperm in the victim does not prove criminal conduct.
3. What does the Court mean when it characterizes the defendant's arguments as "speculative." What specific contexts?

State v. Lindsey
Court of Criminal Appeals of Tennessee, At Jackson
October 29, 2019, Assigned on Briefs at Knoxville; February 14, 2020,
Filed No. W2018-01987-CCA-R3-CD

A Shelby County jury convicted the Defendant, Timothy Lindsey, of aggravated rape for a crime committed in 2005, and he was sentenced to serve thirty years in confinement. On appeal, the Defendant challenges the sufficiency of the convicting evidence and asserts that the ten-year delay in indicting him violated his right to due process. We conclude that the evidence is sufficient to support the verdict and that the Defendant has waived the issue of pre-indictment delay. Accordingly, the trial court's judgment is affirmed and the case remanded for correction of the judgment form.

OPINION

...

Ms. Taylor Templin, an expert in forensic DNA analysis who previously worked at Cellmark as a DNA analyst, conducted analysis on the DNA after Ms. Estes established that the sample contained spermatozoa. Ms. Templin testified that she extracted DNA from the swabs, purified it, measured it, amplified it by making copies, and then used an instrument to obtain data which she interpreted. She testified about the precautions taken to prevent and detect contamination. She stated that for both the oral and miscellaneous sets of swabs, she attempted to separate the epithelial DNA, which could come from skin cells or saliva, from the sperm cell DNA. She testified that such a separation is not always complete and that there is sometimes "carry- over," meaning that some non-sperm DNA will remain in the sperm fraction of the sample and vice versa.

Ms. Templin concluded that the sperm fraction of the DNA she obtained from the swabs labeled "oral" was a mixture of two individuals. She obtained a profile of the major contributor, who was a male. Because she was only able to identify one allele, or location on the DNA, from the minor contributor, she could not come to conclusions regarding the minor contributor to the sperm fraction of the oral swab. She obtained a DNA profile from the epithelial fraction of the oral swab, and this profile belonged to a woman. From the swabs labeled miscellaneous, she obtained a profile from the epithelial fraction, and she concluded the same woman's DNA was present in the epithelial fraction of both sets of swabs. The male donor of the sperm fraction of the miscellaneous swabs could not be excluded as the male donor of the sperm fraction of the oral swabs.

Ms. Templin acknowledged that it was possible that the mixture in the sperm fraction of the DNA obtained from the oral swabs came from two male donors. She stated that this scenario was unlikely, that the minor contributor's allele was consistent with the female DNA profile, and that it was "highly likely" that the minor contributor was the victim because the sample was obtained from her mouth. She acknowledged that DNA degrades over time and that it was possible that the reason she was unable to construct a profile for the minor contributor was due to degradation.

After Detective Carter obtained the buccal swabs from the Defendant, they were submitted to Cellmark for comparison with the sperm recovered from the victim's mouth. Ms. Stephanie Hickey, an expert in forensic DNA analysis, obtained DNA from the buccal swabs and analyzed it. She concluded that the male profile obtained from the sperm fraction of the oral swab and the male profile obtained from the sperm fraction of the miscellaneous swab were consistent with the Defendant's buccal swab. For each set of swabs, the probability of an unrelated individual matching the profile exceeded the world's population many times.

Ms. Hickey acknowledged that her salary depended in part on revenue gained from performing analysis for law enforcement. She clarified that she would get paid even if law enforcement did not get their preferred result. She also acknowledged that the accuracy of her results depended in part on the reliability of the prior testing performed by Ms. Templin. She agreed that her analysis could not determine whether or not a rape had occurred.

The defense presented the testimony of the Defendant's younger brother to establish that the victim's description of her assailant did not match the Defendant's appearance at the time. Mr. Jeremy Taylor testified that he was two years younger than the Defendant that that they grew up in the same household. They shared a room and also shared clothing and shoes. He testified that they both wore a size twelve shoe beginning in high school. Mr. Taylor also stated that the Defendant has always had facial hair and that the Defendant had facial hair in 2005. Mr. Taylor testified he was six feet, three inches tall, and he said he was taller than the Defendant "by just a little bit." The Defendant had always been tall and reached his full height in high school. The Defendant was eighteen or nineteen years old and weighed approximately 190 to 200 pounds in 2005. To cast doubt on the Defendant's identity as the intruder, who used his left foot to kick in the door, Mr. Taylor testified that he and the Defendant were both right-handed. Mr. Taylor acknowledged that he had prior drug convictions and that the Defendant had always had short hair.

The jury convicted the Defendant of aggravated rape. The trial court determined the Defendant was a Range II offender based on two prior aggravated burglaries committed in around the time of the aggravated rape. After applying two enhancement factors, the trial court sentenced the Defendant to thirty years in prison.

ANALYSIS

I. Sufficiency of the Evidence

On appeal, the Defendant challenges the sufficiency of the evidence, arguing that the proof at trial was insufficient to establish his identity as the assailant. We conclude that a rational trier of fact could have found that the State established identity beyond a reasonable doubt.

...

Aggravated rape, as charged here, is the unlawful sexual penetration of the victim by the defendant when force or coercion is used to accomplish the act and the defendant is armed with a weapon or any article used or fashioned in a manner to lead the victim reasonably to believe it to be a weapon. T.C.A. § 39-13-502(a)(1). Sexual penetration includes fellatio.

T.C.A. § 39-13-501(7) (2005). Here, the victim testified that she was forced to perform fellatio at gunpoint.

The Defendant does not dispute that the State established the elements of an aggravated rape, but he disputes the sufficiency of the evidence regarding identity. Identity is an essential element of every crime. State v. Bell, 512 S.W.3d 167, 198 (Tenn. 2015). The identification of the perpetrator of a crime is a question of fact for the jury. State v. Thomas, 158 S.W.3d 361, 388 (Tenn. 2005). In resolving questions of fact, such as the identity of the perpetrator, "'the jury bears the responsibility of evaluating the conflicting evidence and accrediting the testimony of the most plausible witnesses.'" Pope, 427 S.W.3d at 369 (quoting State v. Hornsby, 858 S.W.2d 892, 897 (Tenn. 1993)).

The Defendant points to the disparities between the victim's description of her assailant and his physical characteristics at the time of the offense. The victim described her assailant as between five feet, seven inches and five feet, ten inches tall and weighing between 130 and 150 pounds. She stated he had no facial hair. Although the Defendant's brother testified that in 2005, the Defendant was taller and heavier than the victim's description, that he had facial hair at the time, and that he was right-handed, other evidence, including the victim's description of her assailant as a young African American man with short hair, the size of the shoe prints on the door, and the DNA evidence, supports the conclusion that the Defendant was the offender. The testimony offered by the defense merely constituted a conflict in the evidence, which the jury was free to resolve in the State's favor. We will not revisit the jury's resolution of conflicting identity evidence presented at trial. Pope, 427 S.W.3d at 369.

The Defendant also asserts that the DNA proof was "inconclusive and unreliable." He notes that the sample was not tested for a number of years and might have degraded. However, Ms. Estes testified that degradation would result in an inability to get a full profile or perhaps any profile at all. She stated that degradation of DNA would not change the profile itself. While Ms. Templin testified that there was some possibility that degradation prevented her from obtaining a full profile of the minor contributor to the sperm fraction of the oral swab, she also testified that it was unlikely the minor contributor was another male and "highly likely" that the minor contributor was the victim. In any event, obtaining the profile of the minor contributor would not exculpate the Defendant, whose DNA matched the sperm in both sets of swabs taken from the victim's mouth shortly after the rape.

The Defendant argues that Ms. Templin's testimony regarding the minor contributor to the oral swab and her testimony that the donor of the sperm fraction of the miscellaneous swab could "not be excluded" as the donor of the sperm fraction of the oral swab do not establish proof of identity beyond a reasonable doubt. However, the State's proof at trial was not solely composed of Ms. Templin's testimony. The victim testified that she was forced to perform fellatio at gunpoint and that the assailant's semen was present in her mouth and on her clothing. The victim gave a detailed account of the assault, and law enforcement found physical corroboration of an assault, including the ransacked apartment, the broken doorjamb, the footprint on the door, and the stains on the victim's shirt and carpet. The victim described the assailant as an African American man with a medium complexion and short hair who was between eighteen and twenty-five years old. The shoe print from the victim's front door was the same approximate size as the Defendant's shoe. Two sets of swabs were taken from the victim's mouth shortly after the assault.

Ms. Templin testified that she obtained a DNA profile from both sets of swabs. While the sperm fraction of the DNA from the swabs labeled "oral" was a mixture of two profiles, Ms. Templin testified that the minor contributor, whose genetic material was present only in a slight quantity at one allele and was consistent with the victim's DNA, was most probably the victim. She acknowledged that it was possible that the minor contributor was not the victim but a male. Ms. Templin stated that the donor of the DNA from the sperm fraction of the miscellaneous swab could not be excluded as the donor of the DNA from the sperm fraction of the oral swab. Subsequently, Ms. Hickey performed analysis of the buccal swabs obtained by Detective Carter from the Defendant. She compared the Defendant's known profile to the profiles from the oral and miscellaneous swabs. Ms. Hickey testified that the Defendant's DNA from the buccal swab matched the sperm fraction of the DNA obtained from both the oral and miscellaneous swabs. The probability of an unrelated individual matching the DNA exceeded the world's population many times. We conclude that a rational trier of fact could have found that the State established the Defendant's identity as the assailant. See State v. Toomes, 191 S.W.3d 122, 131 (Tenn. Crim. App. 2005) (upholding conviction when DNA evidence was the only evidence linking the defendant to the rape).

Questions:

1. While there is a natural tendency to accept DNA-Sperm identity as something highly conclusive, does the defendant seek to show despite the conclusions derived, the victim actually did not clearly or objectively identify the perpetrator?
2. How does the Court characterize the identity conclusion on the sperm swaps- dependable, suspect or a certitude?
3. In what manner would a "degradation" argument undermine the quality and reliability of a sperm sample?

4.3 MISCELLANEOUS FLUIDS OR MIXTURES

The analysis of bodily fluids is often a mixture of things such as blood and alcohol in intoxication tests, which is in essence a mixture of two fields: biology and chemistry. Any acceptable conclusion on Blood-Alcohol levels must measure both substances to arrive at a conclusion on intoxication. These conclusions are often supplemented by a series of field texts where the human body's interaction with the known world is measured through a series of field sobriety tests. For example, can the defendant walk on a traffic line, pick up a dime, utter the alphabet or numbers in reverse, or follow a pencil of light, to name just a few. In these cases, fluids are complimented by bodily observations which manifest intoxication.

Visit the Southern Association of Forensic Science's website at https://safs1966.org.

Locate the following case online using your favorite legal research resource, read the case and answer the questions below.

Bragaw v. State
482 P.3d 1023 (Alaska Ct. App. 2021)

Questions:

1. As noted earlier, the conclusion about Blood-Alcohol rates is a combination skill set primarily rooted in finding substances in biological blood. In Bragaw, the Court highlights the "Drug Influence Evaluation" program. How many biological observations, or what points on the biological body are evaluated in addition to forensic tests.
2. When challenged as a non-scientific test that should not be admitted into a court of law, because it fails under the Daubert test, how does the Court resolve.
3. The defendant in Bragaw even highlights this mixture of fields as a basis for rejecting the DRE protocol. He even called an expert witness to trump up doubt about the DRE testing system. At what place does Dr. Means who is the defense expert relay his concerns.

As in most areas of forensic law, there are both substantive and procedural reasons why the evidence fails to be admitted. Hence, even though the scientific challenge may be meritorious, its admission can be denied if not timely filed or past the statute of limitations.

Hanicek v. State
Court of Appeals of Texas, Fourteenth District, Houston
June 14, 2018, Memorandum Opinion Filed
NO. 14-17-00191-CR

MEMORANDUM OPINION

Appellant Walter Jason Hanicek appeals his felony conviction for third-offender driving while intoxicated. See Tex. Penal Code Ann. §§ 49.04(a), 49.09(b)(2) (Vernon Supp. 2017). In his sole issue, appellant challenges the reliability of the procedures used to perform a field sobriety test and blood sample analysis. Concluding this alleged error was not preserved, we affirm.

BACKGROUND

Deputy Eric Becker was on patrol in Fort Bend County, Texas shortly after midnight on December 8, 2015. Deputy Becker was driving his patrol car on the highway when he swerved to avoid a head-on collision with a vehicle traveling towards him in his lane. Deputy Becker turned to follow the vehicle and activated his emergency lights; the vehicle started to pull over but "continued to slow roll on the shoulder for a considerable distance" before coming to a complete stop.

Deputy Becker approached the vehicle, which was driven by appellant. Deputy Becker testified that he "immediately detected the overpowering odor of alcohol" and stated that appellant "was struggling, having some dexterity problems with his fingers retrieving his ID and stuff." Deputy Becker testified that appellant's speech was slurred and that appellant was "unsteady in his posture." Deputy Becker testified that appellant's eyes were "red and glassy, slightly bloodshot." Deputy Becker saw bottles of alcohol in appellant's car; appellant told Deputy Becker that he had consumed "three or four drinks at the time."

Deputy Becker contacted police dispatch and requested a deputy certified to perform standard field sobriety tests. Deputy Mark Anciso arrived at the scene and performed the horizontal gaze nystagmus test (the "HGN test") on appellant. Deputy Anciso testified that appellant's performance on the HGN test was consistent with intoxication. Appellant refused to undergo additional field sobriety tests.

Deputy Becker placed appellant under arrest. Deputy Becker asked appellant to provide a blood sample, but appellant refused. Deputy Becker transported appellant to a hospital in Richmond, Texas and obtained a warrant to perform a blood draw on appellant.

Deputy Becker testified that appellant "was very uncooperative" when he was taken inside the hospital and "physically resist[ed]" attempts to draw his blood. The first blood draw was completed at approximately 4:00 A.M. and secured only a partial tube of blood. The second blood draw was completed at 5:15 A.M. and secured a full sample.

Haley Yaklin, a supervisor at the Texas Department of Public Safety Crime Laboratory, testified regarding the analysis performed on appellant's blood sample. Yaklin testified that appellant's 4:00 A.M. blood sample was tested first in accordance with the laboratory's policy requiring employees "first to select the [sample] that was collected from the individual at the earliest time." Although the 4:00 A.M. blood draw secured only a partial tube of blood, Yaklin testified that the sample was sufficient to perform a blood-alcohol analysis. Yaklin testified that the blood test result showed appellant's blood-alcohol level to be 0.207.

Appellant was indicted for the felony offense of third- offender driving while intoxicated. See Tex. Penal Code Ann. §§ 49.04(a), 49.09(b)(2). After a three-day trial, the jury returned a guilty verdict. The trial court sentenced appellant to six years' confinement. Appellant timely appealed.

ANALYSIS

Appellant's sole issue on appeal challenges the reliability of the procedures used to perform the HGN test and blood sample analysis.1 Appellant asserts that the State failed to prove (1) the HGN test was "properly administered;" (2) the technique for obtaining appellant's blood sample was "properly applied;" (3) the analysis of appellant's blood sample was "properly applied;" and (4) the blood draw "was performed in a reasonable manner in accordance with acceptable medical practices."

Testimony and evidence addressing the HGN test and blood sample analysis constitute scientific evidence, which is admissible if reliable and relevant. Wolfe v. State, 509 S.W.3d 325, 335 (Tex. Crim. App. 2017); see also Adkins v. State, 418 S.W.3d 856, 863 (Tex. App.—Houston [14th Dist.] 2013, pet. ref'd) (test results for blood drawn from appellant are scientific evidence); Plouff v. State, 192 S.W.3d 213, 218 (Tex. App.—Houston [14th Dist.] 2006, no pet.) ("[t]estimony concerning the HGN test is scientific evidence"). Scientific evidence is reliable if "(1) the underlying theory [is] valid; (2) the technique applying the theory [is] valid; and (3) the technique [was] properly applied on the occasion in question." Wolfe, 509 S.W.3d at 335 (quoting Kelly v. State, 824 S.W.2d 568, 573 (Tex. Crim. App. 1992)).

Challenges to the reliability of scientific evidence must be raised in the trial court or they are waived for appeal. Turner v. State, 252 S.W.3d 571, 584 n.5 (Tex. App.—Houston [14th Dist.] 2008, pet. ref'd) (appellant's objection insufficient to apprise trial court he was challenging the reliability of the scientific evidence; appellant therefore "waived any argument regarding reliability"); see also Hopes v. State, No. 14-14-00403-CR, 2015 Tex. App. LEXIS 11468, 2015 WL 6759450, at *3 (Tex. App.—Houston [14th Dist.] Nov. 5, 2015, pet. ref'd) (mem. op., not designated for publication) (appellant challenged reliability of officer's testimony regarding interpretation of gang tattoos; court declined to consider argument because "appellant did not preserve it for our review by objecting at trial"); Murphy v. State, No. 05-09-00772-CR, 2010 Tex. App. LEXIS 3166, 2010 WL 1729341, at *2 (Tex. App.—Dallas April 29, 2010, no pet.) (mem. op., not designated for publication) (appellant asserted on appeal that officer failed to properly administer HGN test; complaint waived because appellant "did not object at trial when this evidence was offered"). To preserve a complaint for appellate review, a party must present to the trial court a timely request, objection, or motion stating the specific grounds for the complaint and obtain a ruling. Tex. R. App. P. 33.1(a); Wilson v. State, 71 S.W.3d 346, 349 (Tex. Crim. App. 2002).

Here, appellant's challenges invoke the third prong of the Kelly test and assert that the procedures underlying the HGN test and blood sample analysis were unreliable. Appellant's reliability challenges were not presented to the trial court and cannot be raised for the first time on appeal. See Turner, 252 S.W.3d at 584 n.5; see also Hopes, 2015 Tex. App. LEXIS 11468, 2015 WL 6759450, at *3; Murphy, 2010 Tex. App. LEXIS 3166, 2010 WL 1729341, at *2. We agree with the State that appellant failed to preserve error on these issues.

Appellant contends that he nonetheless can assert a challenge to scientific evidence "based on the validity of the tests even in the absence of an objection to its admissibility." Appellant cites City of San Antonio v. Pollock, 284 S.W.3d 809 (Tex. 2009), Volkswagen of America, Inc. v. Ramirez, 159 S.W.3d 897 (Tex. 2004), and Coastal Transport Co., Inc. v. Crown Central Petroleum Corp., 136 S.W.3d 227 (Tex. 2004), to support his argument.

Appellant does not cite any authority adopting these cases in the criminal context. Moreover, the cases appellant cites state that an objection in the trial court is necessary to preserve for appellate review a complaint regarding the reliability of scientific evidence. City of San Antonio, 284 S.W.3d at 816-17 ("When a scientific opinion is not conclusory but the basis offered for it is unreliable, a party who objects may complain that the evidence is legally insufficient to support the judgment. An objection is required to give the proponent a fair opportunity to cure any deficit and thus prevent trial by ambush."); Ramirez, 159 S.W.3d at 903 ("A party may raise a properly preserved complaint on appeal that scientific evidence is unreliable and thus no evidence to support a judgment."); Coastal Transport Co., Inc., 136 S.W.3d at 232-33 ("to preserve a complaint that scientific evidence is unreliable and thus, no evidence, a party must object to the evidence before trial or when the evidence is offered" (internal quotation omitted)). By failing to object at trial, appellant waived his complaints regarding the reliability of the procedures used to perform the HGN test and blood sample analysis. See Turner, 252 S.W.3d at 584 n.5; see also Hopes, 2015 Tex. App. LEXIS 11468, 2015 WL 6759450, at *3; Murphy, 2010 Tex. App. LEXIS 3166, 2010 WL 1729341, at *2.

CONCLUSION

Having overruled appellant's sole issue on appeal, we affirm the trial court's judgment.

Questions:

1. Is the Court's rejection of the blood sample analysis based on the flaw or errancy, or even implausibility of a dependable blood-alcohol measure?
2. Even if the defendant were able to challenge the tests employed in this case, and assuming the science defective, would defendant still have been barred as to the challenge on appeal?

Locate the following case using your favorite legal research resource, read the case and answer the questions below.

Washington v. Griffin, 876 F.3d 395 (2d Cir. 2017)

Questions:

1. Few cases contain facts as unsettling as the Griffin decision. The defendant not only committed a wide array of crimes but did so with a distinct brutality. Lay out those facts relating to Victim Number 3 which confirm this conclusion.
2. When the horrid facts of Griffin are analyzed, it is easy to see how critical fluid, DNA, hair and other microscopic evidence is in cases like these. What types of evidence might lead to a forensic identification that would support a conviction of this brutal defendant?
3. The Griffin case has many constitutional challenges. What is the most frequently encountered?

5 Forensic Toxicology

5.1 INTRODUCTION

Questions regarding toxicity and toxic substances are a common issue in both civil and criminal litigation. In this way, just as biology plays a significant role in forensic analysis, the field of chemistry comes in a close second in forensic science because of its many applications. Causes of death due to toxicity, correlations between chemical substances and injury or harm, rates of intoxication or drug influence, to name just a few, are dependent upon chemical identification and analysis. Of recent interest is the draconian impact of fentanyl, a chemical substance that increasingly challenges our culture and our law enforcement, social and human agencies. Toxicology is the field that identifies and delineates the components of all substances but in forensic law and science, stresses the cause and effect relationship between the substances and the harms caused to victims of its toxicity (See Figure 5.1).

5.2 TOXICOLOGY AS EXPERT EVIDENCE

Most toxicology cases insist on some level of expertise beyond that of the lay witness. As an exception, it could be argued that police officers, reading or giving testimony on Breathalyzer results, may be experts on the machine and its operation, but not on how it computes or concludes that a rate or ratio has been computed. For the most part, the legal system expects higher levels of educational, experiential and peer review in the field of toxicology. In the following series of cases, one will readily discern this need for the expert yet at the same time, despite the background and pedigree of the expert, there is no automatic assurance on the admissibility or acceptability of the opinion offered.

Locate the following case using your favorite legal research resource, read and answer the questions below.

Carswell v. Pihl, 999 F.3d 18 United States Court of Appeals for the First Circuit
May 27, 2021, Decided, No. 19-1630

Questions:

1. Why was this operation, coordinated by the U.S. Military, called "Operation Crested Ice?"
2. What toxic substance was involved in the claim for benefits?
3. Is it fair to conclude that the Court ruled that plutonium is not a substance that could cause harm or could be properly labeled "toxic?" If this not be so, and the Court ruled that plutonium is a toxic substance, how could the claimants lose their case?

> Browse the Amerithrax Investigative Summary regarding the Anthrax Letter attacks at https://www.justice.gov/archive/amerithrax/docs/amx-investigative-summary2.pdf.

DOI: 10.4324/9781003267126-5

FIGURE 5.1 Equipment to test letters for anthrax in Amerithrax investigation in November 2001. See https://www.fbi.gov/history/famous-cases/amerithrax-or-anthrax-investigation for details.

Source: FBI Media.

Hall v. Conoco Inc.
886 F.3d 1308
United States Court of Appeals for the Tenth Circuit
April 10, 2018, Filed
No. 17-6086

I. Background

As a child, Ms. Hall had lived near ConocoPhillips's refinery in Ponca City, Oklahoma. Roughly two decades later, Ms. Hall developed a form of leukemia known as "Acute Myeloid Leukemia with Inversion 16." This disease, according to Ms. Hall, resulted from her early exposure to the refinery's emissions of benzene.

In district court, Ms. Hall tried to prove this link through three expert witnesses:

1. Dr. David Mitchell, an air modeler,
2. Dr. Steven Gore, an oncologist, and
3. Dr. Mary Calvey, an epidemiologist.

Dr. Mitchell created an air model to estimate benzene concentrations near where Ms. Hall had lived. Based on Dr. Mitchell's estimates, Dr. Gore

- calculated Ms. Hall's cumulative exposure to benzene and
- used this calculation to opine that benzene exposure had caused Ms. Hall's leukemia.

Dr. Calvey expressed a similar opinion.

ConocoPhillips moved for:

- exclusion of opinion testimony by multiple expert witnesses, including Dr. Gore and Dr. Calvey and
- summary judgment on the issue of causation.

The district court granted the motion to exclude the expert testimony by Drs. Gore and Calvey. In the absence of their testimony, the court also granted summary judgment to ConocoPhillips, concluding that Ms. Hall had not presented sufficient evidence linking her disease to benzene exposure.

II. Exclusion of Expert Testimony

Ms. Hall challenges the district court's exclusion of expert testimony by Drs. Gore and Calvey. We reject this challenge.

A. Standard of Review

The district court has "wide latitude" in deciding whether to exclude expert testimony, and we review the manner in which the district court exercises this gatekeeping function for an abuse of discretion. Bitler v. A.O. Smith Corp., 400 F.3d 1227, 1232 (10th Cir. 2005).

The district court's exclusion of expert testimony is governed by federal law. See Sims v. Great Am. Life Ins., 469 F.3d 870, 879 (10th Cir. 2006). Under Federal Rule of Evidence 702, a qualified expert witness may give opinion testimony if

- the expert's scientific knowledge would help the fact-finder understand the evidence,
- "the testimony is based on sufficient facts or data,"
- "the testimony is the product of reliable principles and methods," and
- "the expert has reliably applied the principles and methods to the facts of the case."

Fed. R. Evid. 702.

Before expert testimony can be admitted, the district court must determine that the proposed testimony is reliable. Daubert v. Merrell Dow Pharms., Inc., 509 U.S. 579, 589, 113 S. Ct. 2786, 125 L. Ed. 2d 469 (1993).

The district court's assessment of reliability is reviewed for an abuse of discretion. Goebel v. Denver & Rio Grande W. Ry., 346 F.3d 987, 990 (10th Cir. 2003). This review includes consideration of whether "the reasoning and methodology underlying the expert's opinion … is both scientifically valid and applicable to a particular set of facts." Id. at 991.

B. Dr. Gore's Testimony

Dr. Gore rendered a differential diagnosis on the cause of Ms. Hall's leukemia. A differential diagnosis "rule[s] in all scientifically plausible causes of the injury and then rule[s] out the least plausible causes until only the most likely cause remain[s]." Id. at 990.

Dr. Gore's first step was to rule in benzene, smoking, and idiopathic causes as potential causes of Ms. Hall's leukemia. Dr. Gore then ruled out smoking as a potential cause, leading him to conclude that benzene exposure had caused Ms. Hall's leukemia. Dr. Gore did not expressly rule out the possibility of idiopathic causes.

The district court assumed that a differential diagnosis could provide a suitable methodology. But the district court regarded Dr. Gore's differential diagnosis as unreliable based partly on his failure to justify ruling in benzene or ruling out "idiopathic causes" of Ms. Hall's disease. Hall v. ConocoPhillips, 248 F. Supp. 3d 1177, 1190-91 (W.D. Okla. 2017) (quoting Chapman v. Procter & Gamble Distrib., 766 F.3d 1296, 1311 (11th Cir. 2014)). This reasoning fell within the district court's discretion.

1. Ruling In Benzene

To use a differential diagnosis, Dr. Gore needed to consider whether he could rule in benzene as a potential cause of Ms. Hall's disease. We may assume, as the district court did, that benzene emissions can cause Acute Myeloid Leukemia with Inversion 16. With this assumption, however, Ms. Hall would still have needed to show that the benzene emissions actually caused her disease. For this showing, Ms. Hall relied on Dr. Gore's quantification of the exposure to benzene.

To quantify the exposure, Dr. Gore used the work of another expert witness, Dr. Mitchell, who had constructed an air model to estimate the highest hourly average concentration of benzene. With this estimate, Dr. Gore quantified the cumulative exposure to benzene based on how long Ms. Hall had lived near the refinery. The result led Dr. Gore to rule in benzene as a potential cause of Ms. Hall's leukemia.

The district court found two flaws in this methodology:

1. Dr. Gore could not reliably use the highest hourly average-emission level to calculate Ms. Hall's cumulative exposure to benzene.
2. Dr. Gore's calculation was based on mistakes involving the extent of Ms. Hall's exposure to benzene.

In finding these flaws, the district court acted within its discretion.

a. The Choice of the Highest Hourly Average- Emission Level

Dr. Gore estimated Ms. Hall's cumulative exposure by taking one of Dr. Mitchell's figures (the highest hourly average-emission level). ConocoPhillips questions the use of this figure, pointing out that Dr. Gore could have used other figures from Dr. Mitchell's air model.

In this model, Dr. Mitchell had estimated benzene emissions over an extended time period. This estimate included the highest hourly average concentration of benzene while Ms. Hall lived near the refinery. Dr. Mitchell took this figure and instructed Dr. Gore to use it for his own calculation.

But Dr. Mitchell admitted that he was not qualified to decide which figure to use in assessing the impact of benzene emissions on human health:

Q. You are not qualified when it comes to analyzing what value—acute, chronic, subchronic—is more relevant for assessing human health risk for benzene or any other chemical. You are not qualified to offer that opinion.
A. I fully agree with that. I'm an air modeler and I'm a meteorologist.
Q. And you don't know which one of those matters most or matters at all for assessing whether or not Samantha Hall developed [acute myeloid leukemia] from benzene?
A. Counselor, that's out of my expertise. I'm not—I'm an air modeler and a meteorologist.

Appellant's App'x at 4944-45. Dr. Mitchell conceded that selection of the concentration level was best left to an oncologist like Dr. Gore. In turn, Dr. Gore acknowledged that he had not independently chosen which figure to use. Instead, he had admittedly relied on Dr. Mitchell's expertise to select the pertinent figure.

Dr. Gore testified that he had called Dr. Mitchell, who provided assurance that the highest hourly average- emission level was the metric used in the industry. Appellant's App'x at 5162. The district court could reasonably consider this assurance an inadequate safeguard of reliability. ...

Without an adequate safeguard of reliability, the district court could reasonably conclude that neither Dr. Gore nor Dr. Mitchell could defend the use of the highest hourly average-emission level. Both expert witnesses had seemingly disclaimed any responsibility for picking this figure: Dr. Mitchell had regarded Dr. Gore as the expert and Dr. Gore had regarded Dr. Mitchell as the expert, leaving no one qualified to choose which concentration level to use.

This circular loop stirred broader concern over the entirety of Dr. Gore's calculation. That calculation was designed to estimate Ms. Hall's cumulative exposure to benzene. If Dr. Gore had calculated the cumulative exposure based on an incorrect figure, the court could reasonably question the decision to rule in benzene as a potential cause.4 As a result, we conclude that the district court acted within its discretion in questioning the reliability of Dr. Gore's decision to use the highest hourly average-emission level.

b. Dr. Gore's Errors and Inconsistencies Regarding the Extent of Ms. Hall's Exposure to Benzene

The district court also identified three of Dr. Gore's errors and inconsistencies about the extent of Ms. Hall's exposure to benzene.

First, Dr. Gore assumed that Ms. Hall had lived near the refinery for eight years. But Dr. Gore was mistaken; Ms. Hall had lived near the refinery for only about four years. As Dr. Gore acknowledged, this error meant that his estimate of the cumulative exposure was roughly double what it should have been.

Second, Dr. Gore vacillated on his own methodology. He initially stated that he had calculated Ms. Hall's exposure by

- using the highest hourly average-emission level from Dr. Mitchell's model and
- assuming that Ms. Hall had been exposed to that level for one hour every day.

But in a later declaration, Dr. Gore stated that he had assumed exposure for 8 hours per day, 5 days per week, 50 weeks per year. At a hearing, Dr. Gore switched back, testifying that his declaration had been mistaken and that he had actually assumed only one hour of exposure per day.

Third, the district court found that Dr. Gore had been inconsistent about what he was calculating. He sometimes said that he was calculating Ms. Hall's exposure to benzene; other times, however, he stated that he was calculating exposure to volatile organic compounds. These inconsistencies mattered because benzene is only one of many volatile organic compounds.

In light of these errors and inconsistencies, the district court could reasonably question the reliability of Dr. Gore's opinion ruling in benzene exposure as a potential cause.

2. Ruling Out Idiopathic Causes

Even if benzene could be ruled in as a potential cause, Dr. Gore's differential diagnosis would have depended on his ability to rule out less plausible causes of Ms. Hall's disease. The district court concluded that Dr. Gore had failed to rule out idiopathic causes, and this conclusion fell within the district court's discretion.

The parties agreed that acute myeloid leukemia is frequently caused by idiopathic sources. For example, Ms. Hall presented expert testimony that 70-75% of cases of acute myeloid leukemia are idiopathic. And Dr. Gore agreed that acute myeloid leukemia is ordinarily idiopathic in origin. The resulting question is whether Dr. Gore needed to consider and rule out idiopathic causes.

The district court answered "yes."

A differential diagnosis is designed to identify the most likely cause of a disease. ... An expert need not consider and rule out every conceivable cause. ... But the expert must still consider all plausible causes and rule out the less plausible ones until only the most likely cause remains. ...

Dr. Gore did not rule out the possibility of an idiopathic cause for Ms. Hall's acute myeloid leukemia. This omission concerned the district court because the evidence had pointed to idiopathic causes in most cases of acute myeloid leukemia, and the district court could reasonably view the failure to rule out idiopathic causes as a fatal error tainting the differential diagnosis. ...

Ms. Hall responds with two arguments:

1. The district court misunderstood the concept of "idiopathic" causes.
2. The Tenth Circuit does not require differential diagnoses to rule out idiopathic causes.

We reject both arguments.

First, Ms. Hall argues that the district court misunderstood the term "idiopathic." She characterizes "idiopathic" as a diagnosis by exclusion: only if all known factors are ruled out, leaving no known plausible factors, can the leukemia be considered idiopathic. Under this view, it is illogical to ask Dr. Gore to "rule out" idiopathic causes because a disease can be considered idiopathic only if all identifiable causes have been ruled out.

Based on this definition of "idiopathic," Ms. Hall regards the district court's reasoning as faulty. To Ms. Hall, Dr. Gore properly ruled in all known risk factors and ruled out the implausible ones until he was left only with benzene. Because benzene remained a plausible cause, the term "idiopathic" would not fit.

But the district court could reasonably take a different approach when analyzing "idiopathic" causes. One can regard "idiopathic" causes as recognition of the medical community's inability to identify all of the causes of a disease. ...

Because idiopathy accounts for more than half of the cases of acute myeloid leukemia, a differential diagnosis could be considered inherently unreliable here. See Tamraz, 620 F.3d at 675 ("[U]nknown (idiopathic) causation ... currently accounts for the vast majority of Parkinson's Disease cases, making it impossible to ignore and difficult to rule out."); Bland v. Verizon Wireless, (VAW), 538 F.3d 893, 897 (8th Cir. 2008).

("Where the cause of the condition is unknown in the majority of cases, [the expert witness] cannot properly conclude, based upon a differential diagnosis, [the plaintiff's] exposure to freon was 'the most probable cause' of [the plaintiff's] exercise-induced asthma."); ...

Given the predominance of idiopathic causes, we conclude that the district court had the discretion to consider Dr. Gore's differential diagnosis unreliable.

Second, Ms. Hall contends that the Tenth Circuit does not require experts to rule out idiopathic causes in conducting a differential diagnosis. For this contention, Ms. Hall relies on Bitler v. A.O. Smith Corp., 400 F.3d 1227 (10th Cir. 2005). Ms. Hall has misinterpreted Bitler. There we observed that an expert need not "'categorically exclude each and every possible alternative cause.'" Bitler, 400 F.3d at 1238 n.6 (quoting Stephen A. Saltzburg et al., Federal Rules of Evidence Manual 702-33

(8th ed. 2002)). This observation does not preclude a district court from questioning the reliability of a differential diagnosis when the cause of a disease is unknown most of the time.

* * *

The district court could justifiably regard Dr. Gore's differential diagnosis as unreliable because of his failure to (1) justify ruling in benzene or (2) rule out idiopathic causes. Thus, the district court did not abuse its discretion in excluding Dr. Gore's opinion based on his differential diagnosis.

...

III. Conclusion

We conclude that the district court did not err:

- in excluding the testimony of Drs. Gore and Calvey and
- in concluding that ConocoPhillips was entitled to summary judgment on causation.

Therefore, we affirm.

Questions:

1. Explain the basis of the claim relating to the petitioner's leukemia?
2. In this case, the claimant's allegation of benzene causing leukemia was neither proven nor was the court happy about Expert Witness's ready to give said testimony. Part of the reason for this was the expert failed to do what?
3. Does the conclusion that a cause of disease is "idiopathic" favor defense or plaintiff's counsel in this case?

Williams v. Mosaic Fertilizer, LLC
889 F.3d 1239
United States Court of Appeals for the Eleventh Circuit
May 14, 2018, Decided
No. 17-10894

...

I.

Ms. Williams was born in 1967 and has lived her entire life at the same residence in Tampa, Florida. The home, located in the Progress Village neighborhood, is approximately three miles from Mosaic's Riverview plant. She alleged that she suffers from G6PD associated pulmonary hypertension, asthma-related restrictive pulmonary function, obstructive pulmonary disease, airway remodeling, lower lung scarring, allergic reactions, side effects from therapeutic treatments for her lung disorders, extreme fatigue, intense abdominal pain, and diabetes. She alleged that chemicals emitted from Mosaic's facility caused, contributed to, or exacerbated these conditions.

According to Ms. Williams, Mosaic's operations in and around the Riverview plant involve the production and handling of a number of chemicals, including sulfuric acid, phosphoric acid, fluoride, and ammoniate phosphates. She averred that Mosaic's production of these substances produces emissions, in the form of dust and particulates, of toxic substances that permeate the ambient air in and around her home and community. Some of these include various types of particulate matter, arsenic, cadmium, chromium, lead, manganese, nickel, phosphorous, and zinc.

In the past, the U.S. Environmental Protection Agency ("EPA") and the Hillsborough County Environmental Protection Commission have promulgated data taken from monitoring stations at or near the Riverview plant showing that sulfur dioxide levels in the ambient air at those stations exceeded the National Ambient Air Quality Standard ("NAAQS") of 75 parts per billion.

Additionally, Hillsborough County at large has been found in violation of the NAAQS standard, and the ambient air at a monitoring station at the Riverview plant has on multiple occasions been found in violation of Florida's more lenient standard of 100 parts per billion. In addition to these sulfur dioxide emissions, monitoring data at a testing site located near the Riverview facility and Ms. Williams' neighborhood showed, on at least one occasion, that the concentration of PM10 respirable particulates, a respiratory irritant, exceeded the national standard of 150 micrograms based on a one-hour average.

Ms. Williams alleged six causes of action under Florida law in separate counts: negligence, gross negligence, strict liability, strict liability failure to warn, strict liability for "prohibited discharge" of pollutants, and medical monitoring and environmental testing.1 To establish general and specific causation, she turned to Dr. Mink. Dr. Mink, an experienced toxicologist, prepared and furnished a summary of his "preliminary expert opinions."

The report contained sixteen pages of analysis, including a description of Ms. Williams' medical background, information on G6PD deficiency, and a set of "preliminary expert opinions" reached "with a reasonable degree of scientific certainty." Those opinions were:

1. Rhonda Williams has been exposed to significant quantities of regulated pollutants and hazardous materials from both direct and fugitive sources as a result of Mosaic's operations including phosphogypsum mining, processing, storage, transportation and waste handling operations over her lifetime residence in Prospect Village, Florida primarily through inhalation and dermal exposures.
2. Rhonda Williams has developed significant adverse health effects as a result of these hazardous exposures including G6PD associated pulmonary hypertension and obstructive pulmonary disease resulting in a diminished quality of life and potentially reduced life span.
3. Rhonda Williams has developed significant adverse health effects as a result of secondary effects from therapeutic agents used to treat her diseases/symptoms resulting from these exposures further diminishing her quality of life and threatening her long-term physical and mental wellbeing.

At the end of the analysis section of the report, Dr. Mink listed fifty-eight references. These consisted of various empirical studies, website references, and regulatory documents. Within the body of the analysis, he cited another eighteen sources. None were pin-cited or otherwise annotated to show which portions supported each conclusion.

After Dr. Mink submitted his preliminary report, Mosaic deposed him. After the deposition, Mosaic moved to exclude Dr. Mink's testimony under Federal Rule of Evidence 702 and Daubert. Mosaic also moved for summary judgment as to all of Ms. Williams' claims. Ms. Williams filed responses in opposition. Without conducting a Daubert hearing, the District Court granted Mosaic's motion to exclude Dr. Mink's opinion testimony. The Court explained its decision thusly:

> [T]he Court concludes that Dr. Mink's expert report and proposed testimony cannot survive a Daubert challenge. Critically, Dr. Mink has neglected the hallmark of the science of toxic torts—the dose response relationship. In and of itself, this is a sufficient basis for excluding his testimony. He also unjustifiably relies on regulatory standards to determine dose, infers facts from studies that contradict his conclusions, fails to consider the background risk for Plaintiff's illnesses, fails to rule out alternative

> potential causes of Plaintiff's illnesses, and renders speculative and conclusory opinions about Plaintiff's exposure to Mosaic's emissions. All in all, it is clear that Dr. Mink has failed to adhere to the methodology expected of toxicologists in toxic tort cases, and he has not demonstrated a reliable basis for his opinions.

In its order granting the motion, the Court comprehensively analyzed Dr. Mink's report and methodology, identifying and explaining its primary concerns and others. Because Dr. Mink was Ms. Williams' sole source of causation evidence, the Court, in the same order, granted Mosaic's motion for summary judgment as to all claims requiring a showing of causation.

...

II.

We review the District Court's grant of summary judgment de novo. Doe v. Sch. Bd., 604 F.3d 1248, 1253 (11th Cir. 2010). We begin with the exclusion of Dr. Mink's opinion testimony. Federal Rule of Evidence 702 makes expert opinion testimony admissible only "if (1) the testimony is based upon sufficient facts or data, (2) the testimony is the product of reliable principles and methods, and (3) the witness has applied the principles and methods reliably to the facts of the case." As interpreted by the Supreme Court in Daubert, this requires the trial court "to act as a gatekeeper to insure that speculative and unreliable opinions do not reach the jury." McClain v. Metabolife Int'l, Inc., 401 F.3d 1233, 1237 (11th Cir. 2005). To properly serve as a gatekeeper, the trial court must perform "a preliminary assessment of whether the reasoning or methodology underlying the testimony is scientifically valid and of whether that reasoning or methodology properly can be applied to the facts in issue." Daubert, 509 U.S. at 593-94, 113 S.Ct. at 2796. When doing so, "the court must consider the testimony with the understanding that the burden of establishing qualification, reliability, and helpfulness rests on the proponent of the expert opinion." McClain, 401 F.3d at 1238 (quotations omitted) (alterations omitted) (quoting United States v. Frazier, 387 F.3d 1244, 1260 (11th Cir. 2004)).

"The trial judge must have considerable leeway in deciding in a particular case how to go about determining whether particular expert testimony is reliable" under Daubert. Kumho Tire Co. v. Carmichael, 526 U.S. 137, 152, 119 S. Ct. 1167, 1176, 143 L. Ed. 2d 238 (1999). Hence, we review a district court's Daubert rulings for abuse of discretion. Gen. Elec. Co. v. Joiner, 522 U.S. 136, 143, 118 S. Ct. 512, 517, 139 L. Ed. 2d 508 (1997). We will affirm unless the court "has made a clear error of judgment, or has applied an incorrect legal standard." Piamba Cortes v. Am. Airlines, Inc., 177 F.3d 1272, 1306 (11th Cir. 1999) (quotations omitted).

Here, the District Court did not abuse its discretion in excluding Dr. Mink's opinions. The Court comprehensively analyzed Dr. Mink's report and deposition testimony, carefully reviewed the literature upon which he relied in forming his causation opinions, and determined that his methodology was undermined by multiple defects.2 We find no error in the Court's analysis and no need to discuss every aspect of its comprehensive assessment anew. However, three of the most significant problems with Dr. Mink's methodology are illustrative: first, his failure to properly assess dose-response with regard to Ms. Williams; second, his failure to meaningfully rule out other potential causes of Ms. Williams' medical conditions; and third, his failure to account for the background risk of her conditions.

With respect to dose-response, we have explained the importance of the dose-response assessment in these sorts of cases before. "When analyzing an expert's methodology in toxic tort cases, the court should pay careful attention to the expert's testimony about the dose-response relationship." McClain, 401 F.3d at 1241. This attention is due because dose-response is "the hallmark of basic toxicology." Id. at 1242. Stripped to its bare essentials, a dose-response assessment estimates scientifically "the dose or level of exposure at which [the substance at issue] causes harm." Id. at 1241.

Here, Dr. Mink conceded he never conducted an independent dose calculation specific to Ms. Williams. Instead, he relied on two academic studies measuring the ambient air concentration of pollutants in the area in which Ms. Williams lived to estimate the dose she received and on the EPA's NAAQS regulatory standards to establish the dose threshold above which Ms. Williams' conditions would likely result from her exposure. As the District Court correctly observed, his reliance on these sources was methodologically problematic in multiple respects. Among the most glaring problems was the fact that both academic studies he relied upon directly contradicted his causation opinions. For example, one study concluded that "phosphate fertilizer plants make minor contribution to the ambient levels of HAP metals compared to other sources for the general population in the Tampa Bay area," and that the air concentration of the various pollutants studied fell hundreds of times below levels that would present health risks to the public. Hsing- Wang Li, et al., Impacts of Hazardous Air Pollutants Emitted from Phosphate Fertilizer Production Plants on Their Ambient Concentration Levels in the Tampa Bay Area, 8 Air Qual. Atmos. Health 453, 453, 464 (2015).

This cuts directly against his opinion that Mosaic's emissions adversely affected Ms. Williams. And although Ms. Williams argues that Dr. Mink relied on the studies' underlying data while disagreeing with their ultimate conclusions, Dr. Mink and Ms. Williams never clearly explained why Dr. Mink reached a different conclusion with regard to Mosaic's contribution to pollution in Ms. Williams' community or why his conclusion was correct and the authors' incorrect until after the District Court ruled on the admissibility of his testimony. Thus, he failed to squarely present the basis for his disagreements to the District Court until after the fact. See Kilpatrick v. Breg, Inc., 613 F.3d 1329, 1341 (11th Cir. 2010) (holding the district court did not abuse its discretion in excluding expert testimony, when the expert "had ample opportunity to identify all of the bases for his conclusions and to explain his methodology in reaching those conclusions" yet failed to do so).

Moreover, assuming for the sake of argument that Dr. Mink's dose estimates were methodologically sound, he failed to demonstrate a scientific basis for concluding that those exposure levels would likely produce, contribute to, or exacerbate Ms. Williams' conditions. For example, he estimated that Ms. Williams was exposed to sulfur dioxide at a long-term concentration rate exceeding 75 parts per billion (though he never reached a more specific number than this), and, based on her purported heightened sensitivity to exposure and the NAAQS standard's calculation to protect sensitive members of the population, assumed that exposure at or above this level was likely to cause her conditions. He based this sweeping assumption on the fact that the 75 parts per billion number is the primary emissions standard set by the EPA for sulfur dioxide in its NAAQS standard. However, this Court has previously explained the potential methodological perils of relying, at face value, on regulatory emissions levels to establish causation. See McClain, 401 F.3d at 1249 (observing that face-value reliance on regulatory dose standards raises a "subtle methodological issue").

The biggest problem stems from the potential difference in purpose between regulatory standards and toxicological dose-response calculations: regulatory standards often build in considerable cushion in order to account for the most sensitive members of the population and prophylactically protect the public (in other words, they are protective), while dose-response calculations aim to identify the exposure levels that actually cause harm (in other words, they are predictive). See id. at 1249-50 (explaining the different calculations and burdens of proof employed by regulatory agencies in setting exposure standards for the general public, as compared to those employed by toxicologists in calculating dose-response).

But Ms. Williams argues that Dr. Mink's facial reliance on NAAQS standards in this case is different, because those standards are predictive. She contends that the EPA's assessments reveal that "exposure to 75 ppb of [sulfur dioxide] causes (not may cause or can cause, but actually does cause) respiratory morbidity," and that the standards result from "dose-response assessments based on human studies." (Emphasis in original). In other words, relevant dose- response assessments are built into the standard. However, the EPA's own regulations require the agency to establish NAAQS levels ensuring "protection of public health with an adequate margin of safety," even if this means setting emissions thresholds at a level at which "the risk is not precisely identified as to nature or degree." 75 Fed. Reg. 35520-01 at *35521 (June 22, 2010). Further, Dr. Mink stated that he relied upon dose- response calculations included in the EPA's Integration Risk Information System ("IRIS"), which the EPA relies upon in setting NAAQS standards. But the EPA has expressly stated that the dose-response assessments in IRIS are not suited to predicting the incidence of exposure-caused disease in humans:

> In general, risk values, such as those on IRIS, cannot be used to predict the actual incidence of human disease or the type of effects chemical exposures may have on humans. This is due to the numerous uncertainties involved in risk assessment, including those associated with extrapolations from animal data to humans and from high experimental doses to lower environmental exposures. The organs affected and the types of adverse effects resulting from chemical exposure may differ between study animals and humans. In addition, many factors besides exposure to a chemical influence the occurrence and extent of human disease.

53 Fed. Reg. 20162-02 at *20163 (June 2, 1988). It is for precisely these reasons that we have cautioned against facial reliance on regulatory standards as a substitute for scientifically rigorous dose-response assessments. And while it is true that the EPA's studies "led to the conclusion that there is a causal relationship between respiratory morbidity and short-term exposure to SO2," the existence of a generalized "causal relationship" was not the operative issue. What mattered was whether the EPA's data provided a reasonably specific calculation as to the exposure amounts required to cause Ms. Williams' conditions upon which Dr. Mink could reasonably rely in forming his opinions.

Though Ms. Williams alleged that, on account of her G6PD, she is more sensitive to exposure than the average member of the general public, Dr. Mink never attempted to quantify how much more sensitive she is. As the District Court observed, at the time it made its ruling, he did "not cite with specificity which studies and peer-reviewed literature demonstrate the increased sensitivity of individuals with G6PD. Consequently, the Court [could] not evaluate those sources and determine whether they are based on reliable methodologies or otherwise support his opinions." Without this, he could not rely on NAAQS standards, because he never adequately established whether her sensitivity to exposure placed her within the class of persons who would likely suffer from exposure at the ranges he estimated, let alone whether the standards were predictive or protective with regard to her.

To be clear, we have never required an expert to "give precise numbers about a dose-response relationship," McClain, 401 F.3d at 1241 n.6, and we do not do so here. But we do require an expert to lay a "reliable groundwork for determining the dose-response relationship." Id. at 1241. Here, Ms. Williams bore the burden of demonstrating to the District Court that Dr. Mink's facial reliance on NAAQS standards was methodologically sound. In light of their protective nature and the EPA's express warnings that those standards—and the IRIS assessments upon which they were formulated—are unreliable predictors of conditions in humans, we are not persuaded that the District Court erred in determining that Ms. Williams failed to meet that burden.

Further, the District Court correctly observed that Dr. Mink's report suffered from methodological errors related to the cause of Ms. Williams' conditions. Dr. Mink failed to meaningfully rule out other potential causes of Ms. Williams' conditions and symptoms. Indeed, one of the studies heavily relied upon by Dr. Mink determined that environmental factors and emissions by other facilities caused the vast majority of pollution in the area in which Ms. Williams lived. Li, supra, at 453. Yet Dr. Mink never addressed how or even if he ruled out those other potential contributors in reaching his conclusion that Mosaic's facility caused Ms. Williams' alleged harms. Nor did he, in his report or his deposition, eliminate or even address factors not related to air quality, such as Ms. Williams' obesity, allergies, lifestyle, exposure to secondhand smoke, or possible genetic predisposition. When asked about how he eliminated other potential causes of Ms. Williams' conditions, Dr. Mink stated:

> Oh, I think we eliminated causes based on their probability. I mean, they were so low in comparison that—that we can eliminate them. Are they totally nonexistent, absolutely not, and I think I testified to that earlier.

However, Dr. Mink never provided the District Court or this Court with any probability studies regarding those potential causes, and his expert report never referenced those probabilities.

Thus, although he made passing references to the purported "low" probability of those other causes, Dr. Mink never provided the District Court with any scientific basis upon which he relied in concluding that the likelihood that various other potential factors caused Ms. Williams' conditions was low enough to reasonably rule them out. The law does not require the District Court to take him at his word. See Joiner, 522 U.S. at 146 ("[N]othing in either Daubert or the Federal Rules of Evidence requires a district court to admit opinion evidence that is connected to existing data only by the ipse dixit of the expert.")

Nor does it appear Dr. Mink meaningfully accounted for the background risk of Ms. Williams' conditions. The background risk "is the risk a plaintiff and other members of the general public have of suffering the disease or injury that plaintiff alleges without exposure to the drug or chemical in question." McClain, 401 F.3d at 1243 (emphasis in original). When asked during his deposition, Dr. Mink did not demonstrate specific knowledge of the general prevalence of Ms. Williams' various conditions:

Q: What is the prevalence of pulmonary hypertension in the population, the general population?
A: I don't recall. It's not large.
Q: What is the prevalence of obstructive pulmonary disease in the population?
A: I don't recall specifically, but it's significant.

These answers do not indicate serious consideration of the background risk. And while his deposition alone may not authoritatively indicate that he failed to account for it, he also failed to address background risk in his report or elsewhere. Thus, the evidence before the District Court at the time it considered Mosaic's Daubert motion does not indicate he meaningfully accounted for background risk.

In conclusion, Dr. Mink failed to meaningfully address other potential causes of Ms. Williams' conditions or even the background risk of those conditions. This fundamental methodological failure undermined the soundness of his causation opinions, and the District Court was therefore right to exclude those opinions as unsound.

…

IV.

We accordingly affirm the District Court's exclusion of Dr. Mink and Ms. Williams' opinion testimony and its subsequent grants of summary judgment in favor of Mosaic.

AFFIRMED.

Questions:

1. Plaintiff alleges a sweeping array of toxic substance exposure due to living in proximity to the plant. What are some of the toxic substances emanating from the plant?
2. Prove True or False: The Agency dose-response relationship tends to be favorable to the plaintiff and make it more likely that an expert's conclusion on toxicity be admitted?
3. How does the appellate court reconcile the trial court's denial of the testimony of Dr. Mink?

McManaway v. KBR, Inc.
852 F.3d 444
United States Court of Appeals for the Fifth Circuit
March 27, 2017, Filed
No. 15-20641

I

In March 2003, the United States Army Corps of Engineers (USACE) hired KBR to help restore Iraq's oil infrastructure following the United States' invasion as part of Project Restore Iraqi Oil (Project RIO). Pursuant to Task Order 3 (part of KBR's contract with USACE) KBR was to begin work on a facility after USACE determined conditions were "benign." Benign conditions meant, among other things, that the site was "cleared of all enemy forces" and "environmental hazards" including nuclear, biological, chemical, and industrial hazards. USACE did not perform an environmental assessment of Qarmat Ali, a water injection facility that KBR was tasked with restoring, and it is unclear whether USACE declared conditions there benign. Nonetheless, USACE (specifically, Task Force RIO) authorized KBR to start work at Qarmat Ali on May 13, 2003, and KBR began work later that month. Plaintiffs are current or former members of the Army National Guard or the British Royal Airforce who provided military protection for KBR at Qarmat Ali.

The company that operated Qarmat Ali prior to KBR's arrival onsite used sodium dichromate, an anti-corrosive agent and hexavalent chromium compound, when injecting water into oil reservoirs for the purpose of forcing oil to the surface. The use of sodium dichromate at Qarmat Ali, particularly the improper storage of the substance, led to air and soil contamination. Plaintiffs contend that they suffered injuries as a result of their exposure to sodium dichromate, a known carcinogen and irritant, while working at Qarmat Ali.

In the district court, Plaintiffs adduced evidence that they were exposed to sodium dichromate. For example, they submitted a Royal Airforce report concluding that "operational activity would suggest" that soldiers working at Qarmat Ali were exposed to sodium dichromate, although the extent of the exposure could not "be satisfactorily quantified." Plaintiffs also submitted a 2008 National Guard memorandum concluding that soldiers at Qarmat Ali "had a high potential for direct exposure," and an October 2003 report from the Army Center for Health Promotion and Preventive Medicine, concluding that before containment of the chemical, "chromium concentrations exceeded (military) health risk screening values," and exposure "[p]robably occurred to some degree prior to containment." Moreover, August 2003 urine and blood sampling of twenty-seven KBR employees who worked at Qarmat Ali showed that four of the tested employees had higher-than-normal levels of chromium in their urine and twenty-three of the tested employees had higher-than- normal levels of chromium in their blood.

Plaintiffs contend that exposure to sodium dichromate was deleterious to their health, both during and after their service at Qarmat Ali. Plaintiff Jody Aistrop observed that "[e]verybody" in his company complained of nosebleeds and rashes and that he experienced such symptoms, as well as gastrointestinal distress. Plaintiff Russell Powell, a medic, observed that he and other soldiers experienced "intense" gastrointestinal problems. Plaintiff Russell Kimberling developed headaches, respiratory problems, diarrhea, skin rashes, and a septal hole. In 2009, Lieutenant Colonel James Gentry died of lung cancer. The Army deemed his death to be "In Line of Duty for Exposure to Sodium Dichromate between June—September 2003."

In 2010, a group of plaintiffs filed suit in the Southern District of Texas, alleging negligence, gross negligence, fraud, and intentional infliction of emotional distress. Plaintiffs claimed that KBR was aware of the sodium dichromate contamination at Qarmat Ali and failed to take appropriate steps to reduce the risk of harm to the soldiers, to inform the U.S. or British military, or to warn the soldiers who were exposed to the contamination. Plaintiffs claimed that KBR's tortious actions caused them to suffer, among other ailments, nose bleeds, skin lesions, physical pain, emotional distress, and death.

In the district court, KBR filed a motion to dismiss on the grounds that the case presented a non-justiciable political question and that Plaintiffs' claims were preempted by federal common law. The court denied this motion.4 KBR also moved for summary judgment claiming, inter alia, that Plaintiffs could not establish that sodium dichromate caused their injuries. The court granted KBR's motion for summary judgment, dismissing Plaintiffs' claims on causation grounds and, separately, granted KBR's motion for summary judgment, dismissing individual plaintiff Lieutenant Colonel James Gentry's claims on similar grounds. The court entered judgment in favor of KBR. Plaintiffs timely appealed.

…

IV

To prove their tort claims against KBR—fraud, negligence, gross negligence, and intentional infliction of emotional distress—Plaintiffs must establish that KBR's conduct caused their injuries. See, e.g., Wheaton Van Lines, Inc. v. Mason, 925 S.W.2d 722, 728 (Tex. App. 1996). The district court found that Plaintiffs' evidence was insufficient to prove causation because their expert's methodology did not adequately explain the connection between their exposure levels and their injuries and Plaintiffs failed to provide sufficient epidemiological support to show that sodium dichromate exposure caused their alleged injuries. On appeal, Plaintiffs argue that the admission of expert testimony supporting causation bars summary judgment, that their expert's "differential diagnosis" is sufficient to establish causation, and that lay testimony is sufficient to establish causation for their acute injuries.

To survive summary judgment, Plaintiffs must establish a genuine issue of material fact as to whether exposure to sodium dichromate caused their injuries "based on a reasonable medical probability and scientifically reliable evidence." Black v. Food Lion, Inc., 171 F.3d 308, 310 (5th Cir. 1999). Plaintiffs must show both that sodium dichromate is capable of causing their alleged injuries ("general causation") and that it did, in fact, cause their injuries ("specific causation"). See Merrell Dow Pharm., Inc. v. Havner, 953 S.W.2d 706, 714 (Tex. 1997). It is not enough that Plaintiffs have produced some expert testimony supporting their position, as "the reliability of expert testimony is … a prerequisite to its legal sufficiency." Abraham v. Union Pac. R.R. Co., 233 S.W.3d 13, 17 (Tex. App. 2007) (citing id.).

Plaintiffs offered expert testimony from Dr. Arch Carson, a medical doctor who specializes in treating individuals "who have been exposed to industrial chemical substances," and Dr. Herman Gibb, an epidemiologist who has published several influential journal articles concerning the health effects of chromium. Gibb opined to "a reasonable degree of scientific certainty" that the Plaintiffs "have increased risks for a variety of health effects as a result of their exposure to sodium dichromate [, including] lung cancer; dermal and respiratory allergic response to chromium; other respiratory effects; and renal, gastrointestinal, and reproductive effects." In forming his opinion, Gibb examined litigation documents, including Plaintiffs' medical records. Gibb said that he did not have air quality measurements or other "quantitative information" that would enable him to make a dose-response assessment—an assessment describing the effects caused by differing levels of exposure. He opined that, given the lack of data, it was "impossible" to estimate each plaintiff's exposure. However, he testified that "the reporting of symptoms and the description of the exposure" indicated that Plaintiffs incurred similar exposure to a cohort of industrial workers he had studied, and that Plaintiffs "had enough exposure to … develop symptoms like irritation and so-forth."

Carson opined that, within reasonable medical probability, many of the Plaintiffs' injuries were caused by exposure to sodium dichromate at Qarmat Ali. Carson detailed three categories of

Plaintiffs' injuries: genetic transformation injuries, characterized by cell damage and an attendant heightened risk of cancer; acute irritation injuries, such as skin irritation, breathing difficulties, eye irritation, nosebleeds, sinus problems, headaches, persistent coughing, and gastro-intestinal irritation; and remote exposure injuries, such as ongoing skin, respiratory and gastrointestinal problems. In arriving at his conclusions, Carson examined litigation documents, physically examined the Plaintiffs, and studied Plaintiffs' medical records.

As part of his methodology, Carson assigned each plaintiff he examined a control number reflecting that plaintiff's self-reported time spent at Qarmat Ali. Carson did not consider the specific dates a plaintiff was stationed at Qarmat Ali, considering only the total time each plaintiff reported being at the site. Carson used these control numbers to put each plaintiff into an "exposure category." Carson gauged whether Plaintiffs' injuries were caused by sodium dichromate based on the assigned exposure categories. He testified, "The exposure category followed directly from the control number, and was only utilized in terms of my determination whether or not the relationship between chromium exposure and symptoms or illnesses was plausible." He agreed his methodology was to "look at people and compare the symptoms that they claimed as compared to the exposure category [he] had put them [in] and reach conclusions based on that." Carson did not create any tables or summaries of these exposure categories or their application.

This exposure category analysis was Carson's only quantitative analysis of Plaintiffs' dosage. He stated that, to his knowledge, there were "no credible" exposure measurements. Carson testified that he did not know the dose of sodium dichromate any particular individual plaintiff was exposed to. Nevertheless, he opined that the level of exposure was "many orders of magnitude or much greater than historical industrial exposures" based on:

the amount of contamination; the type of area it was; the fact that it was extremely dry most of the time, although it did rain occasionally; that it was very windy, that there were recurrence [sic] of windstorms periodically; and the preponderance of acute irritative symptoms that the Guardsmen out there and the KBR personnel experienced.

Carson opined that Plaintiffs' dosage was higher than that received by the subjects of Gibb's study and quoted Gibb's report stating that the contamination at Qarmat Ali was similar to that of a "Superfund site[]."10 However, he testified that he was not aware of any studies that have conclusively shown that there are health effects in humans from exposure to a hexavalent chromium compound, such as sodium dichromate, in an outdoor setting like a Superfund site.

The Texas Supreme Court held in Merrell Dow Pharmaceuticals, Inc. v. Havner that, in the absence of direct evidence of causation, plaintiffs may rely on epidemiological studies11 to prove causation where such studies demonstrate a statistically significant doubling of the risk of the injuries alleged. 953 S.W.2d at 718; see Merck & Co. v. Garza, 347 S.W.3d 256, 265 (Tex. 2011). Moreover, "[t]o raise a fact issue on causation" a plaintiff "must show that he or she is similar to those in the studies." Havner, 953 S.W.2d at 720. Proof that one is similarly situated to subjects in epidemiological studies must "include proof that the injured person was exposed to the same substance, that the exposure or dose levels were comparable to or greater than those in the studies, that the exposure occurred before the onset of injury, and that the timing of the onset of injury was consistent with that experienced by those in the study." Id. To prove causation through epidemiological studies, a plaintiff must provide more than one study that meets this criteria. See Garza, 347 S.W.3d at 266; Havner, 953 S.W.2d at 727. Moreover, "other plausible causes of the injury or condition that could be negated [must be excluded] with reasonable certainty." Garza, 347 S.W.3d at 265-66 (quoting Havner, 953 S.W.2d at 720); see also Cotroneo v. Shaw Env't & Infrastructure, Inc., 639 F.3d 186, 193 (5th Cir. 2011).

Plaintiffs do not argue that their evidence complies with the Havner standard, arguing instead that epidemiological studies are not the exclusive means of establishing causation. However, Plaintiffs do seek to rely on epidemiological studies, citing to, inter alia, Gibb's studies of chromate-production workers who were exposed to hexavalent chromium at a factory. And, where a plaintiff seeks to rely on epidemiological evidence, Texas law requires that the studies show a statistically significant doubling of the risk of developing their alleged injuries. See Garza, 347 S.W.3d at 265 ("Havner holds, and we reiterate, that when parties attempt to prove general causation using epidemiological evidence, a threshold requirement of reliability is that the evidence demonstrate a statistically significant doubling of the risk."); see also Young v. Mem'l Hermann Hosp. Sys., 573 F.3d 233, 236 (5th Cir. 2009).

The studies relied on by the Plaintiffs and their experts do not reflect a statistically significant doubling of the risk of their injuries. One of these studies did not quantify the risk of developing Plaintiffs' chromium- related-acute-irritation injuries at all and the other study did not find a doubling of the risk. In addition to Gibb's studies, Plaintiffs reference National Institute for Occupational Safety and Health and Agency for Toxic Substances and Disease Registry reports concluding that sodium dichromate causes headaches and nose, throat, skin, respiratory tract and gastro-intestinal irritation. Plaintiffs also claim that Carson relied on "numerous scientific articles supporting his contention that sodium dichromate causes various ailments." Plaintiffs did not submit these studies into the record and do not claim that any demonstrates a statistically significant doubling of the risk of any of their injuries. While the reliability of expert testimony is to be viewed in light of the totality of the evidence, "[t]he totality of the evidence cannot prove general causation if it does not meet the standards for scientific reliability established by Havner. A plaintiff cannot prove causation by presenting different types of unreliable evidence." Garza, 347 S.W.3d at 268.

Because none of this epidemiological evidence meets the reliability threshold of Havner and its progeny, it cannot be considered competent summary judgment evidence establishing general causation. See id.; Daniels v. Lyondell-Citgo Refining Co., 99 S.W.3d 722, 730 (Tex. App. 2003). Because the evidence underlying Gibb's and Carson's opinions is unreliable as a matter of Texas law, their testimony is also insufficient to prevent summary judgment. See Havner, 953 S.W.2d at 714.

Plaintiffs argue that, notwithstanding this deficiency, their claims should still survive summary judgment. First, Plaintiffs claim that "a differential diagnosis [excluding other potential causes of injury] constitutes an accepted means of proving causation when it is scientifically accepted that a particular toxin is capable of giving rise to the underlying injury." However, a differential diagnosis is only relevant after general causation has been reliably established "because a differential diagnosis presumes that chemical X can cause condition Y generally, but does not itself so prove." Coastal Tankships, U.S.A., Inc. v. Anderson (In re Estate of Anderson), 87 S.W.3d 591, 609 (Tex. App. 2002). Carson's differential diagnosis does not relieve Plaintiffs of their burden of adducing reliable evidence of general causation. See Johnson v. Arkema, Inc., 685 F.3d 452, 469 (5th Cir. 2012) (upholding exclusion of expert's differential diagnosis where there was no "reliable or relevant scientific evidence" to support the expert's presumption of general causation); Matt Dietz Co. v. Torres, 198 S.W.3d 798, 805 (Tex. App. 2006) (finding differential diagnosis insufficient to establish causation where there was no reliable evidence that "exposure to a specific substance at specific levels [could] cause the injury in question").

Second, Plaintiffs argue that, regardless of the sufficiency of their expert testimony, they have adduced sufficient evidence of causation for their acute injuries on the basis of lay testimony. While Plaintiffs correctly argue that they need not cite a particular case in the district court to rely on it on appeal, Plaintiffs failed to argue in their opposition to summary judgment in the district court that lay testimony was sufficient to support causation. They similarly failed to rely on lay testimony in their motion for reconsideration of the district court's decision on causation. We will generally not countenance arguments not raised before the district court. See, e.g., Keenan v. Tejeda, 290 F.3d 252, 262 (5th Cir. 2002) ("If a party fails to assert a legal reason why summary judgment should not be granted, that ground is waived and cannot be considered or raised on appeal." (quoting Grenier v. Cyanamid Plastics, Inc., 70 F.3d 667, 678 (1st Cir. 1995))). We therefore deem this argument forfeited.

For these reasons, we find that Plaintiffs have not adduced sufficient evidence to avoid summary judgment.

* * *

In light of the foregoing, we AFFIRM the district court's grant of summary judgment and dismissal of Plaintiffs' claims.

Questions:

1. There is little question that soldiers serving in Iraq and similar theatres of war, had exposure to toxic agents. While this is fairly common knowledge, why does the Court reject the cause and effect relationship posed by the expert?
2. Epidemiological studies were not effectively employed by the Plaintiffs nor the experts used to buttress their case. Why are these sorts of studies an alternative to direct causation between the substance and the harm caused.
3. Before commencing work on the water injection facility, an environmental review was conducted. What term applied to the toxins alleged as causing harm?

5.3 MEASURING INTOXICATION: A CHEMICAL AND BIOLOGICAL EXERCISE

The cross-over between both the methods of measuring intoxication and the measures of intoxication can be quite complicated. Aside from the traditional subjective measures of how a person acts, how well the suspect can do in field sobriety tests, how much easier it is to measure intoxication by the machine versus the extraction of blood and finally, whether the law has kept up with the decriminalization trends in drug use, are but a few of the issues that make the measure of intoxication difficult. Add to this, even the breathalyzer machine, which measures alcohol content in the respiratory tract, depends on the use of a chemical regent in order for the machine to measure correctly. In the end, the measure of intoxication depends upon biology and chemistry (See Figure 5.2).

FIGURE 5.2 The U.S. Drug Enforcement Administration (DEA) Los Angeles Field Division, Southwest Border Initiative (SWB) Group 2, along with the Fontana Police Department, seized approximately 3,552 pounds of methamphetamine and 66 kilograms of cocaine in Norco, California. This is the largest seizure of methamphetamine for the DEA Los Angeles Division. Defendants are well aware of these complexities and use them as an aggressive tool of defense.

Source: DEA at https://www.dea.gov/press-releases/2022/10/05/historic-methamphetamine-bust-dea-los-angeles-division.

Dove v. State
Jan. 30, 2020
No. 02-19-00011-CR (Tex. App.)

I.

On the afternoon of August 12, 2018, Officer Cleburne Eardley was dispatched to address a report of a possibly intoxicated driver at a gas station. He found Dove slumped over at the wheel of an idling gas tanker truck and what appeared to be a trail of vomit leading from the ground up to the truck's cab. When Officer Eardley spoke, Dove jolted awake. Officer Eardley had to catch Dove as he unsteadily climbed down from the cab, and his speech was slurred and mumbled. Officer Eardley did not smell any alcohol, but he still suspected that Dove was intoxicated. Seeing how unsteady Dove was, Officer Eardley believed it would be unsafe to conduct field sobriety testing. When paramedics were through treating Dove, Officer Eardley arrested him. As they drove to the jail, Dove repeatedly told Officer Eardley that had taken prescription hydrocodone; Dove asked if the charges would be dropped in light of the fact that he had a prescription.

At the jail, Dove refused consent for a blood draw, and Officer Eardley obtained a blood warrant. Dove's blood alcohol level was .127. He was charged with misdemeanor driving while intoxicated with an enhancement for a prior assault offense.

At trial, the State sought to elicit Officer Eardley's testimony concerning the signs that Dove was intoxicated by narcotics, including his slow movements, constricted pupils, and apparent dry mouth. Dove objected that Officer Eardley was not qualified as an expert in toxicology and should therefore not be allowed to opine on narcotic intoxication. The trial court overruled the objection and allowed this testimony.

Later, the State called Grayson Sterrett, one of the paramedics who examined Dove at the scene. When the State asked Sterrett what symptoms would suggest intoxication by heroin or hydrocodone, Dove again objected that Sterrett had not been qualified as an expert. The trial court overruled the objection and allowed Sterrett to testify that Dove showed symptoms of narcotic intoxication.

Despite Dove's admitted use of hydrocodone and the testimony concerning symptoms of opioid use, the jury was not charged on intoxication by a controlled substance. Instead, the jury was charged only on intoxication by alcohol. The jury found Dove guilty as charged. After finding his enhancement true, the trial court sentenced Dove to 90 days in jail. Dove appeals.

II.

In his first two issues, Dove contends that the trial court abused its discretion by allowing two nonexperts to testify concerning narcotic intoxication. Relying on Smithhart v. State, 503 S.W.2d 283, 285 (Tex. Crim. App. 1973), Dove argues that the topic of narcotic intoxication is solely the province of an expert on toxicology and that lay opinion on the matter is impermissible.

The State argues that even assuming there was error in allowing lay testimony on this subject, it would be a harmless error. We agree with the State.

Because any error is not constitutional, we apply rule 44.2(b). Tex. R. App. P. 44.2(b). That rule requires us to disregard any nonconstitutional error that does not affect appellant's substantial rights. Id. An error that has a "substantial and injurious effect or influence in determining the jury's verdict" affects a substantial right. Haley v. State, 173 S.W.3d 510, 518 (Tex. Crim. App. 2005); King v. State, 953 S.W.2d 266, 271 (Tex. Crim. App. 1997) (citing Kotteakos v. United States,

328 U.S. 750, 776, 66 S. Ct. 1239, 1253 (1946)). Conversely, an error does not affect a substantial right if we have "fair assurance that the error did not influence the jury, or had but a slight effect." Solomon v. State, 49 S.W.3d 356, 365 (Tex. Crim. App. 2001); Johnson v. State, 967 S.W.2d 410, 417 (Tex. Crim. App. 1998).

In determining whether an error affected an appellant's substantial rights, we review the record as a whole, including any testimony or physical evidence admitted for the jury's consideration, the nature of the evidence supporting the verdict, and the character of the alleged error and how it might be considered in connection with other evidence in the case. Motilla v. State, 78 S.W.3d 352, 355 (Tex. Crim. App. 2002). We may also consider the jury instructions, the State's theory and any defensive theories, whether the State emphasized the error, closing arguments, and even voir dire, if applicable. Haley, 173 S.W.3d at 518-19; Motilla, 78 S.W.3d at 355-56.

A person commits DWI when the person is "intoxicated while operating a motor vehicle in a public place." Tex. Penal Code Ann. § 49.04(a). The Texas Penal Code sets out two definitions of "intoxicated." Crenshaw v. State, 378 S.W.3d 460, 466 (Tex. Crim. App. 2012). The subjective definition of intoxicated is "not having the normal use of mental or physical faculties by reason of the introduction of alcohol, a controlled substance, a drug, a dangerous drug, or a combination of two or more of these substances, or any other substance into the body." Id. (quoting Tex. Penal Code Ann. § 49.01(2)(A)). The per se definition of intoxicated is "having an alcohol concentration of 0.08 or more." Id. (quoting Tex. Penal Code Ann. § 49.01(2)(B)).

However, the charge did not instruct the jury on the full definitions of intoxication; omitted from the charge on the subjective definition of intoxication was any mention of intoxication by a controlled substance, a drug, a dangerous drug, a combination of two or more of those substances, or any other substance into the body. Instead, the jury was instructed that it could find the defendant guilty only if it found beyond a reasonable doubt that Dove was intoxicated by alcohol, either in that alcohol impaired the normal use of his faculties or in that his blood alcohol concentration was 0.08 or more. The jury is presumed to have understood and followed the court's charge, absent evidence to the contrary. Id. at 467. Thus, in Crenshaw, the court presumed that the jury convicted the defendant based only on the subjective theory of intoxication that was charged and not on the uncharged per se theory of intoxication. Id. Likewise, we presume that the jury convicted Dove based only on the alcohol intoxication that was charged and placed no reliance on the drug intoxication that was uncharged.

In a certain sense, the charge thus did away with the sole area where harm might reside. A purely obedient jury would not have considered the drug evidence at all in reaching its conclusion that Dove was intoxicated by reason of alcohol alone. Lacking any evidence to the contrary, we are required to presume that the jury was obedient to the charge in this way. See id.

A similar conclusion was reached in Sifford v. State, 741 S.W.2d 440, 441 (Tex. Crim. App. 1987). There, the indictment wrongly alleged two offenses that were not properly subject to joinder, aggravated sexual assault and robbery, and the State introduced evidence of both. Id. at 440. Nonetheless, the State elected to proceed only on the aggravated sexual assault charge. Id. at 441. The Sifford court held that the election cured any harm from the misjoinder or from the introduction of evidence concerning a robbery offense that was never submitted for the jury's consideration. Id. In the same way, here, the State's election to proceed solely on alcohol intoxication would seem to mitigate any harm flowing from evidence relating to a drug-intoxication theory that was never submitted to the jury. Thus, the nature of the charge strongly implies that any error was harmless.

The overwhelming evidence of alcohol intoxication also suggests harmless error. The prosecution introduced testimony and body-camera video demonstrating that Dove exhibited many of the telltale signs of alcohol intoxication: being found unconscious near a pool of his own vomit, only to wake, lurch down from the truck, and slur incoherent responses to Officer Eardley's questions. Analysis showed that Dove's blood alcohol level was .127, more than enough for per se alcohol intoxication. A lab supervisor for the Texas Department of Public Safety testified that because of the high blood alcohol concentration, the lab decided not to even test the blood's drug content, concluding that the State had no need for this alternate form of intoxication evidence. In the same way, the force of this blood and video evidence might have further convinced the jury to put any drug intoxication out of its collective mind and focus only on alcohol intoxication. See McRae v. State, 152 S.W.3d 739, 745 (Tex. App.—Houston [1st Dist.] 2004, pet. ref'd) (op. on reh'g) (holding improper intoxication evidence harmless in light of "more persuasive" blood and video evidence, among other things).

Also of note is that the complained-of evidence was elicited from nonexperts, was cumulative, and was not emphasized by the State. See Bagheri v. State, 119 S.W.3d 755, 763 (Tex. Crim. App. 2003). The State never attempted to qualify the officer or the paramedic as experts in order to strengthen their testimony. Moreover, by the time the State introduced this evidence, the jury had already seen a video in which Dove proclaimed ten times in the space of six minutes that he had taken prescription hydrocodone. Thus, it was Dove himself who gave the first and strongest evidence of drug intoxication; the officer's and paramedic's testimony on this front was duplicative and lesser. See Harrington v. California, 395 U.S. 250, 254, 89 S. Ct. 1726, 1728 (1969) (holding improper evidence harmless because it was, among other things, cumulative of the defendant's own inculpatory statements). And rather than distancing from this topic, Dove revisited the subject of opioid intoxication multiple times during the defense case and closing argument, further diluting the effect of the complained-of evidence. By contrast, the State never mentioned drug intoxication in voir dire or opening statement. During closing argument, the State even downplayed the legal significance of this evidence, saying "this case isn't all about drugs … . All we needed to prove intoxication was this score, a .127." Later, the State again ventured, "We don't need the drugs. This case was not about drugs." Thus, we are fairly assured that this nonexpert, cumulative, and deemphasized testimony could not have had more than a slight effect on the jury's decision. See Bagheri, 119 S.W.3d at 763; Solomon, 49 S.W.3d at 365.

In light of these factors—a charge that offered little opportunity for prejudice; a compelling record apart from the drug evidence; and the State's efforts to minimize this cumulative, nonexpert testimony—we hold that any error was harmless. See Haley, 173 S.W.3d at 518-19; Motilla, 78 S.W.3d at 355-56. We therefore overrule Dove's first and second issues.

…

IV.

We modify the judgment to delete the enhancement and to reflect that Dove did not plead true to any enhancement. We further modify the judgment and the bill of costs to delete the $100 emergency services fee, leaving total costs of $332.10. As modified, we affirm. See Tex. R. App. P. 43.2(b).

Questions:

1. On what basis does the defendant claim his appeal meritorious? Do not assume these arguments are correct. Choose:
   a. That proof of his use of a drug was not provided at trial not did the jury receive jury instructions
   b. That since there was testimony on hydrocone at trial, the failure to instruct on this charge- reversible error
   c. That given the failure to prove the drug intoxication, the intoxication by alcohol should be dismissed.
2. What was the defendant's BAC level and was it satisfactory for a finding of intoxication?
3. True or False: The Texas DWI statute only covers intoxication by alcohol.

Garcia v. State
Number 13-18-00373-CR
(Tex. App. Apr. 9, 2020)

I. BACKGROUND

Kevin Barron, a sergeant with the City of McAllen police department, testified that he conducted a traffic stop around 3:30 a.m. after observing the driver of a vehicle fail to signal a lane change on Nolana Avenue in McAllen, Texas. He approached the vehicle and identified the driver as Garcia. Sergeant Barron smelled a strong odor of alcohol and urine coming from the vehicle. He also noted that Garcia's eyes were red and bloodshot. He asked Garcia whether she had been drinking alcohol, and she responded affirmatively. Upon request, Garcia walked to the rear of the vehicle, and Sergeant Barron observed that her balance was unsteady. When talking to Garcia, Sergeant Barron smelled a strong odor of alcohol emitting from her breath. He then administered standardized field sobriety tests. Garcia showed nystagmus in response to the horizontal gaze nystagmus (HGN) test, an indicator of intoxication. For the walk-and-turn test, Garcia displayed five clues for intoxication—swaying, failure to touch heel to toe, taking the wrong number of steps, stopping while walking, and making an improper turn. Sergeant Barron was unable to administer the one-leg-stand test because Garcia would not cooperate. At this point, Sergeant Barron concluded that Garcia was intoxicated, and he placed her under arrest. Sergeant Barron testified that supervisor vehicles, such as his, are not equipped with a dashboard camera and he was not wearing a body camera, so there was no video of his encounter with Garcia.

Nystagmus is an involuntary rapid oscillation of the eyes in a horizontal, vertical, or rotary direction. Emerson v. State, 880 S.W.2d 759, 765 (Tex. Crim. App. 1994). Horizontal gaze nystagmus refers to the inability of the eyes to smoothly follow an object moving horizontally across the field of vision, particularly when the object is held at an angle of forty-five degrees or more to the side. See Webster v. State, 26 S.W.3d 717, 719 n.1 (Tex. App.—Waco 2000, pet. ref'd). Consumption of alcohol exaggerates nystagmus to the degree it can be observed by the naked eye. Emerson, 880 S.W.2d at 766.

On cross-examination, Garcia's counsel questioned Sergeant Barron on his administration of the HGN test. Specifically, he questioned Sergeant Barron's technique of holding his pen six to eight inches from Garcia's face when looking for nystagmus. Garcia's counsel suggested through his questioning that the appropriate distance was twelve to fifteen inches. Sergeant Barron acknowledged that this might be the case, but he stated he always administers the test by holding the pen six to eight inches from a suspect's face.

Jose Rios, a McAllen police officer, arrived at the scene to transport Garcia to jail. He noticed Garcia stumble a little as she walked toward his vehicle. Officer Rios requested a blood or breath specimen from Garcia, but she declined. The trial court admitted video footage from Officer Rios's patrol vehicle into evidence, as well as two judgments showing Garcia's prior DWI convictions.

Garcia called several witnesses who had been with her in the hours preceding the traffic stop. Each testified that Garcia may have been tired because she had not slept for over twenty-four hours, but they did not see her consume any alcoholic beverages. Samuel Reyes, Garcia's boyfriend, testified that his cat urinated and vomited in Garcia's vehicle. Garcia's counsel argued that this was the reason that Sergeant Barron smelled urine in the car.

The jury returned a guilty verdict. Garcia now appeals.

II. LEGAL SUFFICIENCY

In her first issue, Garcia argues that the evidence was legally insufficient to establish that she was intoxicated.

A. Standard of Review & Applicable Law

The Due Process Clause of the Fifth and Fourteenth Amendments to the United States Constitution requires that a criminal conviction be supported by a rational trier of fact's findings that the accused is guilty of every essential element of a crime beyond a reasonable doubt. Laster v. State, 275 S.W.3d 512, 517 (Tex. Crim. App. 2009) (citing Jackson v. Virginia, 443 U.S. 307, 316 (1979)). This due process guarantee is safeguarded when a court reviews the legal sufficiency of the evidence. Id. Under this review, we consider all of the evidence in the light most favorable to the verdict and determine whether a rational fact finder could have found the essential elements of the crime beyond a reasonable doubt based on the evidence and reasonable inferences from that evidence. Jackson, 443 U.S. at 319; Whatley v. State, 445 S.W.3d 159, 166 (Tex. Crim. App. 2014). Because the jury is the sole judge of the credibility of the witnesses and of the weight to be given to their testimony, we resolve any conflicts or inconsistencies in the evidence in favor of the verdict. Ramsey v. State, 473 S.W.3d 805, 808 (Tex. Crim. App. 2015); Wesbrook v. State, 29 S.W.3d 103, 111 (Tex. Crim. App. 2000) (en banc).

We measure the legal sufficiency of the evidence against the elements of the offense as defined by a hypothetically correct jury charge for the case. Byrd v. State, 336 S.W.3d 242, 246 (Tex. Crim. App. 2011). Such a charge is one that accurately sets out the law, is authorized by the indictment, does not unnecessarily increase the State's burden of proof or unnecessarily restrict the State's theories of liability, and adequately describes the offense for which the defendant was tried. Id. Garcia challenges only the sufficiency of the evidence demonstrating the element of intoxication. As relevant here, intoxication is defined as "not having the normal use of mental or physical faculties by reason of the introduction of alcohol … into the body." TEX. PENAL CODE ANN. § 49.01(2)(A).

B. Analysis

We first address Garcia's contention that Sergeant Barron's testimony concerning the HGN test should not be considered as evidence supporting the jury's intoxication finding because it is unreliable. Garcia does not contend that the HGN test in general is unreliable, but she maintains that Sergeant Barron deviated from the recommended technique when administering the test. However, Garcia did not object at trial to Sergeant Barron's testimony on this basis, and she does not raise an evidentiary issue on appeal. See TEX. R. APP. P. 33.1(a)(1). Therefore, the admissibility of Sergeant Barron's testimony is not before us.

> The Texas Court of Criminal Appeals has taken judicial notice that the scientific theory underpinning the HGN test is sound and that the HGN test, when properly administered, is a reliable indicator of intoxication. Emerson, 880 S.W.2d at 768; see Plouff v. State, 192 S.W.3d 213, 218-19 (Tex. App.—Houston [14th Dist.] 2006, no pet.).

In addition, slight variations in the administration of the HGN test do not render the evidence inadmissible or unreliable but may affect the weight to be given the testimony. See Williams v. State, 525 S.W.3d 316, 324 (Tex. App.—Houston [14th Dist.] 2017, pet. ref'd); Hartman v. State, 198 S.W.3d 829, 840 (Tex. App.—Corpus Christi-Edinburg 2006, pet. struck); Compton v. State, 120 S.W.3d 375, 378 (Tex. App.—Texarkana 2003, pet. ref'd). Garcia's counsel was able to cross-examine Sergeant Barron concerning the proper technique for administering the HGN test. As the trier of fact, it is within the jury's sole province to determine the weight to be given this testimony. See Ramsey, 473 S.W.3d at 808. For the foregoing reasons, we will consider Sergeant Barron's testimony concerning the HGN test in our legal sufficiency analysis. See Clayton v. State, 235 S.W.3d 772, 778 (Tex. Crim. App. 2007) (explaining that a legal sufficiency review of the evidence admitted at trial includes both properly and improperly admitted evidence).

> Garcia presents similar arguments about the reliability of the remaining standardized field sobriety tests, which we reject for the reasons previously mentioned. In addition, we note that an officer's testimony about a suspect's coordination, balance, and any mental agility problems exhibited during the one-leg-stand and walk-and-turn tests are observations grounded in common knowledge. Plouff, 192 S.W.3d at 223; see Emerson, 880 S.W.2d at 763. Therefore, an officer's testimony concerning those tests is considered lay witness opinion testimony. Plouff, 192 S.W.3d at 223; see Emerson, 880 S.W.2d at 763.

Viewing the evidence admitted at trial in the light most favorable to the verdict, we note that the jury could have considered the following evidence in determining that Garcia was intoxicated: (1) bloodshot eyes; (2) unsteady balance; (3) odor of alcohol from the vehicle (4) odor of alcohol on the person; (5) admission to recent alcohol consumption; (6) failed field sobriety tests; (7) refusal to provide a breath specimen; (8) the smell of urine; and (9) Sergeant Barron's opinion testimony that Garcia was intoxicated.

> We decline Garcia's invitation to consider evidence presented at the motion for new trial hearing in our legal sufficiency analysis. See Idowu v. State, 73 S.W.3d 918, 922 n.12 (Tex. Crim. App. 2002) ("Sufficiency of the evidence … must be based upon evidence submitted at the time of trial, not later.").

A defendant's poor performance on the standardized field sobriety tests is evidence that supports a jury's finding of intoxication. Zill v. State, 355 S.W.3d 778, 786 (Tex. App.—Houston [1st Dist.] 2011, no pet.); see Finley v. State, 809 S.W.2d 909, 913 (Tex. App.—Houston [14th Dist.] 1991, pet. ref'd) ("Texas courts consistently uphold DWI convictions based upon the opinion testimony of police officers who observed the defendant's unsatisfactory performance in field sobriety tests."). Further, a witness does not have to be an expert to testify that a person he observes is intoxicated by alcohol; therefore, lay opinion testimony by a police officer in this regard is probative evidence of intoxication. Henderson v. State, 29 S.W.3d 616, 622 (Tex. App.—Houston [1st Dist.] 2002, pet. ref'd); see Annis v. State, 578 S.W.2d 406, 407 (Tex. Crim. App. 1979).

In addition, Texas courts have classified bloodshot eyes, unsteady balance, and the odor of alcohol on a person as objective indicators of intoxication. See Cotton v. State, 686 S.W.2d 140, 142 n.3 (Tex. Crim. App. 1985) (noting slurred speech, bloodshot eyes, odor of alcohol on the person, unsteady balance, a staggered gait, and the odor of alcohol on the breath as evidence of intoxication). Finally, the jury was entitled to consider Garcia's refusal to provide a blood or breath specimen as evidence of intoxication. See TEX. TRANSP. CODE ANN. § 724.061 (providing that evidence of a person's refusal to submit to an officer's request for a blood or breath specimen is admissible in a trial); Griffith v. State, 55 S.W.3d 598, 601 (Tex. Crim. App. 2001) (explaining that sometimes the reason a defendant refuses a request by police for a biological specimen is relevant to the defendant's consciousness of guilt); Hartman, 198 S.W.3d at 834.

Garcia argues that the evidence that she did not drink in the twenty-four hours leading up to the traffic stop "created no other reasonable inference that supports the jury verdict[.]" However, there was conflicting evidence in this regard, including Garcia's admission to Sergeant Barron that she consumed alcoholic beverages and the indicators of alcohol intoxication previously noted. Our standard of review requires that we resolve this conflicting evidence in favor of the jury's verdict. See Ramsey, 473 S.W.3d at 808.

Considering the evidence in the light most favorable to the verdict, we conclude that a rational fact finder could have found beyond a reasonable doubt that Garcia did not have the normal use of her mental or physical faculties by reason of the introduction of alcohol into the body. See Jackson,

443 U.S. at 319; Whatley, 445 S.W.3d at 166. Therefore, the evidence supporting the jury's verdict is legally sufficient. We overrule Garcia's first issue.

…

IV. CONCLUSION

We affirm the trial court's judgment.

Questions:

1. Nystagamus is strictly a test to be conducted by chemists? True or False
2. The Nystagamus test must be perfectly administered for any change in the method or application will result in evidentiary exclusion. True or False
3. Even with minor errors the appeals court upheld the conviction by relying on other distinct forms of proof for intoxication that were part of the Court's record. What were those?

Locate the following case using your favorite legal research resource, read the case and answer the questions below.

State v. Beltran-Chavez, 286 Ore. App. 590, 400 P.3d 927, Court of Appeals of Oregon June 18, 2014, Argued and Submitted; July 6, 2017, Decided, A152983

Questions:

1. Precisely, why did the Court reverse and remand the DWI conviction?
2. True or False: Officer testimony about the state of intoxication from field tests is strictly expert testimony.
3. What does the Court mean when it describes the Breathalyzer evidence as "not overwhelming?"

For another Horizontal-Gaze Test (HGN) case, see: Schultz v. State, 106 Md. App. 145 August 31, 1995.

Challenges to DWI and DUI cases can be quite creative and at times bizarre. One of the judges in the following case calls this particular appeal "a prime example of the Dickensian hair-splitting."

Burnett v. State
September 20, 2017
541 S.W.3d 77 (Tex. Crim. App. 2017)

Burt Lee Burnett was charged with driving while intoxicated after he rear-ended another vehicle. The State alleged that Burnett was intoxicated "by not having the normal use of his mental and physical faculties by reason of the introduction of alcohol, a controlled substance, a drug, a dangerous drug, a combination of two or more of those substances, and any other substance into his body. …" The jury charge included that language in the abstract and application portions of the charge. Burnett objected to including the entire "loss of faculties" statutory definition in either portion of the charge because the evidence showed that, if he was intoxicated, it was only due to alcohol, not anything else. His objection was overruled, and the jury convicted him. He was fined $1,000 and sentenced to 120 days' confinement in the county jail, which was probated for eighteen months.

On appeal, he made the same argument: the jury should have been instructed that "intoxication" only means "not having the normal use of mental or physical faculties by reason of the introduction of alcohol…." The court of appeals agreed, holding that it was error to submit the entire definition because no rational jury could conclude that Burnett consumed intoxicating drugs based on the evidence. In reaching its conclusion, the court of appeals distinguished our decision in Ouellette v. State, 353 S.W.3d 868 (Tex. Crim. App. 2011). We granted the State's petition for discretionary review to decide whether the lower court erred when it distinguished Ouellette from the facts of this case. Because we agree with the analysis of the court of appeals, we will affirm its judgment.

BACKGROUND

Burnett rear-ended a vehicle occupied by Michael Bussey and Nathan Chappa. After the wreck, Bussey saw Burnett get out of his car and "kind of stagger" to his vehicle. When Burnett got there to ask if everyone was okay, Bussey rolled down his window, and he and Chappa caught a "whiff" of alcohol. Chappa said Burnett had glassy and bloodshot eyes, but Bussey only noticed that Burnett was slurring his words. They both thought that Burnett was intoxicated.

Officer Clinton Coapland was the first to arrive on scene. When he made contact with Burnett, Burnett's speech was slurred. He also had to ask Burnett the same questions multiple times. When Coapland leaned towards Burnett to hear him better, Coapland smelled the faint odor of alcohol, but he did not see that Burnett's eyes were glassy or bloodshot. Coapland radioed for backup. Once Officer William Allred arrived, he began to secure the scene while Coapland began his DWI investigation. Coapland asked Burnett if he had been drinking and whether he would consent to taking standard field sobriety tests (SFSTs). Burnett said that he had not been drinking and consented to take the tests. He showed signs of intoxication during the horizontal-gaze nystagmus test (HGN test), the walk-and-turn test, and the one-leg-stand test. Coapland concluded that Burnett was intoxicated and arrested him. Incident to the arrest, Coapland searched Burnett and found twenty white pills and one blue pill in his jacket. Meanwhile, Allred performed an inventory search of Burnett's car and found pills, a laptop, and a Ruger in a bag in the front-seat of the car. He also found a prescription pill bottle in the car. The bottle was not photographed or admitted into evidence. According to Coapland, Allred told him that the bottle was for the medication found in Burnett's coat pocket and that the pills may be hydrocodone. Allred, however, contradicted Coapland, testifying that he did not remember if the prescription pill bottle was for the pills they found.

The State charged Burnett with Class B misdemeanor DWI and alleged that he was intoxicated "by not having the normal use of his mental and physical faculties by reason of the introduction of alcohol, a controlled substance, a drug, a dangerous drug, a combination of two or more of the substances, and any other substance into his body. …"

Motion to Suppress

Burnett filed a motion to suppress, arguing that the officers should not be permitted to testify as to what type of pills they found because they were not drug recognition experts (DREs). The trial court agreed and ruled that the officers were not allowed to testify that they thought the pills were hydrocodone. The next day, however, the issue of the pills came up again. During this discussion, the State told Burnett and the court that there was video footage from the scene during which Coapland, Allred, and Burnett spoke about the pills. The video showed that, after Coapland found the pills in Burnett's jacket, he gave them to Allred, who said that the pills looked like hydrocodone. Allred asked Burnett whether he had a prescription for the medication, and Burnett responded that he did. The State argued that the evidence of Burnett's pill possession should be admitted into evidence as same-transaction contextual evidence. Burnett again objected that the officers were not DREs and so were not qualified to testify about what kind of pills they found or any potential intoxicating effects. He also made other objections that are not relevant here. The trial court ruled that the pill evidence was admissible as same-transaction contextual evidence.

...

ARGUMENTS

The State argues that a jury charge should include the entire "loss of faculties" definition, irrespective of the trial evidence. According to it, the substance that caused a person to become intoxicated is purely an evidentiary issue because the focus is on only whether the defendant was intoxicated, not the intoxicant. For support, the State relies on Judge Cochran's dissenting opinion in Gray v. State, 152 S.W.3d 125, 136 (Tex. Crim. App. 2004) (Cochran, J., dissenting). It also contends that, by not instructing the juries on the entire "loss of faculties" definition, courts are contravening the purpose of the expansive definition adopted by the legislature. By way of hypothetical, it warns that,

> If a criminal defendant were to demonstrate clear evidence of extreme intoxication yet provides a breath sample showing a blood alcohol level considerably below .08, the defendant's own behavior is indicative that some other substance contributed to his intoxication even if the exact nature of the intoxicant is not known or cannot be determined. Under the Eastland Court's rationale, the jury charge in this defendant's case would only be allowed to give an instruction on alcohol; the possibility that the defendant may be intoxicated on "any other substance" may not even be considered.

State's Brief on the Merits at 21.

Burnett responds that, while the State in its charging instrument need only allege that the defendant was "intoxicated" and is permitted to track the language of the entire statutory definition, it is error to give to the jury portions of a statutory definition that are not supported by the evidence. He argues that the rule urged by the State would invite the exact type of speculation that we condemned in Hooper v. State, 214 S.W.3d 9, 15–16 (Tex. Crim. App. 2007), and that such guessing could ensnare thousands of innocent Texans, such as fatigued drivers and those with naturally bad balance, even though they never ingested any substance as required to prove intoxication. He also contends that the State misreads Judge Cochran's dissent in Gray, and that she was "lament[ing] the Court's holding in Garcia which required the State to plead the specific intoxicant, and then allowed a defendant to claim the State proved the incorrect intoxicant. ..." Appellant's Amended Brief on the Merits at 23 (citing Gray, 152 S.W.3d at 136). He also points out that, in Gray, the majority concluded that, although the State specifically pled alcohol intoxication and presented proof that Gray was intoxicated by alcohol, the synergistic-effect instruction given in that case was not erroneous because there was also evidence that Gray took anti-depressants and that those anti-depressants made him more susceptible to becoming intoxicated by alcohol. Here, however, Burnett contends that there is no evidence that he was intoxicated by prescription medication.

ANALYSIS

We agree with Burnett that the State's reliance on Judge Cochran's dissent is misplaced. Our reading of Judge Cochran's opinion shows that she was in fact discussing the State's pleading burden, not instructions in a jury charge. This is evident from the passage directly following the excerpt provided by the State,

> This case is a prime example of the Dickensian hair-splitting that we have allowed ourselves to fall into. Appellant makes no claim that he was not intoxicated at the time he rear-ended a fellow driver. He admits that he was a physically and mentally impaired driver. His contention is simply that he was not intoxicated because of the alcohol he drank; instead, he was intoxicated because of the anti-depressants he was taking. This is a "defense" that is condoned, if not encouraged, by our decision in Garcia which requires the State to allege the precise substance that it thinks caused an impaired driver's intoxication, and then permits that driver to defend against the charge by claiming that he was intoxicated on some other substance.

Gray, 152 S.W.3d at 136 (Cochran, J., dissenting).

We also decline the State's invitation to hold that the entire statutory definition of "intoxicated" should be included in every DWI jury charge regardless of the evidence adduced at trial. The State is correct that the legislature has adopted a broad definition of "intoxicated" that focuses on whether a person is intoxicated and not the agent that caused it. Ouellette, 353 S.W.3d at 869–70. We have explained that the State can use this to its advantage by alleging in its charging instrument that the defendant was simply "intoxicated" or by including the entire statutory definition, instead of pleading the specific intoxicant as was previously required. However, it is the responsibility of the trial court to deliver to the jury a written charge setting forth the "law applicable to the case." TEX. CODE CRIM. PROC. art. 36.14. Part of that duty includes applying the law to the facts of the case. Gray, 152 S.W.3d at 127–28. And, although the trial court is obliged to include in the jury charge statutory definitions that affect the meaning of elements of the crime, Villarreal v. State, 286 S.W.3d 321, 329 (Tex. Crim. App. 2009), the charge must also be tailored to the facts presented at trial. That is, the trial court must submit to the jury only the portions of the statutory definition of "intoxicated" that are supported by the evidence. To do otherwise is error.

The State argues that, under this rationale, the jury was not permitted to consider the possibility that the defendant was intoxicated from "any other substance." We disagree. The jury is permitted to consider whether a defendant was intoxicated from "any other substance" when there is evidence that the defendant ingested a substance that caused him to become intoxicated or there is sufficient evidence for a rational juror to infer such. But, as we will explain, the record here does not support that Burnett ingested a substance other than alcohol.

The evidence here shows that Burnett was intoxicated because he had been drinking alcohol. The witnesses and police could smell it, and Burnett showed signs of intoxication during three SFSTs, which as Coapland testified, is consistent with a person that has lost his mental or physical faculties due to the imbibement of alcohol. Burnett was also observed to have slurred speech and glassy and bloodshot eyes. Police later found hydrocodone pills in Burnett's vehicle, but there is no evidence in this record as to what kind of drug hydrocodone is, whether it can cause intoxicating effects, or whether the symptoms of intoxication Burnett was experiencing were also indicative of intoxication by hydrocodone. These are critical elements that were present in Ouellette but not in this case.

...

We agree with the court of appeals that Ouellette is distinguishable from the facts of this case and that, here, the jury charge was erroneous because it did not apply the law to the facts produced at trial.

HARM

The court of appeals held that Burnett was caused some harm by the erroneous jury charge in this case. Burnett, 488 S.W.3d at 925. According to it, Burnett was harmed because the pill-related evidence became an integral part of the trial even though there was insufficient evidence to submit that theory of intoxication to the jury. Id. We do not review its harm analysis, however, because we did not grant review on that issue. Juarez v. State, 308 S.W.3d 398, 406 (Tex. Crim. App. 2010) (judgment of the court of appeals finding harmful error should be affirmed when this Court agrees that the jury charge was erroneous but did not grant review of the lower court's harm analysis).

CONCLUSION

We affirm the judgment of the court of appeals.

Questions:

1. As to hairsplitting, why does the definition of intoxication cause such consternation?
2. As a result of this statute, defendant claims that his drug usage should have been part of his jury instructions because the statute includes the possibility of being intoxicated in this manner. And the defendant freely admits to drug usage. How successful is this argument?
3. In the end, jury instructions on intoxication must be tailored to what?

5.4 TOXICOLOGY, MEASURE AND MECHANICAL CHALLENGES

Chemistry plays many roles in forensic field work. Field investigators depend upon various machines to identify chemical and toxic materials. Cases that follow manifest this dependence.

> Read the Michigan press release regarding state police halting THC toxicology testing and taking steps to ensure transparency, accuracy after technical issue with CBD at https://www.michigan.gov/mspnewsroom/news-releases/2022/08/31/state-police-halts-thc-toxicology-testing.

Locate the following case, using your favorite legal research resource, read the case and answer the questions below.

United States v. May, Feb. 24, 2021, No. 19-13114 (11th Cir.)

Questions:

1. How does the court sustain a conviction without an actual supply of drugs on a vessel?
2. What rational inference can be drawn from the discovery of 9 trace toxicology findings of cocaine on a vessel? And given this, is the evidence sufficient for a conviction?
3. What measuring tool for chemical identification was challenged and on what basis?

In Breath-Test analysis, the "Intoxilyzer" measures by infrared absorption a breath sample. Using a pulsed Infrared light source, the wavelength frequency is 3.4 and 9.4 microns for ethanol and other interfering substances. The decrease in intensity of the infrared light due to absorption by alcohol in the breath sample is proportional to the amount of alcohol in the breath. While there less chance to challenge the mechanics behind the machine, the processes and methods for taking the sample is a continuous concern.

Alvarez v. State
Feb 28, 2019
571 S.W.3d 435 (Tex. App. 2019)

Brief Background

Appellant does not challenge the sufficiency of the evidence supporting his conviction. The evidence shows that Officer John Mitchell with the Fort Worth Police Department stopped appellant about 2:00 a.m. after seeing appellant driving his car on Mark IV Parkway with no headlights or taillights on. Appellant had bloodshot and watery eyes, and Officer Mitchell could smell alcohol "coming from [appellant's] person or the vehicle." When Officer Mitchell asked appellant if he had drunk any alcohol, appellant responded that he had drunk two drinks at Buck's, a nearby bar. Officer Mitchell performed standard field sobriety tests; appellant demonstrated six of six possible intoxication clues on the nystagmus test, six of eight possible clues for the walk-and-turn test, and one of two clues on the one-leg stand test. Accordingly, Officer Mitchell arrested appellant, read him the statutory DIC-24 warning—pursuant to which appellant agreed to give a breath sample—and drove him to the downtown Fort Worth jail, where appellant gave a breath sample. The two tested samples showed a blood-alcohol concentration of 0.161 and 0.150. A jury found appellant guilty of driving while intoxicated, and the trial judge assessed his punishment at 80 days' confinement and a $1,250 fine.

Breath-Test Results Admissible

In his first point, appellant contends that the trial court erred by admitting the results of his breath test because the State failed to prove that the intoxilyzer operator observed appellant for at least fifteen minutes before beginning the test; therefore, the trial court should have excluded the results as both unreliable and illegally obtained. See Tex. Code Crim. Proc. Ann. art. 38.23.

A person taking a breath specimen on the request or order of a peace officer must be properly certified and must take and analyze the test "under rules of the department" of public safety (DPS). Tex. Transp. Code Ann. §§ 724.001(7), 724.016(a). The court of criminal appeals has held that compliance with this statute, and therefore DPS's rules, controls whether breath-test results are admissible in a DWI prosecution. Atkinson v. State, 923 S.W.2d 21, 23 (Tex. Crim. App. 1996), overruled in part on other grounds by Motilla v. State, 78 S.W.3d 352, 356–58 (Tex. Crim. App. 2002). One of DPS's testing regulations requires a breath-test operator to "remain in the continuous presence of the [tested person] at least 15 minutes immediately before the test and [to] exercise reasonable care to ensure that the subject does not place any substances in the mouth." 37 Tex. Admin. Code § 19.3(a)(1) (2015) (Tex. Dep't Pub. Safety, Techniques and Methods). But "[d]irect observation is not necessary to ensure the accuracy of the test result." Id.

The trial judge must exclude evidence obtained in violation of the law—including breath-test evidence that was obtained absent compliance with DPS's regulations—from the jury's consideration when the defendant so requests. Tex. Code Crim. Proc. Ann. art. 38.23; Atkinson, 923 S.W.2d at 23. When the evidence is inadmissible as a matter of law, the judge should not admit the evidence for the jury's consideration, but when disputed fact issues exist, the question of whether the evidence should be excluded may be tried to the jury. Atkinson, 923 S.W.2d at 23. Appellant did not request an article 38.23 instruction, nor does he complain that the trial judge erroneously failed to give the jury such an instruction. See Shpikula v. State, 68 S.W.3d 212, 223 & n.10 (Tex. App.—Houston [1st Dist.] 2002, pet. ref'd).

The State offered its exhibit 6, the "test record analytical report" from the breathalyzer, through Sara Skiles, a senior toxicologist for the Tarrant County Medical Examiner's office. Although the exhibit itself noted, "Fifteen minute waiting period completed – YES," Skiles did not have personal knowledge of the observation because she did not perform the test. The operator listed on exhibit 6 is Christian Paschal-Hamilton, who also testified at trial. After Skiles agreed on voir dire questioning that a breathalyzer test is not valid unless the fifteen-minute observation period occurred, appellant objected to admission of the breath-test results "on the basis of insufficient proof of observation."

Officer Mitchell testified that according to his police report, he arrived at the jail with appellant at 2:48 a.m. The first time recorded on the intoxilyzer was 3:09 a.m.—the "operational system check." The machine recorded appellant's first test result at 3:11 CST and his second at 3:14 CST. Thus, assuming Officer Mitchell and the machine were keeping the same time, at least twenty-one minutes elapsed from the time appellant arrived at the jail to the time the machine began recording.

Officer Mitchell testified that when he arrives at the jail with an arrestee, he turns the person "over to the booking officers or the … jailers. They handle it from there. And then from them, he's turned over to the intoxilyzer." Officer Mitchell could not recall the time appellant was turned over to the breath-test operator.

Paschal-Hamilton, the breath-test operator, could not recall appellant's specific test because he performs so many. And appellant's test was not recorded on video because the testing room does not have a camera system. But Paschal-Hamilton remained adamant throughout his testimony that he had waited the requisite fifteen minutes because that is his routine and a "mandatory requirement" of his job. When the prosecutor asked Paschal-Hamilton if he had been in appellant's presence for "at least … 15 minutes continuous" before performing the test, he answered, "I always do a 15 minute observation." Paschal-Hamilton further testified that he does "all of [his] transactions the same way" and that he has "very clear instructions to do this. I do it every single time the same way, 15-minute observation, no errors, every time."

Paschal-Hamilton also described the testing process for the jury:

> … [A] person will be generally brought to me, and the arresting officer will generally say I have read the specified document, DIC-24, or I have not. If they have not, we'll do that at that time. If they have, I'll begin a 15-minute waiting period. Once we're verified that, yes, the document has been read and then approval for the breath test has -- has started. 15-minute observation period. We check for -- to make sure that they don't have any gum, tobacco, anything in their mouth, foreign object of any kind.
>
> During that period I'll give an instruction phase. Show them the instrumentation, show them the mouth pieces that we use. It's a clear plastic. Give clear instructions.

The machine would ask whether the fifteen-minute observation period occurred, and Paschal-Hamilton would have to click on the yes option. After that, the machine begins the start-up process, which involves drawing air through the mouthpiece to ensure that there is no interference with foreign substances inside the machine. After that, the machine will prompt the operator to have the test subject exhale into the mouthpiece.

According to Paschal-Hamilton, the booking desk is "[a]pproximately 25 paces" from the testing device, and booking takes "six to seven minutes, maybe. Perhaps." The booking officer does not bring the test subject to him; either the arresting officer does, or Paschal-Hamilton goes to meet the arresting officer. When asked if he knew when the "waiting period" starts, Paschal-Hamilton answered, "[o]nce I verify things and I'm in their presence." He uses the timer on his own wristwatch to calculate the fifteen minutes, but he does not record the time.

Paschal-Hamilton agreed that the time on his wristwatch could have differed from both the time Officer Mitchell used to record his arrival at the jail with appellant and the time on the intoxilyzer. He also agreed that the purpose of the observation is for the test subject's benefit, to ensure that he does not burp, ingest any other substances, or do anything that could cause the breath-test results to be "artificially high."

Appellant argues that the evidence shows it would have been impossible for Paschal-Hamilton to have waited the required fifteen minutes because appellant arrived in the jail's sally port at 2:48 a.m., stayed in booking for at least six to seven minutes, walked into the intoxilyzer room, and was tested beginning at 3:09 a.m. But appellant's argument assumes Officer Mitchell recorded as the jail arrival time the time he pulled into the sally port, not the time he arrived inside the jail; rigidly applies Paschal-Hamilton's general estimation that booking usually takes six to seven minutes, "[p]erhaps"; fails to take into account that the timing devices might not have been synchronized; and fails to consider Paschal-Hamilton's testimony that he sometimes meets the arresting officer at the front of the jail when the arrestee is brought in. Contrary to appellant's assertion, the evidence does not show that it was impossible for Paschal-Hamilton to have observed appellant for fifteen minutes before the test; thus, the trial judge did not abuse his discretion by admitting exhibit 6. See Patel v. State, No. 01-14-00575-CR, 2015 WL 5821439, at *2–3 (Tex. App.—Houston [1st Dist.] Sept. 29, 2015, pet. ref'd) (mem. op., not designated for publication); Serrano v. State, 464 S.W.3d 1, 6 (Tex. App.—Houston [1st Dist.] 2015, pet. ref'd); see also, e.g., Wall v. State, 184 S.W.3d 730, 743 (Tex. Crim. App. 2006) (reiterating that we review a trial court's decision to admit evidence for an abuse of discretion).

We overrule appellant's first point.

...

Conclusion

Because we have overruled all of appellant's points, we affirm the trial court's judgment.

Questions:

1. When administering the breath-test, which is the most correct statement:
   a. That the test must be administered immediately
   b. That the machine operator need no qualification to operate due to its mechanical nature
   c. That the result is not subject to dispute or challenge
   d. None of these states are correct
2. Prove True or False: The Court holds that during any waiting period before administering the test, the officer must personally observe the defendant during the waiting period.
3. Defendant relies on the statute which guides the administration of the machinery. What is the best argument the defendant has under the statute?

> Read the press release from the U.S. Department of Justice regarding falsified breathalyzer results at https://www.justice.gov/opa/pr/former-federal-supervisory-correctional-officer-pleads-guilty-civil-rights-offense-aiding

Hoover v. State
Jan 5, 2017
NO. 02-16-00019-CR (Tex. App. Jan. 5, 2017)

I. BACKGROUND

On June 12, 2014, at "around 11:57 a.m. to 12:04," Rhome Police Officer Chance Garrett heard a broadcast that "a suspicious person" had urinated in a residential front yard and then had driven off in a truck toward the service road of a highway. Shortly thereafter, at approximately 12:04 p.m., Garrett saw a black truck pull up to a stop sign near that service road and turn left after failing to properly signal the turn. Garrett signaled for the truck to pull over, and the driver stopped partially in the intersection. When Garrett approached the truck, he noticed that the driver, whom Garrett identified as Appellant, smelled of alcohol. Garrett had Appellant get into his patrol car and then called the Department of Public Safety for assistance.

When the two troopers—William Aaron Wallace (W. Wallace) and Ralph Ray Wallace (R. Wallace)—arrived, W. Wallace asked Appellant to get out of Garrett's patrol car. He noticed that Appellant "walked out diagonally and stepping across his feet, almost a staggering motion." Appellant's clothes were "disorganized" and he appeared "tired and kind of sluggish." W. Wallace, like Garrett, noticed a strong odor of alcohol on Appellant's breath. Appellant admitted that "he was in trouble for urinating in someone's yard" and that he had drunk three beers both at home and in his truck. Appellant stated that he regularly drank alcohol "pretty much every day" and that he had not eaten anything that day although he had been awake since 2:00 a.m.

When asked, Appellant consented to field-sobriety testing. Appellant failed the horizontal-gaze-nystagmus test, exhibiting six out of six clues of intoxication, and the walk-and-turn test, exhibiting three out of eight clues of intoxication. Appellant told W. Wallace that he could not perform the third test—the one-leg stand. After concluding that Appellant had "lost his normal use of mental and physical faculties" through the ingestion of alcohol, W. Wallace arrested Appellant for DWI. R. Wallace then recited the required DIC-24 warnings to Appellant at 1:15 p.m. Appellant consented to have his blood drawn. The test revealed that at the time Appellant's blood was drawn—1:44 p.m.—Appellant had a blood-alcohol concentration (BAC) of 0.114, which was over the legal limit of 0.08. See Tex. Penal Code Ann. § 49.01(2)(B) (West 2011).

A grand jury indicted Appellant with felony DWI based on his two prior convictions for misdemeanor DWI. See Tex. Penal Code Ann. §§ 49.04(a), 49.09(b) (West Supp. 2016). At trial, Mary Avalos Belli, the forensic scientist who had analyzed Appellant's blood sample, testified that at the time Garrett stopped Appellant, Appellant's BAC would have been 0.15 or 0.17 based on retrograde extrapolation. The jury found Appellant guilty of felony DWI and assessed his punishment at five years' confinement and a $10,000 fine. The trial court sentenced Appellant accordingly, and Appellant was released on an appeal bond. See Tex. Code Crim. Proc. Ann. art. 44.04(b) (West Supp. 2016).

II. EXPERT TESTIMONY

Appellant argues in his first point that the trial court abused its discretion by allowing Belli to give her opinion based on retrograde extrapolation because it was unreliable. See Tex. R. Evid. 702-03. To show reliability, the proponent of scientific evidence must show by clear and convincing evidence that (1) the underlying theory is valid, (2) the technique applying the theory is valid, and (3) the technique was properly applied on the occasion in question. See Kelly v. State, 824 S.W.2d 568, 573 (Tex. Crim. App. 1992). Appellant specifically attacks the technique Belli used to conduct a retrograde extrapolation—she did not understand the "BAC curve"—and Belli's application of retrograde extrapolation to Appellant based on assumed facts. He does not argue that retrograde extrapolation is invalid as a theory.

We review a trial court's decision to admit scientific evidence for an abuse of discretion. See Russeau v. State, 171 S.W.3d 871, 881 (Tex. Crim. App. 2005), cert. denied, 548 U.S. 926 (2006). As such, we will not disturb the trial court's ruling if it was within the zone of reasonable disagreement. Tillman v. State, 354 S.W.3d 425, 435 (Tex. Crim. App. 2011). For the reasons stated below, we cannot so conclude and overrule Appellant's first point.

A. BELLI'S TESTIMONY

During Belli's testimony, she defined retrograde extrapolation as "a calculation where you want to find out what the [BAC] was at a certain time in the past." She explained that this calculation is based on the knowledge that a body absorbs alcohol, the BAC peaks, and then the body begins to eliminate the alcohol—the BAC curve referred to by Appellant. She recognized that there was "some" controversy in the scientific community regarding retrograde extrapolation but stated that it was "accepted … as long as … [the determinative] factors are known" by the expert. Belli testified that she understood the factors that are to be considered in performing retrograde extrapolation: age, body size, amount of alcohol consumed, and when alcohol was last consumed. She stated that the presence of food in the stomach will slow the absorption rate and that a chronic drinker will eliminate alcohol faster than a nondrinker. Belli also admitted that she would need to know the time of Appellant's last drink to determine what his BAC was at the time of the stop based on retrograde extrapolation.

The State then had Belli assume that a person's third drink occurred "sometime between 11:57 and 12:04," that the person was over six feet tall and weighed 160 pounds, that the person had not eaten in ten hours, and that the person's BAC one hour and forty minutes after the stop was 0.114. Belli applied retrograde extrapolation to those facts and concluded that at the time of the stop, the person would have had a BAC of 0.15 if the person was not a regular drinker or 0.17 if the person was a regular drinker, with an error rate of plus or minus .01. In short, Belli testified that based on the facts given to her, Appellant's BAC was higher at the time of the stop than at the time of the blood test.

B. TECHNIQUE APPLYING RETROGRADE EXTRAPOLATION

Appellant objected to Belli's testimony on the basis that Belli did not have sufficient information to apply retrograde extrapolation to Appellant. The trial court overruled Appellant's objection. On appeal, Appellant again asserts that Belli's opinion was unreliable because she opined that Appellant was in the elimination phase at the time he was driving, not at the time of the blood test, which showed she did not understand retrograde extrapolation.

In determining the reliability of retrograde-extrapolation evidence, courts are to consider and balance: "(a) the length of time between the offense and the test(s) administered; (b) the number of tests given and the length of time between each test; and (c) whether, and if so, to what extent, any individual characteristics of the defendant were known to the expert in providing his extrapolation." Mata, 46 S.W.3d at 916-17. In general, an expert must be able to apply retrograde extrapolation with consistency and clarity and must show a knowledge of the "difficulties," "subtleties," and "risks" inherent in the science. Id. at 916.

Belli understood the application of retrograde extrapolation and was able to clearly and consistently testify to its application to Appellant. Belli was aware of the BAC curve and explained its application to Appellant based on several facts surrounding the offense and on Appellant's physical characteristics. These facts included Appellant's height and weight, his history with and tolerance of alcohol, his admission that he consumed three beers at home and in his truck before he was stopped, the fact that he had not eaten anything that day, and the fact that the blood test occurred one hour and forty minutes after the stop. Belli also understood the difficulties associated with retrograde extrapolation and recognized that there was "some" controversy within the scientific community regarding the theory if the factors Belli considered in Appellant's case were unknown. Although Belli stated once that she assumed Appellant was in the elimination phase "at the time of the stop," Belli also explained that "if you have no food in your stomach, it could take anywhere between 20 to 45 minutes to fully absorb that alcohol and then begin eliminating." Viewing Belli's testimony as a whole, Belli's technique was sufficiently reliable such that we cannot conclude that the trial court abused its discretion in admitting it. See, e.g., Bignon v. State, 252 S.W.3d 360, 367-68 (Tex. Crim. App. 2008); Subirias v. State, 278 S.W.3d 406, 410-13 (Tex. App.—San Antonio 2008, pet. ref'd); Kennedy v. State, 264 S.W.3d 372, 377-79 (Tex. App.—Houston [1st Dist.] 2008, pet. ref'd); Washburn v. State, No. 04-99-00927-CR, 2002 WL 31015255, at *2 (Tex. App.—San Antonio Sept. 11, 2002, no pet.) (mem. op. on remand, not designated for publication); cf. Mata, 46 S.W.3d at 917 (finding retrograde extrapolation unreliable because only one blood test given more than two hours after offense occurred and expert knew no personal characteristics of defendant); Veliz v. State, 474 S.W.3d 354, 360-62 (Tex. App.—Houston [14th Dist.] 2015, pet. ref'd) (holding expert's retrograde-extrapolation testimony unreliable because blood was drawn three and a half hours after arrest, expert did not know many of appellant's characteristics, expert believed retrograde extrapolation could be based on knowing only the time of the stop and the time of the blood test, and expert could not explain her calculation with clarity).

C. PROPER APPLICATION OF RETROGRADE EXTRAPOLATION

Appellant also argues that Belli's testimony was unreliable because although there was no evidence of the time Appellant had his last drink, Belli relied on the State's hypothetical and assumed his last drink occurred between 11:57 a.m. and 12:04 p.m. The use of hypotheticals when examining expert witnesses is an accepted practice, and the facts included in a hypothetical may be facts admitted into evidence or facts assumed by counsel in accordance with the theory of the case. Matson v. State, 819 S.W.2d 839, 853 (Tex. Crim. App. 1991). Although there is no evidence establishing the time of Appellant's last drink with precision (and as the State points out, could not be established with unassailable accuracy absent Appellant's cooperation), this assumed fact does not contradict the evidence introduced at trial and fits with the State's theory of the case. See Hearrean v. State, No. 08-13-00338-CR, 2016 WL 3021627, at *8-9 (Tex. App.—El Paso May 25, 2016, pet. ref'd) (mem. op., not designated for publication); Subirias, 278 S.W.3d at 412-13. The State's hypothetical assumed that Appellant's last drink was immediately before he was stopped—between 11:57 a.m. and 12:04 p.m.—which was supported by Appellant's admission that he had been drinking in his truck as well as at home. Belli's application of retrograde extrapolation to Appellant, even though based on an assumed fact, was sufficiently reliable such that we cannot conclude that the trial court abused its discretion by admitting it. See Bigon, 252 S.W.3d at 367 ("[T]he trial court does not abuse its discretion by allowing an expert to base testimony upon an assumption if there is a factual basis for the assumption in the record.").

…

V. CONCLUSION

Having overruled Appellant's points, we affirm the trial court's judgment. See Tex. R. App. P. 43.2(a).

Questions:

1. In cases of repeat DUI/DWI, and when those cases tend to be more egregious, the test results will be assessed from drawn blood. What substance in the blood does the BAC test evaluate?
2. True or False: BAC analysis is certain to capture the most accurate proof of intoxication because it is done immediately upon arrest.
3. When the defendant challenges the scientific reliability of the BAC result, his chief line of argument rests in what theory?

State v. Speer
Jun 17, 2019
A18-1592 (Minn. Ct. App. Jun. 17, 2019)

FACTS

In July 2017, Rochester Police Sergeant Bert Otto stopped a vehicle driven by appellant Brady Edward Speer after observing it cross the fog line several times. When Sergeant Otto spoke with

Speer, he could "smell a slight odor of alcohol coming from him and the interior of the vehicle." When questioned by Sergeant Otto, Speer admitted drinking "two Budweiser beers." Rochester Police Sergeant Terry Pretzloff arrived on the scene and administered three field sobriety tests. After witnessing signs of impairment during testing, Sergeant Pretzloff arrested Speer for DWI. At Speer's request, Sergeant Otto parked Speer's vehicle off of the roadway. While moving the vehicle, Sergeant Otto discovered two open cans of Budweiser beer under the driver's seat.

Sergeant Pretzloff took Speer to the adult detention center to conduct the DataMaster (DMT) breath test. Sergeant Pretzloff situated Speer in the interview room then entered an adjacent room—separated by a glass partition. Speer agreed to take the DMT breath test. The first sample indicated an alcohol concentration of 0.086. The second sample indicated an alcohol concentration of 0.090. The final result of the test indicated an alcohol concentration of 0.08. Sergeant Pretzloff did not observe anything that concerned him about the integrity of the test.

The state charged Speer with two counts of fourth-degree DWI under Minn. Stat. § 169A.20, subds. 1(1), (5) (2016), and possession of an open container under Minn. Stat. § 169A.35, subd. 3 (2016). At the jury trial, Speer objected to Sergeant Pretzloff's testimony and the breath-test results based on foundation. When questioned further by the district court, Speer explained that he objected to the observation period. The district court admitted the breath-test results, but agreed to reconsider admissibility after cross-examination of Sergeant Pretzloff. During cross-examination of Sergeant Pretzloff, Speer renewed his objection. The district court denied the objection and allowed the breath-test results to remain in evidence.

During closing argument, the prosecutor suggested that the DMT training manual, which was not admitted into evidence, did not require "constant visual on the Defendant for the full 15-minute period of time." Speer objected arguing that the statement was improper. The district court sustained the objection and allowed the prosecutor to rephrase the argument. Speer did not object to the rephrased argument. During deliberation, the jury asked whether it could view the procedural guidelines for observation in the DMT training manual. The district court responded that the jury could not because the manual was not admitted into evidence.

...

DECISION

I.

Speer first argues that the district court abused its discretion by admitting breath-test results that lacked foundational reliability. "Evidentiary rulings rest within the sound discretion of the [district] court and will not be reversed absent a clear abuse of discretion. On appeal, the appellant has the burden of establishing that the [district] court abused its discretion and that appellant was thereby prejudiced." State v. Amos, 658 N.W.2d 201, 203 (Minn. 2003) (citation omitted).

A DMT breath test consists of "one adequate breath-sample analysis, one control analysis, and a second, adequate breath-sample analysis." Minn. Stat. § 169A.51, subd. 5(a) (2018). A sample is considered "adequate if the instrument analyzes the sample and does not indicate the sample is deficient." Id., subd. 5(b) (2018). The results of breath testing performed by a fully trained person, "are admissible in evidence without antecedent expert testimony that [such] instrument provides a trustworthy and reliable measure of the alcohol in the breath." Minn. Stat. § 634.16 (2018); see State v. Birk, 687 N.W.2d 634, 637 (Minn. App. 2004) ("The results of [a breath] test are admissible as a trustworthy and reliable measure of breath alcohol if a properly certified operator performs the test."). "Once a prima facie showing of trustworthy administration has occurred, it is incumbent on the opponent to suggest a reason why the test was untrustworthy." Bond v. Comm'r of Pub. Safety, 570 N.W.2d 804, 806 (Minn. App. 1997) (quotation omitted).

The DMT breath test here was adequate because there was one adequate breath sample, a control analysis, a second adequate breath sample, and the instrument did not indicate a deficiency.

See Minn. Stat. § 169A.51, subd. 5. The state presented prima facie foundation for admitting the breath-test results because Sergeant Pretzloff testified that he is a certified operator of the DMT instrument. See Bond, 570 N.W.2d at 807 (noting that the commissioner established prima facie foundation for admitting the test results by demonstrating that a properly trained person performed the test).

Speer challenged the validity of the breath-test results based on the observation period. The driver may challenge the test's reliability by establishing that some irregularity occurred during the observation period, which would render the test result invalid. Kramer v. Comm'r of Pub. Safety, 706 N.W.2d 231, 237 (Minn. App. 2005). "The purpose of the observation period is to preclude the possibility that the testimony may be affected by mouth alcohol, resulting from burping or vomiting." State v. Nelson, 399 N.W.2d 629, 631 (Minn. App. 1987), review denied (Minn. Apr. 17, 1987). A "slight interruption of the observation period or a less than perfect observation does not invalidate the test unless the driver has ingested or regurgitated a substance that affects the results, and the burden is on the driver to present such evidence." Kramer, 706 N.W.2d at 237. "At most, the contention that the observation period was 'faulty' goes to the weight and not the admissibility of the test results." Nelson, 399 N.W.2d at 632.

Sergeant Pretzloff testified that the observation period was a necessary scientific safeguard that must be performed properly, and if not performed properly, the breath-test results are not valid. Sergeant Pretzloff testified that during the observation period, he observes whether the person might be putting something in their mouth, burping, belching, or vomiting. He explained that if the suspect belches or burps, the alcohol vapors can go into his mouth and artificially raise the reading. He testified that, after the "mouth check," the suspect is positioned in an interview cell. He then described the layout of the DMT testing room:

> So if you walk into the room, the machine is directly off to the left against the wall … . And then off to the right side would be where officers would go to be on the other side of the glass partition from the subject that we'd bring into the jail.

He explained, "I am on the other side of a glass wall from them so I'm observing them the entire time until I walk around to get them out of that room and walk right next door to do the DMT."

When asked by Speer's attorney whether he could observe a suspect talking on the phone if he was at the testing instrument, Sergeant Pretzloff responded, "No, because there's a wall between there." Sergeant Pretzloff then testified that Speer was on the phone at the time he started operating the breath-test instrument and thus was not in his "area of observation" at that time. Sergeant Pretzloff agreed that it was an improper observation period. It is not clear exactly how long Speer was outside of Sergeant Pretzloff's area of observation, but Sergeant Pretzloff noted it only takes "two to three seconds" to walk around the glass partition into the interview room.

Moreover, Sergeant Pretzloff testified that he did not observe anything going into Speer's mouth or any indication of vomiting or belching by Speer. He also testified that he did not hear Speer burp. He explained that the DMT instrument has safeguards that indicate if mouth alcohol is an issue, such as indicating an invalid sample, and Speer's breath test did not indicate that mouth alcohol was an issue. He testified that nothing caused him concern about the integrity of the test.

Speer suggests the breath-test results are unreliable because Speer may have burped during the time that Sergeant Pretzloff was not observing him. "A driver will not be successful in challenging [breath-]test results if his arguments rest upon mere speculation that something might have occurred to invalidate those results." Hounsell v. Comm'r of Pub. Safety, 401 N.W.2d 94, 96 (Minn. App. 1987). "To impeach the credibility of breath-test results, a driver must prove that he burped during the observation period and must also demonstrate that the burping actually affected the test results." Junker v. Comm'r of Pub. Safety, 925 N.W.2d 661, 663 (Minn. App. 2019). Speer offered no evidence that he actually burped during the observation period. Furthermore, even if Speer had offered evidence that he burped during the observation period, Sergeant Pretzloff explicitly testified that the

DMT instrument did not indicate that mouth alcohol was an issue during Speer's test administration. Speer offered no evidence to counter this. Therefore, Speer did not "sufficiently demonstrate that the burping, belching, or vomiting actually affected the test results." Id. at 661.

Speer cites McGregor v. Comm'r of Pub. Safety, as one case in which the breath-test results were deemed inadmissible based on improper observation. 386 N.W.2d 339 (Minn. App. 1986). But the observation in McGregor was improper because "the officer had no training, did not know why the driver was to be observed, or what to observe." Id. at 340. This court noted, "The fact [that] the officer worked on her police reports during the observation period is not fatal. We have previously held similar lapses in observation do not invalidate the test, but instead provide the driver with an opportunity to suggest reasons as to why the lapse makes the test unreliable." Id. Distinguishable from McGregor, Sergeant Pretzloff was trained in administering DMT breath tests and explained why the observation period was important and what he observes. Thus McGregor does not require exclusion of the breath-test results here.

Speer argues that Sergeant Pretzloff's testimony alone rendered admission of the test results an abuse of discretion because he "testified that a properly conducted observation period is essential to conducting a proper breath test, and admitted that a proper observation period was not conducted with Speer." But case law establishes that a faulty observation period goes to the weight of the test results, rather than admissibility. See Nelson, 399 N.W.2d at 632 (noting that the faulty observation period goes to the weight of the test results rather than admissibility); Kramer, 706 N.W.2d at 236 (noting that evidence challenging the trustworthiness of the test results affects the weight of the results, not admissibility). As a faulty observation period does not require exclusion of the breath-test results, we conclude that the district court did not abuse its discretion in admitting them.

...

Affirmed.

Questions:

1. True or False: The Datamaster Machine has less reliability than other machines measuring intoxication results.
2. What external factors does defendant claim impact the credibility and integrity of the results?
3. Who has the burden of proof in a challenge of the Datamaster results?

5.5 FORENSIC TOXICOLOGY AND CAUSE OF DEATH

Proving the cause of death, whether in a civil or criminal case or in general trends in the population, heavily depends on forensic toxicology. Cases which highlight the specificity and precise reason for death follow (Figure 5.3).

Locate the following case using your favorite legal research resource, read the case, and answer the questions below.

Bobo v. Tennessee Valley Authority, 2017 855 F.3d 1294 (2017)

Questions:

1. Asbestosis litigation has been a regular visitor to American Courtrooms for the last 75 years. In Bobo, as in nearly all similar cases, there must be proof of sufficient exposure. Where and how did this exposure take place?

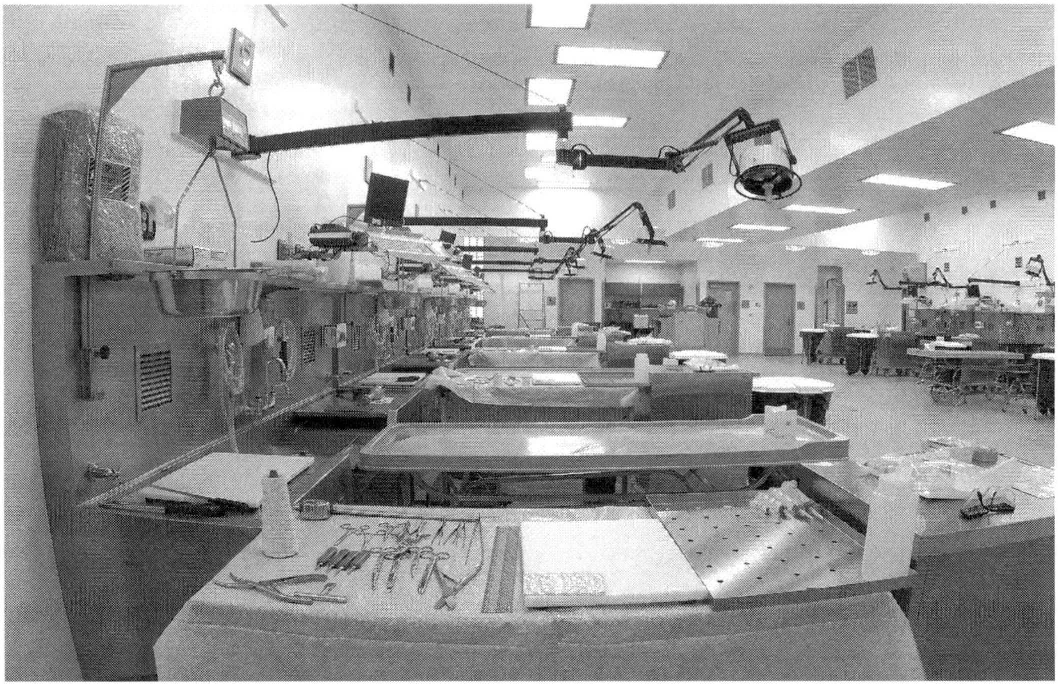

FIGURE 5.3 The Charles C. Carson Center for Mortuary Affairs, at Dover Air Force Base (AFB), Delaware (DE), with 23 state of the art autopsy stations.

Source: National Archives NAID: 6659003.

2. It is not enough for a plaintiff to allege that exposure caused a premature death due to the toxic or hazardous material. The Court relies on another test which is the plaintiff's burden. What is that test?
3. What does the Court conclude concerning the causation in this case?

United States v. Ewing
Aug. 31, 2018
No. 17-5496 (6th Cir.)

Joshua Ewing was convicted of distributing a mixture containing heroin and fentanyl, the use of which resulted in the death of Jeremy Deaton, in violation of 21 U.S.C. § 841(a)(1) and (b)(1)(C). He received a mandatory life sentence. On appeal, Ewing challenges his conviction on three grounds: (1) the district court gave internally inconsistent jury instructions, violating Ewing's due process rights; (2) juror misconduct deprived Ewing of his constitutional right to a fair trial and impartial jury; and (3) there is insufficient evidence to support his conviction and "death results" sentencing enhancement. For the following reasons, we find no error and AFFIRM on the first two grounds; on the third ground, we VACATE Ewing's death results conviction and sentence and REMAND to the district court for entry of judgment and re-sentencing on the lesser included offense of distribution of a controlled substance in violation of 21 U.S.C. § 841(a)(1).

I. BACKGROUND

On May 5, 2016, Joshua Ewing was indicted on one count of distributing a mixture or substance containing heroin and fentanyl, the use of which resulted in the overdose death of Jeremy Deaton, in violation of 21 U.S.C. § 841(a)(1) and (b)(1)(C). Ewing turned down a 15-year plea agreement and proceeded to trial. At trial, the Government introduced evidence demonstrating that Deaton died of an apparent drug overdose on the morning of February 8, 2016. Data from a defibrillator implanted years earlier showed that Deaton's heart was beating over 350 times per minute the morning of the overdose and that the defibrillator shocked Deaton's heart three times between 8:07 and 8:17 a.m. Deaton likely died shortly after the last shock, and his body was discovered in his garage at approximately 10:15 a.m.

Deaton's body was found lying supine on the floor of the garage; one of his sleeves was pushed up, and he had bruising on his arm that was consistent with intravenous drug abuse. An empty syringe and a burnt spoon were located near his body. Residue collected from the spoon tested positive for both heroin and fentanyl; the empty syringe did not contain enough material to provide an identification. John McCarty, the deputy county coroner who responded to the scene, testified that Deaton likely collapsed immediately after using drugs. No leftover drugs were found in the garage, and the officers did not search Deaton's home beyond that area.

Toxicology tests were performed and revealed the presence of narcotics in Deaton's blood and urine. Fentanyl, cocaine metabolites (the compounds produced when the body metabolizes cocaine), and low levels of other drugs were found in Deaton's blood. Fentanyl, fentanyl metabolites, cocaine metabolites, heroin metabolites, and low levels of other controlled substances and their metabolites were found in Deaton's urine. McCarty recorded Deaton's cause of death as acute combined drug (cocaine and fentanyl) toxicity.

Law enforcement officers who responded to the overdose scene were able to recover and access Deaton's cell phone, including text messages between Deaton and a contact identified as "Josh." Further investigation identified Ewing as the owner of the phone associated with the "Josh" number. The Government introduced these text messages as evidence that Ewing sold Deaton heroin laced with fentanyl the night before Deaton's death. The Government also called Timothy Graul, a Lexington Police narcotics detective assigned to the Drug Enforcement Administration task force, as a witness to interpret what some of the text messages meant. Graul testified that the texts were consistent with a heroin sale and that, based on the messages, a transaction had occurred.

The jury deliberated for over seven hours over the course of two days. After twice indicating that they were unable to reach a decision, the jury returned a unanimous guilty verdict. Several days later, the jury foreperson sent a letter to the district court in an attempt to recant her verdict. The letter set forth several specific allegations of misconduct, including bias and misapplication of the law. Ewing moved for judgment notwithstanding the verdict and, in the alternative, for a new trial. The district court denied the motion without a hearing, determining that there was sufficient evidence to support the jury's guilty verdict and that the allegations in the foreperson's letter were inadmissible to impeach the verdict under Federal Rule of Evidence 606(b). Ewing was then sentenced to a mandatory life term due to the "death results" enhancement and a prior felony drug conviction. See 21 U.S.C. § 841(b)(1)(C).

Ewing appeals his conviction, raising three arguments: (1) the court gave inconsistent jury instructions, violating his due process rights; (2) juror misconduct deprived Ewing of his constitutional right to a fair trial and impartial jury; and (3) there is insufficient evidence to support his conviction.

II. ANALYSIS

. . .

C. Sufficiency of the Evidence

Finally, Ewing argues that there is insufficient evidence to support the jury's guilty verdict. Ewing raised this argument before the district court in his motion for judgment of acquittal; we therefore review it de novo. See United States v. Lowe, 795 F.3d 519, 522 (6th Cir. 2015). When evaluating a sufficiency challenge, we view the evidence in the light most favorable to the Government and ask whether it is sufficient to permit any rational juror to find the defendant guilty beyond a reasonable doubt. See id. "Circumstantial evidence alone is sufficient to sustain a conviction and such evidence need not remove every reasonable hypothesis except that of guilt." United States v. Barnett, 398 F.3d 516, 522 (6th Cir. 2005) (citation omitted).

To sustain Ewing's "death results" conviction, the Government must prove: (1) "knowing or intentional distribution of heroin [and fentanyl]"; and (2) "death caused by ('resulting from') the use of that drug." Burrage v. United States, 571 U.S. 204, 210 (2014). To satisfy the second element, the Government must prove that use of the drug distributed by the defendant was "a but-for cause of the victim's death." United States v. Volkman, 797 F.3d 377, 392 (6th Cir. 2015).

Ewing's sufficiency challenge presents three main arguments: (1) there is insufficient evidence that Ewing sold drugs to Deaton on the night before Deaton's death; (2) there is insufficient evidence that fentanyl was a but-for cause of Deaton's overdose death; and (3) even if there is sufficient evidence that Ewing sold drugs to Deaton that night and that fentanyl was a but-for cause of Deaton's death, there is insufficient evidence that Ewing sold Deaton the drugs that caused Deaton's death.

1. Whether Ewing Sold Drugs to Deaton

The text messages exchanged between Deaton and "Josh" indicate that Deaton was interested in buying drugs and that Josh had some "KILLA" to sell—a term Detective Graul interpreted to mean particularly potent drugs. The two men discussed price, quantity, and a meeting place—"BP." Deaton later texted that he was early because he had to stop by an ATM and was also getting gas. Bank records revealed that Deaton withdrew money from an ATM across from a BP gas station that night. Josh later warned Deaton to use less than the "usual amount" due to the drug's potency.

The Government also introduced evidence linking Ewing to the phone associated with "Josh's" number and therefore to the transaction: the contact in Deaton's phone was listed as "Josh," records showed that Ewing owned that phone, and the messages themselves suggest a level of familiarity between the two men and possible past dealings. Graul also testified that drug dealers tend to be "very protective" of their phones and that it was "extremely unlikely" that Ewing would have allowed someone else to access or use his phone. The evidence as a whole is sufficient to permit a rational juror to conclude that a drug transaction actually occurred the night before Deaton's death and that Ewing was the dealer.

…

3. Whether Ewing Sold the Drugs that Caused Deaton's Death

To sustain Ewing's death results conviction, the Government must also prove that the drugs sold by Ewing were the same ones that caused Deaton's death. Ewing contends that the Government did not meet its burden because (1) there is insufficient evidence that Ewing sold heroin laced with fentanyl, rather than cocaine or some other drug, to Deaton; and (2) even if Ewing sold Deaton heroin with fentanyl, there is insufficient evidence that it caused Deaton's death.

The Government introduced evidence that the drug transaction between Ewing and Deaton involved heroin laced with fentanyl. Detective Graul testified about the increasing prevalence of heroin and heroin laced with fentanyl in the area. He explained that he could not remember any case involving pure heroin and that all the heroin he had seen in recent years "ha[d] been cut with fentanyl or something similar." He also testified that the terms, quantity, and price discussed in the text messages between Deaton and Ewing were consistent with a heroin transaction, including heroin laced with fentanyl. For instance, the $170 per gram price is consistent with heroin, which is commonly sold for between $150 and $200 per gram. That testimony also indicated that the drug Ewing sold was probably not another substance, such as cocaine. Heroin is commonly sold in gram or half gram quantities; cocaine, on the other hand, is more commonly referred to in ounces or "eight balls," referring to an eighth of an ounce, and has a fairly consistent price of approximately $100 per gram. Thus, viewing the evidence in the light most favorable to the Government, there is sufficient evidence to permit a reasonable jury to find that the drug sold by Ewing to Deaton was heroin laced with fentanyl.

There is, however, an unexplained gap in the evidence with respect to Deaton's heroin use that calls into question whether the fentanyl that caused Deaton's death was from the mixture of heroin and fentanyl that Ewing sold to him.

The evidence demonstrated that Deaton collapsed immediately after using drugs at about 8 a.m. and died shortly thereafter. Deaton's defibrillator shocked his heart three times between 8:07 a.m. and 8:17 a.m. and recorded a heart rate of over 350 beats per minute. Detective Graul testified that he believed the drug use that caused Deaton's overdose death occurred at approximately 8 a.m., and Deputy Coroner McCarty opined that Deaton "had used the drug and had immediately collapsed." Deaton had a lethal level of fentanyl in his blood, which indicates that he used fentanyl—either alone or mixed with another drug—at approximately 8 a.m., and it was this fentanyl that caused his death.

The jury heard expert testimony from both the Government and Ewing that the presence of a drug or its metabolites in a person's blood indicates whether that person used that drug in the immediate past. According to Ward, "[t]he blood, as you analyze it, will give you a more recent history of ingestion." In contrast, the presence of a drug or its metabolite in the urine shows only past or historical use and, unlike blood test results, does not reveal immediate past use. McCarty testified that drugs found in the urine have "been in the body long enough to move into the urine, which is not immediate"; Ward added that urine tests give "a historical perspective" and that substances ingested two or three days earlier can still show up in one's urine.

Toxicology test results showed that fentanyl and fentanyl metabolites were present in Deaton's blood, along with cocaine metabolites. Heroin metabolites were found in Deaton's urine, but his blood contained neither heroin nor heroin metabolites. The absence of heroin itself in Deaton's blood would not necessarily undermine the Government's case, given the drug's fast half-life, but the absence of any heroin metabolites in Deaton's blood rendered it highly unlikely that he had used heroin "in the several hours before his death." According to Dr. Feola, if Deaton had used heroin within a few hours of his death, "it would be much more likely" that morphine and 6-Monoacetylmorphine, the two tested-for heroin metabolites, would have been present in his blood. The absence of those metabolites rendered it "much more likely that [Deaton's] heroin ingestion was not around the time of [his] death."

The Government did not present any evidence to rebut Feola's testimony or to otherwise explain the absence of heroin or heroin metabolites in Deaton's blood. And the testimony of the Government experts—that urine reveals only historical use, while blood reveals more recent use—supports Feola's conclusion that Deaton had not used heroin in the several hours immediately before his overdose death. It is during that time frame, according to the Government's other evidence, that Deaton used the drugs that caused his death. In the absence of any evidence explaining this inconsistency, the jury lacked sufficient evidence to conclude beyond a reasonable doubt that the heroin mixture sold by Ewing contained the fentanyl that caused Deaton's death.

The other evidence relied on by the Government does not alter this conclusion. The spoon recovered from the workbench near Deaton's body contained both heroin and fentanyl residue, indicating that Deaton had likely used that spoon to heat both drugs. It does not, however, tell us or the jury whether the residue accumulated because Deaton used the spoon to heat a substance that contained both drugs or whether he had used the spoon at some point to heat each drug separately, leaving residue from both. Deaton had a history of drug abuse and evidently had been using the spoon to cook narcotics. He had multiple drugs in his system, including cocaine, which the Government does not contend came from Ewing, so while no evidence of a subsequent drug transaction was presented, Deaton clearly had access to other illicit substances in the days and hours before his overdose death.

It is of course true that the evidence "need not remove every reasonable hypothesis except that of guilt" to support a conviction, Barnett, 398 F.3d at 522 (citation omitted), but the absence of heroin or heroin metabolites in Deaton's blood and the lack of any evidence or testimony to explain its absence leaves us unable to conclude that the jury's verdict as to the death results enhancement is supported by sufficient evidence.

However, as discussed above, there is sufficient evidence that Ewing sold Deaton heroin laced with fentanyl in violation of the lesser-included offense of distribution of a controlled substance. See 21 U.S.C. § 841(a)(1); Burrage, 571 U.S. at 210 n.3 ("Violation of § 841(a)(1) is thus a lesser included offense of the crime charged in count 2 [the unlawful distribution of a Schedule I or II drug that results in death, § 841(a)(1), (b)(1)(C)]. It is undisputed that [the defendant] is guilty of that lesser included offense."); United States v. Burrage, 747 F.3d 995, 998 (8th Cir. 2014) (reversing the defendant's conviction following remand from the Supreme Court, and remanding to the district court to "enter judgment on the lesser included offense of distribution of a controlled substance in violation of 21 U.S.C. § 841(a)(1)" and for resentencing).

We therefore vacate Ewing's death results conviction and remand to the district court to enter judgment on the lesser-included offense of distribution of a controlled substance in violation of 21 U.S.C. § 841(a)(1) and for resentencing. See Rutledge v. United States, 517 U.S. 292, 306 (1996) (noting that "federal appellate courts appear to have uniformly concluded that they may direct the entry of judgment for a lesser included offense when a conviction for a greater offense is reversed on grounds that affect only the greater offense"); United States v. Ward, 37 F.3d 243, 250-51 (6th Cir. 1994) (reversing a continuing criminal enterprise conviction and remanding for resentencing on the merged conspiracy count); see also United States v. Hickman, 626 F.3d 756, 770-71 (4th Cir. 2010) ("Because we conclude that the record contains sufficient evidence to persuade a rational fact finder beyond a reasonable doubt of [the defendant's] guilt on the lesser included offense of conspiracy to distribute one hundred grams or more of heroin, we direct entry of judgment against [the defendant] under Count I of the indictment for conspiracy to distribute and to possess with intent to distribute heroin in the amount of one hundred grams or more.").

III. CONCLUSION

For the reasons explained above, we AFFIRM the district court on Ewing's jury instruction and juror misconduct challenges, VACATE Ewing's death results conviction and sentence, and REMAND to the district court for entry of judgment and resentencing on the lesser-included offense of distribution of a controlled substance in violation of 21 U.S.C. § 841(a)(1).

Questions:

1. In the Toxicology and Autopsy report, what is the cause of death?
2. How did law enforcement connect the specific seller of heroin with the specific cause of death victim?
3. The Court tends view the activity of selling lethal drugs as a strict liability activity. How so?

United States v. Lewis
895 F.3d 1004 (8th Cir. 2018)

Following a three-day bench trial, the district court found Antrell Desharron Lewis guilty of one count of conspiracy to distribute a mixture or substance containing heroin and furanylfentanyl resulting in death and serious bodily injury, in violation of 21 U.S.C. §§ 813, 841(a)(1), 841(b)(1)(C), and 846; and one count of distribution of a mixture or substance containing heroin and furanylfentanyl resulting in death and serious bodily injury, in violation of 21 U.S.C. §§ 813, 841(a)(1), and 841(b)(1)(C). The court sentenced Lewis to concurrent terms of 252 months' imprisonment to be followed by concurrent three-year terms of supervised release.

On appeal, Lewis argues the evidence was insufficient to support the convictions because (1) other individuals might have distributed the heroin laced with furanylfentanyl that caused the death and/or serious bodily injuries; (2) the government failed to establish "but-for" causation due to the intervening act of redistribution by another individual; (3) no conspiracy existed between Lewis and the individual who distributed the drugs to the victims; and (4) the government failed to prove Lewis knowingly distributed an analogue of a controlled substance. Having jurisdiction under 28 U.S.C. § 1291, we affirm.

I. Background

Joshua Manning testified that for about two months prior to March 2, 2016, he had been obtaining heroin from Lewis in Dubuque, Iowa. Manning usually conducted a couple of transactions with Lewis each week involving half gram or gram quantities. He was sometimes joined by his friend Jeremy Nadermann on these trips to Dubuque.

According to Manning's testimony, on March 2, 2016, he contacted Lewis about obtaining five grams of heroin on a "front." Manning asked Nadermann to go with him because Manning did not have a vehicle. Manning intended to give one or two grams of heroin to Nadermann for letting him use a car to make the trip.

Nadermann and Manning joined by two other individuals, Michael Vanamburg and Anthony Kelly, drove to Dubuque for the purpose of obtaining heroin. Manning testified he gave Lewis $200 in proceeds from Manning's sale of previously-fronted heroin. Manning explained at trial that five grams was the largest quantity he had received from Lewis in a single transaction. Nadermann confirmed during his testimony that a meeting was arranged with Lewis on March 2, 2016, to obtain heroin. Nadermann added that there was some discussion about the four of them pooling their money together to obtain a larger amount of heroin, which they would then divide into portions. According to Manning, Lewis agreed to front five grams of heroin to him for $800.

After Manning received the heroin, Nadermann requested the group drive to Jeremy Stierman's residence in Dubuque. Manning testified that he and Nadermann went inside Stierman's apartment while Vanamburg and Kelly stayed in the car. Manning weighed out one or two grams of heroin for Nadermann and a "50 bag" (about 0.2 grams) for Vanamburg. Manning put the rest in a bag. After dividing the heroin into portions, Manning and Nadermann returned to the vehicle to get high. Manning gave Vanamburg the "50 bag." Manning used a spoon and water to prepare a liquid mixture and drew some of it into a syringe for his use, and then gave the spoon to Kelly so he could use what was left.

Manning testified that he believed the substance he received from Lewis on March 2, 2016, was more potent than what he usually received. He described feeling a more intense high. Nadermann testified that he used some of the heroin Manning received from Lewis that evening and experienced a "strong weird feeling" but did not lose consciousness. In a very short period of time after using the heroin inside the vehicle, Manning noticed that Kelly was unresponsive in the backseat. Manning and Nadermann attempted to revive Kelly. When they were unsuccessful, Manning took his portion of heroin back to Stierman's apartment and called 911. During the emergency call, Nadermann advised Manning that Vanamburg, too, was unresponsive.

Before emergency responders arrived on scene, Manning testified that he saw Nadermann throw what Manning believed to be a syringe and Nadermann's portion of the heroin into a snowbank. Emergency responders, suspecting opiate overdoses, testified they administered Narcan and revived both Kelly and Vanamburg. Once alert, Kelly and Vanamburg were transported to the hospital in ambulances. Dr. Joshua Pruitt, an emergency room physician, the deputy medical examiner for Linn County, Iowa, and the chief medical examiner for Cedar County, Iowa, testified at trial. Dr. Pruitt testified that, upon review of the evidence, it was his opinion that Kelly was rendered unconscious because of an opiate overdose and that Kelly was at serious risk of death without medical intervention. Dr. Pruitt also testified that Vanamburg faced the same risks and that Vanamburg would not have been in that situation without the use of an opiate. Dr. Pruitt noted that Vanamburg had been prescribed Oxycodone, but discounted the possibility that Vanamburg's overdose was caused by the prescription medication, as he was using it as directed and the frequency dosage would not cause the effects Vanamburg experienced. Dr. Pruitt opined that Oxycodone was not the cause of Vanamburg's overdose.

During the search of the vehicle, law enforcement officers found a metal spoon containing some residue and a small cotton swab. The items were sent to the crime lab for testing. The lab report indicated the substance on the spoon was heroin and furanylfentanyl. In the rear passenger door pocket, officers found a small plastic baggie with a white rock substance. The lab report indicated the substance in the baggie contained a mixture of heroin and furanylfentanyl. The substance weighed 0.12 grams.

Manning admitted he provided false information to law enforcement about the details of Kelly's and Vanamburg's heroin usage because he did not want to go to jail. Manning and Nadermann were not arrested and were allowed to leave the scene. Manning testified that after they left, Nadermann went back to Stierman's apartment to get Manning's heroin as well as the heroin and syringe Nadermann had thrown in a snowbank. Nadermann denied that he went back to Stierman's apartment. Nadermann testified that he and Manning drove to the hospital to check on Vanamburg and Kelly, but upon arriving decided not to go inside. According to Nadermann, he then parted ways with Manning. Nadermann drove back to Maquoketa while Manning stayed in Dubuque at his mother's house.

Brian Koster, Stierman's co-worker, testified at trial. He testified that during the evening of March 2, 2016, he was at Stierman's apartment. Stierman told Koster that he intended to use drugs that night. Koster was not a drug user and did not see Stierman use drugs that night. Koster was present at Stierman's apartment when two men arrived, went to the kitchen with Stierman for about 15 minutes, and then left. Koster identified one of the men as Manning. Koster was also present when Manning returned to the apartment and announced one of his friends had stopped breathing in the car. Koster was concerned about getting into trouble for being in the presence of drugs, so he asked Stierman if drugs were inside the apartment. Stierman confirmed the presence of drugs. Koster told Stierman he should get rid of them. According to Koster, Stierman left the apartment and came back about one minute later. Koster and Stierman remained in the apartment while emergency responders and law enforcement officers addressed the situation outside. About an hour after law enforcement officers cleared the scene, Koster left Stierman's apartment and drove home.

In the early morning hours of March 3, 2016, Nadermann sent text messages to Stierman. Stierman did not respond to any of the messages. Koster talked on the phone to Stierman when Koster arrived home. He noticed Stierman's state of mind was different and Stierman sounded intoxicated. At 2:44 a.m. on March 3, 2016, almost immediately after the telephone conversation ended, Stierman sent a text message to Koster that stated:

> So I just wanted to get my story straight again in case jerry asks … as far as cops go we're in the clear im just hoping jerry isn't suspicious and at least ill have you as an alibi … just like we talked about me and you were havin a few beers after work when some random dudes just knocked on my door … call me when you can.

Koster testified that "jerry" was Stierman's landlord and he was also Koster's landlord.

At 2:07 p.m. on March 3, Stierman's father arrived at Stierman's apartment to pick up Stierman and take him to work, as they worked at the same manufacturing company and Stierman's father often gave Stierman a ride to work. When Stierman did not come out, Stierman's father drove to work. He became concerned when he learned Stierman did not show up for work. He called Stierman's sister during his break around 7:00 p.m. She drove to Stierman's apartment and found Stierman dead inside his apartment.

Dubuque County Sheriff's Officer Adam Williams testified that a baggie containing a substance was found in a kitchen cabinet in Stierman's apartment and a small amount of white powder was found on the kitchen counter. A lab report identified the baggie contents as 0.14 grams of furanylfentanyl and heroin. The powder on the counter consisted of furanylfentanyl.

Manning testified that the only drugs in his possession on March 2-3, 2016, were those he had obtained from Lewis. Manning acknowledged that Nadermann had "a little bit" of heroin with him when Nadermann picked Manning up, but that the group had stopped during the trip and used Nadermann's heroin before meeting Lewis. Manning testified that he received a call from Lewis a couple days later asking for the money for the heroin that had been fronted. Manning told Lewis that he did not have the money and never paid Lewis for the heroin.

Dr. Julia Goodin, Chief Medical Examiner for the State of Tennessee, who previously held the same position for the State of Iowa, performed an autopsy on Stierman. She offered her expert opinion at trial that furanylfentanyl was the cause of Stierman's death based on the toxicology report, her autopsy findings, and evidence gathered from Stierman's apartment. Because Stierman's urine was presumptively positive for opiates, Dr. Goodin could not rule out the possibility that another opiate was involved as well. Dr. Goodin testified at trial that it was her opinion that Stierman would not have died if he had not used furanylfentanyl.

II. Discussion

Lewis contends the evidence was insufficient to sustain the convictions. He raises several arguments, but his main argument focuses on whether the district court erred when it convicted him on the charged offenses because at most he was guilty of distributing heroin, for which the death/serious bodily injury sentencing enhancement found in 21 U.S.C. § 841(b)(1)(C) would not apply. We consider each of Lewis's arguments in turn.

"We review the sufficiency of the evidence de novo, viewing evidence in the light most favorable to the government, resolving conflicts in the government's favor, and accepting all reasonable inferences that support the verdict." United States v. Trejo, 831 F.3d 1090, 1093 (8th Cir. 2016) (quoting United States v. Washington, 318 F.3d 845, 852 (8th Cir. 2003)) (internal quotation omitted). "Reversal is appropriate only where a reasonable [fact finder] could not have found all the elements of the offense beyond a reasonable doubt." Id. at 1093–94 (quoting United States v. Armstrong, 253 F.3d 335, 336 (8th Cir. 2001)) (internal quotation omitted).

Lewis contends the evidence failed to establish a conspiracy between Manning and him because Manning had no intention of paying for the drugs. It is immaterial whether Manning repaid Lewis or intended to repay Lewis. "To convict a defendant of conspiracy to distribute drugs, the government must prove that there was an agreement to distribute drugs, that the defendant knew of the agreement, and that the defendant intentionally joined in the agreement." United States v. Chavez-Alvarez, 594 F.3d 1062, 1066 (8th Cir. 2010) (citing United States v. Benitez, 531 F.3d 711, 716 (8th Cir. 2008)). "An agreement to join a conspiracy need not be explicit and can be inferred from the facts of the case." United States v. Davis, 826 F.3d 1078, 1081 (8th Cir. 2016) (citing United States v. Slagg, 651 F.3d 832, 840 (8th Cir. 2011)).

There is overwhelming evidence in the record establishing that Lewis and Manning were engaged in a heroin trafficking conspiracy. Manning testified that he had been receiving heroin from Lewis for approximately two months. Manning paid Lewis $200 from a previously fronted drug transaction when he obtained the five grams of heroin that gave rise to the charges in this case. Overwhelming evidence established that Lewis distributed heroin to Manning on March 2, 2016. Lewis agreed to front the drugs to Manning in return for $800. The charged offense was a conspiracy to distribute drugs, not to sell them. There was sufficient evidence for the district court to conclude that Lewis knowingly and voluntarily joined a conspiracy to distribute heroin, and he shared a common purpose with Manning and others.

Lewis's identification of evidence that weighs against the verdicts and his claim that Manning provided "self-serving" testimony are arguments that go to the district court's credibility determinations and weight to be given the evidence. "[W]e will not disturb the district court's reasoned credibility determinations." United States v. Bowie, 618 F.3d 802, 814 (8th Cir. 2010).

The court next turns to the focal point of Lewis's appeal, whether the sentencing enhancement found in § 841(b)(1)(C) is sustainable on the record. Lewis contends that the government failed to prove whether it was the heroin, furanylfentanyl, or combination of heroin and furanylfentanyl that caused the overdoses and death. Expert testimony provided by medical professionals established beyond a reasonable doubt that Vanamburg and Kelly would not have overdosed "but-for" the use of furanylfentanyl and that Stierman would not have died "but-for" the use of furanylfentanyl.

The government was not required to prove that Lewis knowingly distributed an analogue of a controlled substance. We have repeatedly explained that "[a] defendant does 'not need to know the exact nature of the substance in [his] possession, only that it was a controlled substance of some kind.' " United States v. Anwar, 880 F.3d 958, 967 (8th Cir. 2018) (quoting United States v. Morales, 813 F.3d 1058, 1068 (8th Cir. 2016)). Section 841(b)(1)(C) is a sentencing enhancement, not a separate offense. To sustain a conviction under 21 U.S.C. § 841(a)(1) with a serious bodily injury or death enhancement under § 841(b)(1)(C), the government must prove: "(i) knowing or intentional distribution of [an illicit drug], ... and (ii) [serious bodily injury or] death caused by ('resulting from') the use of that drug." Burrage v. United States, 571 U.S. 204, 210, 134 S.Ct. 881, 187 L.Ed.2d 715 (2014). The Supreme Court explained that "where use of the drug distributed by the defendant is not an independently sufficient cause of the victim's death or serious bodily injury, a defendant cannot be liable under the penalty enhancement provision of § 841(b)(1)(C) unless such use is a but-for cause of the death or injury." Id. at 218–19, 134 S.Ct. 881. Following Burrage, the statutory sentencing enhancement in § 841(b)(1)(C) may be proved in two ways: (1) "but-for" cause, or (2) independently sufficient cause. There is sufficient evidence in this case to conclude that the government proved that Lewis knowingly and intentionally distributed an illicit drug and, under the facts of this case, use of that drug was an independently sufficient cause of the overdoses and death. Accord United States v. Allen, 716 F.App'x 447, 450–51 (6th Cir. 2017) (where victim was found with a "cocktail of drugs" in her system, the state medical examiner's testimony combined with the close proximity of a spoon containing fentanyl was sufficient for the jury to conclude that fentanyl was an independently sufficient cause of death). At a minimum, the evidence established that, without the incremental effect of furanylfentanyl in the heroin, Stierman would have lived and Kelly and Vanamburg would not have suffered serious bodily injury. The statutory requirements for the sentencing enhancement set forth in § 841(b)(1)(C) have been met.

Finally, Lewis argues that even if it was proven that he provided the heroin to Manning, who then supplied it to Vanamburg, Kelly, and Stierman, the redistribution constitutes an intervening cause of the injuries and death for which Lewis cannot be held responsible. Nothing in Burrage or the plain language of the statute limits responsibility to only the last person to distribute the drug before the harm occurs.

III. Conclusion

For the foregoing reasons, the judgment of the district court is affirmed.

Questions:

1. Death caused by drug usage and the sales thereof are common and fairly predictable. In Lewis, the challenge of matching victims and their deaths with the drug dealers seems to have been met. How exactly?

2. How does the discovery of a spoon confirm and corroborate the tie between the seller and the victim?
3. What is the Court's resolution of an "enhanced" sentence for causing the death by the distribution of a lethal drug?

United States v. Sica
Jan. 24, 2017
No. 15-3486-cr (2d Cir.)

Defendant Dennis Sica was convicted after a guilty plea of conspiracy to traffic in heroin and fentanyl, causing the deaths of three people, see 21 U.S.C. §§ 841(b)(1)(B), 846, and sentenced principally to 420 months' imprisonment. On appeal, Sica challenges the factual basis for his guilty plea and the procedural and substantive reasonableness of his sentence. We assume the parties' familiarity with the facts and record of prior proceedings, which we reference only as necessary to explain our decision to affirm.

I. Factual Basis for the Guilty Plea

Sica argues that his guilty plea was insufficient to establish that (a) his narcotics sales caused the victims' deaths and (b) such deaths were foreseeable. See Fed. R. Crim. P. 11(b)(3). We normally review a defendant's challenge to his plea's factual basis for abuse of discretion, subject to harmless error, see United States v. Culbertson, 670 F.3d 183, 189, 192 (2d Cir. 2012), but we review only for plain error where, as here, the defendant failed to challenge the validity of his plea in the district court, see United States v. Rodriguez, 725 F.3d 271, 276 (2d Cir. 2013) (citing United States v. Vonn, 535 U.S. 55, 59 (2002)). Sica thus bears the burden of showing (1) an error (2) that is clear and obvious, (3) affecting "substantial rights," and (4) seriously impugning the "fairness, integrity, or public reputation of judicial proceedings." United States v. Wagner-Dano, 679 F.3d 83, 94 (2d Cir. 2012) (internal quotation marks omitted). In the Rule 11 context, substantial rights are affected only if there is a "reasonable probability that, but for the error, [the defendant] would not have entered the plea." United States v. Rodriguez, 725 F.3d at 276. We identify no error here, plain or otherwise.

a. Causation

As to causation, Sica asserts that his guilty plea established only that his drug trafficking was a but-for cause of the victims' deaths, not the sole cause. The argument is defeated by Burrage v. United States, 134 S. Ct. 881 (2014), wherein the Supreme Court clearly and repeatedly stated that the causation element for § 841(b) offenses resulting in death requires only that the defendant's conduct be "a but-for cause of the death," id. at 892 (emphasis added); see id. at 885, 888, 889, 891. In urging otherwise, Sica relies on a decision by a district court outside this circuit that, in a Title VII case, construed § 841(b)(1)(C) to require a showing of sole causation. See Hendon v. Kamtek, Inc., 117 F. Supp. 3d 1325, 1331-32 (N.D. Ala. 2015). We are not persuaded.

In explaining the causation requirement, Burrage distinguished between instances in which the defendant's narcotics trafficking "merely played a nonessential contributing role" in the death, which does not suffice to establish causation, and instances in which the "incremental effect" of the drugs was the "straw that broke the camel's back," which does suffice. Burrage v. United States, 134 S. Ct. at 888. The record fixes this case firmly within the latter category, as toxicology reports and witness accounts provided a sufficient basis to conclude that the narcotics sold by Sica were essential to the victims' deaths, and therefore a but-for cause. Indeed, Sica's counsel appears to have conceded as much at the plea allocution:

> We believe that the government will adduce on its direct case legally sufficient evidence that a jury could conclude that the deaths of those persons was a but/for—the death would not have occurred but/ for the use of the drugs that were distributed as part of the conspiracy.

App'x 168 (emphasis added). Burrage supports the penalty enhancement for § 841(b)(1) crimes in such circumstances.

We therefore identify no abuse of discretion—much less plain error—in the district court's conclusion that Sica's guilty plea provided a sufficient factual basis for the element of causation.

b. Foreseeability

Sica further argues—but also did not challenge at his plea allocution—the absence of a factual basis to conclude that the victims' deaths were foreseeable to him. This argument fails because he cites no controlling authority for a foreseeability requirement, and an error is not "plain" when the urged rule is not "clearly established by the time of the appeal." United States v. Irving, 554 F.3d 64, 78 (2d Cir. 2009). Indeed, Sica's foreseeability argument has been rejected by every court of appeals to have considered the question. See United States v. Burkholder, 816 F.3d 607, 616 (10th Cir. 2016); United States v. Webb, 655 F.3d 1238, 1257 (11th Cir. 2011); United States v. De La Cruz, 514 F.3d 121, 137 (1st Cir. 2008); United States v. Hatfield, 591 F.3d 945, 948, 949 (7th Cir. 2010); United States v. Houston, 406 F.3d 1121, 1124-25 (9th Cir. 2005); United States v. Carbajal, 290 F.3d 277, 284 (5th Cir. 2002); United States v. McIntosh, 236 F.3d 968, 972 (8th Cir. 2001); United States v. Robinson, 167 F.3d 824, 832 (3d Cir. 1999); United States v. Patterson, 38 F.3d 139, 145 (4th Cir. 1994).

We need not ourselves decide the need to establish foreseeability because, even if we were to answer that question in favor of Sica, we would identify no error here, where foreseeability was supported by facts in the record, specifically (1) Sica's experience with potent narcotics, particularly after his prior narcotics convictions; and (2) his continued sale of those narcotics even after learning that they had caused a young woman's illness and, shortly thereafter, a young man's death.

…

4. Conclusion

We have considered Sica's remaining arguments and conclude that they are without merit. Accordingly, the judgment of the district court is AFFIRMED.

Questions:

1. Why does a toxicology report or finding resolve the "but-for" causation debate?
2. How does the Court characterize the toxicology reports when compared to other evidence forms?
3. What impact did the theory of "foreseeability" have upon the Court's conclusion?

6 Forensic Science and Trace Evidence

6.1 INTRODUCTION

The idea of trace evidence and its role in criminal prosecutions seems almost an evidentiary impossibility. By being "trace" it is by nature a small, visible and even invisible evidence source and by nature circumstantial in scope and design. Yet, even though its size may be small, its power to prove and persuade is extraordinary. The term "trace" implies a scant part of the whole, a remnant of the larger evidence. Paint chips, carpet fibers, clothing threads, dog and human hairs, to name just a few, have been the prime evidence in a host of famous cases and despite their size, have been repeatedly declared "sufficient" to uphold a conviction. Trace evidence allows for a reconstruction of events, places and conditions as well as proof of agency and personhood in the investigation and prosecution for criminal acts (Figure 6.1).

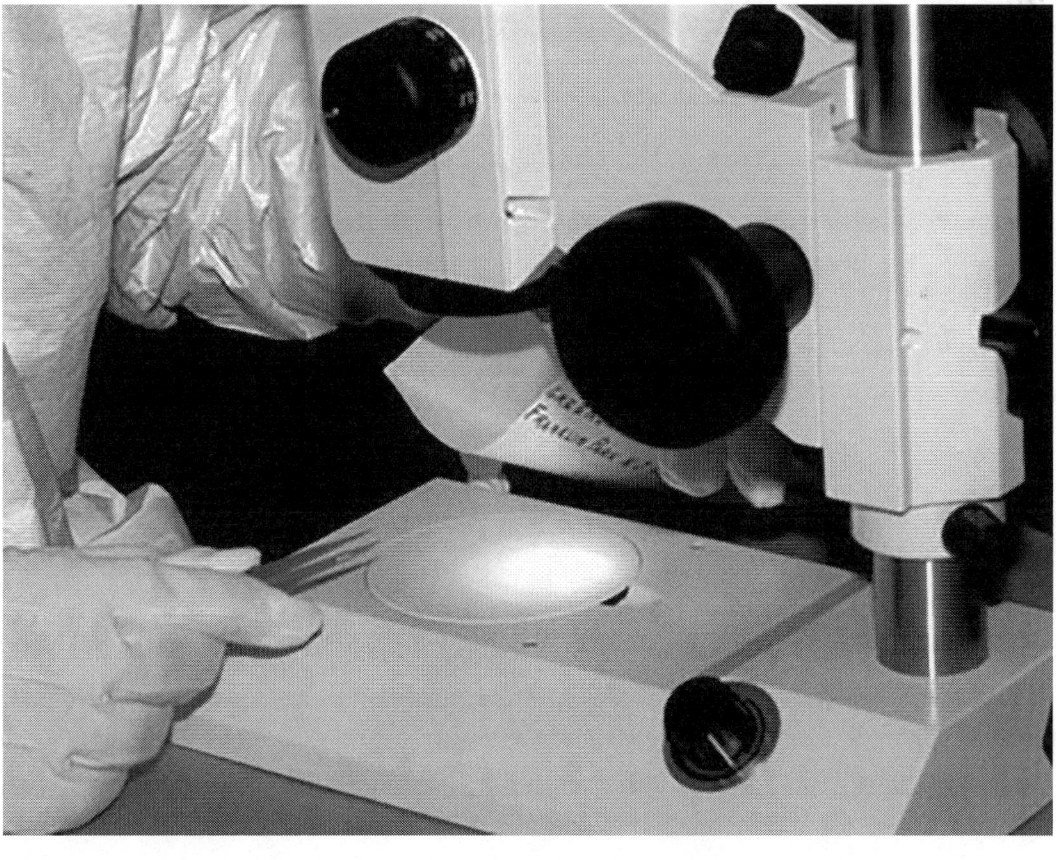

FIGURE 6.1 In the Amerithrax investigation among the analyses performed on the letter and envelope are trace evidence examinations.

Source: FBI at https://archives.fbi.gov/archives/about-us/history/famous-cases/anthrax-amerithrax/forensic-analysis.

DOI: 10.4324/9781003267126-6

Of very recent significance was the discovery of dog hairs in Christopher Kohburger's vehicle: the hairs being matched to the victim's dog who resided in the apartment where these horrid crimes took place.

6.2 TRACE EVIDENCE: TYPES AND METHOD OF MEASUREMENT

As in all forms of forensic analysis, the proponent of trace evidence will have to deal with its collection, preservation, its relevance to the case at hand and finally, the means and method that analyzes and assesses the evidence form. Hence, as in all other forms of forensic evidence, there are both substantive and process questions that need attended to.

State v. Murry
Court of Appeals of Washington, Division Three
January 28, 2020, Oral Argument; June 4, 2020,
Filed No. 35035-5-III

¶1 Roy Murry appeals from convictions for three counts of aggravated first degree murder, one count of attempted first degree murder, and one count of first degree arson. Due to an admitted defect in the charging document, we reverse the attempted murder conviction without prejudice. Because the evidence of identity was sufficient and because the trial court did not abuse its considerable discretion in resolving evidentiary challenges, we affirm the remaining convictions.

¶2 In the published portion of this opinion, we address Murry's Frye challenge and the inadequacy of the attempted murder charging language.

FACTUAL BACKGROUND

¶3 Although extensive evidence was admitted during the lengthy trial, the nature of the appellate challenges counsels we leave more detailed discussion of the voluminous facts to the appropriate argument. Accordingly, there need only be a generalized discussion of the factual background of this case.

…

¶5 On the night of May 25, 2015, Memorial Day, Amanda Constable worked her standard shift at a Spokane hospital and was expected to return home around 12:00 to 12:15 a.m. on May 26. A coworker called in ill and Amanda Constable had to work until 3:38 a.m. to cover. When she finally reached the family home, she discovered that law enforcement had responded to a crime scene.

¶6 The Canfields and John Constable had been murdered. Each had been shot multiple times and their bodies set on fire. Both the house and an outbuilding where Terry Canfield's body was found were burned. The subsequent investigation determined that both gasoline and barbecue lighter fluid had been used as accelerants in multiple areas of the house. Investigators did not identify the ignition sources, but several possible fire starters were located.

…

¶8 Detectives twice interviewed him within four days of the killings. He claimed to have been camping with friends along the Snake River, but declined to name his companions. Extensive efforts ensued to verify the alibi, but no corroborating evidence was located.

...

¶10 The jury found Mr. Murry guilty on the five noted charges and the court imposed the mandatory sentence of life in prison on the three aggravated first degree murder convictions. Mr. Murry timely appealed to this court. A panel heard oral argument of the case.

ANALYSIS

¶11 The published portion of this case addresses two issues. We first consider what is the relevant scientific community for purposes of a Frye analysis. We then turn to the adequacy of the attempted first degree murder charging language.

Frye Community

[1] ¶12 We conclude that the relevant scientific community is not the "criminal forensics community," but is instead the community of experts who are familiar with the use of the technique in question.

¶13 This issue arises from the discovery of strangely shaped nanoparticles on some of the shell casings recovered from the crime scene. William Schneck, a forensic scientist from the Washington State Patrol Crime Laboratory, used a scanning electron microscope (SEM) to examine the casings. The SEM is the most powerful microscope at the lab. He believed the particle might be AccuDure, a firearms lubricant, but his opinion was inconclusive. He therefore sent the samples to MVA Scientific Consultants, a private laboratory in Georgia. Richard Brown of MVA used a transmission electron microscope (TEM) and concluded that the samples were unique, synthetic silicon-based nanoparticles that were consistent with the distinctive component of AccuDure.

...

¶16 A Frye hearing was conducted prior to trial. Mr. Murry alleged that the TEM was not used in the "criminal forensics community" and that, accordingly, Frye precluded consideration of TEM evidence in Washington. Mr. Murry had an expert listen to the Frye hearing testimony, but, ultimately, he did not present any evidence or testimony at the hearing. The State presented testimony from Brown; Schneck; and the developer of AccuDure, Pavlo Rudenko, PhD. Dr. Rudenko used both a SEM and a TEM while developing AccuDure and also hypothesized that he would use a TEM in order to protect his patent should the need arise. He testified that differences between lubricants are discernable under a microscope.

¶17 Mr. Brown, who had used the TEM for 35 years, testified to the history of TEM, a microscope developed during the 1940s that became useful in forensic work in the 1980s due to its ability to distinguish asbestos fibers. He explained that due to the high resolution offered by TEM, it is the most common tool for examining nanoparticles. In addition to forensics, TEM commonly is used by medical device manufacturers and also by the Centers for Disease Control and Prevention to identify viruses.

¶18 Brown explained that TEM is the most powerful microscope for purposes of magnification and resolution. The difference between SEM and TEM is the difference between looking at the surface level of a particle (SEM) or at the atomic level (TEM). Brown also explained that there was no debate in the scientific community concerning use of TEM.

¶19 The trial court rejected the defense effort to classify the relevant scientific community for Frye purposes as the criminal forensic community. Concluding that the more general scientific community was appropriate, the court ruled that the testimony about the AccuDure nanoparticles was admissible.

[9] ¶20 In this court, Mr. Murry reprises his challenge to the trial court's determination of the relevant scientific community for the Frye assessment. One enduring criticism of Frye has been the court's failure to define the scientific community by which to judge the acceptance of novel scientific methods. ... This problem becomes complicated because various overlapping scientific disciplines use the same information and techniques. Id.

¶21 Washington courts have not squarely addressed this issue. A commonly cited answer to this challenge was provided by the Massachusetts Supreme Judicial Court: "the requirement of the Frye rule of general acceptability is satisfied, in our opinion, if the principle is generally accepted by those who would be expected to be familiar with its use." Commonwealth v. Lykus, 367 Mass. 191, 203, 327 N.E.2d 671 (1975).

¶22 We believe the Lykus standard is consistent with the actual application of Frye by the Washington Supreme Court. The mechanics of childbirth injuries were at issue in a recent medical malpractice case. L.M. v. Hamilton, 193 Wn.2d 113, 135, 436 P.3d 803 (2019).

In addition to hearing from the obstetrics community, the court permitted the testimony of a biomechanical engineer despite his lack of expertise with the biomechanics of childbirth. Id. at 138. In the seminal criminal cases that paved the way for use of DNA evidence at trial, the court looked at evidence from experts in multiple disciplines. In the case involving statistical DNA analysis, the court heard from forensic scientists, a university genetics professor, a university genetics researcher, and a university statistics professor. State v. Kalakosky, 121 Wn.2d 525, 542, 852 P.2d 1064 (1993). In the case involving DNA typing, the court heard from a large number of university researchers, geneticists, biochemists, and a statistician, in addition to forensic scientists. State v. Cauthron, 120 Wn.2d 879, 884, 846 P.2d 502 (1993).

¶23 In none of these cases did the experts belong solely to the civil or criminal forensics community. Mr. Murry has not identified a single Frye case where our courts have excluded expert testimony from outside the forensic community. Limiting testimony solely to those who use the science or equipment, instead of those also familiar with the principle, unduly narrows the field to those who favor the science in question. It also discourages innovation by excluding the opinions of cutting-edge researchers who may be demonstrating the utility of a new principle or a device.

[10] ¶24 The Massachusetts standard is consistent with the Washington practice and we adopt it. Accordingly, we hold that scientists familiar with the use of the scientific principle in question constitute the relevant scientific community for purposes of a Frye analysis.

¶25 Here, the trial court heard from scientists familiar with the examination of nanoparticles and properly based its ruling on their testimony. The trial court did not err in determining that examination of nanoparticles by a TEM was accepted in the scientific community familiar with the technology. Accordingly, its ruling is affirmed.

...

UNPUBLISHED ISSUES

¶35 The appeal raises numerous other challenges including the sufficiency of the evidence of identity on all charges as well as a contention that the killer did not take a substantial step toward killing Amanda Constable. We group those two challenges together before turning to look at his evidentiary arguments, another topic that we treat as one. Next, we briefly consider Mr. Murry's challenges to his mental competency and legal financial obligations (LFOs). Finally, we briefly address Mr. Murry's statement of additional grounds (SAG).

Sufficiency of the Evidence

...

¶39 The physical evidence tied Murry to the scene. The most damaging evidence was the AccuDure particles discovered on some of the shell casings. Dr. Rudenko testified that there were only two vials of AccuDure in existence—one belonged to Murry (and was discovered in his car) and the other vial Dr. Rudenko turned over to the WSP Crime Laboratory. Rudenko did not own or use firearms, while Murry was a gun enthusiast who owned numerous weapons. A .22 caliber gun missing from Murry's collection had been used to test AccuDure. The victims were killed by .22 caliber Remington rimfire hollow-point bullets. The same ammunition was found in Murry's car and his residence.

¶40 Traces of a fire starter, Trioxane, were discovered on a headlamp found inside Mr. Murry's car. He gave away his remaining Trioxane supplies shortly after the killings. Investigators believed that Trioxane could have been used to set the fires. Flares were identified as another possible ignition device. Flares also were recovered from Murry's car. The arson was committed by the same person who killed the victims.

¶41 In summary, the killer used AccuDure to lubricate his weapon. Mr. Murry was one of two people to possess that unique synthetic lubricant, and the only one of the two who had a motive to kill the family. He owned the same ammunition as the killer. The gun Murry used to test the AccuDure fired the same ammunition and was missing from his collection. Based on this evidence, the jury could conclude beyond a reasonable doubt that Mr. Murry was the killer and the arsonist. The evidence supported the verdicts.

¶42 In order to convict a person of attempted first degree murder, the evidence must allow the jury to conclude that a defendant intended to commit first degree murder and took a substantial step toward committing the offense. RCW 9 A.28.020(1); RCW 9 A.32.030(1)(a); In re Pers. Restraint of Borrero, 161 Wn.2d 532, 539-40, 167 P.3d 1106 (2007). "A 'substantial step' is conduct strongly corroborative of the actor's criminal purpose." Borrero, 161 Wn.2d at 539.

...

Evidentiary Objections

...

¶57 No fingerprints or DNA were recovered from the shell casings collected at the scene. Mr. Murry's habit of cleaning his ammunition and handling it with gloves explained the absence of any trace evidence. Once again, this was highly relevant evidence.

¶58 The habit evidence was relevant and not prejudicial. The trial court did not abuse its discretion by admitting it.

...

¶90 The third and fourth SAG issues allege that witnesses testified differently than expected and that trace evidence may have been contaminated. Neither of these issues was raised at trial and, therefore, neither is preserved for our review. Also, there is no factual support in the record for the third argument. His eighth argument fails for both of these reasons.

...

¶95 Reversed in part and remanded for further proceedings consistent with this opinion.

Questions:

1. What type of trace evidence was being utilized by the Court? Why is this form of evidence properly categorized as "trace."
2. What other forensic evidence subject matter does the examination of nanoparticles get close to in this examination?
3. What is the nature of the mechanical or procedural challenge to the trace evidence form?

> The Amerithrax report at page 12 describes fibers found in the envelopes, https://www.justice.gov/archive/amerithrax/docs/amx-investigative-summary2.pdf.

The Cordes decision paints a picture of trace evidence and fiber comparisons long before the age of micron microscopes and close scrutiny. At the same time, the case proves why good investigative work is always a central requirement for trace evidence comparisons.

Cordes v. the State
54 Tex. Crim. 204, 112 S.W. 943 (Tex. Crim. App. 1908)

Appellant was indicted for the murder of her child, and upon conviction her punishment was assessed at imprisonment for life.

The facts, in substance, show that appellant and her brother, Eilert Cordes, lived upon a farm in Gonzales County, occupying alone their home. ... Johnnie Hamilton, testified that he slept in the kitchen by the cooking stove in the same house and in an adjoining room to appellant. "That night I heard a baby cry in the night and heard them walking around in the house. I don't think I heard it cry but once that night before I went to sleep. I got up the next morning before sun up and heard it crying in the house next morning. It sounded like it was in the room from me adjoining the kitchen. I don't know who lived in there. I couldn't say what room they stayed in. Nobody else lived in that house but these people. It was about between six and seven o'clock in the morning when I heard the baby cry. I saw Mr. Cordes around the house that morning and I saw him afterwards during the day when he passed me going down towards the creek. He had a bundle under his arm that looked like a croker sack or saddle blanket balled up under his arm, but I couldn't say what it was. When he came back from the creek he had nothing with him. I never heard the baby cry after he carried the bundle off. I stayed there Tuesday, Wednesday, and Thursday, and went to town Friday evening and heard that the baby had been found Saturday night of the same week." A baby was found in the creek, which was some three or four hundred yards from appellant's home.

Dr. Finny Blackwell, in reference to finding the child, testified as follows: "I visited the water hole about two or three hundred yards away from the house of the defendant about the 10th or 11th of August of last year, and then went with Mr. Johnson, the sheriff, on that night to Cordes' house. At the water hole I found something covered up in a gunny sack and blanket very badly decomposed. I came to the place and found Mr. Pittman and another gentleman there and stayed on the bank and waited for the justice of the peace and Mr. Johnson. And they didn't come and I picked the gunny sack up and unwrapped the blanket and found a little girl baby and I examined it and went over it very thoroughly. The body was very much decomposed. The head had been fractured and the ribs had been broken or fractured. The brains were gone, and the membranes were intact. It was either the third and fourth or the fourth and fifth rib that was fractured. In addition the child had a string tied around its neck three times and tight as tight could be, and the back and shoulders of the child were badly decomposed. I pried the ribs open and got hold of the lung and opened it and extracted the lung and introduced two fingers and pulled out a part of the lung tissue, and when I pushed around in it air and water escaped. That air might have come from two or three places. It might have been caused from putrefaction and it might have been caused from natural prevailing conditions. The color of the lung tissue differed; it was light in places, but where it was badly decomposed it was of a brown hue, but I could not say positively just what the color was, looking at it with my naked eye, and from a scientific standpoint. I could not say whether the air had been breathed in there or whether it was the result of putrefaction. The child was well developed and would have weighed about eight or nine pounds. For a living child to have its brains knocked out would kill it sure; the wounds in the side untreated I think would have killed it. I don't think it would have produced immediate death. The string tied three times around its neck would have killed it. A living child would live a very few minutes with that string around its neck. A still born child would have a flat chest and the diaphragm would be way up. In a child that had lived you would find air in its lungs that could not be expelled. If it had been completely submerged for four or five days it would not have decomposed that fast. From my examination of the body and the appearance of the lung tissue I could not state whether that child had ever lived or not. I can state that you would have found those conditions in a child that had breathed, but I must state that results of putrefaction would produce the same results. A child by the act of breathing produces a rounded condition of the chest, and this child's chest was rounded. The lungs were extended; its abdomen was swollen and the skin around its neck had broken where the string was tied and it was decomposed all over. Any part of that child would float. As a medical expert I can not say whether or not that child lived. I can state that the air and extension of that lung would indicate that the child had breathed, but at the same time I believe the results of decomposition in that side would produce the distension just the same. The child's breast was round and distended and the diaphragm was down." The child had been taken out of the water upon which it was floating a short while before the doctor reached the scene. Johnnie Hamilton swore it was two or three o'clock in the evening on Monday that he saw Cordes go down to the creek with a bundle under his arm. … . Various witnesses testify that appellant appeared to be pregnant a short while before the child was found, and at that time did not show any signs of pregnancy. The doctor who examined appellant after she was placed in jail testified that everything indicated that she had given birth to a child a short while before the examination, some seven or eight days. The testimony shows that this will correspond with the date that Hamilton says he heard the baby cry. The sheriff testifies to taking a sack and blanket and a string that were found around the baby's neck and comparing them with blankets and cloth found in the home of appellant, and they correspond in texture and color with said blanket and cloth so found. He says the blanket he took from around the child was dirty and bloody and after washing it, it was of a lighter color than the blanket found in the house, but in texture and other respects corresponded. The figures on the cloth string around the baby's neck corresponded with the figures on the cloth in the home. The sheriff through an interpreter asked appellant if she had given birth to a child prior to her arrest. She denied it. Several bloody clothes and garments were found in the house indicative of child birth. This, in substance, as we glean from the statement of facts is the testimony.

…

Bill of exceptions No. 3 complains that W.W. Johnson, sheriff, was permitted to identify and exhibit to the jury in evidence a portion of a blanket purporting to have been found wrapped around the deceased infant, and which the witness testified was similar to another portion of blanket found at the home of defendant; that the portion shown to the jury and purporting to have been found around the body of the deceased infant, has been washed and was now white in color, while originally when found it was very dirty and bloody and of a dark color, practically the same color of the other portion found at the home of defendant. Appellant objected to the same for the reason it did not show the exact condition of the blanket when found, nor the same color. This objection would only go to the weight of the testimony, and not to its admissibility.

…

Bill of exceptions No. 9 complains of the following: Dr. McCaleb, medical expert, was asked the following hypothetical question: "If you were called upon to examine a dead infant found under the following condition: The body badly decomposed and freshly taken from a hole of water, and you were to unwrap a gunny sack with rocks in it and a blanket wrapped around the body of the child and find that child's head broken, the brains all gone, two of its ribs broken and the membranes of the brain intact, with a string tied three times tightly around the child's neck and you were to prize its ribs open and pull out a portion of the child's lung and you were to press on the lung and air and water came out of it, and the appearance of that lung was light in color in places, and where it was decomposed it was of a brown hue and you were to find the naval string of that child tied with a string, and the appearance of the child showed it to be a fully developed female child weighing about nine pounds and upon pressing on it, when the lung was pulled out of the pleural cavity there were gasses in the lung, and it would spue and bubble, what would be your opinion as to whether that infant lived or not independent of its mother were that child's breast rounded out and its diaphragm dropped down?" To which question the defendant then and there in open court objected on the ground that same was illegal and inadmissible for the reason that said question did not contain all the material facts as to the condition of the child when found; that the doctor was not present and did not hear the witness testify to the condition of the child, and that the question does not give an accurate description of the condition of the child when found, all of which objections were overruled, and the doctor answered, "I think it lived," and said answer of said witness was permitted to go to the jury. In the trial of cases of this character it is permissible to put hypothetical questions to medical experts. In the first place, we wish to say, in substance, the above is practically all, if not literally, of the testimony as to the condition of the child when found. These facts were testified to by Dr. Blackwell, and the State in asking Dr. McCaleb the question practically reproduced the testimony of said Dr. Blackwell. See Burt v. State, 38 Tex.Crim. Rep..

...

On the doctrine of principals the court charged as follows: "Or that by such means and at such time and place Eilert Cordes did unlawfully and with express malice kill said infant, and that when the said Eilert Cordes so killed said infant, the defendant was present and had agreed with the said Eilert Cordes to so kill said infant; or that she was present when the said Eilert Cordes so killed said infant (if he did so), and knowing his unlawful intent, did aid him by acts or encourage him by words or gestures in so killing said infant, you will find defendant guilty of murder in the first degree, and assess her punishment at death or confinement in the penitentiary for life in your discretion." Appellant insists said charge is erroneous in that it did not require her to have express malice at the time she agreed with the said Eilert Cordes to kill said infant nor did it require her to have express malice at the time she aided him by acts or encouraged him by words or gestures in so killing said infant. The charge, we hold, shows clearly a proper presentation of this question. Furthermore, we can not isolate this clause from the balance of the charge, but the court properly charged all the law applicable to the facts of this case, telling the jury that appellant must aid by words, acts or gestures and be present at the time of the commission of the offense, knowing the unlawful intent of the said Eilert Cordes.

Various other errors are alleged in the motion for a new trial of omission and commission in the charge. We have reviewed each and all of same, and must say that the charge, as stated above, is a clear, succinct and proper presentation of all the law applicable to the facts of this case. The evidence in the case, as shown above, shows clearly by circumstantial evidence that the deceased infant came to its death by the direct co-operation, participancy and consent of its mother, appellant, while assisting her brother. That it had an independent existence at the time of its death separate and independent of its mother is manifest from the cries both on the night of its birth and the morning thereafter as testified by the witness Hamilton. It is true the witness Hamilton is thoroughly impeached by divers and sundry witnesses, but this is a matter we can not pass upon. The jury have seen fit to believe him and we are not authorized to disturb the finding of the jury on a question of fact.

Finding no error in the record, the judgment is in all things affirmed.

Affirmed.

Questions:

1. Why does trace evidence become more important than it might other be in the proof of a pregnant woman giving birth? What action by the defendant causes the focus of attention to turn toward trace evidence proof?
2. Without the physical examination, what trace evidence is examined which ties the death of the baby with this defendant?

3. In making a comparison of texture and color of blanket fibers and string, what objection does the defendant pose?

The impact of trace evidence on appeals can be burdensome for a system overwhelmed by litigation. Part of the problem of trace evidence is the mechanisms, the machines needed to measure and evaluate. In a sense it is a higher level due to its minutia. Of course, defendants always claim a host of challenges to trace but side by side with claims of mechanical or measure failures, is the allegation that due process rights are negatively impacted by the defense's lack of equipment and expertise to those tying trace evidence to eventual culpability. It is a tough balancing act.

Locate the following case using your favorite legal research resource, read the case and answer the questions below.

Williams v. State, 251 Ga. 749, 312 S.E.2d 40 (Ga. 1983)

Question:

1. What is the trace evidence at stake in this case?
2. How does the Court rule on the challenge to fiber methodology?
3. What argument is posed by defendant as to crime laboratories and trace evidence?

6.3 FORENSIC SCIENCE AND TRACE PAINT ANALYSIS

Read the DOJ Press release regarding how evidence from a shotgun shattered glass and resulted in missing paint on a gate assisted in a hate crime conviction at https://www.justice.gov/opa/pr/federal-jury-convicts-california-man-hate-crime

Atwood v. Ryan
870 F.3d 1033
United States Court of Appeals for the Ninth Circuit
June 7, 2017, Argued and Submitted, Seattle, Washington; September 13, 2017, Filed
No. 14-99002

Frank Jarvis Atwood was found guilty of kidnapping and first-degree felony murder and sentenced to death. Atwood appeals the district court's denial of his petition for a writ of habeas corpus. We have jurisdiction under 28 U.S.C. §§ 1291 and 2253, and we affirm.

I

Because the facts as found by the Arizona Supreme Court are presumed correct, 28 U.S.C. § 2254(e)(1), the following background relies on the state court's determination of factual issues.

…

After he was released on parole in May 1984, Atwood began traveling across country in his black 1975 Datsun 280Z. Id. at 593, 832 P.2d 593. In August 1984, Atwood met Jack McDonald, who became his traveling companion. Id. Among other stops, Atwood and McDonald visited Bernsienne in Enid, Oklahoma. Id. In mid-September, Atwood and McDonald traveled to Tucson, Arizona. Id. The morning of September 17, 1984, Atwood was seen at De Anza Park, a congregating spot popular among Tucson's transient population. Id. Atwood left the park in the middle of the afternoon. Id.

Around the same time, Sam Hall, a teacher at a Tucson elementary school, noticed a "dark Z car" with California license plates in an alley near the school. Id. at 592, 832 P.2d 593. Hall "described the driver as a man with a medium frame, shoulder-length hair, and a dark beard and mustache," id., a description that fit Atwood at the time, and noted that the man was "making strange gestures and shaking his head," id. at 593, 832 P.2d 593. Because Hall was "somewhat unnerved by the driver's appearance and behavior, he wrote down the car's license plate number." Id.

At around 3:30 p.m., an eight-year-old girl named Vicki Lynn Hoskinson left her Tucson home on a pink bicycle to drop off a card at a nearby mailbox. Id. at 592, 595, 832 P.2d 593. Two teenage boys in the neighborhood passed Vicki as she was riding her bicycle south toward the intersection of Root Lane and Pocito Place, which was only a few hundred feet from the elementary school where Hall had observed the "dark Z car" with California license plates. Id. at 592–93, 832 P.2d 593. The boys also passed a man with "long dark hair, a mustache, and the beginnings of a beard" in a "dark Datsun Z car" heading towards the same intersection. Id. (internal quotation marks omitted). When Vicki did not return home, her mother sent Vicki's sister to look for her. Id. at 592, 832 P.2d 593. Vicki's sister found Vicki's pink bicycle lying in the street near the intersection of Root Lane and Pocito Place, but Vicki was missing. Id.

...

On September 18, Hall told the authorities about his observations, including the license plate number of the car. Id. at 593, 832 P.2d 593. Authorities traced the car to Atwood, id., and agents from the Federal Bureau of Investigation (FBI) contacted his parents for information about his whereabouts, id. at 596, 832 P.2d 593. The FBI also obtained a warrant for Atwood's arrest on kidnapping charges. Id. at 593, 832 P.2d 593.

... Atwood's parents informed the FBI that Atwood was taking his car to Ken Stoepel Ford in Kerrville, Texas for repair work. Id. at 593, 832 P.2d 593.

On September 20, 1984, the FBI arrested Atwood at the garage and impounded his car. Id. ... FBI agents conducted an initial inspection of Atwood's car in Kerrville. Id. FBI Agent Declan Hoffman later testified that, as a part of the inspection, he photographed Atwood's car from numerous angles. (This set of photos is referred to as the Kerrville Suite.) At least one photo included in the Kerrville Suite shows a streak of pink paint on the front bumper of Atwood's Datsun.

... The same day, the FBI agents transported Atwood's car to an FBI facility in San Antonio and obtained a search warrant to conduct a more thorough search of the car. Id. On September 22, 1984, the FBI agents executed the search warrant and observed pink paint on the front bumper of Atwood's car during this search. FBI Agent Edward Burwitz took scrapings of this pink paint from the front bumper, and FBI Agent Hoffman took additional photos. (This set of photos is referred to as the San Antonio Suite.)

On September 27, 1984, Atwood was charged with kidnapping. Id. at 594, 832 P.2d 593. Because the investigation was still ongoing at that time, authorities did not know what additional evidence would be found or whether Vicki was still alive. Id. This uncertainty changed on April 11, 1985, when the skull and bones of a child were found in the desert northwest of Tucson. Id. Through dental records, the Pima County Medical Examiner, Dr. Richard Froede, and a physical anthropologist from the University of Arizona, Dr. Walter Birkby, identified the remains as Vicki's. Id. at 594, 598, 832 P.2d 593. Both experts noted the presence of adipocere on some of the bones during their evaluation. The State's experts could not, however, determine the cause of death from the bones that were found. Id. at 598, 832 P.2d 593.

. . .

State consolidated the kidnapping and murder charges for trial. Id. The Arizona Supreme Court summarized the State's theory at trial as follows:

> Defendant, a convicted pedophile, was cruising [Vicki's] neighborhood in search of a child. He saw [Vicki] riding her bike near the mailbox and followed her on the street as she cut through a field on her way home. The field led to a short street—Pocito—that intersected with Root Lane. Defendant proceeded to Root Lane (where he was seen by the teenagers), intending to abduct [Vicki] as she traveled down Pocito. [Vicki] stopped briefly to talk with a friend who lived on Pocito, and then continued down the street. Defendant's car struck [Vicki's] bike, leaving a pink paint smear on the car bumper, and he grabbed the child. He then proceeded toward northwest Tucson. En route, his car was sighted by three people, all of whom later identified defendant and testified that they saw a young child in the passenger's seat. Defendant took the child to the desert, where he molested and murdered her. He then returned to De Anza Park in central Tucson, where he had been earlier in the day.

Id. at 594–95, 832 P.2d 593.

According to the State's witnesses at trial, laboratory tests established that the pink paint on the front bumper of Atwood's car had come "from the victim's bike or from another source exactly like the bike" and that Vicki's bicycle had nickel particles on it that were consistent with nickel on Atwood's bumper. Id. at 595, 832 P.2d 593. The State's accident reconstruction expert, Paul Larmour, testified that "he found a nearly perfect match heightwise between the contact area on the backside of the bicycle and the [paint] transfer on the bumper," id. (alteration in original) (internal quotation marks omitted), and that "marks on the car's gravel pan were consistent with the theory that it struck the bicycle at a low speed and caused the bike to lodge beneath the car," id.

After a two-month trial, the jury found Atwood guilty of kidnapping and first-degree felony murder. Id. at 591–92, 832 P.2d 593.

. . .

In 1996, Atwood filed his first state habeas petition for post-conviction relief. Atwood raised his Eighth Amendment claim, as well as two other claims relevant here: (1) a claim that his trial counsel was ineffective for failing to further investigate the presence of adipocere and discover the existence of the grave (we refer to this claim as the "adipocere ineffective assistance of trial counsel claim"), and (2) a claim that law enforcement officials (FBI agents and Pima County Sheriff deputies) engaged in misconduct by planting the pink paint found on the bumper of Atwood's car (we refer to this claim as the "law enforcement misconduct claim"). The Arizona Superior Court denied relief on all claims. The Arizona Supreme Court denied review, and the United States Supreme Court denied certiorari.

. . .

. . . The district court stayed the proceedings to allow Atwood to exhaust the new aspects of his law enforcement misconduct claim that had not been presented to the state court. The court denied all of Atwood's other claims on the merits and denied a certificate of appealability for these claims.

. . .

IV

We next turn to Atwood's law enforcement misconduct claim. The state habeas court dismissed without an evidentiary hearing Atwood's claim that FBI agents and Pima County investigators planted pink paint from Vicki's bicycle on the bumper of his car. The state court determined that there was no colorable basis for the claim and that Atwood's theory lacked any "link to provable reality." On appeal, Atwood argues that the state court made an unreasonable determination of the facts, see 28 U.S.C. § 2254(d)(2), by failing to hold an evidentiary hearing on this claim.

A state court does not have to "conduct an evidentiary hearing to resolve every disputed factual question." Hibbler v. Benedetti, 693 F.3d 1140, 1147 (9th Cir. 2012). The "ultimate question" is whether a state court was "unreasonable in holding that an evidentiary hearing was not necessary in light of the state court record." Id. at 1148 (emphasis omitted). We have held that a state court's failure to hold an evidentiary hearing is not unreasonable under § 2254(d)(2)"where the allegations are said to be incredible in light of the record," or where the "petitioner's factual allegations are entirely without credibility." Perez v. Rosario, 459 F.3d 943, 950–51 (9th Cir. 2006). Nor does a state court's failure to hold an evidentiary hearing make its factual findings unreasonable "when the record already before the court is said to establish a fact conclusively," or "there is no likelihood that an evidentiary hearing would have affected the determination of the state court." Id.

...

In order to determine whether the state court here was objectively unreasonable in failing to conduct an evidentiary hearing, we must first explain Atwood's theory of law enforcement misconduct. Under Atwood's theory, Pima County investigators, in collusion with the FBI, removed both bumpers from Atwood's car in San Antonio and transported the bumpers as luggage on a commercial airline to Tucson. These investigators then scraped pink paint off Vicki's bicycle (which was in Tucson at the time) and planted it on the front bumper of Atwood's car. Next, the investigators collected scrapings from the pink paint they had applied to the bumpers, combined these scrapings with scrapings taken by the FBI from an unrelated pink paint smear on the bumper, and substituted the combined sample in the evidence log. The Pima County investigators then used water-soluble paint to cover up the scrape marks created while collecting the sample, transported the bumpers from Tucson back to the FBI's impound garage in San Antonio, and reattached the bumpers to Atwood's car. Once the bumpers were reattached, the agents took photos of the pink paint on the bumpers and then substituted these photos for original photos in the Kerrville and San Antonio Suites to cover up their actions.

Atwood supports his theory by pointing to alleged anomalies in three different suites of photos (the Kerrville Suite, the San Antonio Suite, and a third suite of undated photos taken in Tucson, which we refer to as the Tucson Suite), which he claims raise inferences of law enforcement misconduct.

A

We first consider the alleged anomalies in the Kerrville Suite of photos.

The record establishes that FBI agents conducted an initial inspection of Atwood's car at the garage in Kerrville on September 20, and that FBI Agent Hoffman took photos of the car. Frame 9 of the Kerrville Suite shows pink paint on the passenger side front bumper.

Atwood alleges that Frame 9 was taken in San Antonio (after the FBI tampered with the car to add pink paint from Vicki's bicycle to the bumper, as explained above) and the FBI substituted Frame 9 for the original photo. From this alleged evidence of tampering, Atwood claims, it can be inferred that pink paint from Vicki's bicycle was not present on the bumper of Atwood's Datsun when the FBI took custody of the car in Kerrville.

Atwood points to the following evidence to support this allegation. First, Hill made a digital photo enhancement of Frame 9. Hill claims the enhancement shows that the right side of the front bumper in Frame 9 is "specularly clean and reflective" compared to the left side of the front bumper in Frame 1, which is "grimy, dirty and splattered with dead gnats and other insects." According to Atwood, the differences in the cleanliness of the bumper establish that Frame 9 was taken at a different time than the rest of the Kerrville Suite. Second, Atwood submitted an affidavit from Hulick stating that Frame 9 "does not appear to be a part of the same set of photographs" because it is "in sharper focus and taken from a noticeably different angle than the other photographs." The government gave Atwood sequential negatives of these photos, but Hill stated that "[t]hese images are not 'original negatives' as that term is usually understood." Instead, according to Hill, "they are negatives made from other negatives or photographic positives." (Hulick did not opine on this issue.)

Based on our review of the record, the state court could have reasonably concluded that Atwood's claim that law enforcement had tampered with the Kerrville Suite was not credible. First, on their face, the enhanced photos do not show that the right side of the front bumper pictured in Frame 9 is cleaner than the left side of the front bumper pictured in Frame 1. Moreover, the mere fact that Frame 9 was taken at a different angle than the other Kerrville photos does not raise the inference that it was falsified. Indeed, the majority of the photos in the Kerrville Suite are taken from differing angles. Atwood presents nothing more than "conclusory allegations" that are "unsupported by facts and refuted by the record," which we have repeatedly held is insufficient to entitle a petitioner to an evidentiary hearing. Farrow v. United States, 580 F.2d 1339, 1360–61 (9th Cir. 1978) (en banc).

The state court could also reasonably discount the testimony of Hill and Hulick. The state court's conclusion that Hill's evidence was not competent under Arizona law was not an unreasonable application of Supreme Court precedent nor an unreasonable determination of the facts, given Hill's lack of specialized knowledge. Moreover, Hulick's testimony was merely an inference from the evidence that the court could reasonably reject.

B

Next, we consider Atwood's claim that law enforcement tampered with the San Antonio Suite.

Witnesses for the government provided the following evidence regarding the events in San Antonio. On September 21, 1984, the FBI transported Atwood's car from Kerrville to an FBI impound garage in San Antonio. Two Pima County investigators (McCarter and Dhaemers) flew into San Antonio on a commercial airline the same day. On September 22, FBI Agent Burwitz and four other FBI agents inspected Atwood's car, which had been cordoned off in the FBI's impound garage. FBI agents collected evidence from the car, including scrapings of pink paint from the front bumper. McCarter and Dhaemers were prohibited from touching the car, but observed the FBI activities. The investigations were documented in numerous photos.

According to Atwood, the photos in the San Antonio Suite showing pink paint on the car's bumper were not taken in San Antonio on September 22. Rather, Atwood alleges that these photos were taken later, after the Pima County investigators fabricated evidence in Tucson, returned the bumpers to the FBI impound garage in San Antonio, and reattached the bumpers.

Atwood supports this theory as follows. First, Atwood states that a "close dimensional analysis" establishes that four of the photos (one of which shows pink paint on Atwood's front bumper) were not taken on September 22 with the other San Antonio photos. Atwood next relies on an affidavit from Hill, which states that one pre-scraping photo from the San Antonio Suite, "when enhanced [,] reveals the existence of scrape marks beneath the surface of the paint." Finally, Atwood points to a photo in the Kerrville Suite that shows a smooth bumper cowling, and a photo in the San Antonio Suite that shows the cowling slightly out of alignment. Hulick states that comparing Exhibit 26–1 (Frame 1 taken in Kerrville on September 20) to Exhibit 25–10 (taken in San Antonio) "clearly shows that the bumper and its attached cowling are in a different position from one photograph to another," which she claims supports Atwood's contention that the bumper was removed and reattached. After reviewing the record, we conclude that the state court reasonably determined that Atwood's claim that Pima County investigators tampered with the San Antonio Suite was not credible. Atwood does not explain or support his contention that a "close dimensional analysis" proves that the photographs in the San Antonio Suite were "[c]ounterfeit [a]dditions." Nor does the evidence substantiate Atwood's claim that a water-soluble paint was affixed to the bumper and then removed to enable false pre-scraping photos to be taken. Given Hill's lack of expertise, the state court reasonably discounted his affidavit regarding the existence of scrape marks beneath the surface of the paint, which are not visible in the photo. Although the Kerrville and San Antonio Suites show a change in the alignment of the bumper cowling, the change does not raise the inference that the bumpers were removed in San Antonio, transported to Tucson, transported back to San Antonio, and then reattached.

C

Finally, we consider Atwood's claim that a series of unmarked and undated photos taken on the Pima County Sheriff's loading dock in Tucson (the Tucson Suite) shows investigators planting pink paint from Vicki's bicycle on the bumper of Atwood's car.

According to the undisputed facts, Vicki's bicycle remained in Tucson until law enforcement agents shipped it to the FBI laboratory in Washington, D.C., on September 25, 1984. The bicycle arrived at the FBI laboratory on September 26. Also on September 26, the FBI loaded Atwood's car into a trailer for transportation from the FBI's San Antonio impound garage to Tucson. The trailer arrived in Tucson the next day, and the Pima County Sheriff's Department took custody of the car. In October 1984, the Pima Country Sheriff's Department sent the bumper of Atwood's car to the FBI. The bumper was returned to Tucson in March 1985, and Couser took photos of the bumper when it arrived in Tucson. (This set of photos is referred to as the Couser Suite.)

Atwood contends that an evaluation of the Tucson and Couser Suites establishes that the bumpers and the bicycle were together on the Pima County Sheriff's loading dock in Tucson before March 1985 and also shows that investigators planted the pink paint on the bumper of Atwood's car. To support this theory, Atwood points to one of the Tucson Suite photos showing a car's chrome bumper, and argues that the shiny surface of the bumper reflects a man holding a pink bicycle. According to Atwood, this reflection establishes that the bicycle and the bumpers were together in Tucson at the same time. Although it is undisputed that the bicycle and bumper were together in Tucson in March 1985, Atwood argues that the Tucson Suite was taken earlier, because the bumper looks cleaner in the Couser Suite than in the Tucson Suite, indicating it had been cleaned in the FBI laboratory. Next, Atwood argues that because Vicki's bicycle was shipped to Washington, D.C., on September 25, the Tucson Suite must have been taken before that date. Based on these assumptions, Atwood claims the Tucson Suite gives rise to the inference that investigators shipped the bumpers from Atwood's car to Tucson between September 22 and 25, 1984 to fabricate the pink paint evidence.

Atwood's claims are not credible in light of the record. Even with photographic enhancements, it is not possible to discern the reflection of a man holding a pink bicycle in the car's bumper. Moreover, the resolution, quality, lighting, and angles of the Tucson Suite and Couser Suite are substantially different, making it impossible to conclude that the bumper as photographed in the Couser Suite is cleaner than the bumper as photographed in the Tucson Suite. Because both inferences (the presence of the bicycle and the difference in the cleanliness of the bumpers) are necessary to support Atwood's theory that the Tucson Suite photos were taken before September 25, 1984, the state court could reasonably conclude that Atwood's allegations were not credible and a hearing would not have affected the court's determination.

D

The general implausibility of Atwood's theory further supports our conclusion that the state court was not unreasonable in declining to hold an evidentiary hearing. Under Atwood's theory, FBI agents colluded with Pima County investigators to remove and ship the bumpers to Tucson, insert staged photos, and mix paint samples to be sent to the FBI laboratory, despite the fact that Atwood's car already had a pink paint mark on the bumper. At the time of the alleged fabrication, law enforcement officers could reasonably expect the pink paint found on Atwood's car to match the paint on Vicki's bicycle and therefore would have had no incentive to plant additional pink paint on Atwood's bumper. Moreover, as the district court pointed out, there was no reason for the Pima County investigators, as part of a clandestine operation, to take both bumpers to Tucson (when they needed only one for the alleged fabrication), to check these bulky items as baggage on a commercial airline, or to take photos of the bumpers during the fabrication process. Further, there does not seem to be any reason for the Pima County investigators to apply the pink paint to the bumper, scrape it off, mix the scrapings with scrapings of the other pink paint scraped from the bumper, and cover the scraped bumper with additional pink paint to cover the scrapings. Even more generally, at the time of this alleged misconduct, law enforcement authorities were in the midst of an investigation that might uncover further evidence. Indeed, the officers did not know at that time whether Vicki was alive or dead. Atwood's claim that at this point state and federal officers would have concocted an elaborate plot to fabricate evidence is simply not credible.

In sum, because Atwood's allegations regarding law enforcement misconduct are "incredible in light of the record," and a hearing would not have affected the state court's determination, the state court's failure to hold an evidentiary hearing was not unreasonable under § 2254(d)(2). Perez, 459 F.3d at 950.

...

AFFIRMED.

Questions:

1. The decision lays out some very disturbing facts about the defendant. For justice practitioners, when encountering such people, the temptation to side step usual investigative protocols and professional expectations can be put to the test. Is that possible in this case?
2. Explain what the trace comparison is in this case:
3. Defendant alleges that the paint streak comparisons may be a true match but the results should be disregarded for other reasons. How credible do you think these theories of tampering are?

Flanders v. Commonwealth
Court of Appeals of Virginia July 10, 2018, Decided
Record No. 0486-17-1

Sarah E. Flanders appeals her convictions for felony murder, Code § 18.2-33, and felony hit and run, Code § 46.2-894. She maintains that she was punished twice for the same offense because the murder and hit and run convictions involved the same conduct and there was no causal connection between those offenses. She also maintains the evidence did not prove that she was the criminal agent. Concluding the trial court did not err, we affirm the convictions.

On September 20, 2014, shortly after 5:15 a.m., a woman, later identified as the defendant, drove a red Dodge Durango onto a Dominion Power job site. She told a worker, "somebody needed to call 911. There was a guy laying back there behind the school bleeding from his head." She told the site supervisor, Johnny Burdette, that it "looked like someone had been run over" and "was bleeding to death." She then drove away.

Burdette drove to the school, which was adjacent to the job site. He found a man lying half on and half off the curb of the parking lot. Burdette recognized the man as the person who had walked through the job site about ten minutes earlier. The man was in obvious pain and lost consciousness before rescue workers arrived. He later regained consciousness and identified himself as Rick Pentz, and gave his birthdate. He died about four hours later at the hospital from blunt force trauma.

The police recovered Pentz's cell phone at the scene. It revealed that he had called the defendant earlier that morning. When interviewed, the defendant said that she and Pentz had been friends for six years and had once lived together at her mother's house across the street from the school. The defendant admitted she had been driving the Durango but denied any involvement in Pentz's death. She remarked, "this was crazy and she thought that he was going to make it," after the officers had left her alone in the interview room.

The police found Pentz's blood on the front bumper of the Durango and yellow paint on the inside of the front and rear driver's side tires. That paint was consistent with the paint on the curb where Pentz was found. The defendant's DNA was on the steering wheel and gearshift. The Dominion worker with whom the driver first spoke identified her as that driver. Two days earlier the defendant and Pentz also had a road side altercation which the police had investigated.

The defendant maintains she was convicted twice for the same conduct. She argues the Commonwealth is "seeking to impose two homicide convictions and punishments for a single death" because death was an element of both the felony murder and hit and run offenses. Though she does not so state, her argument is, in fact, a claim that the convictions are a violation of her Fifth Amendment right against double jeopardy.

"[W]e apply a de novo standard of review in determining whether cumulative punishments imposed in a single trial violate the constitutional prohibition against double jeopardy." Payne v. Commonwealth, 52 Va. App. 120, 125, 661 S.E.2d 513, 515 (2008), aff'd, 277 Va. 531, 674 S.E.2d 835 (2009). "Where a legislature intends to impose multiple punishments for the same course of conduct, the imposition of multiple punishments does not violate the Constitution." Id. The first step in the analysis, before applying the test set forth in Blockburger v. United States, 284 U.S. 299, 304, 52 S. Ct. 180, 76 L. Ed. 306 (1932), is to consider "whether 'the legislative intent is clear from the face of the statute or the legislative history.'" Andrews v. Commonwealth, 280 Va. 231, 284, 699 S.E.2d 237, 267 (2010) (quoting Garrett v. United States, 471 U.S. 773, 779, 105 S. Ct. 2407, 85 L. Ed. 2d 764 (1985)).

Code § 18.2-33 comprises part of the codified felony murder doctrine. ... Code § 46.2-894 is an offense contained in the Motor Vehicle Title. The purpose of that statute is "to facilitate accident investigation and to preserve public order." Johnson v. Commonwealth, 14 Va. App. 769, 771, 418 S.E.2d 729, 731, 8 Va. Law Rep. 3619 (1992). Death is not required for a felony conviction. It is clear from the plain language of these statutes that the legislature intended to create two distinct offenses.

In addition to an analysis of legislative intent, Blockburger permits both convictions. The Blockburger test examines "whether each [offense charged] requires proof of a fact which the other does not." Blockburger, 284 U.S. at 304. "In applying this test, the two offenses must be considered in the abstract, rather than in the context of the facts of the particular case being reviewed." West v. Director, Dep't of Corr., 273 Va. 56, 63, 639 S.E.2d 190, 195 (2007).

An examination of the elements of Code §§ 18.2-33 and 46.2-894 reveals that each offense requires proof of a fact that the other does not. Felony murder requires proof that the death occurred during the commission of a felony. Felony hit and run does not require death. It does require proof that the defendant was the driver of a vehicle involved in an accident, that the driver knew or should have known someone was injured in the accident, and that the driver failed to stop and report the accident to the police. Felony murder does not require proof of any of those elements. Each offense contains elements that the other offense does not. Accordingly, the two statutes are not the same offense, and multiple punishments may be imposed for conduct that violates both statutes.

To prove a violation of Code § 18.2-33, the Commonwealth must prove that the defendant killed the victim "'while in the prosecution of some felonious act' or 'in furtherance of the felony.'" Griffin, 33 Va. App. at 424, 533 S.E.2d at 658 (first quoting Code § 18.2-33 (emphasis added); and then quoting King v. Commonwealth, 6 Va. App. 351, 353, 368 S.E.2d 704, 705, 4 Va. Law Rep. 2834 (1988) (emphasis added)). The killing must be within the res gestae of the predicate felony, or "so closely related to the felony in time, place, and causal connection as to make it a part of the same criminal enterprise." Commonwealth v. Montague, 260 Va. 697, 701, 536 S.E.2d 910, 913 (2000) (quoting Haskell v. Commonwealth, 218 Va. 1033, 1043-44, 243 S.E.2d 477, 483 (1978)). "It is not enough that the killing occur 'during' the felony or 'while' it is being committed; something more is required than mere coincidence of time and place." Griffin, 33 Va. App. at 424, 533 S.E.2d at 658.

The trial court found the victim's "death could hardly be categorized as a 'mere coincidence in time and place' with [defendant's] felony hit and run." The evidence supports that finding. The evidence showed that the defendant struck the victim with her Dodge Durango and realized that Pentz was severely injured. She specifically said that it looked like the victim "was bleeding to death." Pentz's death cannot be categorized as a "mere coincidence in time and place." His death was causally related to the defendant's hitting and injuring him with her vehicle before she left without rendering assistance or providing information to law enforcement authorities. The victim's death was "so closely related to the felony in time, place, and causal connection," the death was within the res gestae.

Defendant also argues that the evidence was insufficient to prove that she was the driver who struck Pentz. We review the evidence in the light most favorable to the Commonwealth, the prevailing party below, and reverse the judgment of the trial court only when its decision is plainly wrong or without evidence to support it. …

No one saw the defendant hit Pentz with her vehicle, and the conviction is based on circumstantial evidence. … Such evidence is not "viewed in isolation," because the "combined force of many concurrent and related circumstances, each insufficient in itself, may lead a reasonable mind irresistibly to a conclusion." Brown v. Commonwealth, 54 Va. App. 107, 119, 676 S.E.2d 326, 332 (2009) (quoting Muhammad v. Commonwealth, 269 Va. 451, 479, 619 S.E.2d 16, 32 (2005)).

The evidence established that about ten minutes after Pentz walked through the Dominion Power job site, the defendant drove up in a red Dodge Durango and said that "it looked like someone had been run over" and "was bleeding to death." Pentz's blood was found on the Durango's bumper, and yellow paint from the curb where he lay was on the tires. The defendant admitted she had been driving the vehicle and was identified as the driver. Alone in the interview room, the defendant said that she "thought that [Pentz] was going to make it." The defendant and Pentz had known each other for six years. They had been involved in an altercation two days earlier, and Pentz's cell phone showed that he had called the defendant the morning of the accident. From these facts the fact finder could infer reasonably that the defendant struck Pentz with the Durango.

The circumstantial evidence is sufficient to prove criminal agency, and the trial court did not err in convicting the defendant of both felony second-degree murder and felony hit and run. Accordingly, we affirm.

Affirmed.

Questions:

1. What underlying crime does the trace evidence seek to prove- and in your opinion, has it successfully done so?
2. What type of trace evidence proves or demonstrates that the vehicle and its driver were involved in a hit and run?
3. Prove True or False: The court characterizes paint trace evidence as fully and totally circumstantial yet still finds no insufficiency (Figure 6.2).

FIGURE 6.2 NIST engineer Alan Zheng points to the firing pin impression, one of three marks impressed on the surface of a cartridge case when a gun is fired. The marks, also known as signatures, can act like fingerprints to identify a specific firearm and are key elements in NIST's standard casing, Standard Reference Material 2461.

Source: NIST at https://www.nist.gov/image/firingpinjpg.

United States v. Hayes

United States Court of Appeals for the Seventh Circuit September 6, 2017, Argued; October 3, 2017, Decided

No. 16-3752

Cedric Hayes pleaded guilty to one count of being a felon in possession of a firearm, in violation of 18 U.S.C. § 922(g)(1). He now challenges his sentence, arguing that the district court erred in applying an enhancement to his base offense level under § 2K2.1(b)(4)(B) of the United States Sentencing Commission Guidelines. That section, which applies to offenses involving firearms, states that a four-level enhancement is appropriate where a firearm's serial number has been "altered or obliterated." Hayes also challenges the district court's calculation of his criminal history category.

I. BACKGROUND

On February 2, 2016, a grand jury indicted Hayes on two counts of being a felon in possession of a firearm. On June 28, 2016, as part of a plea agreement, Hayes pleaded guilty to Count Two of the indictment, which specifically charged him with possessing a model AK-47 rifle.

In its recitation of the facts, the plea agreement stated that the serial number on the AK-47 "had been covered by a paint-like substance that prevented the serial number from being visible." It was the government's position that Hayes' offense level should increase by four, pursuant to § 2K2.1(b)(4)(B) of the Guidelines, because the serial number was "altered or obliterated." The plea agreement noted, however, that Hayes disagreed that the enhancement should apply.

Prior to sentencing, the United States Probation Office prepared a Presentence Investigation Report (PSR). The PSR's factual findings repeated the statement from the plea agreement that the serial number was not visible because it was covered by a "paint-like substance." Based on that finding, the PSR recommended that the court apply the § 2K2.1(b)(4)(B) enhancement.

In his sentencing memorandum, Hayes conceded that the AK-47′s serial number was covered in a paint-like substance, but argued that because the serial number was not "physically altered," the enhancement should not apply. As support for that argument, Hayes noted that the Forensic Science Labora-tory of the Bureau of Alcohol, Tobacco, Firearms, and Explo-sives applied a "chemical solvent and light polishing" to the gun, which ultimately revealed the serial number.

On October 17, 2016, the district court held a sentencing hearing. The court acknowledged Hayes' objection, but accepted the position of the PSR and the government that the AK-47′s serial number was "altered or obliterated" for purposes of § 2K2.1(b)(4)(B). The court adopted the Guidelines calculation contained in the PSR, finding that Hayes' offense level was 23, and his criminal history category was five. The resulting Guidelines range was 84 to 105 months' imprison-ment. After hearing from Hayes and addressing the factors to be considered under 18 U.S.C. § 3553(a), the court imposed a sentence of 94 months' imprisonment. Hayes timely appealed.

II. DISCUSSION

A. Base Offense Level Enhancement

Hayes' first argument on appeal is that the district court erred in finding that the AK-47′s serial number had been "altered or obliterated" pursuant to § 2K2.1(b)(4)(B) of the Guidelines. "We review de novo the district court's legal interpretation of sentencing guidelines and review factual findings for clear error." United States v. Harris, 718 F.3d 698, 703 (7th Cir. 2013) (citation omitted).

As an initial matter, Hayes contends that the court improperly accepted as true, without re-questing additional evidence, the fact that the gun's serial number was covered by paint. There is simply no merit to this argument because, although challenging the applicability of the en-hancement, Hayes had conceded the underlying factual basis multiple times. He signed the plea agreement, in which he admitted the AK-47′s "serial number had been covered by a paint- like substance that prevented the serial number from being visible." The PSR also contained the same sentence, and at the sentencing hearing, Hayes expressly told the court that he had no disputes with the PSR's factual findings. Finally, his own sentencing memorandum stated that he "concedes [the AK-47] appeared to be covered in a paint-like substance, which covered the serial number." The district court did not err in accepting this undisputed fact as set forth in the plea agreement, the PSR, and Hayes' own memorandum.

We turn, then, to the issue of whether § 2K2.1(b)(4)(B)'s use of "altered or obliterated" con-templates a serial number that has been covered by a substance that prevents it from being visible. Neither the Guidelines nor their Application Notes provide definitions of "altered or obliterated" or any other helpful guidance. We have, however, previously addressed the parameters of this lan-guage, albeit in an unpublished opinion.

In United States v. Salinas, we held that "a firearm's serial number is 'altered or obliterated' for purposes of § 2K2.1(b)(4)(B) if it has been 'materially changed in a way that makes accurate information less accessible.'" 462 F. App'x 635, 637 (7th Cir. 2012) (quoting United States v. Perez, 585 F.3d 880, 884 (5th Cir. 2009)). The serial number at issue in that case had been filed off, but ultimately was recovered by the Milwaukee Crime Lab. Id. However, because the filing made it "unread-able by the naked eye," we found that it was "altered or obliterated" under the Guidelines. The result in that case was supported by similar holdings from three of our sister circuits, each of which emphasized that the relevant inquiry is whether there has been a material change to a serial number "that makes accurate information less accessible." …

On the record before us, the extent of what we know about the serial number is that it was not visible because it was covered with a "paint-like substance," and that forensic specialists had to use a "chemical solvent" to uncover it. Based on that limited factual detail, we conclude that the AK-47's serial number was materially changed in a way that made the accurate information less accessible. Therefore, the § 2K2.1(b)(4)(B) enhancement should apply.

In reaching that conclusion, we reject Hayes' interpretation of the "altered or obliterated" language, which would limit its application to cases where the serial number has been completely destroyed or where one digit has been changed to another. Instead, we agree with the Ninth Circuit's assessment that "[i]rrespective of how 'obliterated' is construed, 'altered' surely requires a lesser degree of defacement." Carter, 421 F.3d at 912. Hayes argues that a stricter construction is demanded by reading "altered or obliterated" as a single phrase, which, according to Hayes, indicates the Sentencing Commission's intent to limit the enhancement's application to the situations he proposes. That analysis, however, ignores the presence of the word "or," which demonstrates that either term, individually, can serve as a basis for application of the enhancement. Id. at 911 (noting that the language is "presented in the disjunctive" and holding that "even if we were to construe 'obliterated' as Carter urges … the sentence enhancement properly applies so long as the serial number is merely 'altered'").

Hayes' proposed construction also belies the underlying purposes of § 2K2.1(b)(4)(B). In Carter, the Ninth Circuit explained that the legislative purpose of the enhancement was to discourage the use of untraceable weaponry. 421 F.3d at 914 (citing United States v. Seesing, 234 F.3d 456, 460 (9th Cir. 2001)). It clarified, however, that "[t]his purpose is advanced not only by punishing those who possess untraceable firearms, but also by punishing those who possess firearms that are more difficult, though not impossible, to trace … ." Id. Accordingly, any material change that makes the serial number less accessible, thereby making the firearm harder to trace, fits § 2K2.1(b)(4)(B)'s intended definition of "altered or obliterated." Id. at 915-16. Covering a firearm's serial number with a "paint-like substance," the removal of which requires a chemical solvent, surely makes that firearm more difficult to trace. The AK-47's serial number, therefore, was "altered or obliterated" for purposes of § 2K2.1(b)(4)(B).

B. Criminal History Calculation

Hayes' second argument on appeal is that the district court erred in assessing three criminal history points based on a prior state-court conviction for aggravated unlawful use of a weapon. Hayes failed to object to the criminal history calculation in the district court, but the government does not contend that he waived his argument. Typically, where a defendant does not object to his criminal history calculation in the district court, we treat the objection as forfeited and review for plain error. See, e.g., United States v. Brown, 865 F.3d 566, 575 (7th Cir. 2017); United States v. Gill, 824 F.3d 653, 660 (7th Cir. 2016). Under that standard, we will reverse only where: "(1) there is an error; (2) the error is clear or obvious; (3) the error affected the defendant's substantial rights; and (4) the error seriously impugns the fairness, integrity, or public reputation of the proceedings." Brown, 865 F.3d at 575 (citation omitted).

In 2003, Hayes was convicted in Illinois state court of aggravated unlawful use of a weapon, in violation of 720 ILCS 5/24-1.6. Based on that conviction, the PSR recommended, and the district court assessed, three criminal history points in Hayes' Guidelines calculation. After the 2003 conviction, however, both this Court and the Illinois Supreme Court have held that the Illinois statute is unconstitutional. ...

Hayes argues, and the government agrees, that the court erred in assessing criminal history points based on a conviction under an unconstitutional statute. Indeed, the Guidelines themselves instruct that convictions that are ruled unconstitutional are not to be counted in calculating a defendant's criminal history.

U.S.S.G. § 4A1.2, Application Note 6. Thus, we agree that the court erred in assessing criminal history points based on this conviction. ... We have consistently held that this type of error, which leads to an incorrect Guidelines calculation, constitutes plain error on review. Brown, 865 F.3d at 575 (citing Jenkins, 772 F.3d at 1098). Therefore, we must remand for resentencing.

III. CONCLUSION

For the foregoing reasons, we affirm the district court's application of the base offense level enhancement under U.S.S.G. § 2K2.1(b)(4)(B), but REMAND for resentencing based on the improper assessment of criminal history points for Hayes' prior conviction under 720 ILCS 5/24-1.6.

Questions:

1. What is the underlying crime in this prosecution and appeal?
2. How does the use of paint prove intentionality?
3. What other evidence supports the claim that paint's usage implies or infers an illegal alteration of the serial number?

6.4 TRACE EVIDENCE AND GLASS

The use of glass, as a whole or in fragments or chips, is now a well-accepted trace evidence protocol. Aside from matching the composition of glass, there are unique styles of manufacture, color refraction and other specialty characteristics and a bevy of ways to match to specific vehicles (Figure 6.3).

Turner v. State
Court of Appeals of Mississippi
January 28, 2020, Decided
NO. 2018-KA-00945-COA

P1. On June 13, 2018, a Tunica County Circuit Court jury found James Turner guilty of leaving the scene of an accident resulting in death. Turner was sentenced to eight years in the custody of the Mississippi Department of Corrections, with five years suspended and three years to serve. The trial court also ordered Turner to pay a $1,000 fine and $6,500 in restitution.

P2. Turner filed a motion for judgment notwithstanding the verdict (JNOV) or, in the alternative, a new trial. The court denied Turner's motion. On appeal, Turner argues that (1) the trial court erred in denying his two- theory circumstantial-evidence jury instruction; (2) the trial court erred when excluding portions of his statement to investigators; (3) the evidence was insufficient to sustain the jury's verdict; (4) the verdict was against the overwhelming weight of the evidence; and (5) the restitution portion of his sentence was illegal. Finding no error in Turner's first four issues, we affirm Turner's conviction. Because the trial court's order of restitution was not based on any evidence, we reverse and remand on that issue.

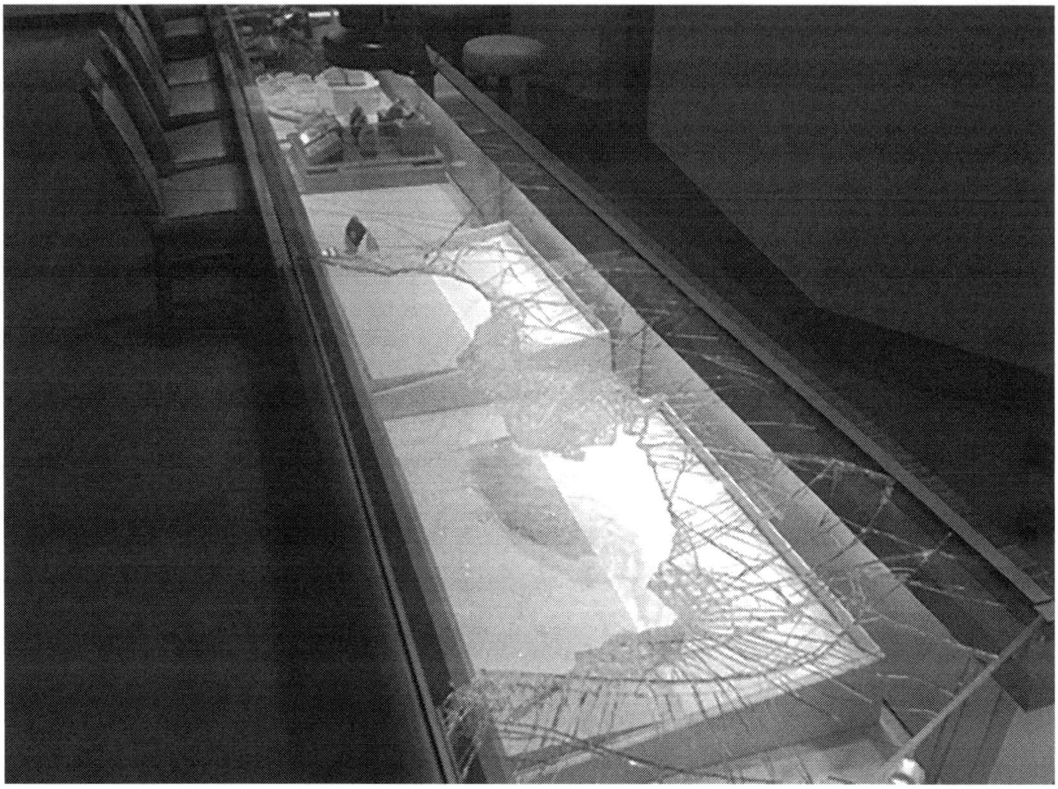

FIGURE 6.3 Shown are the remains of a glass display case after a smash and grab robbery of a jewelry store. See https://www.fbi.gov/news/stories/joint-effort-takes-down-detroit-based-robbery-group for details.

Source: FBI Media at https://multimedia.fbi.gov/item?type=image&id=5163.

FACTS

P3. On the night of August 19, 2014, Samantha McClain Spates was driving down a dark highway in Tunica, Mississippi. As Spates continued down the highway, her vehicle stalled, bringing the car to a halt on the shoulder of the road. She exited the vehicle to investigate the issue. Suddenly, Spates was struck by a passing vehicle. The vehicle persisted forward and left the scene of the accident, leaving Spates injured and alone. Eventually, the police were alerted to the accident and arrived at the scene. Spates was found dead at the scene.

P4. A few miles down the road, a vehicle crashed into a ditch outside a residence. When the resident, Henry Williams, went outside to investigate, he saw a man, later identified as Turner, walking toward him. Turner was walking away from the vehicle in the ditch. As Turner neared Williams, he threw Williams a key and told him to get the car out of the ditch because Turner would return to retrieve the car later that evening. Williams did not get the car out of the ditch. When the police arrived at the scene, Williams described the interaction with Turner, and the police began to investigate the accident contemporaneously with the hit-and-run.

P5. Two days after the incident involving Spates's death, the police verified that the vehicle belonged to Turner, and they brought him to the police station for questioning. After the police obtained a valid waiver from Turner acknowledging his rights, two investigators questioned Turner concerning the night of his accident. After a brief period, the first two investigators were replaced by Lieutenant Katie Johnson. During the interrogation, Turner admitted to driving the silver Jaguar that was in the ditch but did not confess to striking Spates with his vehicle. Additionally, Turner claimed the damage to his vehicle was caused by rocks. The next day, Lieutenant Johnson questioned Turner once more.

P6. As a product of Turner's questioning, the investigators learned that Turner walked from the Williams residence to the home of Monique Walls, Turner's cousin. Walls cooperated with the police and described the events of the evening. Walls testified that on the night in question, Turner arrived at Walls's home, which is located three miles from the Williams residence. Walls recounted that Turner asked for a ride to his home and told her that he had run his car into a ditch. Notably, Turner told Walls that he had just purchased a Jaguar but did not describe the color.

P7. Meanwhile, the police collected and processed the evidence from the two scenes involving Spates and Turner. After the evidence was processed by the police and analyzed by a Mississippi Forensic Laboratory technician, the police determined that Turner's silver Jaguar was the vehicle that struck Spates on that dark night in Tunica. The State subsequently indicted Turner for leaving the scene of an accident resulting in death—a felony.

P8. At trial, Henry Williams testified and identified Turner as the individual who crashed the Jaguar into the ditch, tossed him the keys, and told him to have the vehicle removed.

P9. In addition, the State called Jacob Burchfield, a trace-evidence technician from the Mississippi Forensic Laboratory, to testify, and Burchfield was accepted as an expert in glass examination and comparison. Burchfield described the test he performed on samples from the shattered windshield on Turner's Jaguar and glass fragments from Spates's clothing. Burchfield testified that glass from Spates's clothing was consistent with the known glass samples from the shattered windshield on Turner's Jaguar. Burchfield based his findings on the matching, color-refractive index, and elemental composition of the glass samples, stating that to a reasonable degree of scientific certainty, the glass could have originated from the silver Jaguar. Burchfield confirmed that his results were peer reviewed.

P10. On cross-examination, Burchfield admitted that a definitive match between glass fragments could only happen if there was a physical in-fracture match. Thus, to be one-hundred percent certain, the sample from Spates's clothing would require finding its originating position in the shattered windshield. Even so, Burchfield remained confident in the test results between the two samples because the elemental composition, refractive index, and the color of the samples matched.

...

ANALYSIS

...

1. There was sufficient evidence to support Turner's conviction

P20. Turner argues that the court erred in denying his motion for a JNOV. "A motion for [judgment notwithstanding the verdict] challenges the sufficiency of the evidence presented to the jury." Sacus v. State, 956 So. 2d 329, 334 (¶12) (Miss. Ct. App. 2007). We apply a de novo review to challenges to the sufficiency of the evidence. Sanford v. State, 247 So. 3d 1242, 1244 (¶10) (Miss. 2018). In doing so, "we view the evidence in the light most favorable to the State and decide if rational jurors could have found the State proved each element of the crime." Lenoir v. State, 222 So. 3d 273, 279 (¶25) (Miss. 2017). The issue is not "whether we think the State proved the elements. Rather, we must decide whether a reasonable juror could rationally say that the State did." Poole v. State, 46 So. 3d 290, 293- 94 (¶20) (Miss. 2010).

P21. Turner was charged with the crime of leaving the scene of an accident resulting in death. The State was required to prove to the jury that Turner drove the vehicle that was involved in the accident that resulted in Spates's death. Here, the jury was presented with ample evidence to consider: (a) eyewitness testimony that Turner walked away from the crashed silver Jaguar on the night in question; (b) eyewitness testimony that Turner said he drove his car into the ditch, threw the keys to the eyewitness and told the eyewitness to get the car out of the ditch, all of which occurred just two miles from the accident scene where Spates was killed; (c) damage to his Jaguar which appears to have been caused by striking an object; (d) expert testimony that the glass from Turner's windshield matched the glass fragments found on Spates's clothing; and (e) police testimony concerning the investigation and interrogation of Turner, during which he admitted to owning and driving the Jaguar on the night in question. Additionally, the trial court properly gave a general circumstantial-evidence instruction to the jury. Reviewing the evidence in the light most favorable to the State, rational jurors could have found that the State proved each element necessary to convict Turner. Accordingly, the evidence was sufficient to support Turner's conviction, and the trial court did not commit error when denying Turner's motion for a JNOV.

. . .

CONCLUSION

P29. We find that the trial court was within its discretion to refuse Turner's two-theory circumstantial-evidence jury instruction and to exclude portions of Turner's statement to investigators. We further find that the evidence was sufficient to sustain the jury's verdict and that the verdict was not against the overwhelming weight of the evidence. Because we find the trial court erred in ordering Turner to pay $6,500 in restitution, we reverse the restitution order and remand for resentencing as to restitution only.

P30. AFFIRMED IN PART; REVERSED AND REMANDED IN PART.

Questions:

1. From the facts of this case, is it circumstantially or directly proven evidence that demonstrates defendant's guilt? And what was the crime associated with this decision?
2. How is glass as a trace form utilized in the proof of this case?
3. In the state's testimony concerning the glass fragments, does it assert identicality of the glass fragment to the windshield under comparison.

In re Pers. Restraint of Schirato
Court of Appeals of Washington, Division Two
October 13, 2020, Oral Argument; January 5, 2021, Filed
No. 53913-6-II

¶1 GLASGOW, J. — After a work holiday party, AL went to a bar with some colleagues, including Gregory Schirato. AL went home alone and was intoxicated when she fell asleep. Sometime during the night, Schirato broke a window, entered her home, and sexually assaulted AL. A jury found Schirato guilty of second degree rape and first degree burglary. In this personal restraint petition (PRP), Schirato challenges his convictions or, alternatively, seeks a reference hearing.

¶2 Because AL did not see her assailant, the police conducted an investigation before charging Schirato. The State obtained a warrant to seize the clothes Schirato wore that night. At trial, the State presented evidence that glass fragments found on Schirato's clothing could have been from glass in the door of AL's home that was broken during the burglary. In addition, the State presented evidence that male and female DNA consistent with a mixed profile of Schirato and AL was on a bra that AL wore to bed that night.

¶3 Schirato argues that his constitutional rights were violated because the affidavit in support of the State's search warrant application contained material false statements and omissions made intentionally or with reckless disregard for the truth. Schirato contends that he received ineffective assistance of counsel on eight grounds, including that his counsel failed to move to suppress the glass evidence under Franks v. Delaware.

¶4 None of Schirato's arguments merits reversal or a reference hearing. Schirato's constitutional rights were not violated when the State executed the search warrant because the warrant was supported by probable cause. Likewise, Schirato's trial counsel was not ineffective for failing to file a Franks suppression motion because he would not have prevailed. None of Schirato's other ineffective assistance of counsel arguments entitles him to relief.

¶5 We deny Schirato's PRP.

FACTS

A. Background

¶6 In 2014, AL was on the executive team of a state agency in Olympia. In December, AL went to a work dinner party at a restaurant in Olympia. AL had a cocktail and wine at the party.

¶7 At the party, AL and two other coworkers, including Schirato, took photographs together because they were wearing new shoes. The photographs showed Schirato's brand new brown leather dress shoes, as well as his suit and shirt.

...

¶9 AL testified that she left the bar between 11:30 p.m. and 11:40 p.m. because she felt too drunk. AL recalled swerving on the drive home and struggling to park in her narrow garage. At trial, AL acknowledged that it was a bad decision to drive home.

...

¶14 When AL awoke to her alarm at 6:30 a.m., she expected to see Anderson in her bed, but he was not there. AL then discovered that her bra was unclasped and her underwear was pulled down. AL immediately called Anderson asking if he had been in her house during the night. Anderson told her that he had not been at her house. He asked AL if she had brought someone home, and she said she had not.

¶15 AL went to check the house. The front door was locked as usual. AL went downstairs to her basement where she discovered that the glass window in the basement door was broken, there was glass on the floor and stairs, and the basement door was slightly ajar. AL called 911.

B. Initial Investigation

¶16 Detectives examined AL's house and yard for evidence. They discovered shoe prints in the side and backyard of the house near the basement door, which they photographed and casted. The detectives also collected broken glass from the basement door.

...

C. Glass Evidence

¶28 Johnson sent Schirato's seized clothing to the Washington State Patrol Crime Laboratory. State Patrol forensic scientist Susan Wilson found glass fragments on the clothing. Wilson was not yet qualified as a forensic glass scientist and in early 2015, the State Patrol's laboratory was not equipped for glass analysis. Wilson sent the glass fragments from Schirato's clothing and the glass from AL's broken window to the FBI laboratory. The FBI laboratory concluded the fragments from Schirato's suit were too small to test using its refractive index analysis instrument, one of several forensic techniques used to identify the physical and chemical characteristics of glass to assess whether glass fragments could share a common origin.

¶29 By summer 2016, the State Patrol laboratory was equipped to perform forensic glass analysis and had sensitive instruments capable of handling very small glass fragments. Forensic scientist Daniel Van Wyk had become certified in glass analysis. Van Wyk conducted refractive index analysis on the glass fragments found on Schirato's clothing and on the glass from AL's window. He concluded that the glass fragments in Schirato's clothing could have come from the broken window glass from AL's basement door.

¶30 Schirato's trial counsel hired his own glass expert, Skip Palenik, to review the FBI's and Van Wyk's reports. Palenik wrote in a pretrial report, "the analytical results … and the incomplete and contradictory nature of [Van Wyk's] report do not meet the … rigor required of a scientific glass comparison … to support legal proceedings." Clerk's Papers (CP) at 237. Palenik did not independently test the glass.

¶31 Schirato researched glass evidence and found a 1971 scientific article discussing glass fragments found on men's clothing brought to a British dry cleaner. Richard Woodrow, petitioner's trial counsel, e-mailed the article to Palenik but did not introduce or discuss the article at trial.

D. Trial

¶32 At trial, the State presented evidence that male and female DNA consistent with a mixed profile of Schirato and AL was on the clasp of the bra AL wore to bed on the night of the December 2014 party. The DNA analysis excluded Anderson, AL's boyfriend, as a source. The State also introduced photographs of the shoe prints in AL's yard and the Nordstrom shoes.

¶33 Van Wyk testified that he used x-ray fluorescence, comparison of element intensity ratios, and refractivity analysis to test the glass fragments. He concluded that the glass fragments from Schirato's clothes could have been from the glass in AL's door. Van Wyk explained that his report was peer reviewed by Oregon State Patrol scientists, who found no problems with the substance of his analysis. In response to cross- examination questions about whether the fragments from Schirato's clothing were too small for reliable refractivity analysis, Van Wyk explained that his assessment of whether the fragments were too small was not "based on any set number," but reflected a holistic assessment of the fragment. VRP (Jan. 16, 2018) at 701.

¶34 Palenik testified for the defense. He agreed that Van Wyk used appropriate scientific methods, but found Van Wyk's work imprecise. Palenik criticized Van Wyk for comparing fragments of different sizes and including data with relative standard deviations that were too high. Palenik questioned the validity of Van Wyk's conclusion that the glass found in Schirato's clothes and in AL's basement could have had a common origin. On cross- examination, the State noted that Van Wyk eliminated the data with the highest relative standard deviations. Palenik said this assuaged his criticism somewhat, but stated on redirect, "regardless of how it's put, this data does not have the precision that it should." VRP (Jan. 22, 2018) at 1059-60.

¶35 Van Wyk and Palenik agreed that scientists typically refrain from saying that glass fragments "match" unless they physically fit together. VRP (Jan. 16, 2018) at 688; VRP (Jan. 22, 2018) at 1065. Rather, if there are no indications that the glass originated from different sources and the samples have certain common characteristics, scientists conclude that the glass could have come from the same source.

¶36 In her closing argument, the prosecutor said that scientists typically do not use the word "match" for glass evidence:

We have a comparison of the glass. He wouldn't use the word "match." They don't like to use the word "match." The glass scientist -- neither of the glass scientists wanted to use the word "match." They said "possible" is the best we're going to get. The glass found on the defendant's suit, it's possible it's the same glass found at [AL's] house.

VRP (Jan. 23, 2018) at 1221. But later in her rebuttal argument, the prosecutor argued that the glass from Schirato's clothing "matched" the glass from AL's door. VRP (Jan. 23, 2018) at 1262. She also noted that the jury was to judge the credibility of witnesses and argued that AL "told you what she remembered and did so to the best of her ability. She swore to tell the truth and she did so." Id.

E. Conviction, Appeal, and PRP

¶37 The jury found Schirato guilty of second degree rape and first degree burglary. He filed a direct appeal but voluntarily dismissed it. He then filed this PRP.

¶38 Schirato supports this PRP with his own declaration and two expert declarations, including one from Wayne Fricke, a defense attorney not involved in Schirato's case. Fricke opines that Woodrow provided ineffective assistance of counsel.

¶39 In its response to Schirato's PRP, the State submits an affidavit from the Oregon State Police forensic scientist who peer reviewed Van Wyk's report and concludes that Van Wyk used scientifically accepted procedures and methodologies. The State also submits an extensive declaration from Schirato's trial counsel, Woodrow, who explains that his representation of Schirato was not ineffective and describes the strategic reasons for his pretrial and trial decisions. Woodrow reviews at length the evidence supporting the application for the search warrant, explaining why he believed a motion challenging the warrant's validity would have been denied. He also describes why he did not seek a Frye3 hearing on Van Wyk's glass comparison methods and that his decisions regarding the FBI report and the defense expert were tactical.

ANALYSIS

...

I. FRYE MOTION

¶59 Schirato argues that his trial counsel was ineffective by not moving to exclude testimony about Van Wyk's glass testing under Frye. Because the FBI lab concluded the glass fragments from Schirato's clothing were too small for its refractive index analysis, Schirato argues that Van Wyk failed to employ generally accepted scientific techniques by conducting refractive index analysis on small fragments. We reject Schirato's argument and hold that Woodrow was not deficient because Frye only applies to new scientific techniques and refractive index analysis was not new.

¶60 "Washington uses the [Frye test] to limit expert testimony to principles generally accepted in the scientific community." State v. Murry, 13 Wn. App. 2d 542, 547, 465 P.3d 330, review denied, 196 Wn.2d 1018 (2020). "While Frye governs the admissibility of novel scientific testimony, the application of accepted techniques to reach novel conclusions does not raise Frye concerns." Lakey v. Puget Sound Energy, Inc., 176 Wn.2d 909, 919, 296 P.3d 860 (2013). Under Frye, we look to whether "accepted and reliable mechanisms" exist for implementing a scientific theory. Id. at 920. Where there exist "protocols for assuring reliability, an expert's errors in applying proper procedures go to the weight, not the admissibility, of the evidence, unless the error renders the evidence unreliable." Id.

¶61 Here, refractive index analysis is a well- established technique for forensic glass analysis. Woodrow states in his declaration that he strategically decided not to move to exclude the results of the glass testing under Frye because "[t]esting by glass experts has been occurring for decades. None of the science of examination or interpretation is new or novel. There is no disagreement among scientist[s] in the field of the science, methods or interpretation of glass examinations." Resp., App. II at 10. At trial, Schirato's own expert explained that refractive index analysis was a standard scientific technique. And the Oregon State Police forensic scientist who peer reviewed Van Wyk's report agreed that Van Wyk used "methods generally accepted within the scientific community as reliable for analytical comparison of glass particles." Resp., App. I, at 2. Conflicting expert testimony about the accuracy of Van Wyk's analysis raised issues about the weight the jury would give the evidence but did not implicate Frye.

…

A. Palenik's Testimony

¶67 Schirato argues Woodrow was ineffective by allowing Palenik to agree with the State that the glass on Schirato's suit could have come from AL's door. Schirato also argues that Woodrow was ineffective for failing to have a defense expert retest the glass. We reject these arguments.

¶68 Palenik's testimony did not support the State as Schirato contends. Rather, Palenik steadfastly maintained, consistent with his pretrial report, that Van Wyk's methods were sloppy. And Woodrow used Palenik's report to effectively cross-examine Van Wyk and argue in closing that the jury should reject Van Wyk's conclusions about the glass evidence. On cross- examination, Palenik acknowledged that based on Van Wyk's analysis of the glass, the glass could have come from the same source. And when the prosecutor informed Palenik on cross-examination that Van Wyk had discarded the data with the highest relative standard deviations, Palenik testified that this information reduced his concerns somewhat, but he continued to testify that Van Wyk's imprecision undermined likelihood that the glass fragments could have been from the same source, In closing, Schirato's trial counsel argued that because Van Wyk's methodologies were imprecise, the only valid conclusion the jury could draw was that glass of an unknown source was found on Schirato's clothing.

¶69 Palenik's testimony supported the reasonable defense strategy of casting doubt on the reliability of the scientific data that the State used. … To the extent Schirato argues that Woodrow should have asked Palenik or another defense expert to retest the glass, Woodrow explains that not doing so was a strategic decision because testing might have confirmed the State's conclusions. …

¶70 We hold that Woodrow did not provide ineffective assistance by calling Palenik as a witness and by opting not to have him retest the glass.

…

B. Prosecutor's Comment in Closing Argument that the Glass Fragments "Matched"

¶77 Schirato claims his counsel was ineffective by not objecting to the "false and misleading" comment in the State's closing rebuttal argument that the glass from Schirato's clothing "matched" the glass in AL's window. Br. of Pet'r at 45-46.

¶78 The decision to object is a matter of trial tactics, and "prosecuting attorneys have wide latitude to argue facts and reasonable inferences from the evidence," so long as they do not "make prejudicial statements unsupported by the record." Lui, 188 Wn.2d at 557. This court considers the entire argument in context. …

¶79 In context, the prosecutor's comment that the glass "matched" was not misleading. VRP (Jan. 23, 2018) at 1221. Before inferring that the glass from Schirato's clothes "matched" the glass from AL's window, the prosecutor reminded the jury that although Van Wyk explained why scientists do not use the word "match" for glass evidence unless the pieces physically fit together, Van Wyk concluded that the glass could have been from the same source. See VRP (Jan. 23, 2018) at 1221; VRP (Jan. 16, 2018) at 688. Moreover, Woodrow's declaration states that he strategically opted not to object because he believed the defense had a strong argument that the glass on Schirato's suit could have originated anywhere. And an objection could have led to additional discussion from the prosecutor about Van Wyk's precise results. The decision not to object was strategic.

¶80 We hold that Woodrow did not provide ineffective assistance by failing to object to the State's closing argument inference that the glass evidence matched.

…

CONCLUSION

¶86 We deny Schirato's PRP.

Questions:

1. Prove True or False: The glass comparison was not based on piece reconstruction or match comparisons but on composition and type of glass.
2. What is the most compelling evidentiary inference that can be drawn or derived from the trace glass evidence?
3. What error did the Prosecutor make in closing arguments. Did that error result in a reversal?

6.5 TRACE EVIDENCE AND HAIR

Read the DOJ press release regarding tampering with evidence, including a bag of hair at https://www.justice.gov/opa/pr/four-additional-defendants-charged-assaulting-practitioners-amish-religion-cleveland.

Locate the case below using your favorite legal research resource, read the case and answer the questions below.

Commonwealth v. Chmiel, 240 A.3d 564 (Pa. Sup. 2020) July 14, 2020, Submitted; October 21, 2020, Decided No. 780 CAP

Questions:

1. Similar to glass and other trace evidence forms, what do the proponents of this type of evidence, on comparison, generally not conclude?
2. Where and how were the hair samples acquired? Secondly, does the location of the sample impact its credibility?
3. In other cases of trace hair analysis, the FBI Crime Lab has been severely criticized for its erroneous protocols. How so and and what language?

Locate the following case using your favorite legal research resource, read the case and answer the questions below.

United States v. Butler, 955 F.3d 1052 *; 446 U.S. App. D.C. 247 (United States Court of Appeals for the District of Columbia Circuit November 7, 2018, Argued; April 14, 2020, Decided) No. 17-3080

Questions:

1. Trace analysis of hair can and should some conclusions about similarity in comparison.
2. Comparative hair analysis was labeled?

3. Not only is the means and methods of measuring hair comparisons, according to this decision, the government that employed this forensic technique was found to have done something even worse. What is that?

6.6 TRACE EVIDENCE IN ARSON AND FIRE CASES

Trace evidence is a constant in arson and fire cases. Fires intentionally set usually have some sort of trace element or material that shows point of acceleration or a distinct fire point or setting where the fire originates or the agent that triggered the event (Figure 6.4).

Allstate Indem. Co. v. Dixon

932 F.3d 696 (United States Court of Appeals for the Eighth Circuit April 17, 2019) Submitted; August 1, 2019, Filed

On February 14, 2014, Joseph and Casey Dixon purchased a former resort hotel in West Plains, Missouri, for $120,000 with the intention of turning it into a bed and breakfast (B&B) and marriage retreat. The property contained a 15-bedroom, 12-bathroom structure. The structure was

FIGURE 6.4 On September 17, 2020, at approximately 4:15 a.m., an unidentified arsonist set fire to the administration building of St. Gregory the Illuminator Armenian Apostolic Church in San Francisco, which housed offices, a library and a Sunday school. The building sustained significant fire, smoke and water damage. No injuries were reported. See https://www.fbi.gov/contact-us/field-offices/sanfrancisco/news/press-releases/fbi-san-francisco-offers-50000-reward-in-church-arson

Source: FBI Media at https://multimedia.fbi.gov/item?type=image&id=6117.

insured for $479,857 actual cash value—an amount far in excess of the $120,000 purchase price. On April 12, 2014, the property caught fire and burned to the ground in a little over an hour.

Allstate investigated the fire and concluded that it was set by or at the direction of Joseph Dixon. Allstate paid the mortagees $107,526.29 under the terms of the policy. It then commenced an action seeking a declaratory judgment that the Dixons had violated the intentional acts exclusion of the policy and that Allstate was entitled to recover its payment to the mortagees. The Dixons filed a counterclaim for breach of the insurance contract. The district court1 denied the Dixons' motion for summary judgment. Allstate ultimately prevailed at trial. The Dixons appeal several evidentiary decisions and the denial of their motions for summary judgment, for judgment as a matter of law, and for a new trial. We affirm.

I. Background

The Dixons purchased the subject property located at 791 State Hwy T, West Plains, Missouri, for $120,000. At the time of purchase, they owned an uninsured property in Willow Springs, Missouri, that they ran as a location for marriage retreats. The Dixons' stated purpose for purchasing the West Plains property was to run a B&B and to also use it as another location for wedding retreats. Between the time of purchase and the blaze, no material steps were taken to convert the property into its intended use as a B&B or a wedding retreat location, and the structure sat mostly dormant.

When the Dixons were looking to purchase insurance for the property, they rejected two proposals to insure the property for $120,000 and $300,000 before accepting a proposal to insure the property for $479,857 actual cash value and $60,000 for personal property. The property was insured for nearly four times the $120,000 purchase price. One month after the insurance policy went into effect, the building burned to the ground. The Dixons filed a sworn statement in proof of loss. Shortly thereafter, Allstate began investigating the claim.

At the beginning of the investigation, Allstate retained Terry Decker, a certified fire investigator, to collect data about the incident and the fire scene. Decker was certified as a fire and explosion investigator by the National Association of Fire Investigators in 1997 and as a fire investigator by the International Association of Arson Investigators in 2001. He also was a certified fire investigator through the Missouri Department of Public Safety. As part of maintaining his certifications, Decker was required to engage in ongoing education and training.

Decker investigated and photographed the scene, and reviewed photographs of the house prior to the fire. He interviewed the Dixons, their neighbors, and public fire officials, and also reviewed the local fire department report, 911 logs, and weather history data. Based on that information, Decker created a timeline of events. Joseph Dixon indicated he left the property with a man named Donald Hitch sometime between 7:30 p.m. and 8:00 p.m. Dixon described the sky as getting dark when he left. When the fire was discovered at 9:11 p.m., it was "totally involved" and caused the complete destruction of the building. Decker came up with a number of hypotheses to explain the blaze, including the possibility of accidental, undetected fires burning prior to when Dixon left and the possibility that the fire was burning at or shortly after Dixon left the property. After weighing those hypotheses against the collected evidence, Decker concluded that the fire must have been burning when or shortly after Dixon left the property. As part of his determination, Decker considered weather data that showed that twilight set at 8:06 p.m. on the night in question. Given Dixon's statement that it was getting dark when he left, Decker concluded that Dixon left the property at or slightly after 8:00 p.m. The destruction of the building meant that Decker was unable to come to a conclusive opinion as to the cause or origin of the fire, but that he could not rule out that it was an intentionally-set fire. Decker opined that the fire developed rapidly and progressed at an abnormal rate.

Allstate retained a second certified fire investigator, Robert Wysong, to review the fire given the tight timeline that Decker identified. Wysong was the owner and president of ACS Investigative Services, a fire investigation company in operation since 1991. Wysong was certified as a fire and explosion investigator by the National Association of Fire Investigators. Wysong reviewed Decker's report as well as the sworn statements of the Dixons and Donald Hitch and photographs of the property. Wysong considered a number of hypotheses regarding the blaze, including the possibility of a fire caused by lightning, a fire caused by candle, a fire set in more than one location, and a fire that had been burning more than an hour. Wysong ultimately concluded that the fire must have been burning for more than an hour before discovery or that the area of origin was not a single location.

Allstate's interviews with the Dixons and Donald Hitch did not suggest any alternative issues with electrical appliances that might have led to the blaze. During Casey Dixon's examination under oath, she testified that she thought somebody started the fire. Allstate, relying on the fire expert reports and the Dixons' sizable financial motive to commit arson, determined that the loss was not covered because the fire was intentionally set by or at the direction of Mr. Dixon. Pursuant to the policy, Allstate issued a payment to the mortgagees in the amount of $107,526.29.

Allstate filed a complaint for declaratory judgment on November 20, 2014. On February 26, 2015, Allstate amended its complaint and sought a judgment that the intentional acts exclusion barred coverage for the loss and that Allstate was entitled to recover its payment to the mortgagees of $107,526.29. The Dixons filed a counterclaim alleging that Allstate breached the insurance contract. The district court denied the Dixons' motion for summary judgment on both the complaint and the counterclaim.

Before and during trial, the Dixons objected to the admission of testimony from Allstate's fire experts. Their objections were based in part on whether the fire experts had followed NFPA 921, an authoritative guideline in the field of fire investigation. Neither expert purported to have exhaustively applied NFPA 921. Instead, each suggested that they had used the scientific method and deductive reasoning, as explicitly referenced by NFPA 921. The Dixons argued that NFPA 921 required physical testing that neither fire expert had conducted and that therefore neither expert had used a reliable method. The district court overruled the objections and allowed the experts to testify. In addition to testifying to their certifications and training experience, Decker testified that he had conducted over 2,000 fire investigations, and Wysong testified that he had conducted over 3,000 fire investigations. Each also testified to previous cases in which he had served as an expert fire witness.

The Dixons offered testimony from their own fire expert, Phillip Noah. Noah criticized the opinions offered by Decker and Wysong after conducting a "desktop review" of their work. Noah did not conduct physical testing or visit the scene of the fire itself.

The Dixons filed motions for judgment as a matter of law at the close of Allstate's evidence and at the close of all evidence. Each motion was denied. The jury returned a general verdict in favor of Allstate. In response, the Dixons filed a renewed motion for judgment as a matter of law and for a new trial, $107,526.29 payment to the mortgagees. The district court entered an order granting the motion and entered a judgment against the Dixons for $107,526.29. The Dixons now appeal.

II. Discussion

A. Expert Testimony Regarding the Cause of the Fire

We review the district court's decision to admit expert testimony for abuse of discretion. Peitzmeier v. Hennessy Indus., Inc., 97 F.3d 293, 296, Freem. Ch. 1020 (8th Cir. 1996) (citing Westcott v. Crinklaw, 68 F.3d 1073, 1075 (8th Cir. 1995)). In order for a court's decision to admit evidence to amount to an abuse of discretion, any error must be "clear and prejudicial to the outcome of the proceeding." Torbit v. Ryder Sys., Inc., 416 F.3d 898, 903 (8th Cir. 2005) (citation omitted).

The admission of expert testimony is governed by Federal Rule of Evidence 702 and Daubert v. Merrell Dow Pharms., Inc., 509 U.S. 579, 113 S. Ct. 2786, 125 L. Ed. 2d 469 (1993). "The inquiry envisioned by Rule 702 is … a flexible one." Daubert, 509 U.S. at 594. "Its overarching subject is the scientific validity—and thus the evidentiary relevance and reliability—of the principles that underlie a proposed submission." Id. To properly admit evidence under the rule, district courts must determine "whether the expert is proposing to testify to (1) scientific knowledge that (2) will assist the trier of fact to understand or determine a fact in issue." Id. at 592.

The Dixons challenge the admission of Allstate's fire experts' testimony because the experts did not precisely follow the industry guide NFPA 921, in particular due to a lack of a physical testing. The Dixons' claims related to testing conflate the testing of various hypotheses using the scientific method with physical tests. Where physical testing is not feasible, "observations coupled with expertise generally may form the basis of an admissible expert opinion." Shuck v. CNH Am., LLC, 498 F.3d 868, 875 (8th Cir. 2007). A hypothesis is appropriately tested under the scientific method by comparing all known facts, acquired data (of which physical test results is but a part), and the developed body of scientific knowledge to the hypothesis. Allstate's well-qualified fire experts thoroughly explained their methodologies to the district court and explained why their methods had scientific reliability, even absent physical testing.

The "flexible inquiry" that district courts must conduct under Rule 702 and the different explanations that fire experts might give for their choice of methodologies may lead to differing decisions as to whether to admit different experts' testimonies. For example, in Hickerson v. Dist. Pride Mobility Prods. Corp., 470 F.3d 1252, 1257 (8th Cir. 2006), we upheld a district court's decision to admit an expert's testimony where the expert:

> examined burn patterns, examined heat, fire, and smoke damage, considered this evidence in light of testimony regarding the fire, and identified a point of origin … considered as possible causes of the fire those devices that contained or were connected to a power source and that were located at the identified point of origin … [and] eliminated as possible sources those devices that were not in the area of origin or that were not connected to a power source and contained no internal power source.

In a case involving similar scientific theories, Fireman's Fund Ins. Co. v. Canon U.S.A., Inc., 394 F.3d 1054, 1058-60 (8th Cir. 2005), we affirmed a district court's decision to exclude fire expert testimony where the experts could not show they reliably applied the methodology they articulated and where their interpretations of the meaning of burn pattern evidence dramatically changed between initial report and rebuttal report.

In this case, Allstates' experts "observed the relevant evidence, applied their specialized knowledge, and systematically included and excluded possible theories of causation." See Shuck, 498 F.3d at 875 (allowing opinion of experts who used that approach). Accordingly, the district court did not make a "clear and prejudicial" error affecting the outcome of the proceeding by admitting the opinions.

B. Denial of Motion for Judgment as a Matter of Law

We review the district court's denial of a motion for judgment as a matter of law de novo. McKnight v. Johnson Controls, Inc., 36 F.3d 1396, 1400 (8th Cir. 1994) (citation omitted). We apply the same standards as the district court and ask "whether there is sufficient evidence to support a jury verdict." Id. (quoting White v. Pence, 961 F.2d 776, 779 (8th Cir. 1992)). "Judgment as a matter of law is appropriate only when all of the evidence points one way and is 'susceptible of no reasonable inference sustaining the position of the nonmoving party.'" Id. (quoting White, 961 F.2d at 779).

III. Conclusion

We affirm.

Questions:

1. Part of the reason the Dixon case is so compelling is its lack of trace analysis on burn patterns, charring or seared materials. Indeed, as the Court notes, it uses other non-physical extrapolations to conclude that the fire was intentional set due to its speed. Defendants object on what theory?
2. Experts hired by the insurance company were highly qualified? True or False?

Bradford v. Brown

831 F.3d 902, United States Court of Appeals for the Seventh Circuit April 14, 2016, Argued; August 4, 2016, Decided

No. 15-3706

In 1993 Glenn Bradford was convicted in an Indiana state court of a murder and arson committed in Evansville the previous year, and was sentenced to 80 years in prison, where he remains … .

At 6:35 a.m. on a day a month or so after Lohr's threats and just after Bradford had finished a night shift, he reported a fire at Lohr's house. He told the responders that he'd entered the house to try to extricate her but had been unable to do so and believed she was dead. A firefighter had driven by the house at 6:30 a.m. and seen no signs of fire, but shortly after Bradford reported it the firefighter entered the house, extinguished the fire, and found Lohr's corpse on her bed. He estimated that the fire had started only a few minutes before he arrived. Fire investigators found in Lohr's bedroom an empty gasoline can from which gasoline had been poured onto the mattress on which Lohr's body lay and onto the floor between the mattress and the bedroom door. They inferred that the gasoline had been ignited by someone standing in the bedroom doorway. Lohr, it was discovered, had been stabbed to death and her body had been burned after she died. Her poodle was found dead, also with stab wounds, in the living room, which was just outside the bedroom, but we'll see that the dog was still alive when the fire started. From the conditions of the house and grounds the police concluded that the arsonist-murderer had staged a burglary rather than broken into the house.

Although the exact time of Lohr's death could not be determined, the state believed that she'd been stabbed during the night. Suspicion quickly focused on Bradford, who appears to have concealed his whereabouts on the fatal night and thus could have murdered Lohr sometime during his night shift before returning in the morning to set the fire. He had no alibi for the period from about 11 p.m. to midnight. …

Although only 65 seconds elapsed between when a bank camera revealed Bradford two blocks from Lohr's house and driving toward it and when he reported the fire, an investigator testified at trial that that was enough time for Bradford to have driven to the house, entered, spread the gasoline, and lighted it.

At trial, rival fire experts testified for the state and the defense about whether the fire had to have burned for more than eight minutes, in which event Bradford would have had a solid alibi because the fire was extinguished by 6:43:19 a.m. and he could not have set it before 6:35 given that he was seen by the bank camera two blocks away at 6:34:04 and that it would have taken him about a minute to reach the house from there, start the fire, and then call emergency services (which he did at 6:35:09)—and he'd been on police duty until about 6:30. No gasoline had been found on Bradford's uniform and the police had failed to investigate another possible suspect— ...

Bradford was convicted by a jury of murder and arson, and his conviction was upheld in ... He principally argues in the present proceeding that he couldn't have been the arsonist (and if not, then presumably not the murderer either) because his expert witnesses testified that the fire must have burned for more than eight minutes. ... his principal expert witness in the post-conviction proceedings, Douglas Carpenter, went further, testifying that the fire must have burned for at least 30 minutes before it was extinguished. He based that estimate on his inspection of burned and unburned wood and other materials in the house and on the level of carboxyhemoglobin (COHb) in the poodle's blood—a house fire generates carbon monoxide (CO) that can combine with hemoglobin, a constituent of blood.

Bradford relies on three facts asserted by Carpenter to support his contention that the fire must have lasted more than eight minutes: (1) the time it would have taken the fire to burn through the upper panel of the bedroom door; (2) the time it would have taken for the poodle to have accumulated the amount of COHb found in its blood; and (3) the time it would have taken the fire to consume the mattress on Lohr's bed. These facts don't make a case. Carpenter was unable to prove that the fire would have taken too long to burn through the upper panel of the bedroom door for Bradford to have set it. He conducted experiments under conditions that simulated a maximum temperature near the bedroom ceiling of 350°C. But the fire might well have been hotter than that. He contended that because none of the 12 panes of glass in the front bedroom window was broken, the room temperature near the ceiling (and thus near the top door panel) must have been below 280°C, and at that temperature it would he claimed have taken at least 30 minutes for the bedroom door to char through. He based this temperature claim on a study reported by Vytenis Babrauskas, "Glass Breakage in Fires" 2-3, www.doctorfire.com/Glass Break.pdf (visited Aug. 2, 2016), of the probability of glass breakage at different temperatures. That study had been limited to glass three millimeters thick, and he admitted not knowing the thickness of Lohr's windows, though he testified that three-millimeter glass is "typical."

He claimed that because none of the 12 panes of glass was broken, the temperature near the ceiling of Lohr's room must have been below 280°C. That was the temperature at which the three-millimeter glass in the study on which he relied had a one-in-twelve probability of breaking. But a one-in-twelve probability of breakage at a particular temperature does not mean that if there are twelve panes one of them is certain to break, and Carpenter failed to justify his assumption that the more panes in the unbroken window, the lower the temperature had to have been. Moreover, he had no evidence that all the windowpanes had been exposed to the same temperature as the ceiling and the top panel of the door. In a normal room fire the highest temperature is at the ceiling and the temperature near the floor is much lower. See T. J. Shields, et al., "Performance of a Single Glazing Assembly Exposed to Enclosure Corner Fires of Increasing Severity," 25 Fire & Materials 123, 125-26 (2001).

His testimony that the top panel of the door would have required about 11 minutes to char through even if the fire's temperature had reached (though it hadn't) 600°C (measured just below the ceiling, where the fire would have been at its hottest) misrepresented his own report, according to which the 11-minute estimate was based on the heat flux (the rate at which heat energy passes through a given surface per unit of time) that would have been produced at a near-ceiling temperature of 350°C. And as we said, Carpenter didn't test char-through times for heat fluxes corresponding to temperatures higher than 350°C, even though the fire in Lohr's house may have been hotter than that.

There was also no basis for his conclusion that it would have taken 30 minutes for the fire to have caused the COHb level discovered in the poodle's autopsy. That estimate was based on the rate of carbon-monoxide uptake for an adult human being at rest, and such an estimate is not valid for a fifteen-pound dog who may moreover as we noted have been moving around in the presence of the fire. (What is clear, however, is that the poodle was alive when the fire started, because the only possible source of the elevated COHb found in its blood was inhalation of carbon monoxide (CO) produced by the fire.) Carpenter's own source explained that the uptake rate is significantly higher for smaller bodies and bodies in motion. David A. Purser, "Assessment of Hazards to Occupants from Smoke, Toxic Gases, and Heat," The SFPE Handbook of Fire Protection Engineering 2-116 (4th ed. 2008).

The fire duration that he inferred from the poodle's COHb level was further inflated by his use of carbon- monoxide concentrations measured outside closed doors. The door to Lohr's bedroom was open by about an inch because of clothing jammed into the opening, and so the dog, who was a small distance outside the bedroom, would have breathed more carbon monoxide from the fire raging in the bedroom than had the door to the bedroom been closed.

Carpenter concluded that Lohr's mattress must have been completely consumed before the door charred through, for had it been consumed afterward additional air ventilation caused by the fire burning a hole in the door panel would have interacted with the mattress (fuel for the fire) to cause the fire to reach flashover conditions. Flashover conditions occur when the upper layer of gas in a room becomes hot enough (approximately 500°C to 600°C) to ignite all available combustible surfaces. Since it is undisputed that the room did not reach flashover, Carpenter concluded that the mattress must have been completely consumed before the door charred through. He used studies of rates of mass loss for flexible polyurethane foam (the material in Lohr's mattress) to predict what the mass loss rate would be in a room on fire with the door open one inch. Estimating the maximum duration of the fire as two hours and the minimum as 30 minutes, he concluded that the fire must have been started between 4:30 and 6:00 a.m.

Bradford infers from this that it would have taken two hours for Lohr's mattress to be consumed, and so the fire must have lasted at least that long. But two hours was Carpenter's estimate of the maximum time for the mattress to be consumed. His estimate of the minimum time (30 minutes) if it were correct would exonerate Bradford, but the reliability of that estimate was undermined by the errors, discussed earlier, in Carpenter's testimony and report.

Two officers who appeared at the fire scene three minutes after Bradford had arrived reported that smoke was coming from the eaves of the house. Carpenter argues that smoke couldn't have come from the eaves until after the door had charred through because until then smoke would have gone out through the roof vents rather than sideways through the eaves. After the door charred through, however, the roof vents would be inadequate to vent all the smoke emerging from the bedroom and therefore smoke would go through the eaves as well as the vents. If Carpenter is correct, then, for Bradford to have been the arsonist, the fire had to have started, the door must have charred through, and smoke from the eaves must have appeared—all within three minutes of Bradford's arrival at the house, since responders to the fire arrived three minutes after Bradford and observed smoke upon arrival. But as we've said, the char-through time if the fire was hotter than 350°C is unknown and we also don't know how fast the smoke would have moved through Lohr's house and out the eaves and roof.

It's true that two joggers saw smoke coming from Lohr's house at 6:36:07, about a minute after Bradford called emergency services and two minutes after he was recorded by the bank camera driving toward the house. Carpenter claimed that the joggers' testimony corroborated his conclusion that Bradford could not have been the arsonist, since it is unlikely that Bradford would have been able to set the fire in time for it to emit visible smoke in fewer than two minutes. But this argument does not appear in Bradford's briefs and the state's expert witness testified that smoke would have been visible outside the house within a minute after the fire was set.

Bradford further argues that the evidence that he could have reached Lohr's house and set the fire within 65 seconds (between 6:34:04 when he was photographed by the bank camera and 6:35:09 when he called emergency services) did not account for the time required for the murderer-arsonist to carry the gas can from the back porch to Lohr's room, turn off the electrical breakers, and stab the poodle. Not so. Although Lohr normally kept her gas can on the back porch, there is no evidence that it was there when the arsonist arrived that morning. Had Bradford brought it inside during his visit to Lohr between 11 p.m. and midnight, he could have used it at 6:34 a.m. without having to fetch it from the porch. And he would have needed to turn off the electrical breaker in the morning only if Lohr's neighbor was correct that he saw her light on at 12:30 a.m. That neighbor acknowledged, however, that he was estimating the time and had not been looking at a clock, so Bradford may have turned off the breaker earlier. Nor would it take long to stab a small poodle twice. The jury thus could reasonably find that 65 seconds was enough time for Bradford to arrive at Lohr's house and complete all the tasks that the evidence indicated had been completed by the murderer-arsonist … .

The judgment denying habeas corpus is AFFIRMED.

Dissent by: HAMILTON

HAMILTON, Circuit Judge, dissenting. Petitioner Bradford has come forward with powerful evidence of both his innocence and his trial lawyers' ineffective assistance in dealing with the critical scientific issue—the duration of the fire in Tamara Lohr's bedroom. The State and the courts upholding Bradford's convictions have not yet offered a plausible theory to account for the physical evidence of the fire and the time Bradford supposedly set it.

Instead, to rebut the scientific foundation for Bradford's petition for a writ of habeas corpus, the majority has struck out on its own with some modest criticisms based on the majority's own research and analysis. Those criticisms do not undermine the conclusion that Bradford could not have been the arsonist and murderer. Bradford and his key post-conviction expert also have not had an opportunity to respond to even these modest criticisms. Rather than affirm the denial of relief for Bradford based on these untested critiques, we should order a new trial to test all the relevant evidence. I respectfully dissent.

The scientific issue requires close attention to the timing of Bradford's actions, to the physical details of the fire and the damage it caused, and to objective information about fire behavior. I cannot offer the reader a shortcut, but I can offer a roadmap. Part I lays out the critical facts about the murder and fire: in I-A, the timing of Bradford's movements and the fire, in I-B the State's theory for Bradford's guilt, and then in IC the key physical evidence. Part II turns to the law and how Bradford's otherwise able counsel provided ineffective assistance by failing to present available and credible expert testimony on the critical issue of fire duration. Part II reviews the evidence of fire experts at trial and then the post-conviction case, applies the law governing ineffective assistance claims to those facts, and finally addresses the majority's critiques of the new evidence.

…

D. The Majority's Critique of Carpenter

Carpenter's scientific analysis showing that Bradford could not have set the fire was unscathed by cross- examination in the state courts. The prosecutor focused on trying to show that another expert could have done the same sort of analysis at the time of trial. That is also my point, of course, for it is key to showing the ineffectiveness of relying on Davie. The prosecution tried a few substantive attacks, such as trying to show some uncertainty about the thickness of the window glass, how similar the door samples were for the char-through experiments, and the exact size of Lohr's mattress. Carpenter easily parried those, showing why his estimates were reliable and/or that minor changes in the variables would not change the bottom line conclusion. Not surprisingly, then, the state trial and appellate courts did not try to engage Carpenter's analysis on the merits. In this federal appeal, we find the first attempt by a court to engage Carpenter on the merits. The majority offers some new criticisms based on its own research. None of those criticisms are serious enough to undermine the key conclusion.

The majority's most telling point is that Carpenter's analysis of the probability of glass breakage was wrong, and I agree with the point. A one in twelve probability that a pane of glass will break does not mean that in twelve separate, independent events, one will certainly break.

(In simpler terms, if you flip a coin twice, the probability of heads is one-half on each flip but there is no guarantee that one of two flips will turn up heads.) Carpenter also did not explain why all twelve panes might have been exposed to approximately the same heat flux, given that heat rises. He also could not be sure that the glass in Lohr's bedroom windows was three millimeters thick, though there is no evidence of non-standard glass in this small and modest home. And so on.

The prosecutor did not challenge Carpenter on the probability of the window panes remaining intact. The data in Carpenter's report show that the probability of breakage rises dramatically after about 275° Celsius (530° Fahrenheit). By the time the temperature reaches about 350° Celsius (660° Fahrenheit)—which is not hot enough to produce a char-through time of eight and a half minutes based on Carpenter's tests—the probability of glass breakage rises to about 0.6. (A140) If each of the twelve panes had an independent six in ten chance of breaking, the probability that none would break would be (4/10), or 1.677 × 10 × 10-5—less than two in a hundred thousand. Also, the data refer to glass breakage, meaning that pieces of the glass fall out, further ventilating the fire. Babrauskas, cited by the majority, also says it is fairly well-established that a windowpane of ordinary float glass "tends to crack when the glass reaches a temperature of about 150- 200°C." Carpenter's report indicates that the front bedroom window "showed no signs of cracking." (A138)

In other words, even if we correct the error in Carpenter's probability analysis of the glass, the State still cannot explain how, if the fire burned hot enough to char through the door fast enough to support its theory, all of the front windowpanes remained unbroken.

More fundamental, most of the majority's critiques were not even raised during the cross-examination of Carpenter at the post-conviction hearing. Carpenter never had the opportunity to expand upon his reasoning, to defend his work, or, if there was in fact a significant error, to explain why it might or might not affect his ultimate conclusion.

For example, at page 6, above, the majority attacks Carpenter for having confused floor- and ceiling-level heat flux during his testimony. That apparent error seems to have been limited to his testimony at the hearing. His report did not make that mistake, nor were the report's overall conclusions based on it. His report makes clear that his estimate of the char-through time is based on the intact glass. (A138) It is not unusual even for experts (and lawyers and judges) to misspeak occasionally in such ways. The questioning prosecutor did not identify the point or ask for an explanation.

The majority also criticizes Carpenter for failing to explain why, since heat rises, all the windowpanes would have been exposed to the ceiling level heat flux. One could just as easily say it is not clear why the entire door panel from top to bottom would have been exposed to the ceiling level heat flux, given that the door, like the window, extended down from the ceiling into what were presumably slightly cooler temperatures. More to the point, Carpenter simply was not asked to justify his assumption. We do not know why he made it, whether it was appropriate, or whether changing it would have a significant effect on his conclusion.

In contrast, one such critique was raised on cross- examination—Carpenter's assumption that the window was made of plain three-millimeter glass. He parried the attack easily, noting that tempered glass "isn't traditionally used in residential situations," and that three millimeters is a typical thickness. This seems a reasonable explanation to me, and it suggests Carpenter may well have been able to rebut the majority's other critiques if he had been asked about them.

Carpenter's evidence is not perfect. It depends on certain assumptions and estimates, both because of the passage of time and because the behavior of an uncontrolled fire can be difficult to pin down. But his evidence is powerful. It is the only analysis of the fire's duration based on empirical data, derived from actual tests instead of conjecture, obsolete rules of thumb, or inexact witness guesstimates. His analysis is quantitative and testable. His results appear to be so definitive that minor adjustments for surprisingly thick glass or having the bedroom door open a little farther could not solve the basic problems with the State's theory.

To cause the physical damage within the State's eight- and-a-half-minute maximum burn time, the fire would have had to have been much more intense, consuming the entire mattress in just a few minutes and resulting in broken windows and the flashover that the experts agree never happened. And there is no plausible explanation how the fire in the nearly closed bedroom could have produced, in no more than 90 seconds, a heavy smoke plume visible to the runners more than two blocks away. To explain that smoke plume at all, the char-through time would need to be measured not in minutes but in seconds. And that would mean in turn that the fire would have been ventilated almost from the beginning, contrary to all the fire experts' views, and it would have reached flash- over, which did not happen.

If we are going to apply our own analysis to Carpenter's analysis, we also should not close our eyes to the grisly report from Dr. John D. DeHaan, which focuses on the fire damage to Lohr's body. His analysis, based on the autopsy report and tests that have been done by exposing human cadavers and the bodies of other animals to flames, corroborates Carpenter. DeHaan found that the heavy fire damage to Lohr's body was inconsistent with a fire of only five to seven minutes and was "much more consistent with a fire exposure of approximately 20 minutes to direct flames"—or perhaps to a much longer fire, one of reduced intensity due to the ventilation limits on the fire in this case. (A179)

* * *

To sum up, even if Carpenter's analysis has a couple of minor errors in it, his analysis is the only analysis of the fire's duration with a scientific and empirical basis. His conclusion appears sound, and the State did not show otherwise. If it is correct, it exonerates Bradford. In a new trial, the State would have the chance to attack the scientific analysis on its merits. Perhaps the State would then be able to show that Carpenter's work is not to be trusted. Or perhaps it could present, finally, an expert who could explain in reliable scientific terms how a fire could have done the damage this fire did so quickly without leaving unmistakable signs of a much hotter fire.

Given the state court's conclusion that the analysis Carpenter provided was available at the time of the original trial, I would treat the failure to seek and find such critical expert evidence as ineffective assistance of counsel. In the alternative, I would treat this case as appropriate for an actual innocence grant of habeas corpus. We should order issuance of a writ of habeas corpus directing that Bradford be released or retried.

Questions:

1. Case hinges upon a series of experts and their respective testimonies. Why does the Dissenting Opinion dismiss the testimony of some witnesses and land up accepting only the testimony of expert Carpenter?
2. Why is timing so critical in this case-how fast and for how long a burning took place?
3. What is the outcome for the Defendant's appeal?

Gavitt v. Born

835 F.3d 623 United States Court of Appeals for the Sixth Circuit

August 3, 2016, Argued; September 1, 2016, Decided; September 1, 2016, Filed File Name: 16a0216p.06

Nos. 15-2136/2434

Plaintiff in the action from which these two appeals arise, David Gavitt, was sentenced to life in prison in 1986 after a jury found him guilty of arson and felony murder, charges stemming from a house fire that took the lives of his wife and two daughters. In June 2012, the state court granted Gavitt's unopposed motion for relief from judgment based on newly discovered evidence. The newly discovered evidence is in the nature of advancements in fire science research and investigation methods that tend to impugn some of the evidence on which Gavitt's convictions were based. The judgment was vacated, the charges dismissed, and Gavitt was released from prison.

Two years later, Gavitt brought this civil rights action against numerous city and county entities, prosecutors, law enforcement officials, and investigators who participated in the prosecution against him. He claims that defendants violated his due process rights by intentionally misrepresenting evidence and failing to disclose exculpatory evidence, and that they conspired to deprive him of his rights. All defendants moved to dismiss on the pleadings and the district court granted all but one of the motions, that of the Estate of John DeVries, a Michigan State Police forensic laboratory technician who testified at Gavitt's trial. While the district court identified legal deficits that warranted dismissal of most of Gavitt's claims, it held that Gavitt's claim against DeVries included sufficient factual allegations to state a valid claim for relief.

We now address two appeals stemming from the district court's rulings. First, in No. 15-2136, the Estate of DeVries challenges the denial of its motion to dismiss based on qualified immunity. In short, the Estate contends the district court read Gavitt's claim too generously, failing to recognize that the public record made in state court contradicts his allegations, rendering the claim implausible. Second, in No. 15- 2434, Gavitt challenges the dismissals of his claims against the other defendants, contending that his allegations pass muster at the pleading stage and that he deserves the chance to conduct discovery. For the reasons that follow, we deny relief in both appeals. The Estate's appeal must be dismissed for lack of appellate jurisdiction. As to the second appeal, we affirm, finding no error in the dismissals of Gavitt's claims against the remaining defendants.

I. FACTUAL AND PROCEDURAL BACKGROUND

Each of the district court's substantive rulings includes substantially the same summary of the factual and procedural background. No party having disputed the accuracy of the summary, it is reproduced here:

A. Fire, Investigation, Arrest, Trial, and Conviction

Gavitt survived a March 9, 1985 house fire. His wife and two daughters tragically did not. An investigation was initiated by the City of Ionia Police Department and the Michigan State Police Arson Task Force.

On the morning of March 10, 1985, Defendants Kalman and Fatchett, then-Michigan State Police ("MSP") officers assigned to the MSP Arson Strike Force Unit ("Det./Sgt. Kalman" and "Det./Sgt. Fatchett" respectively), were dispatched to the scene of the house fire to investigate its cause and origin. Based on their initial review of the evidence, they concluded that the fire was incendiary in nature. At 2:30 in the afternoon that same day, Det./Sgts. Kalman and Fatchett summoned Defendant Klein, then-Sergeant with the Ionia Police Department ("Sgt. Klein"), to the burned home, walked him through the evidence at the fire scene that led them to their initial conclusion that the fire was incendiary in nature, and collected evidence that Sgt. Klein then placed in an Ionia Police Department evidence locker. Sgt. Klein then continued his investigation by obtaining evidence from and interrogating Plaintiff David Gavitt ("Gavitt"), and obtaining more evidence from the burned home.

On March 12, 1985, as reported in Det./Sgt. Kalman's March 1985 Report, a meeting was held "for the purpose of reviewing the evidence and determining the course of the investigation." Defendant Prosecutor Gabry is listed as being present. Det./Sgt. Kalman reported that he presented evidence, a discussion was held, and a conclusion reached that Gavitt may have set the fire himself and was unable to save his family once the fire started:

> Undersigned officer explained the burn patterns and also relating [sic] the burn patterns to the burns on the victim. A formal discussion was held on all the evidence obtained and it is the feeling that there is

> strong evidence pointing to the fact that MR. DAVID GAVITT may have indeed set the fire himself and was unable to save his family once the fire started.

Sgt. Klein's March 20, 1985 Report also discussed the March 12th meeting and calls it a "'skull session' starting at/around 8:30 am, ending a short time later." Sgt. Klein does not list Prosecutor Gabry as being present. Rather, he reports:

> Journal Entry: It was on TUES, MARCH 12th, 1985 that this investigating officer, Sgt. Wieczorek and Chief Voet met with the following: City Superintendent Allen Housler, Det/Sgt. JOHN KALMAN and Det/Sgt. JOE DeKRACKER of the Arson Strike Force, MSP Rockford Post, this meeting was an "initial assessment of the case". It should be noted that this meeting was a "skull session" starting at/around 8:30 am, ending a short time later.

On June 10, 1985, a felony complaint was issued, and state criminal charges were brought against Gavitt - three counts of murder, three counts of felony murder, arson, and arson insured property - and he was subsequently arrested. Sgt. Klein was the complaining witness on the criminal complaint.

On June 21, 1985, a preliminary examination hearing was held on the criminal charges brought against Gavitt. District Court Judge James Ward was the presiding judge and Defendant Gabry was the prosecutor. The District Court found that probable cause existed on the charged offenses - murder, felony murder, and arson - but dismissed the insurance fraud charge.

A jury trial was held in the Circuit Court for the County of Ionia. On February 14, 1986, a jury convicted Gavitt on three counts of murder committed in the perpetration of arson (first degree felony murder) and one count of arson to a dwelling place. The one count of arson was dismissed by the Court at sentencing.

On April 18, 1986, Gavitt was sentenced to "imprisonment for life on each of the three counts of murder, to be served concurrently with each other."

B. Innocence Clinic's Post-Trial Motion for Relief, Stipulation, and Dismissal

In September 2011, a motion for relief from judgment was filed on Gavitt's behalf by the University of Michigan Law School's Innocence Clinic, arguing that there was newly discovered scientific analysis of the origin and cause of the March 1985 fire establishing that there was no arson. That motion explained that evidence of actual innocence was only recently discovered because, beginning in 1992, there has been a complete revolution in the field of fire investigation:

> 14. The field of fire investigation has undergone a complete revolution since Mr. Gavitt was convicted in 1986. In 1992, the National Fire Protection Association adopted NFPA 921, the current standard of care for fire investigations, which for the first time put the field of fire investigation on a scientific basis.
>
> 15. In light of the changes in the field of fire investigation, John Lentini - a world-renowned fire investigator who has reviewed all available testimony and evidence in this case - has concluded that there is no basis to conclude that arson was the cause of the Gavitt fire. Mr. Lentini's affidavit is attached to the brief accompanying this motion.
>
> 16. Mr. Lentini's findings are rooted in the crucial concept of "flashover," which was not well understood by the fire investigation community at the time of Mr. Gavitt's trial. During flashover - a phenomenon that takes place when a compartment like the Gavitt living room catches fire - a room becomes so hot that every exposed combustible surface can catch fire.
>
> * * * *
>
> 29. Mr. Lentini's scientific conclusions regarding the origin and cause of the Gavitt fire meet this [newly discovered evidence] test. His findings are based on the new standards of origin and cause investigation, which were not adopted until the early 1990s, at the earliest, years after Mr. Gavitt's trial.

> The evidence refutes all scientific evidence presented at trial, so it is not cumulative. Further, because the prosecution's case relied wholly on the testimony of fire experts, no rational jury could have found Mr. Gavitt guilty of murder if the findings of Mr. Lentini or any competent fire investigator applying modern standards had been presented. This is particularly true when this evidence is considered alongside the new evidence regarding the absence of gasoline on the carpet samples… .
>
> 30. Finally, because the field of gas chromatography was much less precise in 1986 than it is today, trial counsel could not have discovered the new evidence that undermines Mr. DeVries's testimony through the exercise of due diligence, and the fourth and final prong for granting a new trial based on newly-discovered evidence is satisfied.

The supporting brief relied heavily on John Lentini's Affidavit and similarly argued that, because of "significant advancements in the field of fire science and arson investigation," there is newly-discovered evidence that undermines the prosecution's case against Gavitt:

> While the investigation of this case was, perhaps, normal for a fire investigation conducted in the mid-1980s, approximately a decade before scientific principles were first applied to fire investigation, practically all of the investigative methods and conclusions reached by the various fire investigators in this case fail to meet modern standards of accuracy and reliability.
>
> * * * *
>
> The field of fire investigation has undergone a complete revolution since Mr. Gavitt's conviction. John Lentini Affidavit ¶¶ 15-59. Theories that low-burning, alligatoring, pour patterns, depth of char, and temperature and speed of fires can serve as indicators of arson were once unquestioned, but have been completely and unequivocally repudiated by rigorous scientific testing. Id. at ¶¶ 36-59. As such, every indicator of arson relied upon by the prosecution's experts at Mr. Gavitt's trial has been discredited and is understood to be useless in determining the true origin and cause of fires. Id. Many factors once thought to be present only in accelerated fires are now understood to be present in natural fires that have undergone flashover and progressed to full room involvement, a phenomenon that was not understood in 1986[.] Id. at ¶¶ 29-35.
>
> * * * *
>
> Mr. Lentini's finding that Mr. DeVries's testimony about the carpet flame tests was false meets all parts of the standard for newly- discovered evidence. Given that the understanding of flashover is a novel concept in arson science and certainly was not known outside of a very small subset of the scientific community at the time of Mr. Gavitt's trial, the evidence itself and not merely its materiality is newly-discovered. For this same reason, it is clear that the evidence could not have been discovered with due diligence at the time of trial; indeed the concept of flashover would not become widely known and understood by fire investigators until at least a decade after Mr. Gavitt's conviction… .

As Mr. Lentini admitted in his affidavit, at the time that Sgt. Fatchett and the Ionia County Prosecutor's expert, Dr. Edwards, considered the impact of flashover, they had no way of knowing that their generally accepted interpretations of burn patterns would be refuted years later:

> Neither Sgt. Fatchett nor Dr. Edwards considered the impact of flashover and the impact of the burning curtains when they attempted to discern what caused the patterns they observed on the living room floor in the course of their investigation. Nor would such consideration have been expected in 1986, because the state of the art in fire investigation had not come to fully recognize flashover at that time.
>
> * * * *
>
> All of the above testimony [Defendant Fatchett's trial testimony evaluating evidence from the fire and opining that the fire was intentionally set] can be shown by today's standards to have been false and misleading, albeit without malicious intent… .

* * * *

> The State's experts had no way of knowing that their interpretation of the burn patterns at the Gavitt residence was without any scientific validity because, at the time of the trial, those interpretations were, in fact, generally accepted by most fire investigators, including your affiant. The misinterpretation was bolstered by the incorrect laboratory analysis performed by Mr. DeVries. The state's fire investigators "expectations" were not properly "calibrated." They expected the confined fire in the Gavitt residence to behave like an unconfined fire. Because the fire did not meet their expectations of normal fire behavior, they incorrectly determined the fire to be incendiary.

An experienced fire investigator that Gavitt's defense attorney consulted with in 1985 also provided an affidavit which was attached as an exhibit to Gavitt's motion for relief. Mr. Churchwell, like Mr. Lentini, stressed that "the world of fire science is vastly different today than it was in 1985;" that the way he "would have viewed the fire scene in 1985 is completely different from the way" he "would view the same scene today;" and "the advancements in fire science would enable [him] to have far better insights and be wary of false findings today." Mr. Churchwell, like Mr. Lentini, stated that he subscribed to the same beliefs that science has now proven to be false; that he can say now that Gavitt was falsely convicted; but could not have reached that conclusion in 1985-86 when he was consulting with defense counsel:

> Well into the 1980's, the arson investigation profession believed that things like floor burn patterns, low burning, deep charring and alligatoring were automatic indicators of arson: I subscribed to those beliefs at one time as well. But in the 1990s, with a wider understanding of the concept of flashover and the emergence of NFPA 921, the profession grew up and began to embrace the rigors of actual science. Upon doing so, the open-minded among us discovered that the old indicators that we thought were automatic markers of arson were in fact not. This led to the realization that each of us - investigators who had worked in the 1970s and 1980s - had misread many fire scenes, possibly leading to the conviction of innocent people. I have no doubt in my mind that David Gavitt is one such falsely-convicted person. I can say this knowing what I know today, but such a conclusion would have been impossible for me to make in 1985-86 (when Mr. Kolenda consulted with me) because the profession had yet to become enlightened to the errors of the old ways of arson investigation at that time.

Mr. Churchwell also admitted that he "would likely have made the same mistake" as those investigating the Gavitt home fire by failing to give adequate consideration to possible accidental causes of that fire:

> As I know from having worked many similar fires in the 1980s, the fact that obvious sources of ignition and the presence of various fuels (candles, ashtray, oil lamps, paneling, etc.) went largely ignored at the trial is not all that surprising. In those days, fire investigators would look first for pour patterns, alligatoring and other such things, and upon finding them, we would assume the fire was arson - so much so that we'd ignore far more obvious accidental sources of the fire. I have no doubt that the investigators who looked at the fire scene at the Gavitt home made this mistake. As cautious and careful as I always try to be, I would likely have made the same mistake upon seeing the Gavitt fire scene in 1986: Fire investigators simply did not have enough knowledge about the true nature of enclosed (compartment) fires at that time. Today, years later, being wise to the many advancements and the rigors of actual science that have finally come to dominate the arson investigation profession, I can say that the prosecution's experts were blinded by the myths (alligatoring, charred glass, low burning, etc.), and failed to give due deference to far more obvious and likely accidental sources of the fire.

The parties stipulated to a stay of Gavitt's motion, allowing the Ionia Prosecutor's Office time for scientific review of Gavitt's claims. On June 6, 2012, after a thorough investigation, current Ionia County Prosecutor Ronald Schafer, on behalf of the People of the State of Michigan, responded to Gavitt's motion.

Prosecutor Schafer acknowledged that, although "fire investigators" involved in the original investigation "held to a common understanding within fire investigation," it is now known that that "was inaccurate, specifically regarding the recognition and identification of unusual burn patterns in the floor as definitive evidence of ignitable liquid pour patterns. Unfortunately, during that time period many fire investigators did not understand the phenomena of flashover and post- flashover and their effects on the production of unusual burn patterns in floors within compartments." Prosecutor Schafer also acknowledged that Gavitt had satisfied the requirements for a new trial.

> [T]he Peoples [sic] recent investigation confirms the 1985 findings which found the presence of accelerants on the carpet samples introduced at trial can no longer be independently verified as having a presence of ignitable liquids, specifically gasoline. Therefore, the parameters for meeting the legal requirement for a new trial are not disputed.

Despite an admission that "three independent analyses of the evidence suggest there was likely no gasoline on the carpet samples taken out of the Gavitt house in 1985," Prosecutor Schafer emphasized that "there are still a great deal of questions surrounding this case which raise significant questions." "Still today, these unanswered questions linger when looking at the case on the whole, even in light of the acknowledged findings in this response. In fact, this is the exact type of case that would have remained open had it not been prosecuted earlier; the type of case where justice would demand that it stay open." Nonetheless, Prosecutor Schafer acknowledged, "it does not change the fact that fire investigation has advanced in the twenty-seven years since this fire."

Prosecutor Schafer identified some of those fire investigation advances and explained why, in light of those advances, Gavitt cannot be retried.

> In particular, our understanding of flashover, post-flashover and the production of unusual burn patterns in floors, potentially identified as pour patterns, is different today than in 1985. Testing of materials in fire cases has also advanced, with more sophisticated instrumentation and analysis. Consequently, there is new evidence in this case and [Gavitt] is entitled to a new trial. As outlined, based on today's understanding of fire dynamics and the evolved level of fire investigation, this fire incident would likely be classified as undetermined and consequently the People will not be able to retry [Gavitt] … . there is only one thing known with certainty, as of today, this case involves a fire of undetermined origin and cause. Having no laboratory verification of the presence of an accelerant, combined with what the People now know through scientific research and testing regarding flashover and post-flashover compartment fires, and the production of unusual burn patterns in the floor, the determination that an ignitable liquid (gasoline) was used to initiate the fire at the Gavitt residence cannot be verified. As a result, this is a case this office could not charge as arson based on the evidence available today. However, this is also a case that, if it was new today, this office would not close. There are simply too many questions, questions which may never be answered. Ultimately, this remains a case in which the lives of three innocent people were taken by a fire that can only be classified as having an undetermined origin and cause.

On June 6, 2012, the People of the State of Michigan and Gavitt stipulated that Gavitt's motion for relief from judgment be granted, that all charges against Gavitt be immediately dismissed, and that the Court order his immediate release from the custody of the Michigan Department of Corrections.

R. 93, Opinion at 2-11, Page ID 3006-15 (citations omitted) (emphasis in original).

Following his release from prison, Gavitt filed the instant action, alleging that his wrongful conviction was not merely the product of a tragic but innocent misunderstanding of scientific evidence.

…

In sum, the appeal presented in Case No. 15- 2136, challenging the denial of the motion to dismiss the amended complaint against the Estate of DeVries, is DISMISSED for lack of appellate jurisdiction. As to the appeal presented in Case No. 15-2434, the district court's judgment dismissing all claims against the remaining defendants is AFFIRMED.

Questions:

1. Officers assigned to conduct the investigation concluded that the fire was "incendiary" and as a result of this conclusion indicate that defendant set the fire and killed his family. How do these officer reach the conclusion based on evidence observation?
2. Defendant hires his own Expert who completely and totally refutes this conclusion and also observes that historic techniques of measure and discovery of trace evidence proving intentionality in fire are now totally discredited. Explain.
3. What is the Court's resolution of these opposing views in this particular case?

State v. Rose
423 S.C. 382; 814 S.E.2d 529;
Court of Appeals of South Carolina February 5, 2018, Heard; April 11, 2018, Filed
Opinion No. Op. 5551

Paula Reed Rose appeals her convictions of third-degree arson, filing a false police report, burning personal property to defraud an insurer, and making a false insurance claim to obtain benefits for fire loss, for which the trial court sentenced her to a cumulative term of five years' home incarceration with five years' probation. On appeal, Rose argues the trial court erred by (1) refusing to direct a verdict of acquittal on all charges when the State failed to present substantial circumstantial evidence of her guilt, (2) qualifying Investigator Benjamin Cannon as an expert in the origin and causes of fires, and (3) permitting the expert testimony of Investigator Charles Gonzalez regarding the use of his accelerant detection canine at the scene. We affirm.

FACTS

On the morning of July 27, 2012, Paula Rose called 911 to report a burglary. During the call, Rose stated three men were in her garage threatening her and demanding she open a gun safe. While on the phone with the 911 dispatcher, Rose heard a strange noise. She stepped out of her bedroom and realized a wicker couch on her back porch was on fire. The responding authorities found Rose outside worried about her pets who were still inside the home. Rose was covered in soot and wore jeans, a black shirt, and open-toed shoes. The firemen deduced there were two fires—one on the back porch and one on the front porch; however, the only active fire was in the rear of the home.

Shortly after the responders extinguished the fire, Investigator Randy Morgan arrived. The only potential traces of the alleged burglars were some footprints and disposed latex gloves near the front and back porches. Investigator Morgan testified nothing was stolen from the property, but both of the gates and the garage were found open.

During Investigator Cannon's preliminary investigation, he found a red gas can on the front porch and a yellow gas can on the back porch. He noted "the aroma of gasoline on the front and prob[ably] kerosene or diesel-type on the back porch." Investigator Cannon subsequently called his colleague Investigator Gonzalez and his accelerant detection canine Misty to the scene. Misty alerted to multiple areas on the front and back porch.

After taking Rose's statement, Investigator Morgan met with Investigators Cannon and Gonzalez to compare their findings. Based upon various oddities and inconsistencies, such as the location of the fires in relation to the alleged burglary; discrepancies between Rose's 911 call and victim statement; the superficial damage to the safe; the lack of stolen items; and Rose's calm, unaffected demeanor, all three investigators felt the case needed further investigation. Thereafter, they returned to the scene to further investigate the garage and interior of the home.

Describing the damage to the safe, Investigator Cannon testified there were "a dozen or so marks" on the "upper right hand corner" and a few marks around the dial. Investigator Cannon explained the paint appeared chipped but there was no substantial damage to the safe. Looking for an item potentially used by the assailants to attempt to gain entry to the safe, the investigators found a small hatchet sitting on a shelf in the garage that appeared to be undisturbed. The investigators noted the hatchet had paint chips on the blade. While examining Rose's bedroom, Misty alerted to a pair of slippers.

Before trial, Rose moved to exclude evidence of Misty's alerts and the testing results arising therefrom, arguing the use of a canine in this context was unreliable. Relying upon Florida v. Jardines, the trial court ultimately denied Rose's motion, finding testimony regarding Misty's alerts would be admissible if the State established the proper foundation by showing the dog was properly certified through a recognized program. Rose renewed this objection at trial, and the trial court overruled the objection, later stating,

> [A]t the end of the hours of preliminary testimony we had yesterday I don't feel the need to have any further [argument]. I know what I'm gonna hear … my ruling will be … that as long as the appropriate foundation is laid on the admission of the testimony about the … dog then [it will] be allowed … .

At the close of the State's case, Rose moved for a directed verdict on all charges, arguing the State failed to present substantial circumstantial evidence of her guilt. Although the trial court noted its concern regarding the fraudulent insurance claim charge, the court ultimately denied Rose's motion as to all indictments. The jury found Rose guilty of all charges as indicted, and the trial court sentenced her to five years' home incarceration with five years' probation. This appeal followed.

…

LAW/ANALYSIS

I. Directed Verdict

Rose argues the trial court erred in denying her motion for a directed verdict as to all charges because the State failed to show substantial circumstantial evidence tending to prove her guilt. Rose asserts the evidence the State presented fails to rise above vague suspicion. We disagree.

"When ruling on a motion for a directed verdict, the trial court is concerned with the existence or nonexistence of evidence, not its weight." State v. Weston, 367 S.C. 279, 292, 625 S.E.2d 641, 648 (2006). "A defendant is entitled to a directed verdict when the state fails to produce evidence of the offense charged." Id. "On appeal from the denial of a directed verdict in a criminal case, an appellate court must view the evidence in the light most favorable to the State." State v. Stanley, 365 S.C. 24, 41, 615 S.E.2d 455, 464 (Ct. App. 2005). "If there is any direct or any substantial circumstantial evidence reasonably tending to prove the guilt of the accused, we must find that the issues were properly submitted to the jury." State v. Mollison, 319 S.C. 41, 46, 459 S.E.2d 88, 91 (Ct. App. 1995). "Nevertheless, a[n appellate] court is not required to find that the evidence infers guilt to the exclusion of any other reasonable hypothesis." State v. Bennett, 415 S.C. 232, 236, 781 S.E.2d 352, 354 (2016). "[T]he lens through which a[n appellate] court considers circumstantial evidence when ruling on a directed verdict motion is distinct from the analysis performed by the jury." Id. "Accordingly, in ruling on a directed verdict motion where the State relies on circumstantial evidence, th[is] court must determine whether the evidence presented is sufficient to allow a reasonable juror to find the defendant guilty beyond a reasonable doubt." Id. at 237, 781 S.E.2d at 354.

In South Carolina, a person is guilty of third-degree arson if she "wilfully and maliciously … sets fire to, burns, or causes a burning which results in damage to a building or structure." S.C.

Code Ann. § 16-11-110(C) (Supp. 2017). Section 16-11-130 of the South Carolina Code (2015) provides:

> Any person who … wilfully and with intent to injure or defraud an insurer sets fire to or burns or causes to be burned or … aids, counsels, or procures the burning of any … personal property of any kind, whether the property of himself or of another, which is at the time insured by any person against loss or damage by fire is guilty of a felony … .

Additionally, a person is guilty of a felony if she:

> wilfully and knowingly presents or causes to be presented a false or fraudulent claim, or any proof in support of such claim, for the payment of a fire loss … upon any contract of insurance or certificate of insurance which includes benefits for such a loss, or prepares, makes, or subscribes to a false or fraudulent account, certificate, affidavit, or proof of loss, or other documents or writing, with intent that such documents may be presented or used in support of such claim … .

S.C. Code Ann. § 16-11-125 (2015) (emphasis added).

Subsection 16-17-722(A) of the South Carolina Code (2015) provides that "[i]t is unlawful for a person to knowingly file a false police report."

In the case at bar, Rose called 911, alleging three men in her garage were threatening her and seeking entry to her gun safe. During the call, Rose informed the dispatcher she heard a "crackling" noise and subsequently realized a wicker couch on her back porch was on fire. When Benjamin Temple, a good samaritan, stopped to inspect the voluminous smoke he saw from the road, Rose never mentioned the burglary or cause of the fire. In fact, when Temple asked if anyone else was inside the home, Rose replied only her dog remained inside. Further, the investigating authorities were unable to find the three men, and nothing was stolen from the property. Additionally, Investigators Cannon, Morgan, and Gonzalez found the damage to the gun safe to be superficial with most of the shallow marks on the upper right side rather than near the dial. The investigators testified they found the hatchet likely used to make the markings on a shelf that seemed undisturbed. At trial, trace evidence expert Megan Fletcher testified the paint chips found on the hatchet and safe either originated from the same item or an item containing the "same physical and chemical characteristics."

In addition to the fire debris on the front and back porches, Misty also alerted to a pair of Rose's slippers in the bedroom, which were not the shoes Rose wore during the fire. Fletcher testified the slippers tested positive for ignitable liquids. Additionally, Fireman Brandon Barwick testified that in September after the fire at her home, Rose came to the Wade Hampton Fire Department in search of burn records for her property. According to Barwick, Rose stated she wanted the records to provide an explanation for why Misty alerted to her slippers. She explained she previously burned leaves on her property while wearing them. However, Fletcher testified ignitable liquids vaporize quickly, so the products found on the tested items would last "less than a few days," and, according to the Forestry Commission's records, the last controlled burn on the Rose property occurred on May 29—a couple of months before the fire.

Fletcher further testified samples from the jeans and shirt Rose wore during the fire also tested positive for ignitable liquids. Fletcher testified that "for [her] to be able to determine that a[n] ignitable liquid was found, … a liquid would have had to have been deposited on [the collected] items." Fletcher explained the ignitable liquids would have to be in liquid form to be deposited onto an item. Fletcher clarified "ignitable liquids are not found in smoke from a fire scene," and therefore, an ignitable liquid cannot be deposited onto an item, such as clothing, in a gaseous state. Finally, although the State Farm insurance policy only listed Homer as the named insured, Nicholas Gregory, the assigned State Farm representative, testified Rose participated in the claim process, including providing information about how the fire occurred. See § 16-11-125 ("Any person who wilfully and knowingly presents or causes to be presented a false or fraudulent claim, or any proof in support of such claim, for the payment of a fire loss … upon any contract of insurance or certificate of insurance which includes benefits for such a loss … or subscribes to a false or fraudulent account … used in support of such claim, is guilty of a felony." (emphasis added)).

Viewing the evidence in the light most favorable to the State, we find a jury could reasonably deduce Rose fabricated the burglary and caused the fires for the purposes of receiving insurance benefits. … Accordingly, the trial court properly refused to direct a verdict in Rose's favor. …

II. Expert Qualification

Rose argues the trial court abused its discretion in allowing Investigator Cannon to testify as an expert in arson cause and origin. Specifically, Rose maintains the State failed to demonstrate that, at the time of the fire, Investigator Cannon possessed the requisite knowledge, skill, experience, training, or education to allow for his opinion testimony. Rose additionally contends Investigator Cannon's expert testimony was unreliable because he failed to comply with the National Fire Protection Association's guidelines during his investigation of the fires. We disagree.

"The qualification of a witness as an expert is within the sound discretion of the trial court and will not be reversed absent an abuse of discretion." State v. Martin, 391 S.C. 508, 513, 706 S.E.2d 40, 42 (Ct. App. 2011). "An abuse of discretion occurs when the trial court's ruling is based on an error of law or a factual conclusion that is without evidentiary support." State v. Price, 368 S.C. 494, 498, 629 S.E.2d 363, 365 (2006).

"If scientific, technical, or other specialized knowledge will assist the trier of fact to understand the evidence or to determine a fact in issue, a witness qualified as an expert by knowledge, skill, experience, training, or education, may testify thereto in the form of an opinion or otherwise." Rule 702, SCRE. "The expertise, reliability, and the ability of the testimony to assist the trier of fact are all threshold determinations to be made prior to the admission of expert testimony, and generally, a witness's expert status will be determined prior to determining the reliability of the testimony." State v. Tapp, 398 S.C. 376, 388, 728 S.E.2d 468, 474-75 (2012). "There is no exact requirement concerning how knowledge or skill must be acquired." State v. Henry, 329 S.C. 266, 274, 495 S.E.2d 463, 467 (Ct. App. 1997). Further, "[t]here is no abuse of discretion as long as the witness has acquired by study or practical experience such knowledge of the subject matter of his testimony as would enable him to give guidance and assistance to the jury." Id. at 273, 495 S.E.2d at 466. "Once a witness is qualified as an expert, continued objections to the amount or quality of the expert's knowledge, skill, experience, training, or education go to weight of the expert's testimony, not its admissibility." Martin, 391 S.C. at 513, 706 S.E.2d at 42.

We find the trial court properly qualified Investigator Cannon as an expert in arson cause and origin. Although Investigator Cannon possessed less than a year's worth of experience in the arson field at the time of the Rose investigation, the trial court properly accounted in its analysis the breadth of Investigator Cannon's experience and training possessed at the time of trial. At the time of his qualification, Investigator Cannon testified he had previously investigated between eighty and eighty-five arson cases, including cases similar to the Rose investigation, and had previously been qualified as an arson expert in the Thirteenth Circuit. In regards to his education and training, Investigator Cannon testified he had attended a two-week, eighty-hour class at the National Fire Academy in Maryland that was specifically designed for learning the origin and causes of fires and also included a fire pattern certification course. Investigator Cannon testified he also attended "a basic fire investigation course" at the Criminal Justice Academy and had taken 130 hours of "online certified fire investigative classes." Additionally, Investigator Cannon testified he takes approximately forty hours of continuing education classes per year. … Accordingly, the State sufficiently demonstrated Investigator Cannon possessed the requisite experience and training to testify as an expert in arson origin and causes.

Further, Rose's claim that Investigator Cannon's expert testimony was unreliable because he failed to comply with the National Fire Protection Association's guidelines is unavailing. Rose never raised this issue to the trial court prior to its qualification of Investigator Cannon; therefore, Rose failed to preserve this issue for appellate review and any objections subsequent to Investigator Cannon's qualification went to the accorded weight of his testimony. ...

Based on the foregoing, the trial court did not abuse its discretion in qualifying Investigator Cannon as an expert in arson cause and origin ... .

III. Admissibility of Accelerant Detection Canine

Evidence

Rose argues the trial court abused its discretion in allowing Investigator Gonzalez's expert testimony concerning the use of his accelerant detection canine Misty at the scene. Specifically, Rose challenges the testimony detailing Misty's alert to Rose's slippers and the forensic testing results arising therefrom. Rose contends the trial court erred in allowing this testimony because the State failed to establish that (1) the use of an accelerant detection canine was reliable and (2) Investigator Gonzalez possessed the requisite knowledge, training, and skill to properly handle an accelerant detection canine and assess its alerts. We disagree.

Preliminarily, the State contends Rose failed to properly preserve this issue for appellate review. However, we find Rose's pretrial motion in limine and subsequent objections at trial sufficiently safeguarded the issue for our review.

"A trial court's decision to admit or exclude expert testimony will not be reversed absent a prejudicial abuse of discretion." State v. White, 382 S.C. 265, 269, 676 S.E.2d 684, 686 (2009). Although our appellate courts have not yet addressed the admissibility of canine detection evidence in the context of arson investigations, we find our jurisprudence concerning the admissibility of dog tracking evidence instructive. In White, our supreme court held all "[n]on[-]scientific expert testimony must satisfy Rule 702, both in terms of expert qualifications and reliability of the subject matter." 382 S.C. at 273, 676 S.E.2d at 688. However, "[t]here is no formulaic approach for determining the foundational requirements of qualifications and reliability in non-scientific evidence." State v. Chavis, 412 S.C. 101, 108, 771 S.E.2d 336, 339 (2015). "The expertise, reliability, and the ability of the testimony to assist the trier of fact are all threshold determinations to be made prior to the admission of expert testimony, and generally, a witness's expert status will be determined prior to determining the reliability of the testimony." Tapp, 398S.C. at 388, 728 S.E.2d at 474-75.

Investigator Gonzalez testified he and Misty had worked on 142 fire scenes together prior to trial. Regarding the extent of his training and education, Investigator Gonzalez testified he had worked in law enforcement for twenty-one years and had been a part of Greenville County's arson unit since 2005. Investigator Gonzalez testified he received specialized training through the National Fire Academy in Maryland and held numerous certifications in the field of arson investigation.

Regarding Misty's qualifications as an investigative canine, Investigator Gonzalez testified he began "training with her from the beginnin[g] of the imprint [on ignitable liquids] to the end [of] certification." Investigator Gonzalez testified that in January 2011, he and Misty took a six-week accelerant detection training course in Alabama where she was exposed to "a minimum of sixteen different ignitable liquids." While Misty learned to search for and detect various accelerants, Investigator Gonzalez testified he "learned how to know when she is alerting on ignitable liquids and when she isn't alerting ... and how to handle her in workin[g] in fire scenes." After completing the course in Alabama, Investigator Gonzalez testified he took Misty to Austin, Texas where they engaged in "actual fire scene training" with the State Fire Marshall's Office. After becoming additionally certified in Texas, Investigator Gonzalez testified he brought Misty back to South Carolina. Since the original certifications in 2011, Investigator Gonzalez testified he and Misty have both re-certified "every year in the Fall." He explained Misty "certifies each year with the … North American Police Work Dog Association," which is a recognized organization for certification of police canines. Investigator Gonzalez clarified that both "the canine and the handler" re-certify each year. Investigator Gonzalez explained that for Misty to receive her certification, she can neither miss nor falsely alert to any of the tested ignitable liquids. According to Investigator Gonzalez, Misty has not failed any certification process since receiving her original certification in March 2011. He further testified Misty had never falsely alerted in any case.

Pursuant to the framework established in White, we find the State sufficiently demonstrated (1) Investigator Gonzalez possessed the requisite knowledge, training, and skill under Rule 702, SCRE, to qualify as an expert in "the handling of accelerant detection canines" and (2) Misty's alerts were reliable. …

Regarding Rose's challenge to Misty's reliability based on potential crime scene contamination, we find evidence in the record supports the trial court's admission of Misty's alerts. …

Accordingly, the trial court did not abuse its discretion in allowing Investigator Gonzalez's expert testimony concerning Misty's alerts at the scene and the subsequent forensic test results arising therefrom. …

CONCLUSION

Accordingly, the decision of the trial court is
AFFIRMED.

Questions:

1. The Rose decision is a treasure trove of forensic trace evidence. To prove the arson, the prosecution introduced what forms of trace evidence as to proof of fire and its intentional start?
2. Were the defense effort's to disqualify the expert based on a lack if expertise successful?
3. How does the defense seek to discredit the canine used to detect smell of unauthorized accelerants?

7 Documentary Evidence

Documents play a large part in a wide array of litigation, from medical and business records to statements and confessions, from contracts to legal titles such as Deeds. The documentary form memorializes exchanges in the commercial marketplace, resolves potential marital dispute by prenuptial agreements and fosters the law of succession in estates, trusts and wills. The impact of the documentary form is often subject to forensic examination and scrutiny.

The idea of a "document," as the following cases will edify, is no longer a simple matter of a piece of paper. Instead, the list of documentary forms reflects a changing marketplace for not only communication but also for commerce. The concept of an "original," by way of illustration, can be a challenging question for courts and litigants alike (Figure 7.1).

FIGURE 7.1 Forensic testing and analysis of pages from H.R. "Bob" Haldeman papers from Nixon Presidency, [carried out by inter-agency team, under the direction of the National Archives, at the Library of Congress]. Photographer: Earl McDonald—NARA Staff.

7.1 DOCUMENTARY EVIDENCE AND THE "BEST EVIDENCE"

Any time there is a dispute between parties regarding the authenticity or integrity of a documentary form, common sense dictates the evaluation be of the best possible form of that document, which in most cases is the "original." Despite its plain meaning, an original, under our rules of evidence,

DOI: 10.4324/9781003267126-7

may be a duplicate and even a copy. And in some cases, especially when the original has been misplaced or destroyed, the best evidence may not be the original.

Frisby v. Warden
269 So. 3d 371
Court of Appeals of Mississippi May 8, 2018, Decided
NO. 2017-CA-00548-COA

P1. Clifford C. Frisby appeals from the Harrison County Chancery Court's finding that three handwritten documents were enforceable contracts between Frisby and Ferrell Warden for the sale of a home to Warden. The documents provided for credit toward the purchase price of the property in exchange for work performed by Warden. The chancellor ordered specific performance by Frisby and that Warden pay the remaining amount owed. Finding no error, we affirm.

FACTS AND PROCEDURAL HISTORY

P2. Oasis Real Estate Investment, Inc. is a Mississippi Corporation. Frisby is its president and owner.1 In June 2010, Frisby acquired the property through a quitclaim deed from Michael Neill. In 2011, Frisby allowed Warden to move into this property. On February 9, 2015, Frisby filed an eviction action against Warden in justice court. On March 19, 2015, in response to the eviction action, Warden filed a "Petition To Enforce Purchase Contract and Stay Proceedings" against Frisby and Oasis in chancery court. Warden sought enforcement of three handwritten contracts between himself and Frisby for purchase of the property. In support of his petition, Warden attached three handwritten documents that were apparently signed by both parties. The documents detailed work that Warden performed for Frisby, and indicated that Warden's work would count toward the $24,000 purchase price of the property.2 Warden also sought a stay of the justice court eviction proceedings during the pendency of the chancery court proceedings.

P3. On May 12, 2015, Frisby answered Warden's petition with a motion to dismiss, a motion to deny stay of eviction, and a counterclaim for fraud alleging that his signatures on the agreements were forgeries. He further requested monetary damages for fraud, slander of title, and attorney's fees. On July 15, 2015, the chancellor heard Frisby's motion to dismiss and motion to deny stay of eviction, ultimately denying both of Frisby's motions. The chancellor also transferred the justice court eviction action and consolidated it with Warden's enforcement-of-contract action in the chancery court.

P4. On August 2, 2016, and September 8, 2016, the chancellor held a hearing on the consolidated matter.

Warden introduced duplicates of the three alleged contracts and called four witnesses, including Frisby as an adverse witness. Frisby called six witnesses, including an expert in forensic document examination. At the close of the testimony, the chancellor verbally ordered Frisby to specifically perform on the three contracts and ordered Warden, based on the amounts previously paid pursuant to the contracts, to pay $11,300 as full and final payment toward the purchase price of the property. A final judgment reflecting her verbal order was entered on October 4, 2016. Aggrieved, Frisby filed an unsuccessful motion for reconsideration and to set aside the judgment.

P5. Frisby timely appeals from the final judgment and denial of his motion for reconsideration. Frisby raises five issues: (1) whether the chancellor erred in allowing duplicates of the handwritten documents into evidence; (2) whether the chancellor erred in placing the burden of proof on Frisby to prove his fraud allegation; (3) whether Warden carried his burden of proof; (4) whether the chancellor erred in finding that Warden detrimentally relied on Frisby's promise; and (5) whether Frisby presented sufficient evidence to prevail on his fraud claim.

...

DISCUSSION

I. Whether the chancellor erred in admitting the handwritten agreements into evidence.

P7. Frisby asserts that the chancellor improperly admitted the disputed handwritten contracts into evidence. According to Frisby, these documents were duplicates of original handwritten documents that he never signed. Therefore, he argues that pursuant to Mississippi Rules of Evidence 1002 and 1003, the duplicates were inadmissible because there was a genuine question as to their authenticity. In response, Warden asserts Frisby has offered nothing to show that the chancellor abused her discretion nor that any substantial right has been affected. A review of the record indicates that the chancellor, in denying Frisby's motion for reconsideration, explained the handwritten contracts were admitted into evidence pursuant to Mississippi Rule of Evidence 1004(c). We find this determination was not manifest error.

P8. Frisby correctly asserts that pursuant to Rule 1002, known as the best-evidence rule, an original writing is generally required to prove its contents. Further, pursuant to Rule 1003, Frisby correctly asserts that a duplicate cannot be admitted when a genuine issue has been raised about the original's authenticity. However, an exception to the best-evidence rule exists when the party against whom the original would be offered had control of the original, received notice that the original would be subject to proof at trial, and failed to produce the original at trial. M.R.E. 1004(c).

P9. In the present case, Warden introduced duplicates of three handwritten documents into evidence in support of his complaint for specific performance. Frisby initially objected to their introduction, but allowed them to be introduced "for the purpose of this hearing," while still contesting their authenticity. Thus, the hearing proceeded to determine the authenticity of the alleged contracts, with both parties presenting multiple witnesses.

P10. During the hearing, Frisby testified as an adverse witness and explained that he had never seen the three alleged contracts before and that Warden had never been to his office. However, Warden testified that he drafted all three of the handwritten documents "so that [he] could have some kind of documentation on a deal [they] had on the house." Further, Warden testified that while he originally had the original documents, he met Frisby at Frisby's office, where Frisby made copies of the documents and retained the originals, and gave Warden copies. Michael Neill, the previous owner of the property, also testified for Warden. He testified that he deeded the property to Frisby in 2010, and that when he spoke with Warden in 2011 or early 2012, Warden said he was buying the house from Frisby and "doing odd jobs" to pay off the house. Further, Neill testified that he saw Frisby sign a document in 2014, but that he did not know the document's purpose. Neill later testified that he had overheard Frisby and Warden discussing ownership of the house for labor.

P11. As previously mentioned, the admission or suppression of evidence is within the discretion of the trial judge and will not be reversed absent an abuse of discretion. Tunica Cty., 926 So. 2d at 212 (¶5). Further, "the chancellor sits as the fact finder and is the sole judge of the credibility of a witness when resolving factual disputes." Stokes v. Campbell, 794 So. 2d 1045, 1048 (¶11) (Miss. Ct. App. 2001). As such, it was the chancellor's job as trier of fact to determine which version she found more credible. LeBlanc v. Andrews, 931 So. 2d 683, 689 (¶19) (Miss. Ct. App. 2006). The chancellor, after hearing all the evidence, accepted Warden's testimony as the most credible, admitting the duplicates pursuant to Rule 1004(c). Because there was substantial credible evidence in the record to support the chancellor's finding, this Court must accept them. Accordingly, this issue is without merit.

II. Whether the chancellor incorrectly placed the burden of proof of the fraud allegations on Frisby.

P12. Next, Frisby argues that the chancellor erroneously presumed that the disputed documents were authentic, which incorrectly shifted the burden to Frisby to prove his fraud claim by clear and convincing evidence. However, a review of the record indicates that the chancellor, after hearing evidence by both parties, found that the contracts were valid and enforceable and that Frisby failed to establish that the signatures were a product of fraud.

...

P15. Further, prior to her discussion of the contracts' validity, the chancellor explained that Frisby's "expert's opinion [did] not leave the court with a firm conviction that the signatures were forged by the plaintiff." Moreover, the chancellor explained that testimony from two witnesses, Neill and Brandon Wooldridge, evidenced the authenticity of the signatures. Therefore, the chancellor found that Frisby had failed to overcome the presumption against fraud. Having reviewed the record, it is clear that the chancellor understood the burdens required by each party. Therefore, this issue is without merit.

...

CONCLUSION

P22. In summation, the chancellor did not commit manifest error in enforcing three contracts for the sale of land and ordering specific performance. Although the authenticity of the contracts was disputed, the chancellor did not err in admitting them into evidence and proceeding with a trial to determine the authenticity of the signatures. Further, the chancellor, after hearing the testimony and evidence from both parties, did not manifestly err in finding that Warden had proven the validity of the contracts. Additionally, because Frisby asserted a counterclaim for fraud, the chancellor did not err in requiring him to prove his case by clear and convincing evidence. Moreover, the chancellor did not commit manifest error in finding that Warden detrimentally relied on Frisby's promise. Therefore, we affirm the chancellor's judgment requiring Frisby to specifically perform on the parties' three contracts and ordering Warden, based on amounts previously paid, to pay Frisby $11,300 as full and final payment of purchase price of the property.

P23. AFFIRMED.

Questions:

1. Explain the dispute between the parties and why documents are essential to its resolution?
2. What Rule does the court rely upon to evaluate the "copy" rather than original issue which is the preferred evidentiary form?
3. Is there an exception to the rule when an original may not be required?

Locate the following case using your favorite legal research resource, read and answer the questions that follow.

Maley v. Shell W. Exploration & Prod., LP, 203 A.3d 326, J-A11021-18 (Pa. Super. Ct. Dec. 18, 2018)

Questions:

1. What role do documentary evidence forms play in this case?
2. Why did the Plaintiff's, the Maley's, argue that the Best Evidence Rule needs to be applied? Why are they so insistent that the original be needed and the copy or duplicate excluded?
3. What does the Best Evidence Rule seek to inhibit or prevent? And in this case, does the expert document examiner reach that conclusion?

Even a formal, recorded and filed document relative to land ownership or realty can be a fraudulent document. In McNeff, the plaintiff's allege forged signatures on an original deed despite the acknowledgement by a Notary and the regularity of the document itself.

McNeff v. Cerretani
Massachusetts Land Court September 12, 2018, Decided
16 MISC 000327

Plaintiff Ann C. McNeff, as Personal Representative of the Estate of Ralph Thomas Cerretani (Plaintiff), initiated this action on June 15, 2016, seeking to set aside a deed (Deed) purportedly signed by the father of both parties, Ralph Thomas Cerretani (Ralph). The Deed, dated November 14, 2014, described registered property located at 65 Forest Park Avenue in Billerica (Property), purporting to convey it from Ralph to his son (and Plaintiff's brother), Defendant Thomas R. Cerretani (Defendant). Plaintiff claims Defendant forged Ralph's signature on the Deed.

Two days of trial were held in February, 2018. Plaintiff Ann McNeff; Jules McNeff, her spouse; Eileen Page, a handwriting expert; and Claude Turcotte, a New Hampshire notary public, were called by Plaintiff to testify. Defendant Thomas Cerretani; Peter Cerretani, the parties' brother; and Charles P. Antonelli, Jr., a Massachusetts notary public, were called to testify for Defendant. Twenty-three exhibits were entered in evidence, some with multiple parts, including telephone records detailing the parties' cell phone usage on relevant days. The defense orally moved for an involuntary dismissal at the close of Plaintiff's case, which this court denied.

The facts set forth below include the joint material facts from this court's July 13, 2017 order denying summary judgment, the parties' stipulated facts from their joint pre-trial memorandum, facts established at trial through testimony and documentary evidence, and the reasonable inferences drawn therefrom.

1. The Property is registered land known as and numbered 65 Forest Park in Billerica, which is the current residence of Defendant, and Defendant is the owner under Certificate of Title 42396 that issued by the Middlesex North Registry District of the Land Court as a result of the filing of the disputed Deed.

2. The parties' father, Ralph Thomas Cerretani, died on December 9, 2014. He resided at the Property during his lifetime.

3. The parties are brother and sister, and are two of three living children of Ralph. The third sibling is Peter Cerretani. Prior to their father's death, Plaintiff and Defendant had a close relationship with one another.

4. On October 31, 1997, Ralph executed a Last Will and Testament (Will), appointing Plaintiff as Executrix of his estate, leaving "[a]ll of [his] real estate, real, personal or mixed" to "[his] daughter, Ann McNeff, [his] son, Peter Cerretani, [his] son Thomas Cerretani, and [his] son-in-law Jules McNeff."

5. Ralph executed three separate codicils to his Will on February 17, 2009, January 28, 2010, and February 9, 2010, (together, Codicils). The Codicils did not in any way alter or amend the devisees of the Will, or the disposition of the Property. One of the Codicils provided that if Peter Cerretani had not yet repaid his sister, Plaintiff, the sum of forty-thousand dollars lent to Peter by Plaintiff, such sum would be deducted from Peter's share of the estate and given instead to Plaintiff.

...

7. At some point during the final seven or eight years of Ralph's life, Defendant moved into the Property to live with Ralph, who was then approximately eighty-seven years old. Defendant intended to help care for him, pay his bills, and attend Ralph's medical appointments.

…

RALPH'S WILL AND SUBSEQUENT ACTIONS TAKEN BY DEFENDANT

…

13. In July of 2015, Defendant again asked Turcotte to notarize a document. This time it was a deed purportedly signed by Ralph before his death (on November 14, 2014.) Turcotte was unwilling to notarize that document, … Defendant returned with a new original deed that allegedly had his father's signature on the correct line.

14. Mr. Turcotte refused to notarize that deed, and told Defendant he could not because he could not notarize a deed after someone's death.

15. Eventually, the Deed which is the subject of this lawsuit was notarized by a different notary, as set forth below.

…

18. At no point prior to Ralph's death did Defendant tell Plaintiff the house had been transferred into his name despite numerous opportunities to do so. …

24. On July 17th, 2015, at Defendant's request, Attorney Mike Newhouse prepared and executed an "affidavit of no divorce" under G. L. c. 183, § 5B, stating that, at the time of the death of Bertha F. Cerretani (Ralph's wife and the parties' mother) on August 18, 1997, she had not divorced or dissolved her marriage to Ralph T. Cerretani, her surviving spouse. Defendant paid Attorney Newhouse to prepare this document for him.

25. The affidavit was filed with the Registry together with the Deed, which was dated November 14, 2014, a date before Ralph died.

26. The Deed purporting to convey the Property from Ralph to Defendant for consideration in the amount of ten dollars, was likely not signed by anyone on November 14, 2014. It was not registered with the Middlesex North Land Court Registry District until July 20, 2015, in Book 215, at Page 11, and Certificate No. 42396 issued in Defendant's name.

27. Plaintiff first learned of the existence of the Deed when she received the original in the mail from the Registry District sometime in the summer of 2015.

NOTARIZATION OF THE DEED

…

32. On the morning of December 8, 2016, Defendant texted Claude Turcotte the following message: "I am so sorry my sister dragged you into this. I will pay you for your time and your attorney fees. My attorney has the copy of your paperwork that you notarized for me—for my dad. Tell them that is the only paperwork you saw from me and that should be the end of it. You're a good friend. Thank you."

HANDWRITING ANALYSIS CORROBORATES TESTIMONY AND DOCUMENTARY EVIDENCE

33. Eileen Page is a handwriting consultant, trained to testify as an expert in areas of handwriting analysis, and she was qualified as an expert in this matter.

34. As part of her analysis, Ms. Page reviewed known exemplars of signatures of Plaintiff, Defendant and Ralph. She reviewed, among other things, the following documents:

a. Numerous personal checks signed by Ralph;
b. Numerous personal checks signed by Thomas in his father's name, and loan documents;

c. Ralph's Will and the Codicils;
d. Two powers of attorney signed by Ralph; and
e. Court documents, personal checks and certified mail receipts signed by Thomas Cerretani in his own name; and
f. Loan documents, business credit applications, and sales contracts.

35. Ms. Page also examined documents in which the signatories were not known or confirmed, such as the signature on the disputed Deed.

36. Ms. Page incorporated several techniques into her analysis of the above documents. She specifically focused on the formation of the capital letter "T," the middle initial in Ralph's signature; the formation and placement of the capital "C" in Ralph's last name; the formation and place of the last three letters in Ralph's first name ("lph"); letter idiosyncrasies; letter proportions; letter slant; and pressure patterns. Her testimony was aided by enlarged copies of the signatures, used as chalks.

37. Ms. Page focused particularly on letter slant. This required drawing lines over a signature on the upstroke of each letter, from the baseline to the point where the letter changes direction.

38. Ralph's known signatures have letter slants that are mostly forward lines that are parallel and do not intersect. The questioned signatures have intersecting and crossing letter slant lines throughout.

39. Ms. Page concluded the questioned characteristics of the signature on the Deed did not match the known signatures of Ralph, and stated in her expert opinion the Deed was not signed by Ralph.

* * *

Plaintiff alleges the Deed is a forgery, and that Defendant is the one who actually signed it following their father's death. She argues that Defendant first contacted a different notary, Claude Turcotte (Turcotte), for notarization of the Deed following Ralph's death, that Turcotte refused because Ralph was already deceased, and Defendant then approached a different notary, Charles P. Antonelli, Jr. (Antonelli). Plaintiff claims she is familiar with Ralph's signature and does not recognize the signature on the Deed as his. Plaintiff also retained a handwriting expert, Eileen Page (Page), to review and analyze the signatures of various parties involved in this action, and Page concluded that the signature on the Deed did not match that of Ralph, but was signed by Defendant. Defendant argues he is also familiar with Ralph's signature, the signature on the Deed is authentic, and the Deed was properly notarized by Antonelli.

A forged conveyancing instrument is void, and has no effect even if recorded in regular fashion with the registry of deeds. Strother v. Shain, 322 Mass. 435, 437, 78 N.E.2d 495 (1948); Breed v. Gardner, 187.

Mass. 300, 304, 72 N.E. 983 (1904). A notary certificate on a deed, however, is proof presumptive of a valid acknowledgment, unless there is evidence that rebuts or impeaches the notary's testimony. Keville v. McKeever, 42 Mass. App. Ct. 140, 157, 675 N.E.2d 417 (1997); see also McOuatt v. McOuatt, 320 Mass. 410, 413, 69 N.E.2d 806 (1946) (stating "[t]he certificate of acknowledgement furnishes formal proof of the authenticity of the execution of the instrument when presented for recording"). As the party claiming Ralph's signature on the Deed is a forgery, Plaintiff carries the burden of rebutting the presumption of a valid acknowledgment of Ralph's signature on the Deed.

In the instant action, the key witnesses who testified at trial ion Defendant's side, specifically Defendant and Mr. Antonelli, displayed inconsistencies in their testimony, which, on many points, was simply not credible. These credibility issues and inconsistencies, combined with the expert testimony offered by Plaintiff's handwriting expert, have convinced this court that the signature on the Deed does not belong to Ralph. The court finds Plaintiff has presented sufficient clear and convincing evidence rebutting the presumption arising from the signature on the Deed and the the notary certificate. Accordingly, the Deed is void.

...

III. PLAINTIFF'S EXPERT PERSUASIVELY TESTIFIED THE SIGNATURE ON THE DEED DID NOT MATCH RALPH'S KNOWN SIGNATURES, CORROBORATING THE OTHER TESTIMONY

Plaintiff called Ms. Page as an expert witness in handwriting analysis. She reviewed dozens of documents containing Ralph's confirmed signatures, as well as several documents in which Ralph's signature was questioned. Defendant objected to Ms. Page's qualification as an expert witness, arguing Plaintiff failed to demonstrate her testimony met the standard required under Daubert v. Merrell Dow Pharms., Inc., 509 U.S. 579, 113 S. Ct. 2786, 125 L. Ed. 2d 469 (1993) (and, by extension, Commonwealth v. Lanigan, 419 Mass. 15, 641 N.E.2d 1342 (1994)); see Tr. vol. 1, 129:19-25. Massachusetts, however, has long admitted in evidence handwriting comparisons by a qualified expert. ... Additionally, handwriting comparisons are generally considered sufficiently established that any preliminary screening is unnecessary. Commonwealth v. O'Connell, 438 Mass. 658, 662, 783 N.E.2d 417 (2003); Commonwealth v. Murphy, 59 Mass. App. Ct. 571, 576, 797 N.E.2d 394 (2003). After argument regarding Ms. Page's qualifications, this court qualified Ms. Page to testify regarding the handwriting exemplars.

As part of her analysis, Ms. Page reviewed several signatures from Plaintiff, Defendant and Ralph, including numerous originals and copies of checks, letters, loan documents, the Will and Codicils, the Deed, and the durable powers of attorney, and the Authorization. She reviewed signatures purporting to be Ralph's on loan documents for Ally Bank, which Defendant acknowledged he had signed for his father, at his father's direction. She incorporated several different techniques in her analysis, and ultimately opined, among other things, that the letter slant of the questioned signature on the Deed did not match the known signatures of Ralph. Her testimony was compelling, aided by the enlargements of many signatures which she used as chalks.

The court notes that Ms. Page determined the signatures on the Ally Bank loan documents (Exhibit 15) were not signed by Ralph prior to Defendant's confirmation that he in fact had signed them for his father. As Plaintiff pointed out in post-trial briefing, the fact that Ms. Page had an exemplar of Thomas signing Ralph's name was helpful. It created a solid point of comparison from which to [*460] proceed when analyzing other like signatures, including those on the January 1, 2014 Authorization purportedly signed by Ralph in favor of Thomas (allowing him to act on Ralph's behalf regarding certain business matters and use of his credit cards) (Ex. 7), and the Deed itself, along with other documents.

IV. CONCLUSION

After careful examination of the evidence and assessment of the credibility of each witness, and drawing inferences from the testimony and documentary evidence, this court concludes Ralph's signature on the Deed, purportedly signed on November 14, 2014, is a forgery. This conclusion is based in large measure on this court's assessment of the credibility of the witnesses. The testimony of both Defendant and Mr. Antonelli does not hold up and strains reason and credibility in light of the documentary and credible testimonial evidence establishing, among other details: 1) Defendant's failure to advise Plaintiff that the Property had been gifted to him by Ralph in November 2014 or otherwise take any action to assert an ownership claim to the Property, particularly given the frequency of their communication; 2) Defendant's failed attempts to secure a notarization of Ralph's signature on the disputed Deed from Mr. Turcotte after Ralph had died; and 3) the Commonwealth's notary public records and the business records of the Corporate

Connection, from where Mr. Antonelli obtained the tools of the notary trade which were used in connection with the notarization of the signature on the disputed Deed.

On the other side, this court credits the testimony of Plaintiff and also Mr. Turcotte, bolstered by the expert opinion testimony of Ellen Page that the signature on the Deed is not that of Ralph T. Cerretani. Accordingly, this court holds that signature of Ralph T. Cerretani on the Deed was forged and the Deed therefore is a nullity.

Judgment to issue accordingly.

...

ADJUDGED and ORDERED the deed dated November 14, 2014, by which Ralph T. Cerretani purportedly conveyed the Property to Thomas R. Cerretani, filed with the Middlesex North Registry District of the Land Court on July 20, 2015, as Document 283094, was forged and therefore a nullity; it is further

ADJUDGED and ORDERED that Certificate of Title 42396, in Book 215, at Page 11, standing in the name of Thomas R. Cerretani is a nullity; and it is further

ORDERED a copy of this Judgment shall be filed with the Assistant Recorder of the Middlesex North Registry District. A notation of this Judgment shall be made on Certificate of Title 43296 and it will be cancelled and the former Certificate of Title 16594 shall be revived.

Questions:

1. In a case of a fraudulent Deed, the document may be the original but the signatories fraudulent. Here the document analysis must go beyond the original to view similar signatures or the lack thereof on other documents. What types of secondary documents does the challenger to the signature draw comparisons from?
2. What impact does a fraudulent signature have upon the legitimacy of an already filed and recorded Deed?
3. What is the Court's final order regarding this Deed?

Read the DOJ press release regarding the use of handwriting in tax fraud case at https://www.justice.gov/opa/pr/chicago-area-return-preparer-barred-preparing-federal-tax-returns

Commonwealth v. Holland
193 A.3d 1105
Superior Court of Pennsylvania
June 26, 2018, Decided; June 26, 2018, Filed
No. 11 EDA 2017

Appellant, Dewayne Holland, appeals from the judgment of sentence imposed following his bench conviction of attempting to acquire a controlled substance by misrepresentation, fraud, forgery, deception or subterfuge. We affirm.

We take the following pertinent facts and procedural background from the trial court's May 16, 2017 opinion and our independent review of the record. Appellant was arrested on November 17, 2014 after presenting an altered prescription for oxycodone tablets at Walgreens Pharmacy. On April 26, 2016, Appellant waived his right to a jury trial. A bench trial commenced that day.

Amanda Dinmore, Walgreen's pharmacy manager, testified that Appellant presented a prescription for oxycodone. The prescription appeared to be altered, from a quantity of twenty-one to eighty pills. When the Commonwealth showed her a copy, made by the pharmacy, of the original prescription, defense counsel objected, stating, "There's a genuine question here that we're raising about the authenticity of that photocopy." (N.T. Trial, 4/26/16, at 13); (see id. at 12). When questioned by the trial court about the location of the original, Ms. Dinmore stated that she could not remember whether it was in her file, or if she gave it to the police. (See id. at 13-14). When the prosecutor observed that defense counsel was only raising a general objection, but not identifying any actual discrepancies between the original and the copy, defense counsel stated that that was a question of fact for the trial court. (See id. at 15). The court denied counsel's objection.

Thereafter, Ms. Dinmore testified that, when Appellant presented her with the oxycodone prescription, it appeared that the quantity was altered. (See id. at 17). Specifically, she testified that "[t]he quantity said 80 in numbers 8 zero, but underneath in Roman Numerals it said XXI[,] which is 21. So I took a closer look and I could see that it looks like it was 21 and then written over 80." (Id.). On cross-examination, defense counsel presented her with a carbon copy of the prescription from the prescribing doctor's pad. When asked why the "1" on the photocopy of the original given to her, and on the carbon copy from the doctor's pad, appeared to be in slightly different places, Ms. Dinmore explained that the "carbon copy could have slid a little" while the doctor was writing the prescription, causing the number to shift. (Id. at 27). She affirmed that she had not altered the prescription, and that the photocopy was the same as the original she observed on November 17, 2014. (See id. at 26-27).

Dr. Brent Weinerman testified that he wrote the prescription for twenty-one oxycodone pills for the treatment of Appellant's pain. (See id. at 30-31, 35). He stated that he wrote the number twenty-one, and then XXI, the Roman numeral representing the number. (See id. at 34). He explained that he would never prescribe eighty oxycodone pills to anyone, and had not written the number eighty on Appellant's prescription. (See id. at 31-32, 35). When Appellant's counsel asked him about an alleged loop on the "O" on the carbon copy that was not on the photocopy, Dr. Weinerman said, "[T]here's not really a loop. It looks like something was impressed over it[.] [I]t's an 'O'." (Id. at 38).

Appellant testified on his own behalf. He denied altering the prescription. He maintained that the copy provided in evidence by the pharmacy was different from the original he had given them to be filled. (See id. at 46-47). Specifically, the following exchange occurred:

Q. You reviewed copies of the prescriptions that were contained in the discovery?

A. Yes.

Q. Did you notice anything in comparing those prescriptions[,] did you notice anything of concern to you?

A. Yes.

Q. Tell us what that was?

A. On them scripts, okay, is different handwritings.

Q. How do you know?

A. I could see it. I could see it. The "X" is small, on my script, but big on his script. The "O" is cursive on his fax cover sheet, and on his carbon copy that he says is a carbon copy is different. It says ... 11/11/14. But on my script it looks like 11/12/14. It's right here because the police even wrote on their copy that my script says 12, it says it in the paperwork that my script says 12, but on here it says 11. So I reviewed everything. The "T" right here, the "T" right here is crossed but on his it's not crossed. At the top it got an [sic] DEA Roman numerals.

(Id. at 46).

At the conclusion of the bench trial, the trial court convicted Appellant of the aforementioned crime. On November 18, 2016, it sentenced him to a term of incarceration of not less than three nor more than six years. The court denied Appellant's timely post- sentence motion on November 28, 2016. Appellant timely appealed.

Appellant raises one question for this Court's review: "Whether the [c]ourt erred when it overruled a defense objection pursuant to Pa. Rules of Evidence 1002 and 1003 to the admission of written evidence?" (Appellant's Brief, at 6). Specifically, Appellant maintains that, because "[t]he contents of the original prescription form constituted one of the elements of the offense[,]" and he "needed the original prescription form to prove during his testimony that the original and the photocopies contained discrepancies[,]" the trial court abused its discretion in allowing the admission of the photocopy. (Id. at 14). We disagree.

The following legal principles guide our analysis of Appellant's issue.

In reviewing a trial court's ruling on the admissibility of evidence, our standard of review is one of deference. It is firmly established that [q]uestions concerning the admissibility of evidence lie within the sound discretion of the trial court, and [a reviewing court] will not reverse the court's decision on such a question absent a clear abuse of discretion. An abuse of discretion requires:

> not merely an error of judgment, but where the judgment is manifestly unreasonable or where the law is not applied or where the record shows that the action is a result of partiality, prejudice, bias or ill will.

Commonwealth v. Giles, 2018 PA Super 63, 182 A.3d 460, 461-62 (Pa. Super. 2018) (citations and quotation marks omitted).

> [The common-law best evidence] rule is codified in Pennsylvania Rule of Evidence 1002, which provides: "An original writing, recording, or photograph is required in order to prove its content unless these rules, other rules prescribed by the Supreme Court, or a statute provides otherwise." Pa.R.E. 1002. Courts apply the best-evidence rule when the contents of documentary evidence are at issue—that is, if the terms of a writing must be proven to make a case or provide a defense. Thus, Rule 1002 requires that an original writing, recording, or photograph be introduced at trial only if the proponent must prove the contents of the writing, recording, or photograph to prove the elements of its case.

Commonwealth v. Ribot, 2017 PA Super 262, 169 A.3d 64, 67 (Pa. Super. 2017) (citations, emphasis, and quotation marks omitted). However, Rule 1003 provides, "[a] duplicate is admissible to the same extent as the original unless a genuine question is raised about the original's authenticity or the circumstances make it unfair to admit the duplicate." Pa.R.E. 1003 (emphasis added). The Comment to Rule 1003 observes that:

> Under the traditional best evidence rule, copies of documents were not routinely admissible. This view dated back to the time when copies were made by hand copying and were therefore subject to inaccuracy. On the other hand, Pennsylvania courts have admitted copies made by techniques that are more likely to produce accurate copies. For example, when a writing is produced in duplicate or multiplicate each of the copies is treated as admissible for purposes of the best evidence rule.
>
> * * *
>
> … As a result, Pa.R.E. 1003 should tend to eliminate purely technical objections and unnecessary delay. In those cases where the opposing party raises a genuine question as to authenticity or the fairness of using a duplicate, the trial court may require the production of the original under this rule.

Pa.R.E. 1003, Comment (case citations omitted).

> Here, the trial court found that the "minor little differences" about which Appellant testified did not render the two copies of the prescription "totally different[,]" as alleged by Appellant. (Trial Court Opinion, 5/16/17, at 3); (N.T. Trial, at 47). It explained that these inconsistencies are more readily explained by the pharmacist's account of the simple movements of the original or the carbon copy[.] … [T]hey in no way

> posed any genuine questions as to the authenticity and accuracy of the copies, particularly since none of those variations had anything to do with the obvious and undisputed fact that there was an attempt to replace the 21 with an 80. The defense did not dispute that the original prescription contained the 80 when it was presented to the pharmacist[,] but did not when it was written by the doctor and, presumably, given to [Appellant].
>
> (Trial Ct. Op., at 3-4)

We agree with the sound reasoning of the trial court that Appellant failed to raise a genuine question about the authenticity of the photocopy to which Ms. Dinsmore testified. Appellant complained about minor inconsistencies between the photocopy of the original and the carbon copy, which were explained by the copying process. Even more importantly, the alleged discrepancies did not implicate the central issue of the case, whether the prescription Appellant provided to the pharmacist was altered to include an order for eighty oxycodone pills. See Pa.R.E. 1003. Therefore, we conclude that the court did not abuse its discretion when it allowed the admission of the photocopy into evidence. See Giles, supra at *2; Pa.R.E. 1003. Appellant's issue lacks merit.

Judgment of sentence affirmed. Judgment Entered.

Questions:

1. What document is in question? And is that document being used offensively or defensively?
2. Why does the argument for the Best Evidence Rule application not work in this casel?

7.2 DOCUMENTARY EVIDENCE: ALTERATION AND MODIFICATION

The sophistication of fraudsters and forgers has been on the dramatic uprise to the type of documents that are covered under the Rules, as well as new and emerging technical capacities to almost mimic the originals. The range of documents and dishonest tactics applied to them can be quite startling. Consider the unilateral lease alteration that follows.

Rufrano v Yovino
110 N.Y.S.3d 502
Civil Court of the City of New York, Kings County August 23, 2018, Decided
051182-KLT -2015

At issue is whether respondent Michael Yovino ("respondent," "tenant," or "Yovino") subsequently added material terms to a lease that had been signed by petitioner landlord Dr. Frederick Rufrano ("landlord" or "Rufrano"), and whether the three initials of "F.R." (the alleged initials of Rufrano) affixed to these changes were forged. Rufrano claims that he never agreed to the additional material terms of the lease and never affixed his initials to those terms. Yovino claims that Rufrano agreed to these terms during a "second seating" of the lease where he affixed his initials to those terms. As will be described below, the court credits the testimony of Rufrano and his expert that the initials were forged and that Rufrano never agreed to the additional material terms. Since Rufrano never accepted the counteroffer, there was no meeting of the minds as to the final lease agreement and it was not valid or enforceable.

...

Rufrano wanted to establish a term of years lease with the social club after an elderly gentlemen from club fell on the property and sued for damages, resulting in Rufrano losing his insurance coverage. Dr. Rufrano testified that the only way his new insurance company agreed to cover him was to ensure that the tenants carried insurance on the one small commercial space of the building and that Rufrano be named as the secondary insuree under the policy. To that end, Rufrano wrote in all large block lettering on the lease including the lease term of seven years on the first page, and paragraph 28 on the third page which addressed the insurance issue. Rufrano also agreed to Yovino's request to add a provision concerning damages and changing light bulbs, which Yovino wrote and Rufrano initialed on the middle of the first page. This initial is the only one that Rufrano claims he put on the lease. This agreed upon provision and all the other small block writing on the lease were put in by Yovino.

Rufrano recalls writing his signature on the lease twice in front of Yovino on or about December 13, 2013. He did not take the original when he left the premises because the copying machine was broken and Yovino wanted a copy, so he told Yovino to mail him the original with the next rent check. Rufrano explained that he signed the lease twice simply because "Mr. Yovino asked me to sign twice I don't know why and I just did." (2/11 at 39). At the time Rufrano signed the last page, there was no handwritten material around the signature line and he does not think that Yovino signed his name to the lease at that time. The court credits Rufrano's testimony that he never received a copy of the original lease from Yovino until 11 months passed and his attorney served Yovino with a notice of termination. This was the first time that Rufrano became aware that Yovino had altered the original agreement without his consent. Rufrano contends that other than the initials "F.R." that he placed next to Yovino's writing that the "landlord is responsible for all electrical plumbing and changing light bulbs," the other three initials "F.R." placed on the first page of the lease as well as the small handwritten terms were forgeries. Rufrano claims that the first forged initial was next to the name Michael Yovino, and that the second forged initial was next to the small print handwritten term that "Rent will be paid in two parts on third and second portion of rent on the 18th of the month," and that he never agreed to this provision, although he did accept the rent in two $500 parcels. (2/11 at 34-35). The third claimed forged initial was underneath the term "social club" (which Rufrano originally wrote) and by the small print handwritten terms "commercial use & business block 2720 Lot 1, lot 2;" and "connected to side building, usage of yard (lot 2) and 500 Graham Avenue entrance (Access to Yard lot 2)," all of which he never agreed to.

Rufrano testified that he wrote in paragraph 28 on the third page of the lease agreement pertaining to the insurance clause, but did not write in or agree to the clause "For 498 Graham Avenue, Brooklyn … RE: seven year lease for storefront Brooklyn (ground floor and lot 2 yard)" next to paragraph 28. Rufrano explained that it would have been ludicrous for him to agree to this clause since both the storefront and lot two were zoned residential, and permitting commercial use to part of lot 1 and all of lot 2 would merely increase his insurance premiums and expose him to greater insurance liability and thus undermine his interests.

Contrary to Rufrano's testimony, Yovino claims that the lease was signed on December 1, 2013, and that it gave him access to the entire first floor, including the back room, at 498 Graham Street and the commercial lot yard at 500 Graham. Yovino testified that the lease itself was signed in two separate sittings; the first occurred at approximately 12 pm. Thereafter, Rufrano left to collect rent and returned for the second sitting approximately three hours later. However, Yovino's explanations as to what was added into the lease and when it was added are incomprehensible. He first claims that at the conclusion of the first sitting, Rufrano signed and initialed the lease twice, and told Yovino that he could add stipulations. However, Yovino was unclear as to which two initials Rufrano affixed during the first sitting and which he affixed during the second sitting. Nor was he clear as to whether he handwrote the provisions about having access to the whole first floor as well as to the commercial space in the lot during the three hours when Rufrano was not in 498 Graham and supposedly collecting rent or during the second sitting.

…

This finding is confirmed by the testimony of the expert witness presented by the petitioner - Mr. Picciochi. Picciochi testified that a "questioned document examiner" is a forensic scientist who specializes in examination of disputed handwriting and suspect documents. He has worked in that field for over 30 years. He first worked in the NYPD Crime and Laboratory Forensic Investigation division for 20 years and since then has been self-employed. He earned a Master of Science with a concentration in Forensic Science and then underwent three years of extensive training with the Police Department, and supplemented this training through FBI and Secret Service programs. In addition, Picciochi taught forensic science at Hofstra University, John Jay College, and City University of New York. He is a member of the American Academy of Forensic Sciences, the Midwest Assn. of Forensic Scientists, and three other societies. He is also certified by the American Board of Forensic Document Examiners in the field of questioned document examination; certification entails written, practical and oral testing accompanied by continued recertification. Picciochi testified that he has been retained as an expert on handwriting and expert in handwriting on over 100 other cases in various courts.

Defendant's expert, Mr. Curtis Baggett, identifies himself as a "forensic document examiner handwriting expert" (10/27/16 at 67) and claims to have testified in every state in the country in both state and federal court. In addition, he lectures at the "School of Forensic Document Examination at Handwriting University" which is owned by his son. Mr. Baggett studied for two years in Dallas under Dr. Ray Walker to obtain a certificate in document examination; this is his only formal training in this area. However, it is unrefuted that Ray Walker was also disqualified as an expert. Baggett supplements his CV with book publications and appearances on television court programs. While Mr Baggett claims to have testified in both state and federal court, five federal courts found him unqualified because his methodology could not be substantiated On cross examination, upon being presented with these decisions, he responded that they were "a lie" as he was not paid or retained to appear in them. (10/17 at 73). In addition, Mr. Baggett has not been certified by any of the 20 organizations in his field, as noted by a federal district court in Georgia as one of the reason it found his credentials to be paltry in comparison to the other expert. Baggett was dismissive and argumentative when questioned as to whether the American Board of Forensic Document Examiners was the gold standard for handwriting experts in the U.S., stating that it was not for the "6900 people who are not members ..." (Tr. 10/17 at 69).He also did not refute assertions made by counsel for petitioner during cross-examination, that his work has never been peer reviewed and that he has never published in a scientific journal, although he does not think that any of these factors would disqualify him from being a member of any of the recognized document examiner associations.

This court understands that years of experience can be sufficient to qualify an individual as an expert and so qualified Baggett for that reason. However, in assessing how much weight and credibility to give to the testimony of an expert, one must also consider the expert's credentials in his field which is measured by membership in and adherence to guidelines proscribed by respected professional associations. This court finds that Mr. Picciochi 's background and credentials as a "Questioned Document Examiner" are much more substantive and venerable than those of Mr. Baggett. While Picciochi was a member of the "Gold Standard" association, had years of forensic experience in governmental agencies, and kept abreast of the latest trends in the area of expertise, Mr. Baggett seems to have had little formal training and exhibited a cavalier and dismissive attitude towards any and all professional associations in his field.

Furthermore, this Court finds that the methodology and analysis used by Picciochi was much more reliable than that employed by Baggett. Picciochi testified that handwriting identification requires examination of individualistic characteristics of the writing through identifying marks and properties in letter designs, formations, and analysis of the relationship between one letter and another in the movement or manner of execution. This combination of characteristics, which recurs in the writing of any individual, allows document examiners to distinguish between the handwriting of different people. There are three phases to this identification: Mr. Picciochi starts with an independent analysis of the questioned writing by looking for individual characteristics, then separately examines other known writing exemplars, and then compares them side by side to find significant similarities or differences between the two as well as for any limitations in the analysis.

Here, Picciochi first surveyed the original lease and conducted a thorough examination of the four initials in dispute. He then studied Rufrano's "known writings" which consisted of 41 samples of his initials, some of which were samples witnessed and notarized by Rufrano's attorney, such as check ledgers, canceled checks and lease agreements, and finally compared the four initials in question with the known writings. Mr. Picciochi prepared a series of charts to demonstrate the different features between the first three "F.R" initials on the lease which he denominated (Q1, Q2, Q3) and the last one (Q4). He first studied the movements of the characters in the four initials by using a microscope to look at the ink lines of the initials. He then compared the individualized characteristics of the four initials on the lease with the initials from the known samples (denominated K-34), side by side, looking at the style of writing (script versus print), direction and sequence of strokes, manner of the connection between the lines, the curvature of the horizontal strokes in the F, and the balance of the cross stroke of the F (3/11 at 8). He testified with a "high degree of confidence" that initials Q 1-3 were "unlikely written by the true signatory" but that Q4 was written by Rufrano (3/11 at 28).

This conclusion was based upon the "fundamental differences in the individualizing characteristics" of the initials Q1-3 and no significant similarities to the compared initials. (2/11 at 94). Picciochi used a procedure entitled Scientific Working Group for Document Examiners SWGDOC (3/11 at 14) in comparing the initials on the lease to Rufrano's initials on other documents so that the initials could be viewed side to side. Specifically, Picciochi found that the letter "F" in Q1, 2, and 3 were all in print form, whereas Q 4 was cursive, as was the case of all the known initials set forth in K-34 (3/11 at 22-24). He then looked at the formation of the staff (vertical stroke of the letter F) and cap (top horizontal stroke of the letter F). Picciochi found that in Q 1-3 that the first stroke was the staff written from top to bottom, followed by the cap and then the cross stroke, whereas in Q 4 and K34 the writer's first stroke was the cap going from left to right with a small loop and the second stroke went downward forming the staff (3/11 at 18). He again concluded that horizontal stroke and shape of the initials in Q1, 2, and 3 were different from the initials in Q4 and the known exemplars. (3/11 at 24). Picciochi then looked at the connecting stroke between the F and the R. While in Q 1-3, the middle horizontal stroke went from left to right and then downward to form the staff of the R, in Q-4 and K-34, the middle stroke of the F went from left to right upward and then downward to form the staff of the R (3/11 at 21). He also noted that the horizontal stroke of the F was basically straight in Q1-3 whereas in Q-4 and K 34 there was a "very curved like formation to both the cap and the horizontal stroke of the F." (2/11 at 113, 119)(Pl. Ech."8"). Finally, he found that in Q 1-3 the cross stroke was longer on the rights side and shorter on the left side of the staff, whereas in Q-4, there was a "balance," meaning that the cross stroke was about even on both sides of the vertical portion of the staff (Tr. 2/11 at 118-19). Based on all of these differences, Picciochi concluded "to a high degree of probability" that Q1-3 were written by someone other than Rufrano. He also claims that number of compared initials (40) was sufficient in reaching his conclusion.

Baggett testified that the four initials on the lease were almost certainly written by Rufrano. Baggett reached this conclusion, as set forth in his March 2016 letter to Yovino, by viewing a copy of the lease agreement and comparing it to four "known writings" of Rufrano which consisted of copies of four cancelled checks that Dr. Rufrano has signed on the back (7/31/2017 at 6). Baggett enlarged the questioned and known documents by 200% with a copy machine instead of putting each document under a microscope. He used a ruler for measuring the spacing, height and width of the letters. Unlike Mr. Picciochi, Baggett does not believe that a microscope should be used as a primary tool for document examiners; in most cases enlarging the image 200% is adequate to make an accurate determination, because using a microscope is "very tedious" (Tr. 7/31/2017 at 7). Baggett also compared the initials on the front page of the lease to Rufrano's printed handwriting on paragraph 28 of the lease to "match" the initials (7/31/2017 at 33). He reasoned that the four exemplars of Rufrano's signatures were enough to make a determination because the questioned initials match "almost perfectly" to paragraph 28 (Rufrano's confirmed writing). However, he did admit during cross examination that he usually prefers to have between six to ten exemplars. When confronted with more examples of Rufrano's initials, Baggett maintained that this only verified his conclusion (7/31/2017 at 30), that the four exemplars were sufficient because he is the "best handwriting examiner in the United States." (7/31/2017 at 36).

Mr. Baggett relied heavily on the book "Handwriting Identification: Facts and Fundamentals" by Huber and Headrick, which states that "whatever features two specimens of handwriting may have in common, they cannot be considered to be of common authorship if they display but one single consistent dissimilarity whose presence is not capable of reasonable explanation." (Tr. 10/27 at 84-87). Yet, in contradiction to these principles, Baggett justified his conclusion that Q 1-4 were the same, despite the differences between these four sets of initials, because in these differences, there are similarities like angles, shape, and form which make the writing verifiable (12/06 at 65-67). In addition, Baggett maintained that it was not unusual for Rufrano to print his initials three times on the lease and only use cursive once for his initials, because most of the writing in the rest of the lease was in print (12/06/16 at 56).

The court credits Picciochi's opinion that comparing initials to initials is the only proper way to conduct such examination. Comparing initials with full written signatures would taint the results of the examination because of the reflective and subconscious nature in which signatures are written (3/11 at 26). Picciochi also disapproved of this comparison because he needs "comparability;" meaning initials to initials or signatures to signatures. He needed to look at relationships between two letters, and the first and last name in a signature are not juxtaposed in the same way that two initials placed side by side would be. He elaborated on cross examination that the capital letters F and R are connected as initials but are not connected in Rufrano's signature which obviously has letters between the F and R. Finally, Mr. Picciochi testified that comparing the questioned initials with only four signatures would not be a sufficient basis for a document examiner to conclude with a "high degree of probability" who the author of the initials was. Looking at 41 known writings, on the other hand, was sufficient to identify both variations in writing which exist for any given individual and consistent writing patterns.

Based upon the above, this Court finds that three of the four initials on the first page of the lease were forged and that the handwriting next to these initials was not Rufrano's. This, combined with the Court's crediting of Rufrano's testimony over Yovino's testimony, leads to the inexorable conclusion that Rufrano never agreed to give Yovino control and use of the yard at 500 Graham Street or of the hallway/bathroom adjacent to the social club, and that he never authorized Yovino to use these spaces for commercial uses.

...

This Court has already found that Yovino made material modifications to the original lease by adding in the terms "commercial and organization lease," and "commercial use & business" under the words "social club;" and "connected to side building, usage of yard and 500 Graham Avenue entrance access to Yard lot 2." This constitutes a counter offer which Rufrano never agreed to or accepted, and never initialed, thus negating the validity or effect of the lease agreement that was signed by Yovino and Rufrano before the changes were made. Since no lease exists, this Court grants petitioner Rufrano's request for a warrant of eviction and judgment of possession. Should there be any extant issues that arise from this Order, the parties are directed to contact the court within 30 days of receipt of this decision.

This constitutes the Decision and Order of the Court (Figure 7.2).

Questions:

1. In what way does the alteration to the lease between landlord and tenant take place?
2. Compare and contrast the document examiner expert for the plaintiff with that of the defendant.

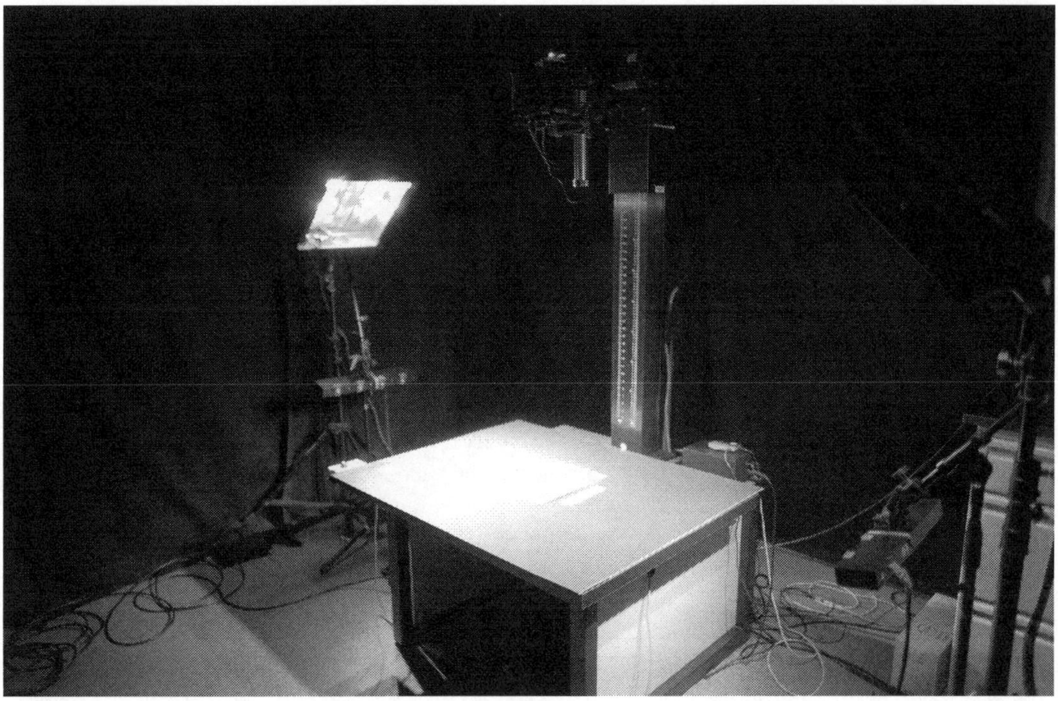

FIGURE 7.2 Forensic testing and analysis of pages from H.R. "Bob" Haldeman papers from Nixon Presidency, [carried out by inter-agency team, under the direction of the National Archives, at the Library of Congress]. Photographer: Earl McDonald—NARA Staff.

3. At times document analysis is dependent on the legal status of the document examined. In this case, was there a legal lease or had it been so changed that there be no contract in a formal sense?

Locate the case below using your favorite legal research resource, read the case and answer the questions below.

United States v. Mallory, 902 F.3d 584, United States Court of Appeals for the Sixth Circuit, August 2, 2018, Nos. 17-3500, 17-3537, 17-3538

Questions:

1. Explain the alteration or intentional modification in this case. What type of document? What type of motive?
2. When analyzing the question of whether the Will's signature was a forgery- it depends on the expert opinion of a handwriting analysis expert. Is it true that the Court admits that handwriting analysis is not "science per se" but the type of technical specialty allowed under Daubert and Kumho?
3. The lower District Court referred to handwriting analysis concerning a signature forgery to not be about "beakers and computers and yardsticks." What does this mean?

Zapata v. State
Court of Appeals of Texas, Third District, Austin
June 19, 2019, Filed
NO. 03-17-00537-CR

Ray Castro Zapata appeals his convictions on two counts of forgery of a financial instrument, one count of theft of property valued at more than $200,000, and one count of money laundering in the amount of $20,000 or more but less than $100,000. See Tex. Penal Code §§ 32.21(d), 31.03(e)(7), 34.02(e)(2). Zapata's offenses involve property belonging to the estate of John Edward Sullivan and began with the forgery of Sullivan's will. A jury assessed Zapata's punishment at: (1) 180 days' confinement in state jail and a $10,000 fine for each count of forgery, with the confinement sentences running concurrently; (2) five years' imprisonment and a $10,000 fine for theft, with a recommendation that the prison sentence be suspended and probated for 10 years; and (3) two years' imprisonment and a $10,000 fine for money laundering, with a recommendation that the prison sentence be suspended and probated for two years. The district court rendered judgments on the jury's verdicts. The district court also ordered Zapata to pay $1.8 million in restitution to Sullivan's estate. In three issues, Zapata contends that: (1) the evidence is insufficient to support his convictions for forgery and theft; (2) the district court erred by admitting his grand-jury testimony into evidence; and (3) the district court's restitution order should be reversed because no evidence supported the initial valuation of the estate. We will affirm the district court's judgments of conviction and order of restitution.

BACKGROUND

The circumstances surrounding Sullivan's death and his purported will resulted in a multitude of criminal prosecutions and civil litigation. We address the facts relevant to Zapata's appeal in some detail, given his challenges to the sufficiency of the evidence supporting his convictions.

Zapata's assistance with Sullivan's bail bond and actions after Sullivan's death

...

Text messages between Zapata and Young showed that Zapata took Sullivan to a meeting with Young on June 3, 2014, the day before Sullivan's body was discovered. ...

Zapata did not testify at trial, but the jury heard evidence that Martinez was with him when they discovered Sullivan's body on June 4, 2014. ... They split up, searching the house.

Inside the bathroom, Martinez found Sullivan undressed and slumped from the toilet into the bathtub, dead. Martinez called to Zapata, who came to the bathroom and photographed Sullivan. Martinez stated that Zapata sent the photo to Sullivan's criminal-defense attorney, Young, and called to confirm that Young had received the photo. But Zapata told a grand jury that he did not send that photo to anyone. While Zapata was speaking with Young, Martinez told Zapata that they needed to get out of there and call the authorities, but Zapata told him to wait and give him some time. Shuffling through papers on Sullivan's desk, Zapata said that he had to find a document with "John Sullivan's signature on it." Martinez testified that Zapata eventually let him call 911, and while he did so, Zapata went into the kitchen and retrieved a book. The book was a missal, a small religious book that looked like a bible. Zapata took the missal, which he told Martinez that Sullivan had given to him, and placed it in his truck before the police arrived.

The jury heard an excerpt of Zapata's testimony at an October 7, 2014 probate proceeding during which Zapata recalled Sullivan saying that he had "written something here" in the missal and "if there's something goes on [sic], it's right here." Zapata said that he took the missal to

Sullivan's attorney with "no idea what was in there." Additionally, Zapata testified to a grand jury that when he dropped off the missal to Young at his office, Young's probate attorney Christianson O. Hartman was already there, although nobody knew that a will was coming.

. . .

Judge Ben Nolen testified that during the June 16, 2014 probate hearing in his court, he placed both Young and Zapata under oath as witnesses. According to Judge Nolen, Zapata testified as to all things contained in the probate application, including that the will was in Sullivan's handwriting, that Zapata had witnessed Sullivan signing the will, and that the will was in the back of a bible that Sullivan owned since childhood. Judge Nolen signed an order on the same day as the hearing, admitting Sullivan's holographic will to probate as a muniment of title and giving all Sullivan's property to Young. Judge Nolen testified that he would not have admitted the will to probate if Zapata had testified—as he did later to a grand jury—that he did not witness Sullivan signing the will and that he was unfamiliar with Sullivan's handwriting.7 Three days after Young received Sullivan's estate, Zapata requested the payoff amount for a real-estate loan from his lender, First Financial Bank.

. . .

Authenticity of Sullivan's will questioned

Texas Ranger Nick Hanna testified that Sullivan's longtime attorney Joe Hernandez contacted him on June 18, 2014, expressing concern about Sullivan's will. Ranger Hanna testified that Sullivan was an educated man, but his purported holographic will contained several grammatical and spelling errors and at least one factual error. Ranger Hanna also testified about his investigation, including that on June 4, 2014, after Sullivan's body was found, eight phone calls took place between Young and Hartman, fifteen phone calls took place between Young and Zapata, and seven phone calls took place between Zapata and Hartman. The calls Zapata had with Young and Hartman that day were deleted from Zapata's phone. But Zapata's phone still had a picture, taken the day before Sullivan's body was found, of a delinquent-rent note handwritten by Sullivan and witnessed by Zapata.

Sullivan's half-sister, Louise Chabot of Worcester, Massachusetts, testified that she and Sullivan saw each other and communicated intermittently over the years, ... She testified that the writing in the alleged will was "not at all" like Sullivan's handwriting because it was in print, every letter "I" was in lower case, and the signature—particularly the way the letter "S" looked like the letter "F"—was not the way that Sullivan typically signed his name. Chabot also noted that the document did not contain any Latin phrase or religious reference, contrary to Sullivan's normal practice.

Sullivan's home-health nurse, Willie Ruiz, testified that he saw Sullivan's signature several times a week over the eighteen-month period that he was caring for Sullivan. Ruiz explained that Sullivan had to sign for every visit that Ruiz made to Sullivan's house and that the signature on the will did not look like Sullivan's because Sullivan's actual signature would have been much larger. Ruiz stated that the language used in the will was "too elementary" and "written at a lower level of intelligence," not the way that Sullivan would have written. Ruiz also stated that he was surprised to learn of Sullivan's cremation because Sullivan "was very much against it."

. . .

First issue: Sufficiency of evidence supporting forgery and theft convictions

In his first issue, Zapata contends that the evidence is insufficient to support his convictions for forgery and theft. He also contends that without proof of his involvement in the will's forgery: (1) both of his convictions for forgery fail; (2) there is no "unlawful appropriation" to support his conviction for theft; and (3) there are no "proceeds of criminal activity" to support his conviction for money laundering.

Applying a legal-sufficiency standard, we consider the evidence in the light most favorable to the verdict and determine whether "any rational trier of fact could have found the essential elements of the crime beyond a reasonable doubt." Jackson v. Virginia, 443 U.S. 307, 319, 99 S. Ct. 2781, 61 L. Ed. 2d 560 (1979); Nisbett v. State, 552 S.W.3d 244, 262 (Tex. Crim. App. 2018). "Beyond a reasonable doubt, however, does not require the State to disprove every conceivable alternative to a defendant's guilt." Ramsey v. State, 473 S.W.3d 805, 809, 811 (Tex. Crim. App. 2015). We defer to the jury's resolution of conflicts in the evidence, weighing of the testimony, and drawing of reasonable inferences from basic facts to ultimate facts. Isassi v. State, 330 S.W.3d 633, 638 (Tex. Crim. App. 2010). We apply the same standard to direct and circumstantial evidence. Id.

Circumstantial evidence is as probative as direct evidence in establishing a defendant's guilt, and circumstantial evidence can alone be sufficient to establish guilt. Nisbett, 552 S.W.3d at 262. Each fact need not point directly and independently to the defendant's guilt if the cumulative force of all incriminating circumstances is sufficient to support the conviction. Id.; ... Further, the trier of fact may use common sense and apply common knowledge, observation, and experience gained in ordinary affairs when drawing inferences from the evidence. Acosta, 429 S.W.3d at 625.

1. Forgery

A person commits an offense if he forges a writing with intent to defraud or harm another. Tex. Penal Code § 32.21(b). "Forge," in relevant part, means "to alter, make, complete, execute, or authenticate any writing so that it purports to be the act of another who did not authorize that act." Id. § 32.21(a)(1)(A)(i). Forgery is a state jail felony if the writing is or purports to be a will. Id. § 32.21(d). Zapata contends that there is no direct evidence that the will was fake and that no one who looked at the will indicated the handwriting was his.

However, the jury had evidence providing context for Zapata's actions after Sullivan's death, in addition to evidence that Sullivan did not write the alleged will,12 but rather that Zapata did, including:

- immediately after the discovery of Sullivan's body, Zapata delayed Martinez's call to 911 until after Zapata called Young;
- shortly after the discovery of Sullivan's body, Zapata had fifteen phone calls with Young and seven phone calls with Hartman, but those calls were deleted from Zapata's phone;
- Zapata testified to a grand jury that he had only one phone call with Young before calling the authorities and that Young did not call him back;
- during one such phone call with Young, Zapata shuffled through papers on Sullivan's desk and stated that he had to find a document with Sullivan's signature on it;
- Zapata removed a missal from Sullivan's house and placed it in his truck before police arrived;
- Zapata testified that he did not get the missal from the house until after the police were there;
- Zapata did not notify police that he took the missal from the house;
- Zapata kept the missal overnight before delivering it to Young;
- Before the alleged will was probated, Zapata confirmed to Young that "Clearing House of all papers is complete" and that "Shredding is in full swing in Sweetwater"
- Zapata had a picture on his phone, taken the day before Sullivan's body was found, of a delinquent- rent note that Zapata had witnessed and Sullivan had handwritten;
- Sullivan was an educated man, but the alleged will inside the missal contained several errors in grammar and spelling;
- Zapata testified to a grand jury that Sullivan did not sign the alleged will;

- Zapata had previously testified to Judge Nolen that Sullivan did sign the alleged will;
- Zapata testified to a grand jury that he would not be able to recognize Sullivan's handwriting;
- Zapata had previously testified to Judge Nolen that he did recognize Sullivan's handwriting;
- Zapata testified to a grand jury that he did not recall his testimony to Judge Nolen about seeing Sullivan writing in the back of the missal;
- Zapata also testified to a grand jury that "there could be somebody else that could have written it [the alleged will], but I don't think it would be John Young";
- distinctive characteristics of Zapata's handwriting were present in the alleged will, including: loops in the lower case letter "p," use of the capital letter "R" between lowercase-lettered words; use of a capital letter "D" at the end of lower-case-lettered words; linking the lower-case letters "iv" in a way that appeared to be a "w" and crossing pairs of the lower case letter "t" with one crossbar;
- forensic analysis found indications that Zapata may have written the body of the alleged will;
- according to Chabot, Manning, Quinn, and Hernandez, the handwriting and signature on the alleged will did not appear to be Sullivan's;
- according to Ruiz, the signature on the alleged will did not look like Sullivan's;
- according to Hernandez, the handwriting and signature appeared to be Zapata's; and
- when Martinez told Zapata, "This is a forgery. There has been a crime committed here," Zapata did not deny the accusations but responded, "[L]et the authorities handle it."

Further, during their deliberations, the jury had all the exhibits admitted into evidence, including the alleged will and numerous handwriting samples from Sullivan and Zapata. We conclude that the combined and cumulative force of all the evidence at trial, viewed in the light most favorable to the jury's conviction, was sufficient to allow a rational jury to find each element of forgery beyond a reasonable doubt. …

…

CONCLUSION

We affirm the district court's judgments of conviction and its order of restitution.

Questions:

1. What differences might occur when assessing handwriting or forgery in a holographic will?
2. What do the handwriting experts conclude about the style of handwriting in this holographic will?
3. What criminal felony most aptly fits this particular fraud?

7.3 DOCUMENTARY EVIDENCE: UNIQUE AND DISTINCT CHARACTERISTICS

Document examiners, if qualified and of sufficient experience, find anomalies in documents all the time. Things like ink differences, signature variance, pressure applied from the writing instrument and general patters of writing which manifest significant differences are all common critiques. In the end, writing style, instrumentation and characteristics of penmanship all can be compared and contrasted. When something looks out of the ordinary, it usually is.

Read the FBI press release on new study of accuracy of handwriting analysis at https://www.fbi.gov/news/press-releases/fbi-laboratory-publishes-major-handwriting-analysis-study

Milner v. Lewis
United States District Court for the District of Connecticut
April 14, 2017, Decided; April 14, 2017, Filed
CIVIL ACTION NO. 3:14-cv-1544 (VLB)

This Plaintiff, Kyron Milner ("Plaintiff" or "Milner"), brings this action claiming that his rights to procedural due process were denied when certain Department of Corrections employees ("Defendants")1 failed to give him prior notice of a disciplinary hearing after which he was designated to administrative segregation. The Court scheduled the case for trial, the parties selected a jury and the case was unable to proceed on the first day evidence was scheduled to be introduced due to the parties' failure to complete discovery and prepare for trial in accordance with the Court's scheduling and other orders and accommodations. For the reasons that follow, the Court declares a mistrial in the interest of judicial integrity and fundamental fairness to the Plaintiff; appoints counsel to represent the Plaintiff; and reopens discovery for limited purposes as more fully stated below.

I. Background

…

The parties agree that according to the Administrative Directives of the Department of Corrections, the date when Plaintiff was formally notified of the November 6, 2013 hearing should be reflected on a Notification of Hearing form ("Notice"). DOC Admin. Dir. 9.4 § 12.B. However, there are three non-identical versions of the Notice: one photocopy was kept within the Department of Corrections' exclusive custody; one duplicate original was kept in Plaintiff's "inmate master file," maintained by DOC, but accessible to Plaintiff upon request, and another duplicate original was kept by Plaintiff, but accessed by DOC officials on at least one occasion when Plaintiff was transferred to Administrative Segregation. [Dkt. 54-1 at 8; 4/11/2017 oral argument on evidentiary issues.] The three versions of the Notice bear inconsistencies, and both parties rely on versions of the Notice to support their positions. When Defendants moved for summary judgment, they submitted the photocopied Notice within DOC's exclusive control, which indicates Plaintiff received and signed the Notice on November 1, 2013, five days before the hearing. [Dkt. 19-12.] Plaintiff submitted with his opposition to summary judgment the two other versions of the Notice, which indicate Plaintiff did not receive or sign the Notice until November 13, 2013—seven days after the hearing. [Dkt. 33-2 at 13, 15.]

Due to the discrepancies between the Notification of Hearing forms, the Court found disputed issues of material fact and denied summary judgment as to whether Plaintiff received sufficient notice before the hearing to determine his administrative segregation. [Dkt. 40 at 21.] As a predicate to that procedural due process claim, the Court found a genuine issue of material fact as to whether Plaintiff held a constitutional liberty interest in avoiding administrative segregation due to the length of time he remained there and the nature and severity of conditions he experienced during that confinement as compared to the conditions in general population. Id. at 16.

…

Defense counsel advised the Court that the Plaintiff had not identified to her the exhibits he intended to introduce. In response, the Plaintiff stated that he intended to introduce at trial the exhibits he filed in opposition to the Defendants' motion for summary judgment and that to assure clarity he would have his counselor forward these exhibits to Defendants' counsel. The Court prepared exhibit binders consisting of the exhibits filed by the Plaintiff in opposition to summary judgment. Later the Clerk of the Court received from Plaintiff's Department of Corrections counselor documents which were not included as exhibits in opposition to the motion for summary judgment with no indication of their intended purpose. As they were in the possession of the Clerk of the Court these documents were uploaded to the docket. [Dkt. 53.]

On Saturday, April 8, 2017, after 10:00 p.m., with the presentation of evidence to begin on Tuesday, April 11, 2017, Defendants filed a Motion to Dismiss pursuant to 28 U.S.C. § 1915A. [Dkt. 54.] In their Motion, Defendants argued that the Notice which was kept in Plaintiff's inmate master file was altered to artificially reflect that Plaintiff signed the Notice on November 13, 2013, after the date of the hearing, rather than November 1, 2013. Id. at 5. The motion to dismiss relied upon and had attached thereto a report of a forensic examination of one of the three hearing notices. As discussed below, the forensic analysis did not comport with the Court's Scheduling Order, the rules of procedure or the rules of evidence.

...

Defendants' newly disclosed expert report describes the forensic examination conducted on April 5, 2017 by the State of Connecticut's Department of Emergency Services and Public Protection Division of Scientific Services. [Dkt. 54-7.] The report states that Forensic Science Examiner-Analyst Greg Kettering and Forensic Science Examiner Technical Reviewer Lisa Ragaza examined the ink used to write the date on the original Notice. Id. at 2-3. The examiners state they found three discrepancies between the number 3 in the date 11/13/13 and the other digits: dissimilar ink chemical properties, dissimilar pen pressure, and an absence of ink line striae present on the other digits. Id. at 2. Defendants argue the report concludes that "someone used the same writing instrument to author everything in the day column with the exception of the '3'" which was "authored with a different writing instrument." [Dkt. 54-1 at 7.] Defendants conclude from this that Plaintiff fraudulently altered the original Notice kept in his inmate master file to state he did not receive notice of the administrative hearing until November 13, 2013, when in fact he received notice on November 1, 2013. Id. at 7-8.

Defendants moved the Court to dismiss the case, arguing that the report demonstrates that there is no question of fact that Plaintiff received notice five days before his hearing. Defendants argue that in light [*10] of this, the Amended Complaint is frivolous and should be dismissed under 28 U.S.C. § 1915A and Campos v. Smith, 418 F. Supp. 2d 277 (W.D.N.Y. 2006). Id.

Defendants also submit that Plaintiff committed a fraud on the Court and committed perjury, and encourage the Court to take appropriate actions. Id.

II. Analysis of Defendants' Motion to Dismiss

a. Defendants' Frivolity Claim

Under 28 U.S.C. § 1915A, where a "prisoner seeks redress from a governmental entity or officer or employee of a governmental entity ... the court shall identify cognizable claims or dismiss the complaint, or any portion of the complaint, if the complaint ... is frivolous, malicious, or fails to state a claim upon which relief may be granted."

In support of their motion, Defendants cite Campos, 418 F. Supp. 2d at 277. In that case, an inmate offered into evidence a document which was inherently unreliable because it referred to events which had not occurred as of the date it was purportedly authored. Id. at 279. The document was also maintained in the exclusive possession of the party who proffered it. Id. Because the document was falsified on its face, the Court dismissed the complaint. Id. at 279. From this,

Defendants conclude Plaintiff's Amended Complaint should also be dismissed because the original Notice kept in the inmate master file was altered.

The case at bar is distinguishable from Campos. ... Here the document was not in the exclusive possession of the Plaintiff. Indeed, none of the documents was in the exclusive possession of the Plaintiff. Two of the documents were in the possession of both Plaintiff and Defendants at various times (Plaintiff's version of the Notice having been given to him by the DOC and the version of the Notice which was placed in Plaintiff's master file), and one version which was maintained exclusively by the DOC, Defendants' employer. In addition, there are also irregularities in the Defendant's recordkeeping which preclude dismissal. The Defense admitted that the version of the Notice maintained in the Plaintiff's master file was initially missing, and was ultimately located in another section of the master file, out of chronological order. [Dkt. 54-1 at 6; 4/11/2017 oral argument on evidentiary issues.] Although Defendants offer evidence that the Plaintiff accessed his master file and alteration of the document may have advanced the interests of the Plaintiff, Defendants do not offer conclusive evidence that the document was in the master file when the Plaintiff accessed the file, or that Plaintiff altered the document.

Nor does the advocate investigation report dated November 5, 2013 provide conclusive evidence that Plaintiff altered the document. [Dkt. 54-12.] Plaintiff signed the report on November 5, 2013, the day before the hearing at issue, but the report states only that Plaintiff will file a written statement "at hearing." Id. at 2. It does not indicate that Plaintiff had notice of the November 6, 2013 hearing before November 5, 2013, and DOC administrative directive 9.4 entitled Plaintiff to two business days' notice. In addition, the fact that Plaintiff challenged the outcome of the hearing to Commissioner Arnone but did not allege that he was not given adequate notice at that time is persuasive, but not dispositive. At best, the Defendant has raised a genuine issue of material fact. In addition, Defendants admit that following the hearing, Plaintiff was transferred to an administrative segregation unit in a different Department of Corrections facility. They further acknowledge that upon the transfer of an inmate, the property of the inmate is packed and searched by Department of Corrections officials before they are returned to the inmate. Therefore the record indicates that the Plaintiff did not have exclusive possession of the original which was given to him to keep. Here, unlike Campos, the document offered by the Plaintiff is not the only document, and it is not facially altered. In addition, no documents were in the exclusive possession and control of the Plaintiff. While Defendants offer strong circumstantial evidence from which a jury could reasonably conclude that the Plaintiff altered a document, it is not irrefutable and thus genuine issues of material fact remain to be decided by a jury.

...

III. Conclusion

Judicial integrity and fundamental fairness outweigh the need for judicial efficiency, and demand that the parties be afforded another opportunity to fully prepare to prosecute and defend this case. For these reasons and based on the reasoning outlined above, the Court hereby: (1) declares a mistrial; (2) orders the Clerk to appoint counsel for the Plaintiff; (3) closes the case pending the appearance of counsel for the Plaintiff; (4) grants the Plaintiff leave to reopen the case on the later of (a) 365 days after the date of this Order, or (b) 182 days (6 months) after the date counsel files an appearance on behalf of Plaintiff; (5) reopens discovery as it relates to the documents marked as Plaintiff's Exhibits 29, 17, 18, 40, and 43 during the first day of trial, April 11, 2017, as well as any other exhibit Plaintiff sought to offer into evidence on April 11, 2017 which Plaintiff did not timely identify in accordance with the Scheduling Order [Dkt. 24], including expert discovery on the authenticity of such proposed exhibits; (6) orders counsel for Defendants to initiate and conduct a Rule 26(f) planning meeting with Plaintiff's counsel and to request a Rule 16 conference with the Court within 35 days after the date Plaintiff's counsel files an appearance.

IT IS SO ORDERED.

Questions:

1. Documents are sometimes part of a larger file applicable to a case or adjudication. In Milner, there are processing documents, such as notice forms, which the Court concludes were tampered with. What impact does the tampering have upon the claimant?
2. Documentary analysis by qualified document examiners found 3 discrepancies which were out of the ordinary. Relay those findings.
3. Due to the significance of these documentary file alterations, what is the Court's remedy and decision?

Locate the case below using your favorite legal research resource, read the decision and answer the questions that follow.

Moses v. Somerset Cmty. Servs., Court of Special Appeals of Maryland, March 26, 2020, Filed No. 2864, September Term, 2018

Questions:

1. In five deeds for property given eventually to grandchildren, a challenge was made as to the inconsistency of?
2. How do the experts in this case evaluate the quality and consistency of signatures on these 5 documents?
3. In this case, there is a battle of two expert conclusions. How does this impact the Court's final decision?

One of the most famous handwriting and documentary analysis cases involved the kidnapping and killing of Charles Lindbergh's baby in the early 1930s. There are many critics of the decision of guilt that exist even today.

State v. Hauptmann
115 N.J.L. 412, 180 A. 809
Court of Errors and Appeals of New Jersey
June 20, 1935, Argued; July 15, 1935, Submitted on Supplemental Briefs; October 9, 1935,

...

Point XI is based upon language of the charge on the subject of expert evidence respecting the handwriting of letters claimed to have been written by defendant. The language of the charge is as follows:

"A very important question in the case is, did the defendant, Hauptmann, write the original ransom note found on the window sill, and the other ransom notes which followed? Numerous experts in handwriting have testified, after exhaustive examination of the ransom letters, and comparison with genuine writings of the defendant, that the defendant, Hauptmann, wrote every one of the ransom notes, and Mr. Osborne, Senior, said that that conclusion was irresistible, unanswerable and overwhelming. On the other hand, the defendant denies that he wrote them, and a handwriting expert, called by him, so testified. And so the fact becomes one for your determination. The weight of the evidence to prove the genuineness of handwriting is wholly for the jury." The sentence beginning "numerous experts" and ending with "overwhelming" is selected for specific attack, which seems to consist of the proposition that the court "unduly and erroneously emphasized the value and effect of handwriting testimony by referring to and quoting from the testimony of the elder Mr. Osborn" (here follows the sentence in question). We fail to see in this anything but legitimate comment. Further, under the same point, the refusal to charge requests Nos. 1, 2 and 3 is urged for error. They are as follows:

"1. All doubts respecting the competency of the opinion of experts in handwriting, based upon mere comparison, as evidence, have been removed by the statute. (Rev. 381, § 19): but it must be esteemed proof of low degree. Very learned judges have characterized it as much too uncertain, even when only slightly opposed, to be the foundation of a judicial decision. Gurney v. Langlands, 5 Barn. & Ald. 185; Doe v. Suckermore, 5 Ad. & El. 751; 1 Greenl. Ev., § 80, note (2); Stark. Ev. 173, note (e). 30 Equity 201. Mutual Benefit Life Insurance Co. v. Brown.

"2. It has been said by an eminent law writer with respect to the evidence of handwriting experts that this kind of evidence is most unsatisfactory, very inconclusive, most unreliable and of the lowest probative force. Moore on Facts, § 615.

"3. A jury is not bound to accept the opinion of an expert as to disputed handwriting, even when uncontradicted. It should consider it as an argument rather than as proof, and to make allowances for all the disturbing influences by which the judgment of the expert may be moved. 1 Whart. Ev., § 722."

These are copied verbatim in order to make it plain that Nos. 2 and 3 are extracted from legal text books, and No. 1 from an opinion by the late Vice- Chancellor Van Fleet in 1878.

The refusal of No. 2 is clearly supportable on the ground that it does not pretend to be a direct proposition of law, but merely a quotation from a text writer. No. 3 we think was adequately covered, if correct, by the broad instruction leaving the weight of evidence on handwriting to the jury. Request No. 1 is taken, as we have said, from the case of Mutual Benefit Life Insurance Co. v. Brown, 30 N.J. Eq. 201. As to this, it should be noted in the first place that the case came to this court (32 N.J. Eq. 809) which did not adopt the language of the vice-chancellor, but went no farther than to say (page 812), "the opinions of experts, however skillful they may be, are weaker in degree of certainty than the direct evidence of the subscribing witness, who has sworn to the genuineness of both signatures, when he proved the execution. There is nothing in the evidence to show that he had any interest or motive to swear falsely, or to impugn his veracity." This is very far from saying that the expert evidence is, generally and without qualification, "proof of low degree." In this case, as the trial judge pointed out, numerous experts in handwriting had testified that the defendant wrote every one of the ransom notes, and only one for the defendant denied this. The history of the legitimizing of this class of evidence, i.e., comparison of handwriting, is most interesting, but cannot be considered in extenso here. See Wigm. Ev.,

§§ 1991 et seq. But it may well be noted, that its legal status was at least very doubtful until it was authorized by statute; in England in 1854, and in this state for the first time, apparently, in the Revision of 1874. See Rev. 1877, p. 381, § 19; Comp. Stat., p. 2226, § 20. In Gordon's Case, 50 N.J. Eq. 397, Chancellor McGill, sitting as Ordinary, said (at p. 422): "Handwriting is an art concerning which correctness of opinion is susceptible of demonstration, and I am fully convinced that the value of the opinion of every handwriting expert as evidence must depend upon the clearness with which the expert demonstrates its correctness. * * * (page 423). Without such demonstration the opinion of an expert in handwriting is a low order of testimony," &c. Neither of the two statements is binding on this court. There was no request to charge the latter; and as to the former, which was requested, we are at least clear that without the qualifying clause of demonstration, the defendant was not entitled to have it charged. There was in fact ample demonstration by the experts for the state; and plainly it was satisfactory to the jury, to whose decision the judge properly left it. 22 C.J. 728; Com. v. Williams, 105 Mass. 62, 68. We find no error in the refusal of the first request.

Point XII also brings up refusals to charge, and is in several subdivisions. The first deals with the charge on reasonable doubt. It was fully and correctly charged. The present complaint is that the court did not add a sentence contained in the opinion of the Supreme Court in State v. Hadley, 113 N.J.L. 335, 337. The trial court in that case had charged that "if the state has failed to prove its charges, and the defendants have caused you to entertain and believe that such a doubt actually exists," &c. The Supreme Court naturally and properly said: "The duty was upon the state to prove the defendants guilty beyond a reasonable doubt, and no duty was cast upon the defendants to cause the jury to entertain a doubt as to their guilt." But that was applicable and applied to that particular situation, and the court was very far from holding that any proper charge on reasonable doubt must include the latter phrase. The accredited formula was used in this case and no such addition was called for.

...

Point XIV. Refusal to strike out testimony of Kelly and Maish. The proposition is that both were allowed to testify as experts and neither was qualified.

Kelly, a state trooper, sworn, said he specialized in identification. Had examined the ladder for finger and other marks; that he placed it against the house in three sections but found no marks where the top rested; that he tried again with two sections and found markings the width of the ladder and one and one-half to two inches above the end when set in the mudholes; the marks were "right with the ladder. Q. What kind of marks were they? A. Ladder marks." Mr. Reilly moved to strike this answer out as a conclusion. The motion was denied and exception taken. The witness went on to testify that the marks were very plain -- a scraping mark in the white paint -- the gray of the stone showed through; also that there were white marks on the top ends of the ladder -- "I would say paint."

The objection was to a conclusion. It was not that Kelly had not been qualified as an expert. Assuming the technical impropriety of the statement that they were "ladder marks," the other testimony put the jury in a position to form their own conclusion, and they had the ladder in evidence.

As to Maish. He had made the baby's thumb guard and said it would not rust. On cross, his authority as an expert was seriously impaired; but the testimony was in and was completed; the witness was dismissed, certain exhibits were marked, the court adjourned for the day and reconvened next morning after Maish had presumably gone home to Wyoming, Ohio, and then counsel for the first time moved to strike the testimony out. The court denied the motion on the merits. It is sufficient to say that it came too late. Moreover, the motion made the next day was to strike out the Maish testimony in toto. Part of it was clearly competent; and for that reason also the motion was properly denied.

...

Point XVI. That the verdict was against the weight of evidence.

Defendant's argument on this point deals principally with the alleged lack of credible evidence tending to show his presence at the scene of the crime on the day of the kidnapping or at any other time.

It is maintained that the testimony of Rossiter, Whitted and Hochmuth on this crucial issue was either willfully false or the product of morbid or irresponsible minds; and that the trial judge laid undue stress thereon to the great prejudice of defendant; that not a single readable fingerprint was found in the nursery room or on any of the equipment therein, or on the ladder, the chisel or any of the ransom notes; that, on the contrary, Dr. Hudson, defendant's expert, found fingerprints but none of defendant's; that the handwriting evidence was unreliable; that the testimony of the wood expert, Koehler, was likewise unworthy of credit in the determination of an issue so grave; that Colonel Lindbergh's Identification of the defendant by his voice was entitled to little weight; that the evidence of the payment of the ransom, and its possession by defendant, bore only upon the crime of extortion, and that, in any event, the evidence that $2,790 of the ransom money was deposited in the Federal Reserve Bank at the time the gold notes were called by one "J. J. Faulkner," whose handwriting on the deposit slip was concededly not that of Hauptmann, was a most persuasive circumstance; and that the testimony of Condon and the remaining witnesses was on the whole of such a character, in its material aspects, as not to furnish adequate ground for a verdict of guilty.

Defendant's reply brief points out that the evidence of Mrs. Lindbergh established that "friendly hands" lifted the child from the crib; that "the safety pins still held the covers to the mattress," and that "there was no bloodstains or evidence of haste, anywhere." The query is put, "how did the kidnapper know] that the southeast window led into the nursery or that it was the only window in the room that could not be locked?" And it is said that, with a thirty-five-mile gale blowing, the defendant did the highly improbable thing of "risking his life" on this frail ladder, and of placing "it against the side of the house without leaving his trail imprinted in the soft earth, climbed the ladder to next to the top rung forty-eight inches below the window sill, and with his left hand opened the shutters in a thirty- five-mile gale, raised the window without the use of a jimmy or instrument, entered the room alone and obtained the baby." Attention is directed to the absence of evidence of an "outcry or alarm" when the baby was lifted from the crib and carried out the window, as alleged, and the fact that a stein resting on the side of the window sill was not disturbed; that the gale did not disturb the note left by the kidnapper; that the closed window and shutters found after the kidnapping demonstrated that "it did not happen that way; that the window was never opened or closed; that the baby never came out of the window and certainly could not have met its death under that window or in the room without leaving some indication somewhere showing where its skull was fractured."

Moreover, it is claimed that there was affirmative credible evidence of an alibi both as regards the night of the commission of the crime and the payment of the ransom money.

Our conclusion is that the verdict is not only not contrary to the weight of the evidence, but one to which the evidence inescapably led. And discarding the testimony of the witness classified as irresponsible by the defendant, the result is the same. From three different and, in the main, unrelated, sources the proofs point unerringly to guilt, viz., (a) possession and use of the ransom money; (b) the handwriting of the ransom notes; and (c) the wood used in the construction of the ladder.

(a) Bills totalling $14,600 were found hidden in defendant's home in the Bronx, New York, beneath the floor of the garage, and in the rafters of the dwelling house. In a hidden recess in one of the rafters containing some of the bills was found a small revolver. On a board in a closet of the house was found, in defendant's handwriting, the address and telephone number of Condon. And we have the return of the sleeping suit by the man to whom the ransom money was eventually paid.

The explanation of the source of this money offered by defendant was incredible, and we find not the slightest evidence to corroborate it. The defendant's handling of the money makes clear his guilty connection with the enterprise. He was without funds before the payment of the ransom money, and in comparative poverty; he worked at his trade apparently, in the light of his subsequent conduct, because it was his only means of livelihood. After the payment of the money, he refrained from work entirely and speculated in Wall Street on a comparatively large scale. His actions thereafter are persuasive of guilt. It is clear his new found prosperity was evidenced long before he says he found the money claimed to have been left by Fisch. It is inconceivable that Fisch, if he had this money, would have left it in defendant's custody in the manner claimed, particularly when he was planning to leave for Germany on what seemed to be his last trip. He was afflicted with tuberculosis and died shortly after returning to his homeland.

From the foregoing it is deducible, to a moral certainty beyond a reasonable doubt, that he collected the ransom money, and was therefore the kidnapper, particularly in view of --

(b) the identification of the handwriting on all the ransom notes, including the one left on the window sill of the nursery window, as that of Hauptmann. The peculiarities of expression and spelling common to all these notes and admittedly genuine writings by Hauptmann. It is likewise an inescapable inference that the writer of the ransom note on the window sill entered the nursery and perpetrated the crime.

Moreover, we have the chisel found at the scene of the crime. It was of a size that is ordinarily a part of a carpenter's tool chest, but was missing from Hauptmann's set found in the Bronx home.

(c) Part of the wood used in the construction of the ladder was taken from Hauptmann's Bronx home. It is significant that long before Hauptmann's arrest, the wood, or part of it, was traced to a Bronx lumber yard near his home by the expert Koehler. And there was evidence that Hauptmann limped for some days after the kidnapping.

It is clear the ladder was used to reach the second story. Colonel Lindbergh heard the sound of breaking wood that evening, and the ladder was found broken. The witness Sweeny, by actual test, demonstrated that the entry of the second story window and the descent therefrom with the child could be accomplished with the ladder.

There is no significance in the selection by the kidnapper of the unlocked southeast window. The shutters, because of warping, could not be completely closed, and it is fair to assume that the intruder selected this unlocked window by pure chance, or when he found the other windows barred.

And the claim that entry was not made through this window had nothing to support it. Muddy footmarks were found on the window sill and nursery floor, and a suitcase placed on the floor under the window. Footprints were found outside the window.

The absence of defendant's fingerprints had no significance; it is clear that this self-confessed and apparently experienced criminal would naturally guard against the making of these tell-tale marks of his visit.

There was much more evidence tending to establish his guilt. It was a circumstantial case with the evidence pointing to guilt from so many directions as to leave no room for a reasonable doubt.

...

The judgment of conviction is affirmed (Figure 7.3).

Questions:

1. First what documentary evidence does the state rely upon in its prosecution? Second, how are these documents used to show uniqueness and individual identity?
2. Is it fair to argue that handwriting comparisons seeking to show a separation from unique patterns to the individual is neither settled science nor binding on the jury as a scientific given?
3. The Hauptmann case has a wide variety of dilemmas and these conundrums still generate doubt in this case nearly 90 years later. Aside from the handwriting analysis, there is another form of comparative analysis in this case. Can you describe where this occurs?

United States v. Al-Farekh
United States Court of Appeals for the Second Circuit
April 16, 2020, Decided
18-943-cr

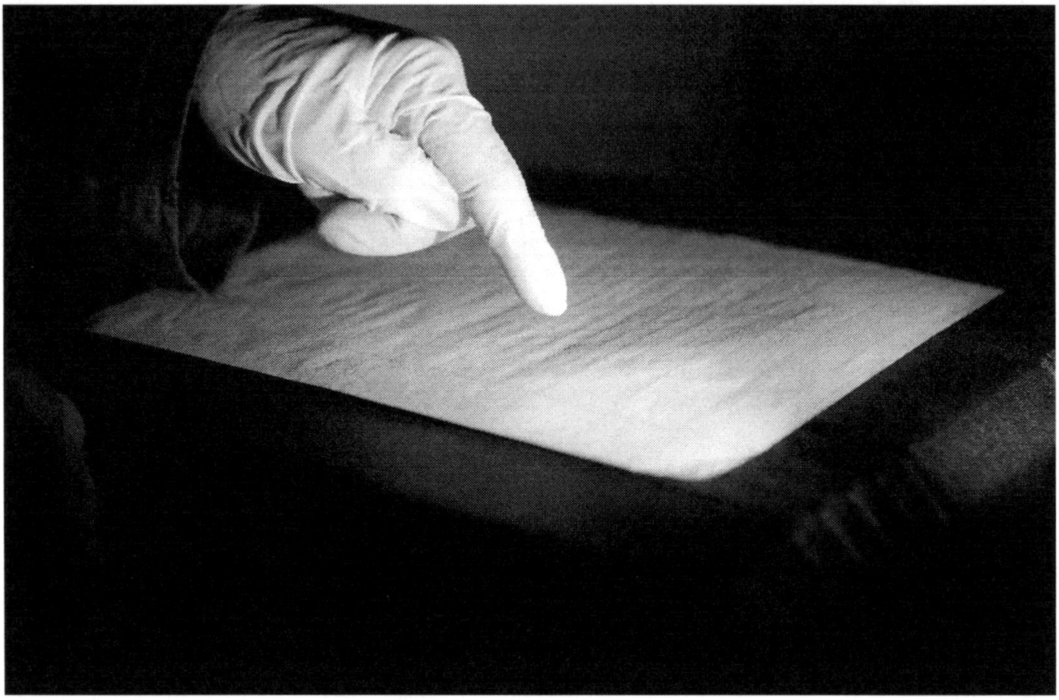

FIGURE 7.3 Haldeman Papers being analyzed at TIGTA Lab. Photographer: Earl McDonald, NARA Staff.

Source: National Archives NAID: 210374965.

Defendant-Appellant Muhanad Mahmoud Al Farekh ("Al-Farekh") appeals from a judgment convicting him, following a jury trial, of: use of explosives; conspiracy to murder U.S. nationals; conspiracy to use a weapon of mass destruction; conspiracy to use a weapon of mass destruction by a U.S. national; conspiracy to bomb a U.S. government facility; conspiracy to provide, attempt to provide, and provision of material support to terrorists; and conspiracy to provide, attempt to provide, and provision of material support to the Foreign Terrorist Organization al-Qaeda. The District Court sentenced Al- Farekh principally to 45 years' imprisonment.

On appeal, Al-Farekh challenges a series of evidentiary rulings, as well as the District Court's denial of his request to declare a mistrial. In addition, Al-Farekh challenges the substantive reasonableness of his sentence. We assume the parties' familiarity with the underlying facts, procedural history of the case, and issues on appeal.

In an opinion filed simultaneously herewith, we reject Al- Farekh's challenges to the handling of the Government's motions filed pursuant to the Classified Information Procedures Act, the admission of an out-of-court photo identification of Al-Farekh, and the limitation on Al-Farekh's cross-examination of the Government's fingerprint examiner. We now address the remainder of Al-Farekh's arguments and conclude that the judgment of conviction and sentence should be affirmed.

I. The District Court's Evidentiary Rulings

"We review a district court's evidentiary rulings under a deferential abuse of discretion standard, and we will disturb an evidentiary ruling only where the decision to admit or exclude evidence was 'manifestly erroneous.'" United States v. McGinn, 787 F.3d 116, 127 (2d Cir. 2015) (quoting United States v. Samet, 466 F.3d 251, 254 (2d Cir. 2006)). Where we find an abuse of discretion, "vacatur is required unless we are 'convinced that the error was harmless beyond a reasonable doubt.'" United States v. Mejia, 545 F.3d 179, 199 (2d Cir. 2008) (quoting United States v. Reifler 446 F.3d 65, 87 (2d Cir. 2006)).

In determining whether an erroneous admission was harmless, we consider: "(1) the overall strength of the prosecutor's case; (2) the prosecutor's conduct with respect to the improperly admitted evidence; (3) the importance of the wrongly admitted testimony; and (4) whether such evidence was cumulative of other properly admitted evidence." United States v. Gomez, 617 F.3d 88, 95 (2d Cir. 2010) (quotation marks and citation omitted).

A. Written Communications by Co-Conspirators Imam and Yar

Al-Farekh argues on appeal that the District Court erred when it admitted into evidence an e-mail by Ferid Imam and two letters by Maiwand Yar, Al-Farekh's co- conspirators. According to Al-Farekh, these statements are irrelevant and contain hearsay that does not fall into any hearsay exception.

On review, we conclude that the District Court did not abuse its discretion in admitting the written communications by Imam and Yar. We do so for substantially the reasons given by the District Court in its thorough September 13, 2017 Decision and Order granting the Government's motion in limine to admit Imam's e-mail and Yar's letters. See Appellant's App'x ("App'x") at 49-55.

B. Al-Farekh's Handwritten Letters

We also reject Al-Farekh's argument that the District Court abused its discretion by admitting the handwritten letters that were found in a USB drive that was handed to an agent of the Federal Bureau of Investigation in Afghanistan. Al-Farekh contends that these letters should have been excluded because they were not authenticated.

We disagree. Although the Government did not present evidence regarding the circumstances surrounding the seizure of the USB drive, Federal Rule of Evidence 901(b)(4) permits authentication based on "[t]he appearance, contents, substance, internal patterns, or other distinctive characteristics of the item, taken together with all the circumstances." The Government satisfied Rule 901(b)(4) here. For example, the letters were signed in Arabic text "Abdullah" or "Abdullah al-Shami," the kunya that the Government witness Sufwan Murad, a former al-Qaeda collaborator, attributed to Al-Farekh. The Government also presented expert testimony that there were considerable similarities between the handwriting in the letters in the seized USB drive and the known samples of Al-Farekh's handwriting. Finally, the content of the letters—namely, the author's desire to wage violent jihad against the United States and the fear for his safety as a leader in al-Qaeda's external operations division—are consistent with Murad's description of Al-Farekh. In light of the totality of the circumstances, the District Court did not abuse its discretion in finding that the proof of authentication was sufficient to pass the relatively low bar for authentication of evidence, see United States v. Al-Moayad, 545 F.3d 139, 172-73 (2d Cir. 2008), and that any remaining questions as to the reliability of the letters go to their evidentiary weight, not their admissibility, see App'x at 39-40.

C. The Testimony of Professor Lorenzo Vidino, the Testimony of Evan Kohlmann, and the Video of a Controlled Detonation

We similarly reject Al-Farekh's challenge to the admission of: (1) expert testimony by Professor Vidino, Director of the Center on Extremism at George Washington University, on the absence of a single jihadist profile and the routes commonly used to travel to join al-Qaeda in the Middle East; (2) testimony by fact witness Evan Kohlmann, founder of a company that collected online information on al-Qaeda and disseminated threat intelligence reports to clients, summarizing certain lectures of Anwar al-Awlaki, as well as the excerpts from certain jihadist materials; (3) a one-minute, 40-second-long video depicting the controlled detonation of a vehicle-borne improvised explosive device, the type of bomb that was used in a terrorist attack against a U.S. military base in Afghanistan and on which Al-Farekh's fingerprints were found. We find no error in the admission of this evidence and conclude that the District Court's analysis under Rule 403, which is entitled to considerable deference on appeal, see United States v. Greer, 631 F.3d 608, 614 (2d Cir. 2011), was not "manifestly erroneous," McGinn, 787 F.3d at 127 (quoting Samet, 466 F.3d at 254).

Even if we were to assume, for the sake of argument only, that the District Court erred in admitting the testimony of Professor Vidino, the testimony of Kohlmann, or the detonation video, any such error would have been harmless in light of, among other things, the strength of the Government's case against Al-Farekh. …

D. Limitations on the Cross-Examination of Sufwan Murad

Al-Farekh also disputes the District Court's decision to permit Murad to testify using a pseudonym and to limit the scope of the cross-examination to preclude Al-Farekh's counsel from eliciting testimony about Murad's identity, country of origin, nationality, location, and ongoing cooperation with authorities. Al-Farekh contends that this decision violated his rights under the Confrontation Clause.

The Confrontation Clause protects a criminal defendant's right to cross-examine the prosecution's witnesses. … That right, however, is not absolute. Trial judges have "wide latitude … to impose reasonable limits … on … cross-examination based on concerns about, among other things, harassment, prejudice, … or the witness' safety." Delaware v. Van Arsdall, 475 U.S. 673, 679, 106 S. Ct. 1431, 89 L. Ed. 2d 674 (1986); accord United States v. Vitale, 459 F.3d 190, 195 (2d Cir. 2006). Although we generally review a limitation on the scope of a defendant's cross-examination of a Government witness for an abuse of discretion, when the limitation directly implicates the defendant's constitutional rights under the Confrontation Clause, we review the ruling de novo. …

We find no error in the District Court's evidentiary ruling. As a threshold matter, Al-Farekh does not present any evidence undermining the Government's reasonable assertion that the safety of Murad, a former al- Qaeda collaborator who assisted in counterterrorism investigations, may be in fact jeopardized in the absence of the District Court's protective order.

Moreover, the narrowly tailored limitations on Al- Farekh's cross-examination are consistent with limitations that this Court has upheld as appropriate to protect the safety of witnesses. …

Al-Farekh was aware of Murad's identity in advance of the deposition, as the Government did not seek to limit its disclosures to Al-Farekh on this subject. And Al- Farekh was permitted to explore Murad's alleged biases and motivations for lying during his cross-examination. Accordingly, we conclude that the District Court's limitation on the cross-examination of Murad's testimony was reasonable and did not violate Al- Farekh's rights under the Confrontation Clause.

II. The Denial of Al-Farekh's Request for a Mistrial

Al-Farekh challenges the District Court's denial of his motion to excuse one of the jurors and, in the alternative, for a mistrial. Al-Farekh moved for a mistrial after the first day of deliberations, following an incident in which Juror 4 reported to the District Court that he had heard that Al-Farekh's father had boarded an elevator with another juror, but that Juror 4 was not aware of what, if anything, the father had said to the other juror.

We review the District Court's denial for abuse of discretion. … To determine if the presumption is properly rebutted, we must consider the "(1) the nature of the information or contact at issue, and (2) its probable effect on a hypothetical average jury." Id. at 169.

On review, we conclude that the District Court did not abuse its discretion in refusing to excuse Juror 4 and to grant Al-Farekh's request for a mistrial. Juror 4's access to extra-record information in this case—hearing that Al-Farekh's father boarded an elevator in the courthouse with a juror—is "harmless." Id. at 168. Juror 4 had no knowledge about any conversation between Al-Farekh's father and any jurors. Moreover, the District Court took several important steps to remedy any potential prejudice, including conducting individualized inquiries of all jurors and alternate jurors and issuing two instructions to the jury that were consented to by both the Government and Al-Farekh. More importantly, Juror 4 indicated to the District Court that the reported extra-record information will not affect his ability to be a fair juror in the case.

Under the circumstances presented, in light of the innocuous nature of the extra-record information, the District Court's reasonable and precautionary measures to minimize the risk of prejudice, and the juror's assurances to remain impartial, we conclude that the Remmer presumption was properly rebutted and thus find no abuse of discretion. ...

III. Substantive Reasonableness of the Sentence

Al-Farekh also challenges the substantive reasonableness of his 45-year prison sentence, which we review "under a deferential abuse-of-discretion standard." United States v. Yilmaz, 910 F.3d 686, 688 (2d Cir. 2018) (citation omitted). Our review of a sentence for substantive reasonableness is "particularly deferential." United States v. Broxmeyer, 699 F.3d 265, 289 (2d Cir. 2012). We will set aside a sentence as substantively unreasonable only if it is "so shockingly high, shockingly low, or otherwise unsupportable as a matter of law that allowing [it] to stand would damage the administration of justice." Id. (internal quotation marks omitted).

Al-Farekh's challenge to his sentence is meritless. Contrary to Al-Farekh's assertion, the District Court specifically considered the letters from Al-Farekh's friends and family and Al-Farekh's own letter. Notably, the letters submitted on Al-Farekh's behalf describe the young teenager who went to college at the University of Manitoba prior to his departure for Pakistan in 2007. These letters provide little, if any, insight into the mindset or character of the adult who traveled to Pakistan to join a terrorist organization to perpetrate violent attacks against military and civilian personnel. Moreover, Al-Farekh's own letter was "not an enthusiastic acceptance of responsibility or expression of remorse." App'x at 565.

In light of the severity of Al-Farekh's conduct, we conclude that a 45-year sentence of imprisonment is simply not substantively unreasonable.

CONCLUSION

We have reviewed all of the arguments raised by Al- Farekh on appeal and find them to be without merit. For the foregoing reasons, we AFFIRM the judgment of the District Court.

Questions:

1. Defendant Al-Farekh, a known terrorist, appeals his conviction on the basis of documentary and handwriting claims. What documents does he dispute as not being accurate, nor in the alternative not authenticated by one who knows the style and content of his writings?
2. This case also brings to light another factor on expert testimony as to handwriting comparisons or document examination. An expert may or may not be necessary because in some jurisdictions a lay witness can match or conclude similarities in style and manner of writing. How could that happen in this case?
3. The defense appeals his sentencing conviction saying it too severe. How does the court respond?

United States v. Foust
989 F.3d 842 United States Court of Appeals for the Tenth Circuit
March 2, 2021, Filed
No. 19-6161

Defendant-Appellant Justin Foust appeals from his conviction on six counts of wire fraud, and one count each of aggravated identity theft and money laundering. He was sentenced to 121 months' imprisonment and three years' supervised release. Briefly, Mr. Foust's company, Platinum Express, LLC, submitted false and fraudulent invoices to its customer, Chesapeake Energy Corporation ("Chesapeake"). Chesapeake identified more than $4.5 million that it had paid out on these invoices. Mr. Foust did not deny that the invoices were improper and that Platinum Express had not performed the work. But he denied that he had forged the signatures and employee identification numbers of Chesapeake employees. A handwriting expert testified otherwise regarding invoices associated with Chesapeake employee Bobby Gene Putman. The jury convicted Mr. Foust on the wire-fraud and aggravated-identity-theft counts associated with these invoices.

On appeal, Mr. Foust argues that the district court abused its discretion by allowing the handwriting expert to testify at trial. He contends that (1) the government did not adequately show that the expert's methodology was reliable and (2) the handwriting expert used unreliable data in reaching his opinion. This court has jurisdiction under 28 U.S.C. § 1291, and we affirm.

Background

Mr. Foust was a production foreman at Chesapeake where he operated oil and gas wells. In 2011, he left Chesapeake to start Platinum Express, ... A few years later, Chesapeake employees discovered the fraudulent invoices submitted by Platinum Express. Chesapeake began investigating the matter and told Mr. Foust that it was going to exercise its contractual right to examine Platinum Express' computers and business records. But while Chesapeake investigators were travelling to the Platinum Express office, Mr. Foust told them that someone had broken into the office and stolen two computers. The investigating sheriff's deputy believed the break-in was staged by Mr. Foust. Chesapeake decided to turn the investigation over to the FBI.

When the FBI initially talked to Mr. Foust, he blamed the fraudulent invoices on Mr. Lucas, Platinum Express' general manager, and Ms. Lucas, the office manager. However, the agents could not connect the Lucases to the fraudulent invoices, so they determined that Mr. Foust was likely involved. The FBI learned, among other things, that: Mr. Foust had access to and controlled the Platinum Express account; all of the checks from Chesapeake were traced into the Platinum Express account and the Fousts used that account for personal expenses; the Fousts received business profits; and Mr. Foust was knowledgeable of Chesapeake's practices due to his previous employment there.

Prior to trial, Mr. Foust requested a Daubert hearing to determine whether the government's handwriting expert, Arthur Linville, would be allowed to testify. Mr. Linville is an experienced, board-certified forensic document examiner and was initially retained by Chesapeake during its investigation. During the hearing he explained his methodology, which consists of comparing the known writing with "exemplars of the suspect's writing." 2 R. 15. He looks for common characteristics between the exemplars as well as any unexplained differences. When comparing exemplars, Mr. Linville considers their "[q]uantity, quality and comparability." Id. at 29. He first determines the number of exemplars needed for a comparison, which depends on the range of variation in an individual's handwriting. He considers whether an exemplar was written "in the normal course of business" and prefers "relatively contemporaneous" writings "within a year or two" of each other. Id. at 28-29. However, Mr. Linville explained that the necessary timing can vary noting that change in handwriting over time is "somewhat overstated." Id. at 43-44. Mr. Linville follows American Society for Testing and Materials ("ASTM") standards but conceded at the hearing that they are "pretty basic" and not "hard-and-fast rules." Id. at 57-59. Finally, he mentioned that studies have found that forensic document examiners had a less than 1% error rate, while lay people had a 6.5% error rate.

While explaining his analysis of this case, Mr. Linville sorted the invoices by the type of forgery, which included "cut and paste" forgeries and "freehand" forgeries. Id. at 18. The fraudulent invoices associated with another Chesapeake employee (Jeff Willis) were cut-and-paste forgeries. Mr. Linville could not provide an opinion on these forgeries and the jury ultimately hung on the counts associated with them. On the other hand, the invoices with Bobby Gene Putman's signatures were freehand forgeries. Mr. Linville testified that the invoices with Mr. Putman's signature "absolutely were not Mr. Putman's signature" but were written in the natural hand of the writer—i.e., the author did not try to recreate the signature. Id. at 26-27. He was able to rule out the Lucases because of the different style of numbers and the quality of penmanship. As to Mr. Foust, Mr. Linville compared the invoices to exemplars of Mr. Foust's writing from 2002, 2011, and 2017. Mr. Linville opined that Mr. Foust forged Mr. Putman's signature because of similarities in pictorial appearance, skill, and other unique characteristics in the numbers.

Next, the district court laid out the standards it would apply under Rule 702 and summarized Mr. Linville's testimony. The court concluded that Mr. Linville was using an accepted methodology and reliably applied the methodology to the facts of the case. Although the court noted that handwriting comparison looks like "black magic" to the "untrained eye," it was still able to look at the samples and determine whether there were facts supporting Mr. Linville's opinion. Id. at 70-71. Therefore, the court concluded that Mr. Linville's testimony was admissible.

Discussion

...

On appeal, Mr. Foust does not challenge Mr. Linville's qualifications but instead raises two arguments aimed at the reliability of his methodology. First, he argues that the government did not meet its burden of establishing that Mr. Linville's methodology was reliable under the Daubert/Kumho Tire test. Second, he argues that Mr. Linville did not use reliable data—thus, making his method unreliable—because the exemplars were not sufficiently contemporaneous. We disagree and conclude the district court did not abuse its discretion in admitting the testimony.

A. Mr. Linville's Methodology

Beginning with Mr. Foust's broader argument, he contends that the prosecution failed to establish that Mr. Linville's methodology was reliable. In Daubert, the Supreme Court highlighted a number of considerations relevant to this inquiry: (1) whether the theory can be tested; (2) whether it is subject to peer review and publication; (3) the known or potential error rate; (4) the existence and maintenance of standards; and (5) the general acceptance in the relevant scientific community. 509 U.S. at 593-94. However, this list is not exclusive, and the test for reliability is flexible. Kumho Tire Co., Ltd. v. Carmichael, 526 U.S. 137, 150, 119 S. Ct. 1167, 143 L. Ed. 2d 238 (1999). Handwriting analysis is primarily an experience-based expertise, as opposed to science- based, which could make some Daubert factors less relevant than others. See id. Still, "some of Daubert's questions can help to evaluate the reliability even of experience-based testimony." Id. at 151.

Turning to the factors, we begin with whether Mr. Linville's method can be (and has been) tested. Although it appears that testing of handwriting comparison "mostly falls short of the rigors demanded by the ideals of science," it is still subject to less rigorous forms of testing. ... Testing may be done through "criminal investigation[and] court proceedings," as well as certification and proficiency exams. Id. Here, Mr. Linville has undergone years of training, has been certified by organizations of forensic examiners, and has rendered expert opinions in hundreds of cases. Therefore, this factor provides some support for admissibility.

Furthermore, the general-acceptance factor weighs in favor of admissibility. As the district court noted, "based on his description of how his science is practiced and the way it has apparently been practiced for a good many years, my conclusion is [] he is using accepted methodologies." 2 R. 70. Mr. Linville testified that handwriting analysis is a comparative process and two fundamental principles underlie the process: (1) no two individuals' handwriting is the same and

(2) an individual's own handwriting is never exactly the same. He has also received training through forensic organizations and federal agencies, which demonstrates some consensus in the field. Although, as Mr. Foust argues, acceptance by unbiased experts is always better, that does not mean this factor cannot support admission. See Baines, 573 F.3d at 991. And given the widespread acceptance of handwriting comparison through the years, we think this factor supports admissibility in this case. ...

On the other hand, the standards, peer-review, and error-rate factors do not necessarily support admission. Mr. Foust was correct to point out that Mr. Linville's testimony regarding peer review and error rates was lacking. Mr. Linville stated that he participated as a guest editorial board member on a peer-reviewed forensic journal; however, there was no testimony regarding whether his own methodology had been peer reviewed. Mr. Linville also testified generically that forensic examiners have a less than 1% error rate while lay people have a 6.5% error rate, but it is not clear that the study concerned the process of identifying the author. While there may be available evidence regarding peer review and error rates, it was not adequately presented at the Daubert hearing to support admission.

Finally, Mr. Linville follows the ASTM standards for his analysis, but he testified that these guidelines were "pretty basic" and not "hard-and-fast rules." 2 R. 59. Much like fingerprint analysis, handwriting comparison relies a lot "on the subjective judgment of the analyst," which may cut against admissibility. Baines, 573 F.3d at 991. With that said, the nature of handwriting comparison as an experience-based expertise lends itself to greater reliance on subjectivity when compared to science-based expertise. So while this factor does not necessarily support admission, we think it has less relevance in the specific context of handwriting comparison. ...

On balance, our review of the Daubert factors provided mixed results, however we recognize that they are "meant to be helpful, not definitive." Id. at 151. This understanding is particularly important because handwriting comparison is not a traditional science, and the Daubert factors do not always correspond perfectly. See id. at 150. During the hearing, the district court heard extensive testimony about Mr. Linville's methodology and analysis in this case and observed that whether the expert used accepted methodologies really had not been questioned. 2 R. 70. We recognize that there has been criticism of handwriting expertise in both the courts and academic literature. ... However, given our standard of review, the district court did not abuse its discretion in finding Mr. Linville's methodology reliable.

B. Mr. Linville's Underlying Data

Mr. Foust next argues that Mr. Linville's opinion is based on faulty data that renders his methodology unreliable. Specifically, he highlights Mr. Linville's use of exemplars outside his one-to-two-year range for contemporaneousness, thus violating his own methodology. We disagree.

To start, Mr. Foust overstates Mr. Linville's testimony. When discussing the timing of exemplars, Mr. Linville said, "it would be nice if they were written relatively contemporaneous in time with the questioned writing, and that can be within a year or two preceding or following the documents at issue." 2 R. 29. He noted that this was his "personal preference" and it was a view "expressed in text that [he] own[s]." 2 R. 43. He elaborated further that the need for contemporaneous exemplars varies, explaining that change in handwriting is "somewhat overstated" but there can be circumstances where very recent exemplars are needed. 2 R. 43-44. Mr. Foust's suggestion that the two- year cutoff is a strict rule under Mr. Linville's methodology is not supported by the record.

Furthermore, Mr. Linville relied primarily on the 2011 and 2017 exemplars that were two-and-a-half months and eight-months outside the two-year range. Mr. Foust's quarrel about the two-year cutoff goes to the weight, not the admissibility of Mr. Linville's testimony. It is "for the jury to evaluate the reliability of the underlying data, assumptions, and conclusions." In re Urethane Antitrust Litig., 768 F.3d 1245, 1263 (10th Cir. 2014). Not every issue raised about an expert's opinion requires the testimony to be excluded. Instead, many of those concerns can be addressed through "[v]igorous cross-examination, presentation of contrary evidence, and careful instruction on the burden of proof." Daubert, 509 U.S. at 596. Indeed, Mr. Foust used much of his cross-examination to discredit the exemplars' reliability. The district court also instructed the jury that it is not required to accept opinion testimony and that it is free to give the testimony as much weight as the jury thinks it deserves. 1 R. 320.

Because we believe this issue concerns the data's reliability—as opposed to the methodology's reliability as a whole—Mr. Foust's reliance on a nonprecedential case, Crew Tile, is inapt. ... In that case, we concluded that the district court erred by admitting a handwriting expert's testimony where the expert failed to complete the verification step of her methodology. Id. at 797-98. We reasoned that by skipping that step the expert had "completely changed a reliable methodology" or "misapplied that methodology." Id. at 797 (alteration and citation omitted). Importantly, we noted that the district court did not conduct a Daubert hearing and the party failed to show that the methodology was still reliable without the missing step. Id. at 797. Here, the district court conducted a Daubert hearing where Mr. Linville testified precisely about how the importance of contemporaneous exemplars can vary. Mr. Linville did not skip a step or change his methodology.

AFFIRMED.

Questions:

1. When evaluating the credentials of handwriting and document experts, the court concludes that Education is more important than Experience. True or False"
2. What three criteria and measures are central to reaching any reliable conclusion regarding the similarity of handwriting samples?
3. In general the timing of exemplars as compared to the actual document or handwriting need be at any time or preferably "contemporaneous?"

7.4 DOCUMENTARY EVIDENCE: INKS, IMPRESSIONS AND COMPOSITION

Forensic examiners in the world of documents and handwriting analysis often delve into the types of papers, inks employed from printers or writing utensils, impressions and watermarks and a host of other secondary, albeit circumstantial, evidence regarding documentary evidence. Sometimes authorship or agency in the authorship can be proven by these secondary forms of proof. Thus, can printer's ink be compared? Or the weight and composition of paper products and the like?

> Read the DOJ's press release about handwriting in notes taken during CIA meetings, used in espionage case at https://www.justice.gov/opa/pr/former-cia-officer-sentenced-conspiracy-commit-espionage

Locate the case cited below using your favorite legal research resource, read the case and answer the questions below.

People v. Moore, 6 Cal. App. 5th 73, Court of Appeal of California, Third Appellate District, November 29, 2016, Opinion Filed C075231

Questions:

1. Proof of murder sometimes relies on documentary evidence. Explain some examples that emanate in People v. Moore?
2. How does the prosecution use these documents, or prove that these documents were created or used by the actual murderer?
3. What other forensic evidence buttressed the documentary conclusions?

People v. Pinkney
316 Mich. App. 450 *; 891 N.W.2d 891
Court of Appeals of Michigan July 26, 2016, Decided
No. 325856

Defendant, Edward Pinkney, was convicted by a jury of five counts of election forgery, MCL 168.937, but acquitted of six counts of making a false statement in a certificate-of-recall petition, MCL 168.957. He was sentenced as a fourth-offense habitual offender, MCL 769.12, to concurrent prison terms of 30 to 120 months. He appeals as of right his December 15, 2014 judgment of sentence. We affirm.

I. BACKGROUND

Defendant's convictions arise out of the unsuccessful recall efforts against the mayor of Benton Harbor, James Hightower. Hightower was elected mayor in 2011 …

On October 23, 2013, James Cornelius filed a petition seeking the recall of Hightower as the mayor of Benton Harbor. According to the petition, Hightower was to be recalled "for voting no to not allow the city income tax to be placed on the 11.5.13 ballot." Carolyn Toliver, the elections administrator for the Berrien County Clerk's Office, testified that the recall petition was accepted, and "a clarity factual hearing" on the petition was scheduled for November 6, 2013. At that hearing, the Berrien County Election Commission approved the recall petition's language.

…

On January 8, 2014, defendant returned to the Berrien County Clerk's Office with a stack of signed recall petitions, but Toliver was unable to accept the recall petitions from anyone but the sponsor, i.e., Cornelius. Consequently, defendant contacted Cornelius, who came to the clerk's office and submitted the signed recall petitions to Toliver that same day. In total, there were 62 signed petitions containing 728 signatures submitted on January 8. While more than 300 of the submitted signatures were disqualified for various reasons, 402 signatures were certified by Toliver. Therefore, the recall petitions were accepted, and a recall election was scheduled for May 6, 2014.

However, the recall election was never held. After Toliver called for the recall election, Hightower raised concerns with her regarding the authenticity of the dates on the recall petitions that were submitted. Detective- Sergeant David Zizkovsky with the Berrien County Sheriff's Department took possession of the 62 recall petitions and studied them for irregularities. Specifically, Zizkovsky identified several dates next to corresponding signatures that appeared to have been altered. Consequently, he took the 62 petitions to the Michigan State Police Crime Laboratory, and he and Detective-Sergeant Mark Goff, a forensic document examiner with the laboratory, who was admitted "as an expert regarding opinions in the area of forensic document examination" at trial, decided to examine 10 recall petitions in detail.

Of the 10 recall petitions that were examined, 5 are of significance, and they are numbered in the record as Petitions 1, 6, 18, 19, and 38. Each of these petitions was circulated and signed by defendant. Goff testified that each petition was examined using photomicroscopy, which he described as "tak[ing] pictures through microscopes," and a "Video Spectral Comparator," which he described as "a flashlight with a group of colored lenses and then some more filters that filer out that light." Using this equipment, Goff was able to look for differences in ink color and variations in the type of pen used. Goff also used an electrostatic detection device, which he testified "detects impression in paper." As it relates to Petition 1, Goff expressed concerns over the authenticity of the dates on lines 2 through 6. Specifically, it appeared as though the dates were originally written as November 8, 2013, but subsequently changed to November 9, 2013. This determination was based on a change in ink color—i.e., from a black ink with a reddish hue to a dark black ink—and a difference in the burr striations. As it relates to Petition 6, Goff expressed concerns over the authenticity of the dates on lines 1 and 2. Specifically, it appeared as though the dates were originally written as November 7, 2013, but subsequently changed to November 9, 2013. Again, this was determined based on a change in ink color, i.e., from one red ink to a different red ink. As it relates to Petition 18, Goff expressed concerns over the authenticity of the dates on lines 5 and 6. Specifically, it appeared as though the dates were originally written as November 8, 2013, but subsequently changed to November 18, 2013. According to Goff, the "1" in "18" was written with a different ink than the remaining numbers in "11-18-2013." As it relates to Petition 19, Goff expressed concerns over the authenticity of the dates on lines 5 through 10. Specifically, it appeared as though the dates were originally written as November 8, 2013, but subsequently changed to November 18, 2013, in the same fashion as those on Petition 18. Finally, as it relates to Petition 38, Goff expressed concerns over the authenticity of the dates on lines 1 through 12. Specifically, it appeared as though the dates were originally written as November 8, 2013, for lines 1 through 4, but subsequently changed to November 18, 2013, and as though the dates were originally written as November 8, 2013, for lines 5 through 12, but subsequently changed to November 28, 2013. Again, the "1" in "18" and the "2" in "28" were written with a different ink than the remaining numbers in the dates. Goff also testified that he was able to find impressions from lines 6 through 10 on Petition 19—i.e., five of the six lines that were altered—on Petition 38, which he described as an indication that Petition 38 was underneath Petition 19 when the alterations at issue were made. Additionally, Zizkovsky testified that between March 29, 2013, and April 13, 2014, he watched a news story that included still photographs of the recall petitions, including Petition 38. In the still photographs of Petition 38, Zizkovsky noticed that the dates were written as November 8, 2013; however, the dates on Petition 38 were written as November 18, 2013, when they were examined.

As it turns out, these alterations to the dates proved crucial. The signatures on these recall petitions remained valid for only 60 days. See MCL 168.961(2)(d). That is, if the signers signed the recall petitions more than 60 days before they were filed, their signatures would be deemed invalid. The parties agree that the 60-day period commenced in this case on November 9, 2013. Therefore, it appeared that the alterations were made in an attempt to validate otherwise invalid, untimely signatures. Numerous individuals who signed the recall petitions were called to testify, but they largely testified that they did not recall whether they were responsible or whether someone else was responsible for the alterations to the dates.

The record reflects that defendant was familiar with the recall-petition procedures, including the applicable time limitations such as the 60-day rule. According to Toliver, defendant had sponsored recall petitions for three members of the Benton Harbor School Board in September 2013, had received multiple information packets outlining the applicable time limitations such as the 60-day rule, and had the applicable time limitations such as the 60-day rule explained to him when he received the information packet on each occasion. The prosecution also presented the testimony of Sharon Tyler, the Berrien County Clerk, who oversees Toliver. According to Tyler, between

August and October 2014, defendant submitted 12 recall petitions against her in connection with what she described as the "recall petition drive" against Hightower.

Defendant was eventually charged with five counts of election forgery, MCL 168.937, and six counts of making a false statement in a certificate-of-recall petition, MCL 168.957.

...

After an eight-day jury trial, defendant was convicted ...

II. ANALYSIS

On appeal, defendant argues that his convictions must be reversed for four reasons. First, defendant argues that his convictions must be reversed because MCL 168.937 does not create the substantive offense of election forgery. Second, he argues that his convictions must be reversed because the prosecution presented insufficient evidence to support his convictions. Third, defendant argues that his convictions must be reversed because the trial court erroneously instructed the jury that it could convict him under an aiding-and-abetting theory. Finally, defendant argues that his convictions must be reversed because other-acts evidence was admitted against him in violation of MRE 404(b), his constitutional right to free speech, and his constitutional right to due process. Because we disagree in each respect, we affirm defendant's convictions and sentence.

A. MCL 168.937

Defendant argues on appeal that his convictions must be reversed because MCL 168.937 does not create the substantive offense of election forgery. We disagree.

Defendant was convicted of election forgery under MCL 168.937, which provides as follows:

Any person found guilty of forgery under the provisions of this act shall, unless herein otherwise provided, be punished by a fine not exceeding \$1,000.00, or by imprisonment in the state prison for a term not exceeding 5 years, or by both such fine and imprisonment in the discretion of the court.

...

B. SUFFICIENCY OF THE EVIDENCE

Defendant also argues on appeal that his convictions must be reversed because the prosecution presented insufficient evidence to support his convictions. Specifically, he claims that the jury could not have concluded that he altered the dates on the recall petitions with the requisite intent beyond a reasonable doubt. We disagree.

This Court reviews a defendant's challenge to the sufficiency of the evidence de novo. People v Cline, 276 Mich App 634, 642; 741 NW2d 563 (2007).

"In determining whether the prosecutor has presented sufficient evidence to sustain a conviction, an appellate court is required to take the evidence in the light most favorable to the prosecutor." People v Bosca, 310 Mich App 1, 16; 871 NW2d 307 (2015) (citation and quotation marks omitted). "The standard of review is deferential and this Court 'is required to draw all reasonable inferences and make credibility choices in support of the jury verdict.'" People v Powell, 278 Mich App 318, 320; 750 NW2d 607 (2008), quoting People v Nowack, 462 Mich 392, 400; 614 NW2d 78 (2000). In this case, because the Michigan Election Law does not define the substantive offense of election forgery, we turn to the common-law definition of forgery, People v Riddle, 467 Mich 116, 125; 649 NW2d 30 (2002), which is, as stated earlier, "a false making ... of any written instrument with intent to defraud," Nasir, 255 Mich App at 42, n 2 (citation and quotation marks omitted). "Because intent may be difficult to prove, only minimal circumstantial evidence is necessary to show a defendant entertained the requisite intent." People v Harverson, 291 Mich App 171, 178; 804 NW2d 757 (2010); ... Indeed, circumstantial evidence, alone, may be sufficient to prove all elements of an offense. People v Unger, 278 Mich App 210, 223; 749 NW2d 272 (2008). Relatedly, "[a]lthough motive is not an essential element of the crime, evidence of motive in a prosecution … is always relevant." Id. When a defendant's "sufficiency argument is couched in terms of the trial court's having denied his motion for a directed verdict," "we only consider the evidence presented by the prosecution up to the time the motion was made." Powell, 278 Mich App at 320, n 1, citing People v Gould, 225 Mich. App. 79, 86; 570 N.W.2d 140 (1997).

Applying those rules to this case, we conclude that the prosecution presented sufficient evidence for the jury to find defendant guilty of violating MCL 168.937 beyond a reasonable doubt prior to defendant's directed-verdict motion. The evidence presented at trial portrayed defendant as the leader in the recall efforts at issue. Defendant, who had previously sponsored several recall campaigns in Benton Harbor, asked Cornelius to sponsor this specific recall petition because he was a Benton Harbor resident; successfully obtained the recall petitions from Toliver; circulated 33 of the 62 recall petitions that were submitted; advocated on behalf of the recall campaign at city commission meetings, BANCO meetings, and generally throughout the community; attempted to file the 62 recall petitions at issue; and demonstrated animosity toward Hightower in various ways during the period at issue. Having filed several recall petitions in the past, defendant was admittedly familiar with the 60-day rule regarding the signatures' validity, and he was undisputedly in control of the petitions when he attempted to file them on his own as well as when Cornelius filed them in his presence. The testimony that was presented indicated that the only dates altered on these recall petitions were those that fell beyond the 60-day period, November 7 and 8. Those dates were altered to instead reflect November 9, 18, or 28, all dates that fell within the 60- day period. This evidence, while circumstantial, constituted sufficient evidence to support the jury's guilty verdicts when viewed in a light most favorable to the prosecution.

Defendant's primary argument on appeal relies on his claim that he was merely found in possession of the altered recall petitions, but he was not actually responsible for the alterations themselves. However, a review of the record reflects that defendant's convictions were supported by substantially more than his mere possession of the recall petitions. As discussed earlier, the evidence presented portrayed defendant, who was familiar with the recall-petition process, as having motive to alter the recall petitions to bring them within the 60-day period. While possession of the forged recall petitions alone may have been insufficient to support a conviction under MCL 168.937, this possession of the forged recall petitions, coupled with his demonstrated animosity against Hightower, his knowledge of the 60-day rule, and his motive to alter the dates to bring them within the 60-day period, constituted sufficient evidence, when viewed in a light most favorable to the prosecution, Bosca, 310 Mich App at 16, for a rational trier of fact to find defendant guilty of violating MCL 168.937.

…

III. CONCLUSION

In sum, we conclude that MCL 168.937 does create the substantive offense of election forgery, that the prosecution presented sufficient evidence to support defendant's convictions, that the jury was correctly instructed, and that the trial court did not erroneously admit other-acts evidence. We therefore affirm defendant's convictions and sentence.

Affirmed.

Questions:

1. What documents were scrutinized and why?
2. Explain some of the irregularities discovered by document investigators who reviewed a sample?
3. Why are the dates on the documents crucial to the state's election forgery case?

State v. Saunders

Court of Appeals of Arizona, Division One June 14, 2016, Filed

No. 1 CA-CR 15-0416

P1 Christopher Lee Saunders appeals his conviction of armed robbery, a Class 2 felony. For the reasons that follow, we affirm his conviction and resulting sentence.

FACTS AND PROCEDURAL BACKGROUND

P2 Saunders entered a sandwich shop and placed a plastic bag and a hand-written note torn from a spiral notebook on the counter. The note read, "I have a gun, put the money in the bag and step away, nobody dies." Before the owner had a chance to read the note, however, Saunders pulled out what appeared to be a gun, and the owner gave Saunders $200 from the cash register. Saunders left the restaurant on a white BMX bike with black rims and black handlebars.

P3 Responding officers interviewed the restaurant owner and reviewed restaurant surveillance video footage. After identifying Saunders as a suspect, officers obtained and executed a search warrant for his apartment. Officers seized a black hand brace, a pair of men's dark sunglasses, a backpack, a long-sleeved black shirt and a white and gray BMX bicycle with black handlebars, all of which were consistent with the surveillance footage. Officers also seized a small spiral- bound notebook from inside the backpack; in the notebook was a page of paper from which a fragment had been torn.

P4 At trial, a forensic document examiner testified about paper-matching and handwriting analyses he conducted on the stick-up note. The document examiner opined that the fragment of a sheet of paper on which the note was written had been ripped from the spiral-bound notebook found in the backpack in Saunders's apartment. He also testified that there were indications Saunders wrote the note but he could not conclusively identify Saunders as the writer. A jury found Saunders guilty, and the superior court sentenced him to 15.75 years' imprisonment and $200 in restitution. Saunders timely appealed; we have jurisdiction pursuant to Arizona Revised Statutes sections 12-120.21(A)(1) (2016), 13-4031 (2016) and -4033(A)(1) (2016).

DISCUSSION

A. Legal Principles

P5 Saunders argues the superior court should not have admitted the document examiner's testimony because his opinions did not comport with the requirements of Arizona Rule of Evidence 702. That rule governs admission of expert testimony, and states:

> A witness who is qualified as an expert by knowledge, skill, experience, training, or education may testify in the form of an opinion or otherwise if:
>
> a. the expert's scientific, technical, or other specialized knowledge will help the trier of fact to understand the evidence or to determine a fact in issue;
> b. the testimony is based on sufficient facts or data;
> c. the testimony is the product of reliable principles and methods; and
> d. the expert has reliably applied the principles and methods to the facts of the case.

P6 We review the superior court's admission of expert testimony for an abuse of discretion. State v. Salazar- Mercado, 234 Ariz. 590, 594, ¶ 13, 325 P.3d 996 (2014). The superior court judge serves as "gatekeeper," determining whether the proposed expert testimony is relevant and admissible, but "it is the province of the jury to determine the weight and credibility of the testimony[.]" State ex rel.

Montgomery v. Miller, 234 Ariz. 289, 298, ¶¶ 19-20, 321 P.3d 454 (App. 2014) (quoting Ariz. R. Evid. 702 cmt. to 2012 amendment). "We review the interpretation of a court rule de novo." Salazar-Mercado, 234 Ariz. at 592, ¶ 4.

B. Paper-Matching Analysis

P7 Saunders argues that, in violation of industry standards, the paper-matching analysis was conducted after the note had been subjected to other forensic testing. In support of his argument, Saunders points to guidelines for the analysis of paper-matching issued by the American Society for Testing and Materials ("ASTM"). That publication, titled "Standard Guide for Physical Match of Paper Cuts, Tears, and Perforations in Forensic Document Examinations," states: "The results of prior storage, handling, testing, or chemical processing (for example, for latent prints) can interfere with the examination of certain characteristics. Whenever possible, document examinations should be conducted prior to any chemical processing. Items should be handled appropriately to avoid compromising subsequent examinations." The document examiner testified the note was slightly discolored when he examined it because it had been already tested for latent fingerprints. Saunders argues that by examining the note after it had been tested for prints, the document examiner failed to follow the ASTM guidelines, thereby rendering the testimony inadmissible under Rule 702(d).

P8 Rule 702(d) requires the superior court to determine whether an expert has reliably applied the applicable standards in formulating his or her opinion. See State v. Bernstein, 237 Ariz. 226, 229, ¶ 13, 349 P.3d 200 (2015). But the court is not required to exclude expert testimony based on any slight variation from standards. "Rule 702 contemplates that expert testimony can be 'shaky' yet admissible...Errors in the application of a generally reliable methodology, therefore, should not serve to exclude evidence unless they are so serious as to render the results themselves unreliable." Id. at ¶¶ 14-15.

P9 Contrary to Saunders's argument, the ASTM guidelines suggest but do not require that paper-matching testing be performed before the evidence is subjected to any other analyses. Although the guidelines caution that other testing may affect the paper-matching analysis, "not all errors in the application of reliable principles or methods will warrant exclusion." Id. at ¶ 14. More importantly, the examiner here testified that the discoloration of the paper did not affect his analysis: "[A]s far as the tear match examination, the previous latent print processing did not hinder that exam whatsoever."

P10 Saunders also argues that the paper-matching testimony should have been excluded under Rule 702(a) because the testimony was not helpful to the jury. Saunders contends that because the examiner admitted that a lay person could have conducted the same analysis, the testimony did nothing to aid the jury in understanding the evidence. Contrary to Saunders's contention, however, the document examiner explained that although a lay person could have put the two pieces of paper together "like a jigsaw puzzle[,]" he also testified he used special equipment and expertise in performing his analysis. The document examiner explained that he used a high-magnification camera that allowed him to look at the individual tears and match up the paper fibers. The examiner's testimony involved specialized knowledge and aided the jury in determining whether the note had come from the notebook recovered at Saunders's apartment.

P11 For these reasons, the superior court did not abuse its discretion in admitting the paper-matching testimony.

C. Handwriting Analysis

P12 Saunders further argues insufficient data supported the document examiner's handwriting analysis as required by Rule 702(b). He also argues the examiner's opinion was too vague to be helpful to the jury as required by Rule 702(a).

P13 The document examiner explained that in analyzing whether the writing on the note belonged to Saunders, he used a nine-point scale, ranging from one ("conclusive elimination") to nine ("conclusive identification") with five ("inconclusive") in the middle. After comparing the note to a sample of Saunders's handwriting, the document examiner opined there were indications the two samples were written by the same person, concluding the identification was a six on the nine-point scale. At trial, the document examiner explained that the limited sample size and overwriting of letters on the note were the most significant factors limiting his ability to make a conclusive identification. Saunders argues that the examiner's inconclusive opinion, undermined by the limited sample size and overwriting, should not have been admitted because it was neither helpful to the jury nor supported by sufficient data.

P14 Notwithstanding Saunders's arguments, sufficient data supported the document examiner's handwriting analysis. The examiner testified, "[T]he amount of writing on the questioned note was limited but there was still enough there to do an examination and a comparison between the questioned note and the known writing I had of Mr. Saunders." Although the examiner's final conclusion was less than conclusive, he was able to point out similarities between the note and Saunders's writing sample. As the court noted when it denied Saunders's motions to exclude the testimony, "The concerns noted go greater to the weight of the evidence, to be determined by the trier of fact, rather than to its admissibility."

P15 Because the document examiner had sufficient data to conduct the handwriting analysis and because the testimony aided the jury in deciding whether Saunders had written the stick-up note, the superior court did not abuse its discretion in admitting the testimony. As the superior court found, Saunders's arguments concerning the certainty of the examiner's conclusion went to the weight the jury could give to the examiner's testimony, not to its admissibility.

CONCLUSION

P16 For the foregoing reasons, we affirm Saunders's conviction and sentence.

Questions:

1. Documents come in a wide variety of forms. What document is examined and offered up as proof of criminal agency?
2. The prosecution also seeks to prove agency by other means. Can you elaborate on the one that deals torn paper?
3. What challenge to the torn paper theory is posed by the defendant?

8 Forensic Evidence

Ballistics

8.1 INTRODUCTION

The field of ballistics is broadly defined as the study and analysis of firearms, their manufacture and design characteristics as well as the projectiles employed and the residue left behind. Firearms, depending on make and manufacturer, display their own unique characteristics and while it is impossible to fully replicate ballistic evidence, there can be, in controlled conditions, enough similarities to lead an evaluator to a match. In some cases, the weaponry has unique markings and rifling which cause markings and striations that are exclusive to the weapon under evaluation. Other ballistic categories include types of power, gas and grease by-products, shot and cotton wadding remnants, and other signature characteristics for the weapon under evaluation (see Figure 8.1).

8.2 FORENSIC EVIDENCE: GUNSHOT RESIDUE

People v. Castillo
Court of Appeal of California, Second Appellate District, Division One
March 23, 2021, Opinion Filed
B301982

Defendant Hector Castillo was convicted of one count of first degree murder, one count of shooting at an occupied motor vehicle, and two counts of unlawful possession of a firearm. The jury found true gang enhancement allegations relating to all four offenses, and it also found true firearm enhancement allegations pertaining to the murder and shooting at an occupied motor vehicle convictions. The trial court sentenced Castillo to an aggregate prison term of 50 years to life, with a minimum parole eligibility date of 15 calendar years. The lower court also imposed, but stayed, certain prison terms, including two 10-year gang enhancements.

On appeal, Castillo alleges claims of prosecutorial misconduct, the erroneous exclusion of evidence supporting his theories of imperfect self-defense and third-party culpability, instructional error, the erroneous denial of his motion to bifurcate the gang allegations from the trial of the underlying substantive offenses, insufficiency of the evidence supporting the gang allegations, and cumulative error. Castillo further contends that the trial court erred in imposing the 15-year minimum parole eligibility date and the two 10-year gang enhancements.

Castillo's prosecutorial misconduct claims fail because he has not demonstrated that he was prejudiced by the People's actions. We also reject his challenge to the trial court's evidentiary ruling and his claim of instructional error because his characterizations regarding the excluded evidence giving rise to those claims are unsupported by the record. His challenges to the denial of his bifurcation motion and to the sufficiency of the evidence fail because the People offered evidence showing that Castillo shot a rival gang member in territory claimed by Castillo's gang in order to defend that territory and avenge the murder of Castillo's brother, who was a member of his gang. Castillo's cumulative error claim does not warrant reversal because, aside from his allegations of

DOI: 10.4324/9781003267126-8

FIGURE 8.1 Close-up view showing measurements of shell casing located on the sidewalk outside the grocery store where the January 8, 2011, shooting by Jared Lee Loughner occurred in Tucson, Arizona. This casing had blood on it, and it was marked as evidence item 26.

Source: FBI Records: The Vault at https://vault.fbi.gov/2011-tucson-shooting/2011-tucson-shooting-media/dsc0262.jpg/view.

prosecutorial misconduct, we explicitly pass upon and reject the merits of his claims of trial error. Lastly, although we accept the parties' concession that the two 10-year gang enhancements should be stricken, we disagree with Castillo's contention that the trial court's imposition of the 15-year minimum parole eligibility date was unauthorized. We thus affirm the judgment as modified.

PROCEDURAL BACKGROUND

On September 28, 2016, the People filed an information charging Castillo with one count of murder, in violation of Penal Code section 187 (count 1); one count of shooting at an occupied motor vehicle, in violation of section 246 (count 2); one count of unlawful possession of a firearm, in violation of section 29805 (count 3); and another count of unlawful possession of a firearm, in violation of section 29815, subdivision (a) (count 4). With regard to counts 1 and 2, the information alleged that Castillo personally and intentionally discharged a firearm causing great bodily injury for the purposes of section 12022.53, subdivision (d). Additionally, the information averred that Castillo committed all four offenses for the benefit of, at the direction of, or in association with a criminal street gang with the specific intent to promote, further, or assist in criminal conduct by gang members. Castillo later pleaded not guilty to all counts.

On February 21, 2019, the jury found Castillo guilty of first degree murder and of all other charged offenses, and found the gang and firearm enhancement allegations to be true.

On October 22, 2019, the trial court sentenced Castillo to an aggregate prison term of 50 years to life. The sentence is comprised of a term of 25 years to life on count 1, with a consecutive sentence of 25 years to life imposed on that count pursuant to section 12022.53, subdivision (d);

the court imposed, but stayed pursuant to section 654, a 10-year concurrent enhancement on count 1 pursuant to section 186.22, subdivision (b)(1)(C). Pursuant to section 186.22, subdivision (b)(5), the trial court ruled that Castillo could not be paroled until he had served a minimum of 15 years of his prison sentence. The court also imposed prison terms for counts 2, 3, and 4, but stayed each of those sentences pursuant to section 654. Of particular note for the instant appeal is the sentence imposed for count 2, which was a 7-year prison term, along with: a 10-year enhancement pursuant to section 186.22, subdivision (b)(1)(C); an indeterminate life sentence pursuant to section 186.22, subdivision (b)(4); and a 25-year enhancement pursuant to section 12022.53, subdivision (d).

Castillo timely appealed the judgment.

FACTUAL BACKGROUND

This part summarizes relevant portions of (a) testimony offered by the People's witnesses, and (b) the defense's theory of the case.

1. The People's Evidence

A. The Shooting and Its Aftermath

On March 21, 2015, Rosario Marban lived at a home on Eastlake Avenue in Los Angeles (Marban residence) with his family, including his son, Alberto Marban. A family gathering was taking place at the house that evening. When Oscar Rivera arrived at the Marban residence, he walked through an archway and saw two persons sitting on chairs outside the house; Rivera later made his way to the backyard, as did Rosario. While they were in the backyard that evening, Rivera and Rosario heard gunshots.

Christian Lepe was driving on Minnesota Street near Eastlake Avenue when he heard gunshots and saw a car that had flipped over. Lepe parked his vehicle and saw a man inside the other car who was not moving.

At 6:54 p.m., two police officers responded to a shooting reported at an address located near the Marban residence. After arriving at that location, one of the officers saw a white sport utility vehicle (SUV) that had flipped over onto its passenger side. A Lincoln Heights gang member named David Cardiel was the sole occupant of the SUV. Cardiel had sustained multiple gunshot wounds to his face and body, and was pronounced dead at the scene. Police found no firearms on Cardiel.

Officers thereafter canvassed the area north of the scene of the shooting. The officers subsequently asked the occupants of the Marban residence to exit the home, and then conducted a sweep of the dwelling for suspects.

One of the officers went upstairs to Alberto's bedroom, where the officer found Castillo lying face down on a bed. Castillo, who was breathing heavily and sweating, was taken into custody.

A detective entered a bathroom in the home, examined a trash can, and found therein a pair of jeans, a Los Angeles Dodgers baseball cap, and a damp shirt that had been turned inside out. Inside the pocket of the jeans were three identification cards that bore Castillo's name, including a California identification card. The detective used the same gloves to remove each item of clothing from the trash can.

As Alberto saw the police remove Castillo from the Marban residence, Alberto noticed that Castillo was wearing Alberto's shirt. At trial, Alberto claimed that although he had seen Castillo on the morning of March 21, 2015, Castillo did not come to family gathering. Alberto further claimed that he did not give Castillo permission to wear Alberto's shirt or enter his bedroom. Alberto also insisted that the clothing found in the bathroom trash can did not belong to him, and that he did not place those articles of clothing in the trash can.

An officer conducted a protective sweep of the backyard of the Marban residence, and found a rifle on the hill by the fence line. The magazine contained four live bullets, and the chamber contained one bullet. A glove was found under the rifle. An opening along the fence led to a neighbor's property. Police also "recovered three cartridge casings, two of which were on the landing between two staircases in front of the residence, a Monster energy drink can in the front yard, and a red spray paint can in a planter."

One of the officers investigating the shooting noticed that the words "Eastlake 1" and "Swifty" appeared in red paint on a wall across the street from the Marban residence. The officer testified that he could smell paint in the air in the vicinity of the wall, and that this smell indicated that someone had recently spray-painted the graffiti.

B. Forensic Evidence

The shell casings found at the scene of the shooting were fired from the rifle found in the backyard of the Marban residence. The weapon was a 7.62 by 39- millimeter caliber rifle.

Gunshot residue (GSR) is a group of microscopic particles that are formed when a firearm is discharged. GSR is comprised of three elements: lead, antimony, and barium. A "characteristic particle" contains all three aforesaid elements. If a characteristic particle is found on a person, then there are four possible explanations for that phenomenon: (1) that person discharged a firearm, (2) the individual was in the general vicinity of a firearm being discharged, (3) the person touched something that had GSR on it, or (4) something with GSR thereon touched that person. For instance, "when a police officer handcuffs someone, [that officer] could be a source of introducing … a characteristic particle onto someone's hands" Additionally, an "indicative particle" typically contains only two of the three elements found in GSR, and could "possibly come from … environmental sources" other than the firing of a gun.

One characteristic particle was found on Cardiel's left hand. A characteristic particle was also found on Castillo. Although no GSR was detected on Alberto, six indicative particles were found on him.

Detectives obtained DNA samples from Castillo and Alberto. DNA swabs were taken from the rifle, hat, shirt, and jeans seized on March 21, 2015. The rifle's trigger, trigger guard, wood stock, wood grips, and bolt were swabbed for DNA.

Alberto was the major contributor of the DNA found on the hat. Castillo was the major contributor of the DNA found on the shirt and jeans, and he was the single contributor of the partial DNA profile found on the rifle. The People's DNA examiner testified that he could not ascertain which part of the rifle the partial DNA profile came from, and that the DNA could have come from any part of the weapon that had been swabbed. Although the examiner could not say with certainty that direct human contact with the rifle is the only means by which the DNA could have been left thereon, he testified that "the easiest and most common way that DNA transfers … is by direct, primary handling of an item."

The random-match probability of the profile from the rifle was one in a hundred billion unrelated people.

…

DISPOSITION

The judgment is modified as follows: (1) We strike the 10-year gang enhancement imposed on count 1 under Penal Code section 186.22, subdivision (b)(1)(C); and (2) we strike the 10-year gang enhancement imposed on count 2 under Penal Code section 186.22, subdivision (b)(1)(C). As modified, the judgment is affirmed.

Upon the issuance of our remittitur, the trial court is directed to prepare a corrected minute order consistent with the views expressed in this opinion, amend the abstract of judgment, and send certified copies of the amended abstract of judgment to the Department of Corrections and Rehabilitation. The clerk of the court is directed to send a copy of the opinion and remittitur to the Department of Corrections and Rehabilitation. (See Cal. Rules of Court, rule 8.272(d)(2).)

Questions:

1. When searching for gunshot or powder residue, what three elements does the examiner try to identify?
2. Are there alternative explanations for why residue would be discovered and hence may not link the residue to the particular actor?
3. Prove True or False: Residue, according to this case, is only legitimate if found on the hands, arms or other exposed skin. Other locales are too remote.

Estate of Loury v. City of Chicago
United States District Court for the Northern District of Illinois, Eastern Division
February 25, 2021, Decided; February 25, 2021, Filed
Case No. 16-cv-4452

Plaintiff Tambrasha Hudson, the administrator of the following reasons, the Court, in its discretion, denies Officer Hitz's motion.

Factual Background

On April 11, 2016, defendant Officer Hitz and non- defendant Officer Richard Riordan were on duty as Chicago police officers when they responded to a call of "shots fired" that went out over the police radio. As the officers were driving southbound on Homan Avenue, individuals informed them that a person in a black sedan traveling northbound on Homan Avenue had fired the shots. The officers then followed the black sedan. After the sedan turned west on Roosevelt Road, Officer Riordan initiated a traffic stop. When the sedan came to a stop, the decedent Pierre Loury, who was sixteen- years-old at the time, exited the front passenger door and ran northbound through a vacant alley. Officer Hitz immediately pursued him. After Loury ran through the alleyway, he reached a fence and began to climb over it. Shortly thereafter, Officer Hitz fired two shots at Loury resulting in his death.

Expert Opinions

Hudson's expert, Ronald Scott, provides the following opinions regarding the gunshot residue ("GSR") found on Loury:

> The presence of GSR particles on Pierre Loury could have originated from his being in the presence of a gunshot including two gunshots fired by Officer Hitz.
>
> The presence of GSR particles on Pierre Lowry could have been transferred through the contact of Officer Riordan handcuffing him after having had his firearm out. Any GSR present on Officer Riordan's firearm either from prior shooting (i.e. range qualification), and the presence of GSR in his holster would likely be on his hands which then contaminate the hands of Pierre Loury through direct contact.
>
> The failure to properly preserve the hands at the scene prior to his movement to the hospital is an egregious violation of evidence preservation and protection and constitutes contamination.
>
> Therefore, the GSR results are based on failure to properly protect and preserve evidence, likely contamination resulting therefrom, and it cannot be excluded that the GSR results are from the gunshots of Officer Hitz or by transfer contact from Officer Riordan while handcuffing the victim.
>
> (R. 201-1, 8/15/17 Scott Expert Report, at 19)

Scott's Qualifications

Scott has testified as an expert witness over 300 times in the areas of firearms ballistics, shooting reconstruction, and shooting dynamics, among other topics, in both federal and state courts … As an independent forensic consultant, he has provide services to the United States Military, engineers, insurance companies, attorneys, prosecutors, authors, and architects. Scott has master's degrees in Management, Business Administration, and Criminal Justice. His bachelor's degree is in Law Enforcement and he has an associate's degree in Science. Throughout his career, Scott has engaged in extensive technical and forensic training, including training in crime science investigation, firearms, and ballistic materials research and testing. Scott also served in the United States Army where he received training in small arms and explosive ordinance reconnaissance.

Scott was employed as a police deputy by the Massachusetts State Police ("MSP") for more than 25 years before retiring and becoming an expert in firearms, ballistics, and police shootings. During his time with the MSP, he served in the "Ballistics Section" and was the commanding deputy of the MSP's ballistics laboratories. In that role, Scott conducted, supervised, and trained personnel in forensic investigations, shooting reconstruction, and the dynamics involved in shooting incidents. In addition, Scott was a member of the MSP Firearms Review Board, which evaluated departmental deputy-involved shooting incidents. Scott has investigated over 400 police-involved shootings and has personally conducted thousands of forensic investigations, attended approximately 700 postmortems, and was trained in the forensic interpretation of gunshot wounds from a shooting reconstructive perspective.

Also, Scott has been involved in a number of forensic science organizations, including the Association of Firearms and Toolmark Examiners. He was a voting member of the American Academy of Forensic Sciences Standards Consensus Board on Firearms and Tool Marks. Scott has given lectures and presentations at Harvard University, Northeastern University, Boston University, American Academy of Forensic Sciences, as well as at police academies.

LEGAL STANDARD

Rule 702 and Daubert require district judges to act as gatekeepers to ensure that proposed expert testimony is both reliable and relevant. Timm v. Goodyear Dunlop Tires North America, Ltd., 932 F.3d 986, 993 (7th Cir. 2019). When determining reliability, the Court's role is to assess whether the expert is qualified in the relevant field and to examine the methodology he used in reaching his conclusions. Id. The Court's focus is on the expert's methodology, not his conclusions. Kopplin v. Wisconsin Central Ltd., 914 F.3d 1099, 1104 (7th Cir. 2019). To be relevant, expert testimony must "help the trier of fact to understand the evidence or to determine a fact in issue." Fed.R.Evid. 702.

A district court's gatekeeping evaluation of expert testimony does not take the jury's place in deciding the issues of accuracy or credibility. Clark v. River Metals Recycling, LLC, 929 F.3d 434, 438 (7th Cir. 2019). Indeed, once the district court determines that "the proposed expert testimony meets the Daubert threshold of relevance and reliability, the accuracy of the actual evidence is to be tested before the jury with the familiar tools of 'vigorous cross-examination, presentation of contrary evidence, and careful instruction on the burden of proof.'" Lapsley v. Xtek, Inc., 689 F.3d 802, 805 (7th Cir. 2012) (quoting Daubert, 509 U.S. at 596). The expert's proponent has the burden of establishing the admissibility of his opinions by a preponderance of the evidence. Varlen Corp. v. Liberty Mutual Ins. Co., 924 F.3d 456, 459 (7th Cir. 2019).

ANALYSIS

Officer Hitz first argues that Scott is not qualified to offer his opinions regarding gunshot residue. Specifically, Officer Hitz takes issue with the fact that Scott does not have a college degree in the physical sciences like chemistry or physics. Officer Hitz also points out that Scott does not list gunshot residue analysis in his areas of expertise on his curriculum vitae and that he has no education or experience in the area of forensic science.

Pursuant to Rule 702 and Daubert, courts ask whether an expert is qualified to answer specific questions, not just whether an expert is generally qualified. United States v. Truitt, 938 F.3d 885, 889 (7th Cir. 2019). And, this expertise comes not just from formal education, but can be based on knowledge, skill, training, or education. United States v. Brown, 973 F.3d 667, 703 (7th Cir. 2020). When law enforcement officers testify as experts, they usually bring a wealth of experience and make important connections for jurors based on their specialized knowledge. See United States v. Wehrle, 985 F.3d 549, 553 (7th Cir. 2021).

Under this standard, whether Scott has a college degree in a physical science is not dispositive nor is the fact that Scott did not specifically list an expertise in gunshot residue analysis on his curriculum vitae. Meanwhile, Officer Hitz's broad statement that Scott has no education or experience in the area of forensic science simply ignores Scott's 40 plus years of ballistics and weapon training and experience, including that he conducted and supervised forensic investigations.

The Court thus turns to whether Scott has sufficient expertise involving gunshot residue analysis and whether the gunshot residue analysis results were based on the failure to properly protect and preserve evidence. The Court first notes that Scott has testified about gunshot residue analyses in other cases, including in the criminal matters, Arizona v. Ramone Trammel, Maricopa County Court, and West Virginia v. Charles Kennedy, McDowell County Court. Similarly, Scott has testified about the spoliation and alteration of evidence in the civil lawsuit, Scott v. Bullets & Burgers, Clark County, Nevada. Moreover, Scott's extensive experience in crime scene investigations coupled with his firearm and ballistics expertise underscores his knowledge in the area of evidence collection and gunshot residue. Although Scott has not worked as a chemist in a laboratory examining the gunshot particles expelled from a firearm, he has abundant experience and training to discuss the results of gunshot residue testing.

Next, the Court examines Officer Hitz's argument that Scott did not use reliable reasoning or a scientific methodology in forming his opinions. Scott's reasoning and "methodology" in coming to his conclusions is based on his observations in relation to the specialized experience that he has garnered over the last 40 years. ... In other words, it is not essential that law enforcement experts follow a scientific method in forming their opinions. ...

Here, Scott based his gunshot residue opinions on documentation concerning the crime scene and the Illinois State Police's ("ISP") Gunshot Residue Testing documents, including the swabbing of Loury's hands and the ISP's final report. He also reviewed photographs taken at the hospital after Officer Hitz shot Loury. After analyzing these reports, Scott connected them to his opinion by describing the absence of documentation on the preservation of Loury's hands from contamination prior to the collection of the gunshot residue. His analysis was independent and not merely a regurgitation of the ISP's documents or report. Scott's expert opinions are thus reliable.

Officer Hitz next contends that Scott's opinion testimony will not be helpful to the jury because it "mirrors" the ISP chemist Ellen Chapman's report, and thus would be cumulative. Although Chapman concluded that Loury either discharged a firearm, had contact with a gunshot residue-related item, or had both hands in the environment of a discharged firearm, Scotts' opinions are more developed and include the opinion concerning the proper preservation of the gunshot residue evidence. As such, his opinions will be help the jury make important connections.

Last, the Court rejects Officer Hitz's argument that the preservation of evidence opinion would be prejudicial to him because he does not sufficiently explain how it would be unfairly prejudicial under the standard set forth in Rule 403. Shipley v. Chicago Bd. of Election Comm'rs, 947 F.3d 1056, 1063 (7th Cir. 2020) (perfunctory and undeveloped arguments are waived).

Conclusion

Based on the foregoing, the Court, in its discretion, denies Officer Hitz's motion to exclude the gunshot residue expert testimony of Ronald Scott.

Questions:

1. Defense claims that the residue was not originally part of the defendant's person but transferred. How so?
2. Using Daubert, the prosecution claims the expert lacks a college degree and because of this, cannot be qualified as an expert. How does the Court resolve this challenge?
3. Is the Court's finding dispositive of innocence or guilt, or is its ruling limited to the admissibility of evidence alone?

> Read the story about the Reservation murders of FBI Agents Jack Coler and Ronald Williams which discusses the ballistics evidence in the case at https://www.fbi.gov/history/famous-cases/resmurs-case-reservation-murders.

State v. Daughtridge
789 S.E.2d 667
Court of Appeals of North Carolina
February 24, 2016, Heard in the Court of Appeals; August 2, 2016,
Filed No. COA15-1160

Travis Lamont Daughtridge ("Defendant") appeals from his convictions for first-degree murder and possession of a firearm by a felon. On appeal, he contends that the trial court plainly erred by allowing the admission of (1) an investigator's testimony concerning Defendant's demeanor; and (2) opinion testimony from a medical examiner that the victim's death was a homicide rather than a suicide. After careful review, we conclude that Defendant received a fair trial free from prejudicial error.

Factual Background

The State presented evidence at trial tending to establish the following facts: In 2011, Simeka Daughtridge ("Simeka") lived with her three children at her mother's house … On 26 August 2011, Simeka married Defendant, …

On 30 October 2011, while Defendant was at Linda's house, Defendant and Simeka began arguing in Simeka's bedroom. The door was shut, and they were alone together in the room. …

Approximately 10-15 minutes after Defendant and Simeka began arguing, Simeka's eldest daughter heard a gunshot from the direction of Simeka's room and observed Defendant run out of the room a few seconds later. Simeka's son also heard a "loud noise" and the sound of shattering glass coming from Simeka's bedroom. He too saw Defendant run out of the room several seconds later.

Defendant, upon seeing the children, yelled: "[Y]our mom shot herself." … Defendant stood in the doorway for several seconds and then fled from the house.

Officers … responded to the scene at approximately 2:00 p.m., …

Upon examining Simeka's bedroom, law enforcement officers discovered a .9 millimeter Kel-Tec …

… Defendant returned to Linda's house. He then told one of the officers that Simeka had shot herself.

… while at the police station he voluntarily surrendered his clothing for gunshot residue ("GSR") analysis.

…

The State also offered the testimony of David Freehling ("Freehling"), an expert in the field of GSR testing, who testified that while Simeka's hands and clothing had tested negative for GSR, Defendant's t-shirt, hooded jacket, and jeans all tested positive for GSR with one particle of GSR found on each of these three articles of clothing.

…

In addition, Defendant introduced testimony from his own GSR expert, Robert White, who testified that he would typically expect more than three particles of GSR to be present on the clothing of an individual who had fired a gun. Finally, Defendant presented the testimony of Dr. Christina Roberts, an expert in forensic pathology, who stated that she was unable to determine whether Simeka's manner of death was homicide or suicide.

The jury found Defendant guilty of both charges.

…

Rule 702 now provides, in pertinent part, as follows:

> (a) If scientific, technical or other specialized knowledge will assist the trier of fact to understand the evidence or to determine a fact in issue, a witness qualified as an expert by knowledge, skill, experience, training, or education, may testify thereto in the form of an opinion, or otherwise, if all of the following apply:
>
> 1. The testimony is based upon sufficient facts or data.
> 2. The testimony is the product of reliable principles and methods.
> 3. The witness has applied the principles and methods reliably to the facts of the case.

N.C.R. Evid. 702(a).

In McGrady, the Supreme Court discussed in detail the implications stemming from the amendment to Rule 702.

> Rule 702(a) has three main parts, and expert testimony must satisfy each to be admissible. First, the area of proposed testimony must be based on scientific, technical or other specialized knowledge that will assist the trier of fact to understand the evidence or to determine a fact in issue. This is the relevance inquiry discussed in both Daubert and Howerton. As with any evidence, the testimony must meet the minimum standard for logical relevance that Rule 401 establishes. In other words, the testimony must relate to an issue in the case. But relevance means something more for expert testimony. In order to assist the trier of fact, expert testimony must provide insight beyond the conclusions that jurors can readily draw from their ordinary experience. An area of inquiry need not be completely incomprehensible to lay jurors without expert assistance before expert testimony becomes admissible. To be helpful, though, that testimony must do more than invite the jury to substitute the expert's judgment of the meaning of the facts of the case for its own.
>
> Second, the witness must be qualified as an expert by knowledge, skill, experience, training, or education. This portion of the rule focuses on the witness's competence to testify as an expert in the field of his or her proposed testimony. Expertise can come from practical experience as much as from academic training. Whatever the source of the witness's knowledge, the question remains the same:

> Does the witness have enough expertise to be in a better position than the trier of fact to have an opinion on the subject? The rule does not mandate that the witness always have a particular degree or certification, or practice a particular profession. But this does not mean that the trial court cannot screen the evidence based on the expert's qualifications. In some cases, degrees or certifications may play a role in determining the witness's qualifications, depending on the content of the witness's testimony and the field of the witness's purported expertise. As is true with respect to other aspects of Rule 702(a), the trial court has the discretion to determine whether the witness is sufficiently qualified to testify in that field.
>
> Third, the testimony must meet the three- pronged reliability test that is new to the amended rule: (1) The testimony must be based upon sufficient facts or data. (2) The testimony must be the product of reliable principles and methods. (3) The witness must have applied the principles and methods reliably to the facts of the case. These three prongs together constitute the reliability inquiry discussed in Daubert, Joiner, and Kumho. The primary focus of the inquiry is on the reliability of the witness's principles and methodology, not on the conclusions that they generate. However, conclusions and methodology are not entirely distinct from one another, and when a trial court concludes that there is simply too great an analytical gap between the data and the opinion proffered, the court is not required to admit opinion evidence that is connected to existing data only by the ipse dixit of the expert.

McGrady, 368 N.C. at 888, 787 S.E.2d at 8-9, 2016 N.C. LEXIS 442, at *14. (internal citations, quotation marks, and brackets omitted and emphasis added).

The Supreme Court's analysis in McGrady—which the trial court did not have the benefit of at the time of Defendant's trial—makes clear that trial courts must now perform a more rigorous gatekeeping function when determining the admissibility of opinion testimony by expert witnesses than was the case under the prior version of Rule 702. Here, Dr. Duval's opinion that Simeka's death was a homicide as opposed to a suicide appears to have been largely—if not entirely—based on his interpretation of non- medical information conveyed to him by law enforcement officers who were involved in the investigation of Simeka's death.

Q. Did you take into consideration statements made by witnesses at the scene as far as the circumstances related to the moments before and after Simeka Daughtridge was shot?

A. Yes.

Q. And could you please tell the jury the information that you considered as far as the circumstances of the moments before and after she was shot?

A. It was relayed to me by law enforcement that eyewitnesses stated that there was some sort of verbal altercation occurring in the bedroom in which the decedent was in. A loud sound or a pop or something analogous to a gunshot was heard and then a person was seen to emerge from the room.

Q. And the person that was seen to emerge from the room, do you know who that was?

A. I believe it was described as the decedent's boyfriend.

Q. And were you informed as to what he had told the police as far as whether or not that was consistent [sic] what other witnesses observed?

A. I believe that again, according to information provided to me from law enforcement, was [sic] that the decedent's boyfriend claimed that he was not in the room with the decedent at the time that the gunshot was heard.

Q. As a result of all of the information that you took into consideration, did you form an opinion as to whether or not Simeka Daughtridge was the victim of homicide?

A. Yes.

Q. And based on all the information that you were provided as well as the testing that you performed yourself, what was your expert opinion as to whether or not she was the victim of homicide?

A. In my opinion the manner of death is homicide.

Dr. Duval further testified during cross-examination as follows:

Q. Okay. A few other questions. The information you had received was from law enforcement only, correct, as far as what the --
A. To my recollection.
Q. -- as to the cause and manner of death?
A. Yes.

...

Q. Okay. Outside of what law enforcement told you, is there anything about the wound itself that would indicate that it could not have been self-inflicted?
A. No.

We believe Defendant has raised legitimate concerns about the admissibility of Dr. Duval's opinion testimony on the issue of whether Simeka's death was a homicide. Clearly, Dr. Duval would have been qualified to provide an opinion that Simeka's death was a homicide—rather than a suicide—if that opinion was based on the type of evidence that was within his area of expertise as an expert witness in the field of forensic pathology. However, based on our review of the trial transcript, it is difficult to escape the conclusion that his opinion on this specific issue was based not on such medical evidence but instead on statements from law enforcement officers about the results of their investigation—information that bore little, if any, connection to his own observations stemming from his autopsy of Simeka.

The State has failed to adequately explain how Dr. Duval was in a better position than the jurors to evaluate whether the results of the officers' investigation were more suggestive of a homicide than a suicide. Therefore, based on the principles set out in McGrady, his opinion failed to pass muster under the new test governing the admissibility of expert witness opinion testimony that is now required in light of the 2011 amendment to Rule 702.

However, we are not convinced that the trial court's error in allowing Dr. Duval to give this opinion rose to the level of plain error. This is so for several reasons.

First, the results of the GSR testing are inconsistent with the theory that Simeka committed suicide. The State's evidence established that had Simeka, in fact, shot herself, GSR would have been present on her hands. However, no GSR was discovered on her hands during forensic testing. During direct examination, Freehling offered the following testimony:

Q. And if somebody was to have shot themselves and died as a result of that gunshot wound shortly after shooting themselves, would you expect there to be gunshot residue on their hands?
A. It's depending on the caliber of the weapon and if the body was touched or the hands were touched or anything that could lead to the lost [sic] of particles. You would expect to find particles on the hands of someone that shot themselves if it was as is. But then there's factors that lead to particle loss which could be the caliber of the weapon, if EMTs touched the body or while transferring to the hospital, anything [sic] of those factors could lead to particle loss.
Q. Now, you talked about caliber. A .9 millimeter would you consider that a small caliber?
A. That's higher caliber. The smaller calibers were [sic] little to no gunshot residue submitted are typically .22 caliber and .25 caliber.
Q. And as far as you had talked about medical personnel, if there's no medical intervention as far as cleaning of the hands or medical procedures on the hands, that would not then affect a loss of gunshot residue; is that correct?
A. That's correct.

Freehling also testified that "[GSR] expands up to a few feet from the weapon. So the further away you get from the weapon, the less likely you are to have gunshot residue on you." He further testified as follows:

Q. And as far as if there's gunshot residue mixed in with, say, blood can you still detect that gunshot residue?
A. Yes.

Significantly, Simeka's blood-stained hands were not disturbed following the shooting as reflected by a Duke University Medical Center report prepared by Dr. Catherine Lynch, who stated therein that "[Simeka] was prepped for autopsy/police investigation with bags placed on her hands and was not wiped clean per my request." Moreover, when Simeka's body was delivered to Dr. Duval for autopsy, he observed that "I saw what appeared to be dried blood stains on [her] hands."

Consequently, the preservation of Simeka's hands in an undisturbed state for forensic testing and the total lack of any GSR on them forecloses the possibility that she shot herself. Furthermore, Simeka was shot with a .9 millimeter high caliber handgun as opposed to a smaller caliber handgun that—as Freehling noted—might not leave appreciable traces of GSR.

Defendant's clothing, conversely, tested positive for the presence of GSR despite his assertion during his interview with Investigator Sole that he had not been in Simeka's room at the time of the shooting. Given Freehling's testimony that GSR only travels "three to four feet ... maximum" from a fired .9 millimeter gun and that the "cloud is only in the air for a matter of seconds" before the particles fall to the ground, the fact that three separate pieces of Defendant's clothing—his t-shirt, jeans, and hooded jacket—all tested positive for GSR clearly indicates that he was, in fact, present in Simeka's bedroom within several feet of the Kel-Tec .9 millimeter weapon at the time she was shot.

Second, both of Simeka's children who testified at trial stated that Defendant exited Simeka's room after they heard the gunshot. No witness at trial who was present in the house at the time of the shooting testified to the contrary.

Third, the fact that Defendant made no effort to tend to Simeka and actually left the residence prior to the arrival of law enforcement officers or emergency medical personnel serves as circumstantial evidence of Defendant's guilt. ...

Finally, any prejudicial effect from the erroneous admission of Dr. Duval's opinion on the manner of Simeka's death was mitigated during cross-examination by Defendant's trial counsel:

Q. If you had been told that a victim, that a deceased person was suicidal, had attempted suicide, and in fact had the day before told her children, mom, was not going to be with you much longer, would you have considered that in determining whether something was a homicide or suicide?
A. Sure.
Q. But if you're not told that, you can't consider it in making your determination?
A. Yes.
Q. Were you told any of that in this case?
A. Not to my recollection.

...

Q. Do you know if the victim in this case ever expressed thoughts of killing herself?
A. I have no idea.
Q. Did you talk to the family and ask them about that?
A. No.
Q. Did you ask the police if she had any of that?
A. I don't know.

...

Q. But at the time, you never even considered the self inflicted angle because it was never communicated to you?
A. No.
Q. And there's nothing about the autopsy itself per her wound that would tell you there's no way she could have done this herself?
A. That is the shortcoming of an autopsy.
Q. But that wound, and that type, where it was, and everything else, you've seen wounds like that that are self-inflicted?
A. Yes.

Furthermore, as noted earlier, the jury heard Dr. Duval explicitly admit during cross-examination that his opinion that Simeka did not commit suicide was based entirely on non-medical information he received from law enforcement officers.

Q. Okay. A few other questions. The information you had received was from law enforcement only, correct, as far as what the --
A. To my recollection.
Q. -- as to the cause and manner of death?
A. Yes

...

Q. Okay. Outside of what law enforcement told you, is there anything about the wound itself that would indicate that it could not have been self-inflicted?
A. No.

Thus, as a result of defense counsel's cross- examination of Dr. Duval, the jury was expressly told that in forming his opinion as to Simeka's manner of death (1) he had not been made aware of any of Defendant's evidence suggesting Simeka had a motive to commit suicide; (2) he instead relied exclusively on the information the officers had related to him about their investigation; and (3) nothing from his analysis of Simeka's body during the autopsy shed light on whether the death was a homicide rather than a suicide. These admissions would have led a reasonable juror to place Dr. Duval's stated opinion on the manner of death in its proper context.

For these reasons, we conclude that the admission of Dr. Duval's opinion testimony did not amount to plain error. ...

Conclusion

For the reasons stated above, we conclude that Defendant received a fair trial free from prejudicial error.

Questions:

1. How does this court perceive whether or not a person qualifies to be an ballistics expert? Is it a rigid or more flexible standard? Is it consistent with the "helpfulness" test as espoused by Daubert?
2. The State refutes the claim by defendant that this was a suicide and not a homicide. What role does Gunshot Residue play in this conclusion?
3. How did the police investigators assure the integrity of the victim's hands when delivering the body for an autopsy?

Steele v. United States
United States District Court for the Southern District of New York
June 4, 2021, Decided; June 4, 2021, Filed
15 CR 836(VM); 20 Civ. 1151 (VM)

On October 28, 2016, petitioner Joseph Steele ("Steele") was convicted of being a felon in possession of a firearm, in violation of 18 U.S.C. §§ 922(g)(1) and (2). Now before the Court is Steele's pro se motion to vacate, set aside, or correct his sentence pursuant to 28 U.S.C. § 2255. (See "Motion," United States v. Steele, No. 15 CR 836 (the "Criminal Docket"), Dkt. No. 84; Steele v. United States, No. 20 Civ. 1151 (the "Civil Docket"), Dkt. No. 1.) The Government opposes the Motion. ("Opposition," Criminal Docket, Dkt. No. 89, at 9.) For the following reasons, the Motion is DENIED.

I. BACKGROUND

A. PROCEDURAL HISTORY

On February 19, 2016, Steele was charged with one count of being a felon in possession of a firearm, in violation of 18 U.S.C. §§ 922(g)(1) and 2. The indictment alleged that Steele had three predicate felonies subjecting him to a sentencing enhancement under the Armed Career Criminal Act, 18 U.S.C. § 924(e)(1). The charges stemmed from an incident that took place on October 10, 2015 in which Steele pulled out a gun and fired it during an argument with three other men. On October 28, 2016, following a four- day jury trial, Steele was convicted of this charge. Steele was sentenced to 180 months' imprisonment, followed by five years' supervised release.

1. The Government's Case

According to the Government's case, Steele pulled out a gun and fired it during an argument with three other men on the corner of West 175th Street and Macombs Road in the Bronx, New York. The Government's evidence included eyewitness testimony that a man in a burgundy jacket pointed a gun at a man wearing a black sweater and a gunshot sounded after the eyewitness turned away. Surveillance video was also introduced, which showed a man in a burgundy jacket brandishing an object in his left hand and pointing it toward the feet of a darkly clothed man in front of him, followed by what appeared to be a muzzle flash from the object.

New York Police Department ("NYPD") Officers Burgos, Pineda, and Guzman drove to the scene of the shooting in response to the eyewitness's 911 call. There, Officer Burgos saw a group of three or four men, with one man standing slightly apart from the others clutching something inside his jacket pocket. When Officer Burgos ordered the men to stand still, the man holding something inside his jacket pocket -- who the Government contended was Steele -- fled. Officer Burgos gave chase and observed a black object in Steele's left hand. During the pursuit, he saw Steele make a swinging motion across his body with his left hand followed by a metallic clatter coming from the line of cars parked along the sidewalk.

A second group of officers, including Officer Perdomo, intercepted Steele further up the block. After Steele was under control, Officer Burgos returned to the spot where he had heard the metallic clatter and found a loaded .380 semiautomatic pistol between the curb and the tire of a parked car. Officer Guzman testified that he found a spent .380 shell casing near the corner of West 175th Street and Macombs Road. Officer Pineda corroborated Officer Burgos's account and added that Steele had been wearing a burgundy jacket, which he subsequently vouchered at the precinct. The officers' testimony was also corroborated by surveillance footage. Furthermore, a ballistics expert testified that the .380 casing had been fired from the same .380 semiautomatic pistol recovered by the officers that night, approximately twenty feet from where Steele was arrested.

DNA testing of the firearm was conducted, and criminalist Asako Ishii testified that Steele's DNA was not detected on the gun (though other peoples' DNA was). The criminalist testified about how skin cells and the DNA contained therein may be left on surfaces, noting that some people shed skin cells more than others, some surfaces like metal retain DNA poorly, and environmental factors like application of physical force to the surface by wiping or brushing may affect whether an individual's DNA persists on the surface.

2. Evidence Not Introduced

As relevant to the instant motion, three pieces of evidence were not introduced at trial. First is the testimony of Vanessa Martinez, a Forensic Criminalist with the NYPD Police Laboratory, as well as the NYPD Gunshot Residue report Martinez prepared. Before trial, Steele moved to admit this report. Steele sought to introduce the report through the hearsay exception for "a record or statement of a public office." The report concluded that no gunshot residue had been detected on the burgundy jacket worn by Steele.

Upon further discussion with Martinez, the Government learned that Martinez used a low power microscope and not a Scanning Electron Microscope ("SEM"). A SEM is the industry standard tool for identifying gunshot residue because a low-powered microscope can adequately identify residue only on victims' clothing or surfaces actually struck by a bullet. The NYPD laboratory does not use a SEM on clothing due to resource-allocation issues. The Government also learned from Martinez that gunshot residue is not commonly found on the shooter unless the shooter's clothes were burned or singed by the gunfire, and certain materials are more efficient than others at retaining residue. Moreover, environmental conditions and the act of running can easily brush gunshot residue from clothing. The passage of time can also reduce the likelihood that gunshot residue will persist on a surface.

The Government opposed the introduction of the gunshot-residue report because it contained expert opinion that had not been admitted under Rule 702; because it was unreliable in that it failed to disclose serious deficiencies concerning the circumstances of the analysis it purported to carry out; and without the necessary explanation and context from Martinez, admission of the report would mislead the jury and was therefore more prejudicial than probative. The Court agreed that the report was not admissible without the testimony of Martinez as well, a ruling which was affirmed by the Second Circuit. United States v. Steele, 729 F. App'x 47, 49–50 (2d Cir. 2018).

Ultimately, the defense called no witnesses, not even Martinez, even though she was present at trial. Defense counsel explained that he did not do so because "[l]ooking over the record as it came in, I had Ms. Martinez here today, we made a strategic decision not to call her because the record, as I just read it to the Court, allowed me to make that argument about gunshot residue" as part of a larger argument about the inadequacy of the investigation (for failure to do testing at the scene). (Trial Tr. at 304.) The Court rejected the defense's attempt to argue that an inadequate investigation can be inferred based on the failure to do gunshot-residue testing at the scene. The Court concluded that such an inference would be unfair for two reasons. First, the two witnesses the defense sought to rely on, Officer Burgos and Detective Fox, were not experts. Second, the surveillance footage showed officers pursuing Steele and the other individuals present drifting away, so there was no opportunity for the officers to conduct further testing on other individuals at the scene.

The jury began deliberations on October 27, 2016. One of the jury's requests was for clarification on whether the burgundy jacket had been checked for gunpowder residue. While defense counsel argued that the jury should be referred to Officer Burgos and Detective Fox's testimony, the Court ultimately informed the jury that "the trial record contains no evidence in testimony or any exhibit that states whether or not the jacket was checked for gun powder residue," with the closest testimony coming from Officer Burgos, who said that the jacket "was sent to the lab. I don't know what kinds of testings were done on it." (Trial Tr. at 362–63.)

B. THE PARTIES' ARGUMENTS

In his Section 2255 motion, Steele alleges that he was denied effective assistance of counsel when his trial counsel (1) failed to call Vanessa Martinez, a Forensic Criminalist with the NYPD Police Laboratory, to testify; (2) failed to call Officer Perdomo to testify about a glove that Perdomo had recovered from Steele following Steele's arrest; and (3) failed to introduce the glove into evidence. Steele contends that had defense counsel called Martinez, the report finding no residue on his jacket could have been introduced. (Motion at 5.) Steele also contends that introduction of the glove, and the related report finding that his DNA was not identified on the glove, would have proven that Officer Perdomo had planted the glove. Steele believes this evidence was exculpatory and would have changed the outcome of his trial in light of the jury's question about whether gunshot residue was found on the jacket, and because casting doubt on Officer Perdomo's credibility would have been enough to sow reasonable doubt of Steele's guilt.

The Government responds that Steele received effective assistance of counsel and that any alleged errors were not prejudicial. The Government argues that Martinez's testimony would have exposed that the gun-residue test was scientifically flawed, so its conclusion was worthless. Similarly, the Government argues that the testimony of Officer Perdomo regarding the glove and the DNA results would have harmed, not helped, Steele's defense. Because Officer Perdomo was not involved in the collection of the key, direct evidence, such as the surveillance video and the firearm or shell casing, the relevant evidence would not have been affected. Moreover, the DNA analysis only excluded Steele as the primary contributor of DNA on the firearm, but there were multiple unidentified secondary DNA contributors, so Steele could have been a secondary contributor. Given this possibility, the evidence would likely have harmed Steele's defense, because it would have provided an explanation for why Steele's DNA was not found on the firearm. For these reasons, the Government argues that defense counsel's actions were sound strategic decisions that the Court should not second guess, and that in any event, the failures were not prejudicial.

In his reply memorandum of law, Steele argues that Martinez's testimony would have been nonetheless valuable in furthering his inadequate-investigation defense, because the fact that a purported expert had used a scientifically flawed test suggests that the investigation was subpar. Steele also contends that the DNA report of the glove excluded him as a contributor of the DNA samples found on the glove. Steele further argues that the failure to call experts and admit the scientific exhibits at issue constitutes ineffective performance under United States v. Nolan, 956 F.3d 71 (2d Cir. 2020), and Bell v. Miller, 500 F.3d 149 (2d Cir. 2007).

...

II. DISCUSSION

...

Nor is there a substantial likelihood or reasonable probability of a different outcome had the jury been presented with Martinez's testimony and the gunshot- residue report finding no residue on Steele's jacket. Countervailing evidence would have mitigated any exculpatory effect of this finding. For instance, the Government would have elicited testimony from Martinez to contextualize the gunshot-residue test results and explain that an unreliable methodology was used, thereby casting doubt on the results. In addition, the Government would in all likelihood have elicited testimony from Martinez that gunshot residue is not captured on certain fabrics easily and can be brushed off by running. Given the type of jacket Steele was wearing, and because the Government's evidence showed that Steele had been running after the shot was fired and before he was apprehended, Martinez's testimony would have provided multiple plausible reasons for why no gunshot residue was found on Steele's jacket.

Steele argues in his reply memorandum of law that testimony that an unreliable test had been performed would have bolstered his argument of an inadequate investigation. Steele contends that "[i]f it was proved that the expert NYPD laboratory examiner had actually used the wrong test on the defendant's jacket, then the defense could have offered the jury reasonable doubt based on the mishandling of the evidence." (Criminal Docket, Dkt. No. 98, at 3.) Such a conclusion is possible, but the Court is not persuaded that it is substantially likely. A SEM was not used due to resource-allocation issues, so it is not at all clear that a jury would have concluded that the evidence was mishandled simply because a SEM was not used.

Moreover, Steele fails to account for the weight of the other evidence supporting the Government's case. In considering whether a petitioner was prejudiced by trial counsel's errors, the strength of the Government's case must also be considered. See Henry, 409 F.3d at 66. Courts must "bear[] in mind that a verdict or conclusion only weakly supported by the record is more likely to have been affected by errors than one with overwhelming record support." Id. at 67 (internal quotation marks and citation omitted). Thus, the Circuit has considered the prosecution's case to be weak when the only evidence connecting the defendant to the crime is a single identification from an arguably suggestive lineup, see id. at 66, or when absent potentially unreliable eyewitness testimony, the probative value of the remaining evidence is "limited at best," see Nolan, 956 F.3d at 82.

In the case at hand, the Government adduced multiple other pieces of evidence probative of Steele's guilt. For example, an eyewitness testified that he saw a man in a burgundy jacket point a gun, and the Government established that Steele was wearing a burgundy jacket. Additionally, while Steele was being chased, Officer Burgos saw a black object in Steele's left hand and observed Steele make a swinging motion across his body as he fled, followed by a metallic clattering sound. Officer Burgos then found a firearm near where he heard this sound. Furthermore, the shell casing that was fired by the firearm found by Officer Burgos was also found near where Steele was arrested. There was also corroborative surveillance video.

While some of this evidence is circumstantial, it is nonetheless strong enough for a jury to conclude that:

> (1) Steele possessed a firearm and had thrown it aside while fleeing from the NYPD; and (2) the firearm had been discharged prior to Steele's flight. The weight of this evidence -- particularly when considered in conjunction with the weak, and even damaging, nature of the unintroduced evidence at issue -- critically undermines Steele's argument of prejudice. Because Steele cannot demonstrate prejudice, his claim for ineffective assistance of counsel must fail.

III. ORDER

Accordingly, for the reasons stated above, it is hereby

ORDERED that petitioner Joseph Steele's motion to vacate, set aside, or correct his sentence pursuant to 28 U.S.C. § 2255 (United States v. Steele, No. 15 CR 836 (the "Criminal Docket"), Dkt. No. 84; Steele v. United States, No. 20 Civ. 1151 (the "Civil Docket"), Dkt. No. 1.) is DENIED.

The Clerk of Court is hereby directed to mail this Order to Joseph Steele, Register Number 73353–054, at F.C.I. Hazelton, P.O. Box 5000, Bruceton Mills, WV 26525 and note service in the docket.

Because Steele has failed to demonstrate a denial of a constitutional right, a certificate of appealability will not issue.

SO ORDERED.

Questions:

1. Prosecution and Defense both had to deal with how the NYPD tested for residue and the equipment employed to do the testing. How so and with result?
2. Given the Court's rejection of the reasonable doubt argument by defendant, is the Court satisfied that there is sufficient evidence to uphold the conviction?

United States v. North
United States District Court for the Northern District of Georgia, Atlanta Division
November 17, 2017, Decided; November 17, 2017, Filed
1:16-cr-309-WSD

This matter is before the Court on Defendant Jeff North's ("Defendant") Motion to File Out of Time ("Motion to File") [94] and Motion to Preclude Gun-Shot Residue Analysis Opinion Evidence (the "Daubert Motion") [94.1].

I. BACKGROUND

On May 26, 2016, a grand jury in the Northern District of Georgia returned a three-count indictment [13] charging Defendant with Carjacking in violation of 18 U.S.C. § 2119(1) (Count 1); Discharging a Firearm During a Federal Crime of Violence in violation of 18 U.S.C. §§ 924(c)(1)(A)(i), (ii), and (iii) (Count 2); and Possession of a Firearm by a Convicted Felon in violation of 18 U.S.C. §§ 922(g)(1) and 924(e) (Count 3). The Indictment alleges that, on or about March 23, 2015, Defendant shot Johnny Dansby and stole his vehicle.

On March 23, 2015, swabs from Defendant's hands were submitted to the Georgia Bureau of Investigation, Division of Forensic Sciences for a gunshot residue ("GSR") analysis. The sealed samples were later analyzed by Microanalyst Alexander Covin. Mr. Covin completed an official report detailing the method of analysis, results, and conclusions. ([97.1]). Mr. Covin's primary trainer and Manager and Acting Director of the Trace Evidence Section, Michael McCarriagher, independently reviewed the evidence, report, and all associated documentation. (Id.).

The GSR report was provided to Defendant at his arraignment hearing on September 9, 2016. The report states that the samples taken from Defendant's hands were tested for the presence of particles characteristic of GSR. It details the test method used ("scanning electron microscopy/energy dispersive x-ray spectroscopy") and summarizes the results (the examination "revealed three particles characteristic of GSR"). It also summarized the analysts' opinion that the examination of the samples:

> revealed the presence of particles characteristic of [GSR]. This supports the possibility that the individual discharged a firearm, was in close proximity to a firearm upon discharge, or came into contact with an item whose surface bears GSR.
>
> ([97.1])

On September 20, 2017, the Court ordered [69] that this case be placed on the Court's December 5, 2017 trial calendar. The Court further ordered that the parties file, by October 16, 2017, motions in limine and motions to exclude evidence or testimony.

On October 20, 2017, the Government officially noticed Mr. McCarriagher and Mr. Covin as experts who would testify in the area of gunshot residue and provided copies of their CVs. The Government also provided backup notes, data, and other information. (See [97.4]).

On November 13, 2017, Defendant filed his Motion to File, [94] in which he moved for leave to file the Daubert Motion beyond the October 16, 2017, deadline. The Defendant attached his Daubert Motion [94.1] to the Motion to File. In his Daubert Motion, Defendant argues that the Government's disclosures related to the GSR analysis fail to comply with Rule 16 of the Federal Rules of Criminal Procedure. He also seeks discovery and a hearing regarding the admissibility of Mr. McCarriagher's1 testimony under Rule 702 of the Federal Rules of Evidence.

By docket entry on November 13, 2017, the Court ordered that the Government respond to the Motion to File by noon on November 16, 2017.

On November 16, 2017, the Government responded to the Daubert Motion. (See [97]).

On November 17, 2017, Defendant filed his Reply [100].

II. DISCUSSION

The Daubert Motion is currently before the Court. Defendant seeks a hearing on the admissibility of Mr. McCarriagher's testimony regarding the presence of gunshot residue on Defendant on the grounds that the Government's expert is not qualified to offer his opinion and the opinions are otherwise inadmissible under Federal Rule of Evidence 702. Defendant also requests discovery on the tests used, including "a summary as required by Rule 16 of the Federal Rules of Criminal Procedure."

A. Legal Standard

Rule 16(a)(1)(G) of the Federal Rules of Criminal Procedure "imposes specific disclosure requirements on the government with regards to expert witnesses that the government plans to utilize at trial." United States v. Holland, 223 F. App'x 891, 893 (11th Cir. 2007). The rule provides, in pertinent part, that, "[a]t the defendant's request, the government must give to the defendant a written summary of any [expert] testimony that the government intends to use." Id. The "summary" provided by the government must include the expert witness's "opinions, the bases and reasons for those opinions, and the witness's qualifications." Id. The commentary to the Rule further provides that the government's summary "should cover not only written and oral reports, tests, reports, and investigations, but any information that might be recognized as a legitimate basis for an opinion … ." Id., Advisory Comm. Notes, 1993 Amendment.

Expert opinion testimony is governed by Rule 702 of the Federal Rules of Evidence and Daubert v. Merrell Dow Pharmaceuticals, Inc., 509 U.S. 579, 113 S. Ct. 2786, 125 L. Ed. 2d 469 (1993). Under Rule 702, expert testimony is admissible if: (1) the expert is qualified to testify regarding the subject matter of his testimony; (2) the methodology that the expert used to reach his conclusions is sufficiently reliable; and (3) the expert's testimony will assist the trier of fact in understanding the evidence or in determining a fact at issue. United States v. Scott, 403 F. App'x 392, 397 (11th Cir. 2010) (citing United States v. Frazier, 387 F.3d 1244, 1260 (11th Cir. 2004)) (en banc); Fed. R. Evid. 702. The Government has the burden to meet each of the admissibility requirements. See Scott, 403 F. App'x at 397–98.

The second prong requires the district court to make a preliminary determination on whether the expert's methodology is reliable. Scott, 403 F. App'x at 397. In Daubert, the Supreme Court provided a non-exclusive list of factors for the district court to consider:

> (1) whether the expert's theory can be and has been tested; (2) whether the theory has been subjected to peer review and publication; (3) the known or potential rate of error of the particular scientific technique; and (4) whether the technique is generally accepted in the scientific community.

Frazier, 387 F.3d at 1262. These factors are only general guidelines, and the trial judge has "considerable leeway in deciding in a particular case how to go about determining whether particular expert testimony is reliable." Kumho Tire Co., Ltd. v. Carmichael, 526 U.S. 137, 152, 119 S. Ct. 1167, 143 L. Ed. 2d 238 (1999).

A court may conduct a hearing on a Daubert motion, but one is not automatically granted. …

B. The Government's Disclosures Comply With Rule 16

The pretrial disclosure provided by the Government includes a clear summary of the method of analysis, results, and opinions. The GSR report was provided to Defendant at his arraignment hearing on September 9, 2016. It discloses that the evidence analyzed was a "[s]ealed GSR collection kit identified as containing samples from the hands of Jeff North." ([97.1]). The report concluded that "three particles characteristic of GSR" were revealed on the sample, supporting the analyst's opinion that "the individual discharged a firearm, was in close proximity to a firearm upon discharge, or came into contact with an item whose surface bears GSR." (Id.). It further states that the "method of analysis" used to arrive at this conclusion was by "scanning electron microscopy/energy dispersive x-ray spectroscopy (SEM/EDS) and analyzed for elemental composition and particle morphology." (Id.). The report notes that the "evidence, report and all associated documentation have been reviewed by primary trainer, Michael McCarriagher." (Id.). The Government also provided copies of Mr. McCarriagher's and Mr. Covin's CVs. (See [97.2–3]). Finally, the Government supplied thirty additional pages of notes and underlying data. (See [97.4]).

The Government met its pretrial disclosure obligations under Rule 16. United States v. Campbell, No. 1:04-CV- 0424-RWS, 2006 WL 346446, at *1 (N.D. Ga. Feb. 13, 2006) ("Rule 16 does not mandate a comprehensive recitation of every nuance and detail that will make up an expert's testimony, or which may be drawn out on cross-examination.").

C. Expert Discovery and a Daubert Hearing Are Unnecessary

Defendant requests discovery and a hearing on the admissibility of expert opinions related to the GSR report. A hearing is not required every time a party raises a Daubert objection. …

Here, the Government has proffered sufficient information to allow this Court to qualify Mr. McCarriagher in his area of expertise. It set forth how it plans to present Mr. McCarriagher's qualifications, training, experience, and the methodology employed in conducting his analysis and reaching his conclusions. (See [97] at 7–8). The Court is satisfied that Mr. McCarriagher has the requisite experience and training to offer opinions on the presence of GSR. "Over the course of [Mr.] McCarriagher's ten-plus years as a forensic scientist, he has conducted more than 1,200 GSR analyses and he has testified in court and been qualified as a GSR expert approximately eighty times." ([97] at 7).

Applying the Daubert criteria, the Court finds that Mr. McCarriagher's opinion is reliable. Defendant does not cite any authorities or other information that the GSR analysis is unreliable, non-scientific, or that it does not have broad acceptance in the forensic community. Defendant mentions "[t]hree reports issued in 2008, 2009, and 2016 … have called into question the reliability of many forensic 'sciences' previously admitted without much doubt." ([100] at 2–3). Defendant does not demonstrate that these reports cast cognizable doubt on the reliability of Mr. McCarriagher's method.2 Defendant also does not rebut other cases submitted by the Government in which courts have admitted expert testimony regarding GSR testing similar to that which it intends to be offered at this trial in this case. …

To the extent that Defendant seeks to attack the credibility and accuracy of the results of the GSR analysis, these matters can be the subject of "vigorous cross examination, presentation of contrary evidence, and careful instructions on the burden of proof." Daubert, 509 U.S. at 596.

III. CONCLUSION

For the foregoing reasons,

IT IS HEREBY ORDERED that Defendant's Motion to File Out of Time [94] is DENIED AS MOOT.

IT IS FURTHER ORDERED that the Defendant's Motion to Preclude Gun-Shot Residue Analysis Opinion Evidence [94.1] is DENIED.

SO ORDERED this 17th day of November, 2017.

Questions:

1. Why was proof of residue relevant to the crime alleged in this case?
2. In your view, should the challenge to the expertise and qualification of the GSR expert be upheld or denied?
3. Defendant argues for a Daubert hearing? Can you discern any legitimate rationale for this argument and does the Court agree with this conclusion?

8.3 FORENSIC EVIDENCE: PROJECTILE TYPE AND COMPOSITION

Precisely what makes up a projectile, whether a bullet or shotgun shell, its composition as to type of powder, packing and powder, jacket design and imprints all lead to comparative analysis in the world of ballistics.

Locate the case below using your favorite legal research resource, read the case and answer the questions.

More v. State, 880 N.W.2d 487, Supreme Court of Iowa May 20, 2016, Filed No. 14–1623

Questions:

1. The More decision is a fascinating example of how courts agree with a defendant's basis for the appeal, and in this case, as the opinion unfolds, it seems as if the Court will have little other choice but to grant the post-conviction relief. What was the chief challenge posed by defendant?
2. In this case, what do the experts testify to as to CBLA?
3. Prove True or False: The Court concurs with the challenge and even admits to a host of errors. Even so, it affirms the conviction.

Another common approach in ballistic's analysis is to compare the imprints from a weapon's firing pin, barrel, from its grooves and landings that are embedded and impressed upon manufacture, with the impression or sometimes labeled toolmark impressions on the casings, the projectile and the base of the casing. For a long time, experts were too often saying they found exact matches, but this is less likely than once thought for unless there is distinguishing characteristics within the rifling or other impression point, the best that can be said is a similarity or an elimination rather than perfect matches (See Figure 8.2).

People v. Gadlin
Court of Appeal of California, First Appellate District, Division One
November 7, 2018, Opinion Filed
A149764

Defendant Gregory D. Gadlin was convicted of first degree murder with personal use of a firearm (Pen. Code,1 §§ 187, subd. (a), 12022.7, subd. (a), 12022.53, subds. (b), (c) & (d), 12022.5,

FIGURE 8.2 This cartridge case shows the three distinct marks, or 'signatures,' impressed on its surface when it was fired by a gun: the firing pin impression (FP), the breech face impression (BF) and the ejector mark (EM). NIST's standard casing ensures that forensic labs can accurately match signatures on shells found at crime scenes with images in a national database.

Source: NIST at https://www.nist.gov/image/12pml028cartridgeheadlrjpg.

subd. (a)) and possession of a firearm by a felon (§ 12021, subd. (a)(1)). On appeal, defendant contends the first degree murder verdict is not supported by substantial evidence under either felony murder based on robbery or premeditation and deliberation. He also argues the trial court erred by admitting unreliable toolmark identification testimony, giving a flight instruction, and imposing improper victim restitution award and costs.

We conclude the verdict is supported by substantial evidence. Nor do we find any error in the court's decision to allow the expert testimony on toolmark identification, the issuance of a flight instruction, or the victim restitution award. We do, however, find the probation investigation fee should be stricken because there is insufficient evidence of defendant's ability to pay. Accordingly, we strike the $250 probation investigation fee but otherwise affirm the judgment.

I. BACKGROUND

Evan Meisner met defendant shortly before he moved out of a home he shared with a former girlfriend. …

Around midnight on the evening of the planned sale, Meisner informed his housemates he was going to make the sale and would return in 15 or 20 minutes. Meisner placed his open beer in the refrigerator, walked to his bedroom, and emerged carrying a basket full of laundry. He left the house driving his truck. Approximately eight calls between Meisner's phone and a cell phone used by defendant were made between 12:13 a.m. and 1:29 a.m. When Meisner did not return after about an hour, his roommates made multiple calls to Meisner, all of which went unanswered.

Around 1:40 a.m., Willie Carter and Rodney Fisher, both of whom resided in the apartment complex neighboring Meisner's former home, reported hearing a gunshot. Carter did not think much of it. Fisher went outside to check on his car, which he had parked in the driveway of Meisner's former home with Meisner's approval. While checking on his car, he saw Meisner's truck parked behind his vehicle. He also testified he saw an individual, who he believed was defendant, run and jump over the fence separating Meisner's yard from a park that abutted the property. Fisher immediately returned to his apartment after seeing defendant.

The defense offered contradictory testimony regarding that evening. Defendant's stepson and wife, who live with defendant, both testified they never heard a gunshot and did not believe defendant left their apartment. Defendant's wife also testified she answered defendant's phone when it rang in the middle of the night and called it a few times to see whether it was a girl calling defendant. However, in a prior statement to the police, she acknowledged she did not know whether defendant left because she takes sleep medication that "knocks [her] out." She also informed police that after she fell asleep that evening she did not wake up again until the morning.

The next morning, the property owner of Meisner's former home arrived to meet a "window guy." She was surprised to see Meisner's truck and another car parked in front of the house because it was supposed to be vacant. The security door was ajar and the door unlocked. The property owner saw Meisner lying face down and called 911. The house had no signs of forced entry.

Meisner was pronounced dead when police and paramedics arrived. He had been shot once in the left side of the neck at close range and died immediately. His body was face down on the floor between the kitchen and a back bedroom. Blood spatter was located on the lower part of the door jamb and the door, and the evidence indicated Meisner had been shot where he was found. A single bullet casing from a Winchester nine-millimeter Luger was located in the bedroom behind Meisner's legs. His pants had been pulled down below his buttocks and one shoe was removed. The marijuana and Meisner's wallet, truck keys, and cell phone were not located in the house.

The same morning Meisner's body was found, defendant arranged to sell his cousin a large quantity of marijuana for $300, claiming it was worth $600. The marijuana was of poor quality and packaged in small bags inside a larger see-through bag. Defendant testified he had received the marijuana from a neighbor.

Approximately one week later, defendant was arrested on unrelated charges and a parole violation. The evening of his arrest, defendant made a telephone call from the booking area to a childhood acquaintance. Defendant requested that his acquaintance go to his car and remove something located under a box in the trunk. He stated it was "very important." The acquaintance did not follow defendant's instructions. The next day the police located defendant's car. The car appeared to be locked although one window may have been slightly ajar. The police discovered a loaded Taurus nine-millimeter gun in the car's trunk under a box. The box contained personal papers and other materials belonging to defendant. Defendant testified the gun did not belong to him, but rather was placed there by a local gang, and he was warned not to touch it. The toolmarks unique to the firearm matched those on the bullet fragment recovered from Meisner's body and the casing recovered from the crime scene.

Defendant was convicted of first degree murder and possession of a firearm by a felon. This timely appeal followed.

II. DISCUSSION

Defendant raises various challenges to his conviction. First, defendant contends the first degree murder conviction is not supported by sufficient evidence. Second, defendant argues the trial court erroneously admitted unreliable expert testimony regarding toolmark identification. Third, defendant challenges the trial court's instruction on flight. Finally, defendant disputes the victim restitution award and the order to pay a probation investigation fee. We address each argument in turn.

...

A. Firearm Toolmark Identification

Defendant contends the court erred in failing to exclude a firearms expert from testifying, or, at a minimum, limiting the scope of that testimony on toolmark identification, because such testimony was not reliable or accepted by the general scientific community. As a result, defendant contends admission of that testimony violated evidentiary law and his right to due process. We disagree.

1. Relevant Factual Background

Todd Weller was an Oakland Police Department criminalist, who compared the cartridge and bullet fragments with the firearm recovered from defendant's car. Defendant anticipated the prosecution calling Weller as an expert in the field of firearms examination and filed a motion in limine to exclude his toolmark identification testimony. Defendant argued a 2009 report issued by the National Research Council of the National Academy of Science, entitled Strengthening Forensic Science in the United States: A Path Forward (hereafter 2009 NRC report) constituted a material change in the prevailing scientific opinion on toolmark identification. Because of changing scientific perceptions of toolmark identification, defendant argued such an analysis is no longer reliable and is inadmissible.

At the hearing on defendant's motion, the prosecutor noted Weller would not be providing a statistical analysis of whether the toolmarks from the gun matched those on the casing and fragments. However, the prosecutor argued Weller should be allowed to state his opinion of certainty. The court ruled percentages of certainty would not be appropriate, but deferred ruling on "how he couches his words as to whether he's confident or not." The court indicated it would rule on any objections made during Weller's testimony as appropriate.

At trial, Weller was presented as an expert criminalist qualified to analyze and give an opinion on the comparison of firearms, bullets, and casings. During voir dire of Weller's qualifications, defense counsel vigorously questioned Weller on the 2009 NRC report and the reliability of toolmark identification analyses. Weller acknowledged toolmark identification is based on an individual's visual comparison, and there is no computer model that can perform the analysis. Weller also acknowledged there is no industry standard regarding how many points of similarity are required to find a match between a gun and a casing or bullet. However, Weller stated if another examiner reviewed the material, he was "confident that other examiner [sic] would get the exact same result I obtained. These types of tests have been studied by our profession, they're called validation studies, and they've been done for decades, a long time." The court accepted Weller as an expert in firearms identification and analysis of firearms, bullets, and casings.

Weller proceeded to testify for the prosecution. He described the manufacturing process of firearms to explain how two guns of the same make and model would leave distinct microscopic toolmarks. He then testified, based on various marks, both striations and impressions, that the firearm recovered from defendant's vehicle fired the casing and bullet collected from the crime scene. Weller testified he was "completely confident" in his conclusion. In discussing the similar markings, Weller explained, "So part of the training of a firearms examiner is to purposefully look at what we call known matches. These are samples that I know come from the same firearm, and known non-matches, samples that I know come from different firearms. And in non-matching cases, this level or amount of agreement is unexpected. It's considered a practical impossibility. Whereas this level of agreement is what I do expect to see in matching casings. So this is consistent with or congruent with the matching situation and is not what I would expect, or is unreasonable to expect from a non-matching situation." No objections were raised by defense counsel to any of Weller's testimony.

On cross-examination, defense counsel repeatedly probed the certainty of Weller's findings. Weller acknowledged the same gun could shoot two bullets and those bullets could have different markings that are not identifiably a match. Defense counsel also questioned Weller about the 2009

NRC report and instances in which examiners reach opposite conclusions. Weller conceded the accuracy of his conclusions "really ultimately depends on [him], on how good [he is] and consistent and experienced [he is] because [he is] the one who's making the determinations of how and what is an identifiable mark."

2. Analysis

Defendant concedes firearms leave toolmarks on bullets and examination of those toolmarks may provide some relevant information. But he argues toolmark identification is "unreliable, misleading, and discredited by the scientific community" and therefore inadmissible under People v. Kelly (1976) 17 Cal.3d 24, 130 Cal. Rptr. 144, 549 P.2d 1240 (Kelly).

"In [Kelly, supra,] 17 Cal.3d 24, [the Supreme Court] held that evidence obtained through a new scientific technique may be admitted only after its reliability has been established under a three-pronged test. The first prong requires proof that the technique is generally accepted as reliable in the relevant scientific community. (Id. at p. 30.) The second prong requires proof that the witness testifying about the technique and its application is a properly qualified expert on the subject. (Ibid.) The third prong requires proof that the person performing the test in the particular case used correct scientific procedures. (Ibid.)" (People v. Bolden (2002) 29 Cal.4th 515, 544–545, 127 Cal. Rptr. 2d 802, 58 P.3d 931.)

"'Kelly is applicable only to "new scientific techniques." [Citations.]' (People v. Leahy (1994) 8 Cal.4th 587, 605, 34 Cal. Rptr. 2d 663, 882 P.2d 321.) It '"only applies to that limited class of expert testimony which is based, in whole or part, on a technique, process, or theory which is new to science and, even more so, the law." [Citation.]' (Ibid.) As stated by the Leahy court in discussing People v. Stoll (1989) 49 Cal.3d 1136, 265 Cal. Rptr. 111, 783 P.2d 698, 'by reason of the potential breadth of the term "scientific" in the Kelly/Frye4 doctrine, the courts often refer "to its narrow 'common sense' purpose, i.e., to protect the jury from techniques which … convey a '"misleading aura of certainty."' [Citations.]" (49 Cal.3d at pp. 1155–1156.)

According to Stoll, a technique may be deemed "scientific" for purposes of Kelly/Frye if "the unproven technique or procedure appears in both name and description to provide some definitive truth which the expert need only accurately recognize and relay to the jury." (Id. at p. 1156, italics added.)' (People v. Leahy, supra, 8 Cal.4th at p. 606.)" (People v. Mitchell (2003) 110 Cal.App.4th 772, 782–783, 2 Cal. Rptr. 3d 49 (Mitchell).)

"Thus, Kelly analysis is limited to situations where it will 'forestall the jury's uncritical acceptance of scientific evidence or technology that is so foreign to everyday experience as to be unusually difficult for laypersons to evaluate. [Citation.] In most other instances, the jurors are permitted to rely on their own common sense and good judgment in evaluating the weight of the evidence presented to them. [Citations.]' (People v. Venegas (1998) 18 Cal.4th 47, 80, 74 Cal. Rptr. 2d 262, 954 P.2d 525.)" (Mitchell, supra, 110 Cal.App.4th at p. 783.)

"'Once a published appellate decision has affirmed admission of a scientific technique, the technique's general acceptance is established as a matter of law. Further hearings on general acceptance are unnecessary "at least until new evidence is presented reflecting a change in the attitude of the scientific community."' (People v. Jones (2013) 57 Cal.4th 899, 937, 161 Cal. Rptr. 3d 295, 306 P.3d 1136.)

Defendant does not argue toolmark identification testimony has not previously been recognized. Nor can he reasonably do so. Toolmark identification evidence has been admitted in California for over 60 years. (See People v. Godlewski (1943) 22 Cal.2d 677, 685, 140 P.2d 381.) Rather, defendant cites the 2009 NRC report as evidence of a change in the attitude of the scientific community. He contends toolmark identification is no longer generally accepted or reliable within the scientific community.

We recognize toolmark identification has come under increased scrutiny in recent years. But the 2009 NRC report, while criticizing the subjectivity of the field, does not call for abandonment of the field. We note such testimony has continued to be admitted in both California and federal courts, and our Supreme Court in People v. Cowan (2010) 50 Cal.4th 401, 113 Cal. Rptr. 3d 850, 236 P.3d 1074 (Cowan), reaffirmed the admissibility of such testimony even after publication of the 2009 NRC report. Indeed, defendant has not cited a single case prohibiting such testimony. Rather, some courts have only placed limits on the degree of certainty an examiner can express in a match. … Other courts, however, have continued to admit toolmark identification testimony without limitation. … In light of such recognition, we cannot conclude the scientific community has rejected toolmark identification. Accordingly, Kelly, supra, 17 Cal.3d 24 is not applicable.

Moreover, even if the scientific community did dispute the reliability of toolmark identification, Kelly only applies to "scientific" methods. (People v. Leahy, supra, 8 Cal.4th at p. 606.) Defendant argues toolmark analysis is a scientific technique, because jurors cannot "competently evaluate for themselves whether 'this gun fired this bullet.'" While defendant acknowledges the California Supreme Court found otherwise in Cowan, supra, 50 Cal.4th 401, he attempts to distinguish that opinion from the current case because he "made a strong showing that toolmark identification carries the weight of science." We find this attempt to distinguish Cowan unavailing.

Cowan resolved the exact issue defendant now seeks to relitigate. There, the Supreme Court addressed the supposedly novel technique of comparing a pistol barrel with recovered bullets by making a cast of the pistol barrel to assist in the comparison. (Cowan, supra, 50 Cal.4th at pp. 468, 470–471.) The defendant did not claim "that either the technique of ballistics comparisons or the technique of identifying tool marks using molds made of elastic material is new" but objected to the combined technique and argued a Kelly hearing was required. (Cowan, at p. 470.) The Supreme Court disagreed. The court emphasized neither technique was "'so foreign to everyday experience as to be unusually difficult for laypersons to evaluate.'" (Ibid.) The expert's procedure in Cowan "merely 'isolate[d] physical evidence'—specifically, the pattern of lands and grooves and associated imperfections on the inside of the Colt pistol's barrel, as well as the corresponding markings on the recovered bullets—'whose … appearance, nature, and meaning [were] obvious to the senses' of the lay jurors." (Id. at p. 471.) "Although there was some dispute about whether the method [the expert] used produced a cast of the barrel 'without tampering or alteration' due to possible bubbling or shrinkage of the Mikrosil, that possibility was fully explored on cross-examination and the jury had the opportunity to weigh its effect on the validity of [his] conclusions." (Ibid.)

Under Cowan's reasoning, testimony matching bullet casings and firearms based on toolmarks is testimony lay jurors can readily understand. (Cowan, supra, 50 Cal.4th at p. 471.) The 2009 NRC report does not alter this analysis because it does not opine on what qualifies as a scientific determination under Kelly. The Supreme Court has, however, ruled on this issue and we are bound by such precedent.

Finally, we disagree that Weller's toolmark identification analysis should be found unreliable under Sargon Enterprises, Inc. v. University of Southern California (2012) 55 Cal.4th 747, 149 Cal. Rptr. 3d 614, 288 P.3d.

1237. In Sargon, the court held an expert "'must provide a reasonable basis for the particular opinion offered, and that an expert opinion based on speculation or conjecture is inadmissible.'" (Id. at p. 770.) Here, Weller provided a reasonable basis for his opinions—namely, his observation and analysis of the gun, casing, and bullet fragments. Defendant's disagreement with Weller's conclusions does not trigger the concerns flagged in Sargon.

Barring a definitive rejection of toolmark analysis by the scientific field or binding authority to the contrary, the trial court properly admitted Weller's testimony.

…

III. DISPOSITION

The judgment is modified to reflect the $250 probation investigation fee is stricken. The clerk of the superior court is directed to prepare a new abstract of judgment with this modification and to send a certified copy thereof to the Department of Corrections and Rehabilitation. In all other respects, the judgment is affirmed.

Questions:

1. Impressions on casings, projectiles are caused by what process?
2. Instead of perfect match, what can an expert testify to when making ballistic comparisons in the type of case?
3. Part of ballistics is the measurement of impressions- toolmark- that is by the mechanical operation of the weapon. How effective was the defense argument that "toolmark" analysis is not longer credible.

Visit the ATF's website of Historic cases at https://www.atf.gov/our-history/blast-past.

The characteristics and condition of a projectile is a story unto itself. The journey each projectile takes impacts its alteration through time and space.

State v. Hendel
4 Idaho 88, 35 P. 836
Supreme Court of Idaho
February 19, 1894, Decided

HUSTON, C. J.--The defendant John Hendel was charged jointly with one Louis Roder, by information filed in the district court for Alturas county, with the crime of murder in the killing of one Leander Holstrom, on the seventeenth day of March, 1893. Upon their arraignment, defendant Hendel, on his own application, was given a separate trial. On said trial, at the June term, 1893, of said district court, defendant Hendel was convicted of murder in the second degree. A motion for new trial was made on his behalf, which was denied by the court, and from the order denying said motion, and from the judgment of final conviction, this appeal is taken.

The case comes to this court upon a bill of exceptions. The errors assigned are seven in number, and are as follows: "1. The court erred in overruling appellant's motion for the appointment of an elisor to take charge of the jury; 2. The court erred in permitting J. S. Waters, Esq., counsel for Louis Roder, to interpose an objection to, and to be heard in argument upon, the request by appellant's counsel, upon the trial, that the bullet taken from the vest worn by deceased be submitted to a microscopical and chemical expert examination and analysis; 3. The court erred in overruling the application made by appellant's counsel, upon the trial, for leave to have made a microscopical and chemical examination of the bullet found in the vest of deceased, and of the substance or substances adhering thereto; 4. That the court erred in overruling the objection of appellant's counsel, upon the trial, and in permitting the witness D. W. Figgins to testify as to his belief in relation to the bullet found in the vest of deceased; 5. That the court erred in overruling the objection of the appellant upon the trial, and in permitting the witness Thomas Miser to testify as an expert in answer to the following question, put to him by the counsel for the state: 'What is your opinion as to whether that bullet [state's exhibit No. 8] was fired from a pistol?'--and to the following: 'What is your opinion as to whether that [bullet found in vest] has been fired from a pistol?' 6. The court erred in overruling the objection of appellant, made at the trial, and in permitting the witness Seaborn Jones to testify as an expert as to the appearance and condition of a bullet, then in evidence, and as to whether it had been fired from a weapon--gun or pistol; 7. The court erred in denying defendant's motion for a new trial." We will consider the errors assigned in the order in which they are presented in appellant's brief.

As to the first error assigned: The refusal of the court to appoint, on application of defendant, an elisor to take charge of the jury during the trial of the case. We think the action of the court was correct. The application did not come within the provision of the statute applicable to such cases. (Rev. Stats., sec. 1887.) The granting or refusing of the application was within the discretion of the court, and it seems to us it was properly exercised in this case.

As to the second assignment of error by the court: In permitting counsel for appellant's codefendant to be heard in opposition to appellant's request that a certain bullet, offered in evidence by appellant, should be submitted to analytical and microscopical examination. While the point presented is somewhat novel, we think it was largely a matter of discretion with the trial court, and we cannot say, upon an examination of the record before us, that the action of the court even tended to prejudice the case of appellant. The facts, as shown by the record here (and it is conceded it is only a part of the evidence), are substantially as follows: The appellant was a keeper of a saloon or beer hall in the town of Hailey, Alturas county. On the night of the 16th of March, 1893, the deceased and three others had been drinking beer at appellant's place, until about 4 o'clock on the morning of the 17th, when as is quite usual in such cases, the debauch culminated in a quarrel between appellant, his codefendant, Roder, who was at the time in the employ of appellant, on the one side, and deceased and his three companions, on the other, during which quarrel two shots were fired, one by appellant and one by his codefendant, Roder. There is some evidence of a third shot being fired, but it hardly arises to the dignity of a suspicion. One of the shots took effect on the body of deceased, entering between the seventh and eighth ribs, on the right side, striking and tearing away a piece of the eighth rib, one-fourth of an inch in width, and one-half an inch in length, glancing upward, and passing through the lungs and through the body of the heart, and out through the left breast, between the sixth and seventh ribs. At the time the inquest was being held, and after the clothes had been removed from the body of deceased, a bullet was found upon the floor, in close proximity to the place where the clothing was removed from the body. This bullet was of size carried by the pistol used by appellant on the night, or rather the morning, of the homicide, to wit, a "British Bulldog" pistol, of 44-caliber. Another bullet was also found at that time, which had passed through a portion of the partition of the room, about eight feet from the floor. This latter bullet was of the size carried by the pistol had and used by appellant's codefendant, Roder, on the night of the homicide, to wit, a 45-caliber Colt's revolver. It is conceded that appellant and Roder each fired one shot at and in the direction where deceased were removed, it was found that the bullet that had passed through him had made a hole in the lining of his vest, at a point thereon corresponding with the place of exit of the bullet from his body. Search was made for the bullet, and one was found, as above stated, on the floor near where the clothes had been taken from the body of deceased. No other bullet was found at that time. The bullet found on the floor, as stated, bore unmistakable evidence of having been fired from a gun or pistol, and also of having come in contact with some hard substance, which had battered the point, and scraped or torn some of the substance from the side of it. This bullet was of the size and fitted the pistol of appellant. But now appears the peculiar and disturbing feature of this case. When the clothing of deceased was brought before the magistrate on the preliminary examination of the defendant (it does not appear from the record when the preliminary examination took place, but it was presumably some time after the inquest, which was held on the 17th or 18th of March; it is not exactly clear which of these two days), upon an examination of the vest of deceased by the sheriff, in whose custody the clothing of deceased had been from the time of the inquest to the time of the preliminary examination, a bullet was found in the lower corner of the left side of the vest, between the outside material and the lining; and this bullet, so found, fitted and was the size carried by the pistol used by appellant's codefendant, Roder, on the occasion of the homicide. Upon the production of this bullet by the sheriff (called on the part of the defendant), who took it from the vest of deceased at the time of the preliminary examination of defendant Hendel, and who had retained it in his possession since that time, counsel for defendant Hendel moved the court to permit a microscopical and analytical examination to be made by a competent expert of the substance or substances adhering to the bullet found in the vest of deceased, for the purpose of determining its true composition. To this motion, one J. S. Waters, an attorney of said district court, and who, it seems, was counsel for appellant's codefendant, Roder, objected, claiming that, as the production of this bullet was an attempt to fasten the alleged homicide upon his client, he insisted that the bullet should be retained in its then condition until the trial of his client, informed against jointly with the defendant on trial, was had. The permitting of Waters, aforesaid, to be heard upon said motion, is assigned as error by counsel for appellant. We are unable to see any error in this ruling of the court. The defendants had been informed against jointly. Hendel, upon his application, had been awarded a separate trial. It was, it seems to us, the duty of Roder's counsel, not only to be present at the trial of Hendel, but to vigilantly watch and scrutinize the conduct of such trial, to the end that none of the legal rights of his client should be sacrificed or jeopardized in the interest of Hendel. We think the district court committed no error in permitting the counsel of Roder to be heard on the said motion of appellant, either as amicus curiae, or in behalf of his client, but that the action of the court was consonant with, and calculated to effect the ends of justice.

The third specification of error presented by appellant is based upon the refusal of the court to permit the "microscopical and chemical examination of the bullet, and of the substance or substances adhering thereto." Aside from the objections urged by the counsel for Roder against the granting of this request, which, of themselves, were sufficient, in our view, to warrant the action of the court, neither the court nor the law ever intended that the granting of a separate trial to one of two or more defendants, jointly indicted, should be used by the defendant first placed on trial as a means by which he could sacrifice his codefendants, or do aught to impair their just and legal defense. But it does not appear from the record that there was any substance or substances adhering to the bullet which required the invocation of either chemical or microscopical aid to define its character. One witness testified: "There is something in the creases of this bullet, but I cannot say what it is. The apex of the bullet is slightly battered, and is marked with the imprint of the cloth--warp of the cloth--on the end of the bullet." And this was all the testimony appearing in the record upon the subject. Neither of the medical witnesses testified to the existence of anything on the surface of the bullet at all resembling, or intimating an appearance of, "blood stains, or human tissues." So far as the record before us shows, the blood and tissue theory has no more support in the evidence than does the theory of the third shot, with which it is intimately allied. The medical witnesses stated that it would take from one to two weeks to procure the test required. We think, under the circumstances, it was within the discretion of the trial court to grant or refuse the application on the part of appellant, and that, in refusing it, the court committed no error.

The fourth specification of error is predicated upon the ruling of the court in permitting D. W. Figgins, one of the medical witnesses, "to testify as his belief in relation to the bullet found in the vest of the deceased." It was a theory sought to be established by the defense that three shots were fired--two by Roder and one by Hendel; that it was one of the shots fired by Roder that killed the deceased. To maintain this theory, the defense called Dr. N. J. Brown, who testified as follows, in regard to the bullet found upon the floor near where the clothing was removed from the body of deceased: "Under all the conditions of the injuries inflicted, and the manner in which the deceased was dressed, this bullet, from its appearance, did not pass through the body of the deceased." On being shown the bullet taken from the vest of deceased, said witness testified in regard to the same as follows: "I believe this bullet could have produced all the effects upon the body of the deceased, as I have observed them, and as they have been testified to here in court. This bullet has marks on the point that resemble the mesh work of weaving of cloth, and the edges are turned over from force of some kind, striking some substance, or some substance striking it." Unsatisfactory as this evidence would seem to be, either from a legal or a logical standpoint, it was enough that it was offered on the part of the defense, and was permitted to go to the jury, and it was but common justice to permit the prosecution to meet it in rebuttal by the testimony of Dr. Figgins, who testified as follows: "A bullet striking a rib would be battered, in my opinion. I do not believe that the bullet taken from Holstrom's vest is the bullet which killed Lee Holstrom. I have examined that bullet, both with the naked eye and with a magnifying glass. There is the appearance of the web or mesh of the cloth on that bullet, while I believe that the bullet which caused the death of Lee Holstrom was battered, more or less, and would not have the meshes of the cloth stamped on it. The meshes of the cloth, if stamped on the bullet, would have been destroyed by contact with the rib. In my experience, as a surgeon. I never saw a bullet that had been fired into bodies that had any grease substance left upon it. The burning of the powder would melt the grease." Without commenting upon the relative value of the testimony of these two expert witnesses, it is enough to say that the ancient conundrum of "Who should decide when doctors disagree?" seems to have been settled by the jury in this case, at least. Several other witnesses were called by the defense in rebuttal, who testified to the same effect as Dr. Figgins--that the appearance of the bullet taken from the vest of deceased showed it had never been fired from a gun or pistol--giving various reasons for this conclusion. Notwithstanding counsel for the defendant had opened the door to this class of testimony by introducing that of Dr. Brown, he now contends that it was error for the court to permit the same class of testimony in rebuttal. A careful examination of the authorities constrains us to believe that the question as to the admissibility of expert testimony must depend largely, if not entirely, upon the circumstances of each particular case; and where it is apparent, not only that no injustice has been done, and that the exigencies of the case seem to warrant or require the admission of such testimony, for a proper and satisfactory presentation of the facts to the jury, such admission is not error.

It is objected by counsel for the appellant that the bill of exceptions does not show that any venue was proven upon the trial. While it is apparent from the record that there was great remissness on the part of counsel who represented the state in the settlement of the bill of exceptions, we think there is sufficient therein to show that the venue was proven. Besides, it was conceded by counsel for appellant, before the argument of the case was entered upon, that the record contained only a portion of the evidence.

The objection that the verdict is contrary to the evidence is not well taken. In addition to the evidence already commented upon, no less than four witnesses (Field, McGarnish, Farrel, and Saunders) testified positively that Hendel admitted to them, and in their presence, that he killed Holstrom, and undertook to justify the killing by saying that deceased had knocked him down, and had knocked his wife down, and he had to kill him. Hendel's attempt to explain away these admissions was, as evidenced by the verdict, a failure. Finding no reversible error in the record, the judgment of the district court is affirmed.

Questions:

1. How many bullets were fired during this altercation and where were these bullets found?
2. Why did the expert witness Dr. Brown have his suspicions about the bullet which allegedly passed through the victim's body?
3. What does the Court say about projectile review under a microscope?

Wiley v. State
Court of Special Appeals of Maryland
January 15, 2020, Filed
No. 989, September Term, 2018

Appellant Gregory Leon Wiley was convicted of the murder of Cornell Antonio Yarbour in 1997. The prosecution's theory was that Mr. Yarbour was killed because he had stolen from

Thomas James Anderson, a local drug dealer, and that appellant was ordered by Mr. Anderson to kill Mr. Yarbour.

At trial, the prosecution used comparative bullet-lead analysis (CBLA), a forensic investigation technique which has since been discredited, to establish a connection between Mr. Anderson and Mr. Yarbour's murder. Without that connection, appellant contends that he would not have been convicted. He filed a Petition for a Writ of Actual Innocence on February 28, 2017, which the Circuit Court for Baltimore County denied. On appeal, he asks:

> Did the trial court abuse its discretion in denying appellant's petition for a writ of actual innocence on the ground that newly discovered evidence discrediting CBLA did not create a substantial or significant possibility that the result at trial may have been different?

For the following reasons, we answer that question "no" and affirm the judgment of the circuit court.

FACTUAL AND PROCEDURAL BACKGROUND

Mr. Yarbour was found dead near appellant's father's residence on March 12, 1992. Found nearby was a crack cocaine pipe, fake cocaine, a bloody rock, some tire impressions. Three .32 caliber bullets were recovered from his body. In the course of the State's investigation, a search warrant was executed at Mr. Anderson's address. There, police recovered an empty black shoulder holster, an "owe sheet," cocaine, and multiple kinds of ammunition, none of which was .32 caliber. During the search, a girlfriend of Mr. Anderson's, Debra Campbell,2 entered the residence and was arrested and charged for narcotics offenses. When interviewed, she claimed not to know who was responsible for the murder of Mr. Yarbour.

Later that month, police seized a Nissan Stanza vehicle registered to Mr. Anderson with tires that left tracks similar to those found near Mr. Yarbour's body. Inside the vehicle were .32 caliber bullets, an ankle holster, and cocaine paraphernalia. A search of appellant's house yielded documents referring to "coke."

Appellant was charged with murder in September of 1996, and trial began in March of 1997.

The Trial

Ms. Campbell testified that appellant's involvement in Mr. Anderson's drug operation included "beat [ing] people up" when they owed money. A few days before the murder, her home was broken into and drugs and money were taken out of a box locked in her bedroom. Because Mr. Anderson suspected Mr. Yarbour of the break-in, she, Mr. Anderson, and appellant drove in Mr. Anderson's Nissan Stanza to search for him. Mr. Anderson carried a shotgun; she carried a .25 caliber handgun belonging to Mr. Anderson; and appellant carried Mr. Anderson's .32 caliber revolver.

After Mr. Anderson spoke with a "guy that claim[ed] to be [Mr. Yarbour's] brother," they went to Ms. Campbell's residence. There, Mr. Anderson asked appellant to kill Mr. Yarbour in exchange for $500 in drugs and money. According to her testimony, Mr. Anderson:

> wanted [appellant] to kill him with the [.]32 because he didn't like that gun as much. [Mr. Anderson] wanted [appellant] to take [Mr. Yarbour] in the car, pick [Mr. Yarbour] up at [his] house, and just fake like they were friends … but not kill him in the car. Because of the blood.

When appellant returned to Ms. Campbell's home hours later and said that he "did it," Mr. Anderson "looked so happy." When she was asked, "How did he say he did it?" she recounted:

> [Appellant] drove [Mr. Yarbour] down a street … I remember asking him why so close to your brother's house I remember near water … He said that they were supposed to get high and they got out of the car. He had him get out of the car.

> When [Mr. Yarbour] wasn't paying attention, [appellant] pulled the gun out and [Mr. Yarbour] thought he was playing and asked him, you know, what are you doing, stop playing.
>
> And then, [appellant] pulled the trigger one time, and shot [Mr. Yarbour] in the stomach, and [Mr. Yarbour] asked him to stop. And when he asked him to stop, he knew he couldn't, so he just shot him four more times.
>
> [Appellant] didn't believe he was dead, and he picked up a brick or a rock and was—because he was still moving—and hit him in the head with it, and he thought he was still alive after that, even when he was back at the apartment.

Ms. Campbell added that she disposed of the gun used by appellant by throwing it in the water at Rocky Point Park.

She denied having been offered any special consideration regarding the outstanding California and Maryland drug charges, but she admitted that the sentencing judge in Maryland would take her "truthful[ness]" into account in deciding her sentence and that she hoped that her testimony in this case would help her.

Donald Ray Bobbitt, an admitted drug dealer and user, worked for Mr. Anderson in 1992. Only after his car was impounded with drug paraphernalia did he contact the police with information relevant to this case, saying that he did not "want to be an accessory to a murder." Police set up a wiretap in the wall of Mr. Bobbitt's motel room, but he was unable to coax Mr. Anderson and appellant into the room to get more information from them. Compounding that failure, Mr. Bobbitt informed police that he had bought drugs instead of paying his motel bill, and needed the police to pay for his room. He denied being offered a deal in exchange for the information he provided.

Mr. Bobbitt testified that appellant was also a drug dealer who did "odds and ends" for Mr. Anderson, such as "snatch[ing] people up" who owed money. According to Mr. Bobbitt, appellant told him, in May of 1992, that "he killed [Mr. Yarbour]" by shooting him "three or four times in the chest," and that appellant was afraid Mr. Anderson would "turn [him] in for what he done."

Mr. Bobbitt recounted that, in March of 1992, Mr. Anderson admitted that he had "had [Mr. Yarbour] taken care of" by appellant who "shot [Mr. Yarbour] three or four times in the chest, beat him, and … took vials of fake cocaine and put it on his person." Mr. Anderson also told him that Ms. Campbell had disposed of the gun. According to Mr. Bobbitt, Mr. Yarbour was murdered because he had broken into an apartment and stolen money and cocaine belonging to Mr. Anderson.

Shirley Ann Scott, another girlfriend of Mr. Anderson's, was living with him at the time of the murder. She testified that appellant told her that he shot Mr. Yarbour three or four times and then beat him.

John Henley, who had been a drug runner for Mr. Anderson, testified that he was Mr. Yarbour's "get-high brother,"3 and that he had seen appellant "beat people up" at Mr. Anderson's direction. A few days before the murder, Mr. Yarbour told Mr. Henley and appellant that he planned to visit Mr. Anderson's house and that he knew "where [to] get some serious money and drugs." Later that night, when he and Mr. Yarbour were both at a friend's apartment, Mr. Yarbour had a bag containing money and drugs that were in the same type of containers that they would get from Mr. Anderson. Mr. Henley and Mr. Yarbour "[got] high off the crack cocaine" that was in the bag.

When Mr. Henley left the friend's apartment, he encountered a Nissan Stanza driven by a "white female" named Debbie, with Mr. Anderson and two other men as passengers. Mr. Anderson, who was holding a sawed-off shotgun, asked Mr. Henley about Mr. Yarbour's whereabouts. Mr. Henley denied having seen him. He stated that he did not see appellant, but he could not identify or describe the other male passengers.

Mr. Henley acknowledged lying to police by providing a fake name and date of birth. He denied receiving any consideration for his testimony, and claimed that he testified because Mr. Yarbour had a son who "was left behind."

At the time of the murder, Sherry Lee Scarbath lived with Mr. Yarbour and their son. Ms. Scarbath testified that the night before Mr. Yarbour's body was found, appellant came to their residence, and then the two men left together. That was the last time she saw Mr. Yarbour alive.

The medical examiner testified that Mr. Yarbour suffered five or six gunshot wounds as well as lacerations and blunt force trauma to the head. Additionally, the State presented evidence that tire tracks found near the body matched one of the tires on Mr. Anderson's Nissan Stanza.

Agent Ernest Peele of the Federal Bureau of Investigation (FBI) testified as an expert in CBLA. He explained that the technique involved determining the ratios of certain elements within particular lead bullets and then comparing their composition. In this case, he compared a total of three fired bullets with twenty unused bullets from a box of cartridges found inside Mr. Anderson's vehicle. He concluded that the fired bullets and the twenty compared bullets "[had] the same composition as if all of these bullets were in the same box at one time," and that "they were made at the same time."

During closing argument, the prosecutor referenced Agent Peele's CBLA testimony:

> Ultimately you can say [Agent] Peele is right, the bullets in the body probably came from this box, bullets manufactured on the same day It makes it probative, it makes it worthwhile evidence, and it makes [Ms.] Campbell correct and consistent.

After two days of deliberations, the jury found appellant guilty of first-degree murder and use of a handgun in the commission of a crime of violence.

The Petition for Writ of Actual Innocence

The Circuit Court for Baltimore County held a hearing on appellant's Petition for Writ of Actual Innocence on February 21, 2018. Counsel for appellant and the State agreed that the CBLA evidence was "newly discovered evidence." The State, however, argued that it did not create a substantial or significant possibility of a different result. Counsel for appellant, pointing out that Ms. Campbell, Mr. Bobbitt, and Mr. Henley all had "credibility problems," argued that the State relied on Agent Peele's CBLA testimony to corroborate the testimony of questionable witnesses.

The circuit court, in its April 2, 2018 memorandum opinion, denied appellant's Petition for Writ of Actual Innocence. The court found that the evidence discrediting CBLA was "newly discovered evidence" that could not have been discovered in time to move for a new trial. But it rejected the argument that it created a substantial possibility of a different result:

> After reviewing the trial transcript, considering the testimony and argument made at the Writ of Actual Innocence hearing, and considering relevant case law, this Court finds that Petitioner's allegation that the denigration of CBLA evidence at the time of trial would have created a substantially different outcome is meritless. In reviewing the transcript, it is clear that Petitioner's conviction was supported by ample witness testimony. This Court finds that even considering the unreliability of CBLA evidence as conceded by the State, the witness testimony presented supports the State's position that the Petitioner does not meet his burden as required by Section 8-301 of Maryland's Criminal Procedure Article. It is evident that the CBLA evidence did not have any substantial impact on Petitioner's conviction, as noted in defense counsel's closing at trial.4 Although the CBLA evidence is newly discovered evidence that could not have been discovered within the time to move for a new trial, this Court finds that the denigration of the CBLA evidence does not create a substantial possibility that the outcome of Petitioner's trial would be different.

This timely appeal followed.

STANDARD OF REVIEW

We review the denial of a Petition for a Writ of Actual Innocence for an abuse of discretion. Smith v. State, 233 Md. App. 372, 411–12, 165 A.3d 561 (2017). In doing so, we apply the well-recognized "before and after" test and ask whether there would be a "substantial or significant possibility" of a different outcome had the new evidence been presented to the jury. Ward v. State, 221 Md. App. 146, 169, 108 A.3d 507 (2015) (citing Yonga v. State, 221 Md. App. 45, 108 A.3d 448).

DISCUSSION

Contentions

Appellant contends that the court improperly denied his petition for a Writ of Actual Innocence, based on "ample witness testimony" supporting appellant's convictions, for two reasons. First, by "failing to recognize that the prosecutor used the CBLA evidence to establish the credibility of the State's critical witness, Ms. Campbell." As he sees it, that evidence was particularly important because it was scientific evidence corroborating Ms. Campbell's testimony that appellant killed the victim with Mr. Anderson's .32 caliber gun. In support, he cites Clemons v. State, 392 Md. 339, 372, 896 A.2d 1059 (2006), where the Court of Appeals stated that "[l]ay jurors tend to give considerable weight to 'scientific' evidence." (quoting Reed v. State, 283 Md. 374, 386, 391 A.2d 364 (1978)). And second, by "plac[ing] undue emphasis on the defense's theory of the case as expressed in closing argument," and "discounting the State's burden to prove the offenses beyond a reasonable doubt."

The State responds that the hearing court acted within its discretion in denying the petition because there was overwhelming witness testimony and forensic evidence as to appellant's involvement in the murder of Mr. Yarbour. It contends that "[e]xcising Agent Peele's testimony" from the trial would not create a substantial likelihood of a different result because all of the admitted forensic evidence—"the tire tracks, the bullets, and the bullet composition"—implicated Mr. Anderson and not appellant. But, for the jury to convict appellant, "it had to credit the testimony of multiple witnesses that said [Mr.] Anderson ordered appellant to kill [Mr.] Yarbour" and that Agent Peele's testimony "was irrelevant to that determination." The State adds that defense counsel's closing was not the reason the hearing court found that the "exclusion of the CLBA evidence did not create a substantial likelihood of a different outcome." Instead, the hearing court viewed it as an "accurate summary of the significance of the CBLA evidence" and "did not unduly emphasize it."

Analysis

Section 8-301(a) of the Criminal Procedure Article ("CP") provides:

> A person charged by indictment or criminal information with a crime triable in circuit court and convicted of that crime may, at any time, file a petition for writ of actual innocence in the circuit court for the county in which the conviction was imposed if the person claims that there is newly discovered evidence that:
>
> (1)(i) if the conviction resulted from a trial, creates a substantial or significant possibility that the result may have been different, as that standard has been judicially determined[.]

See also Md. Rule 4-332(d)(8). The petitioner bears the burden of proof.5 CP § 8-301(g); Md. Rule 4-332(k).

The newly discovered evidence must support the claim of actual innocence. Smith v. State, 233 Md. App. 372, 413, 165 A.3d 561 (2017). In other words, that the "petitioner did not commit the underlying crime for which he or she was convicted." Smallwood v. State, 451 Md. 290, 316, 152 A.3d 776 (2017). But it "need not definitively prove his or her innocence." Smith, 233 Md. App. at 413. "That the newly discovered evidence does not definitively exonerate appellant, or may be countered by other evidence, goes to the weight of the evidence." Id.

We have held previously that evidence undermining the validity of CBLA evidence is newly discovered evidence. Ward v. State, 221 Md. App. 146, 149, 108 A.3d 507 (2015). And when, as in this case, the CBLA evidence was used to link appellant to the murder, it "speaks to" his actual innocence. Smith, 233 Md. App. at 413. A hearing court considering an actual innocence petition "look[s] back to the trial" to "determine whether the newly discovered evidence created a substantial or significant possibility that the result may have been different." McGhie, 449 Md. at 511.

In this case, the hearing court found that "the denigration of the CBLA evidence [did] not create a substantial possibility that the outcome of [appellant's] trial would be different" after summarizing Agent's Peele's testimony and then detailing other witness testimony and physical evidence directed at his guilt:

> [Ms.] Campbell's testimony included being present when [Mr. Anderson] hired Petitioner to kill the victim, being present when Petitioner returned from the murder, wiping Petitioner's fingerprints off of the gun, and hearing Petitioner acknowledge and describe the murder.
>
> The State presented additional evidence at trial. [Mr.] Bobbitt, a frequent drug customer of the [Mr. Anderson] who hired Petitioner to commit the murder, testified that Petitioner confessed the murder to him. Multiple other witnesses corroborated Petitioner's use of [Mr. Anderson's] car, association with [Mr. Anderson], and location on the night of the crime. Alex Mankovich, a Maryland State Police Officer, testified that the car Petitioner used left tire marks found in the mud next to the victim's body.

Appellant contends that the CBLA evidence was critical to "corroborat[ing] Ms. Campbell's testimony that Mr. Yarbour was likely killed with .32 caliber bullets from the box found in [Mr.] Anderson's car." He argues that the hearing court failed to consider the prosecutor's use of the CBLA evidence to bolster the credibility of the State's key witness, Ms. Campbell, who otherwise could have been discredited. In his view, because Ms. Campbell's credibility was "explicitly connected to the reliability of the CBLA," its exclusion would create a significant possibility of a different outcome.

Ms. Campbell was, of course, an important witness in the State's case. She was the only witness who testified to what happened on the night of the murder from start to finish. She testified that shortly after it happened, appellant described the murder to her in graphic detail: that he drove Mr. Yarbour to where his body was found for the purpose of getting high; that when they got out of the car, he pulled the gun out and Mr. Yarbour thought he was playing around; that he shot Mr. Yarbour once in the stomach and then four more times; and, because he did not believe Mr. Yarbour was dead, he picked up a brick or rock and hit Mr. Yarbour in the head with it. And Ms. Campbell also admitted to disposing of the murder weapon by throwing it in the water at Rocky Point Park.

The prosecutor did seek to use the CBLA evidence to bolster Ms. Campbell's testimony:

> Agent Peele is an agent of the FBI and he is trained in chemical analysis … . [H]e says to you … look [ing] at the three bullets that are coming out of the body, and I do the same analysis on those bullets as I do the bullets that are there this box that came out of [Mr. Anderson's] car.
>
> What I say to you is simply this; they are so similar in the amounts of these trace elements that we look for what I can say to you that bullets that came out of the body within a reasonable degree of scientific certainty came from this box or the box manufactured by the same company on or about the same day.
>
> * * *
>
> [U]ltimately you can say [Agent] Peele is right, the bullets in the body probably came from this box, bullets manufactured on the same day that have survived eighteen years. It makes it probative, it makes it worthwhile evidence, and it makes [Ms.] Campbell correct and consistent.

(Emphasis added). On the other hand, the CBLA evidence was not the only evidence advanced by the prosecutor to support Ms. Campbell's credibility. The prosecutor also pointed to: her testimony about Mr. Anderson's drug organization and its consistency with Mr. Henley's and Mr. Bobbitt's testimony; her testimony about the types of weapons that Mr. Anderson possessed and the ammunition found in Mr. Anderson's car; her testimony about Mr. Anderson's motive to kill Mr. Yarbour; and the consistency between her testimony about where Mr. Yarbour could be found and Ms. Scarbath's testimony about where appellant found Mr. Yarbour.

And, in addition to the CBLA evidence and the testimonial evidence of Mr. Henley and Mr. Bobbitt, her testimony that appellant drove Mr. Anderson's vehicle to commit the offenses was supported by tire track evidence. And the medical examiner's findings that Mr. Yarbour suffered five or six gunshot wounds as well as lacerations and blunt force trauma to the head was not significantly inconsistent with her testimony of what appellant told her.

Ms. Campbell testified that at the time of the murder, appellant was employed by Mr. Anderson as an enforcer, and that his duties included collecting debts by "beat[ing] people up" if they failed to pay. Mr. Henley and Mr. Bobbitt both corroborated that testimony. Ms. Campbell recounted that someone broke into her apartment and stole drugs and money, that Mr. Anderson suspected it was Mr. Yarbour, that Mr. Anderson wanted him "taken care of" by appellant, and that she and appellant accompanied Mr. Anderson to search for Mr. Yarbour. Mr. Henley corroborated aspects of that account by testifying that he saw Mr. Anderson, shotgun in hand, looking for Mr. Yarbour in the company of Ms. Campbell.

In sum, three witnesses, Ms. Campbell, Mr. Bobbitt, and Ms. Scott, implicated appellant as the one who shot Mr. Yarbour. Each of them testified that appellant told them that he killed Mr. Yarbour. Ms. Campbell testified that appellant killed the victim, at Mr. Anderson's request, in exchange for $500 in drugs and money. Mr. Bobbitt testified that Mr. Anderson had admitted to ordering appellant to murder Mr. Yarbour. He also testified that appellant admitted to shooting Mr. Yarbour "three or four times in the chest" and then beating him up. Ms. Scott, a girlfriend of Mr. Anderson, testified that appellant told her that he shot Mr. Yarbour three or four times. And Ms. Scarbath, the mother of Mr. Yarbour's son, testified that appellant came to their house and left with Mr. Yarbour on the night of the murder and that was the last time she saw him alive.

Appellant correctly points out that Ms. Campbell, Mr. Bobbitt, and Mr. Henley had credibility problems. All of them were users and traffickers of illegal drugs and had, in the past, been connected to Mr. Anderson's narcotics operation. And, at trial, the credibility of each was attacked by defense counsel. Ms. Campbell denied knowing who was responsible for the murder when she was first arrested. She agreed to testify only after receiving immunity for her cooperation, and admitted that she hoped to receive a favorable disposition in her pending drug case. Mr. Bobbitt only contacted police after drug paraphernalia was found in his vehicle, and relied on police to pay his overdue rent. Mr. Henley was confined in a super-maximum prison, and had a history of lying to the police. All of this was heard by the jury, and, in the end, it was for the jury to assess and weigh their credibility. …

Appellant also contends that, in reaching its conclusion, that the hearing court "plac[ed] undue emphasis" on defense counsel's suggestion that Mr. Anderson killed the victim. In closing, defense counsel argued:

> The question really comes down to three people in this case. It doesn't come down to whether the .32 caliber bullets came from that box or not, because where is that box found? It is found in [Mr.] Anderson's house. Where is the shotgun found? In [Mr. Anderson's]'s house. Where is the handgun found? In [Mr. Anderson's] car. What is found in my client's house? Essentially nothing.

Appellant posits that he "may well have made a different argument if the CBLA evidence had been excluded."

We are not persuaded that the hearing court relied on defense counsel's closing to find no substantial likelihood of a different outcome. The court stated "it is evident that the CBLA evidence did not have any substantial impact on [appellant's] conviction," because the CBLA evidence pointed more to Mr. Anderson than it did appellant—as summarized in defense counsel's argument—and "[e]ssentially nothing" related to weapons and ammunition was found in appellant's possession. Alternatively, appellant's conviction rested on the testimony of witnesses and the other evidence presented.

In short, the hearing court did not err or abuse its discretion in determining that omitting CBLA from the evidence in this case would not create a substantial or significant possibility of a different outcome.

JUDGMENT OF THE CIRCUIT COURT FOR BALTIMORE COUNTY AFFIRMED; COSTS TO BE PAID BY APPELLANT.

Questions:

1. CBLA-Comparative Bullet-Lead Analysis is a now discredited forensic method which primarily does what?
2. What makes CBLA as a testing protocol so troublesome and essentially discontinued by the FBI in 2004.

   In a way, general conclusions are fine but specific finding of identical bullets a tougher sell.
3. Prove True or False: Even though CBLA is discredited and discontinued, at the time of appellant's trial, CBLA was regularly admitted without hesitation. Hence the Court had little choice but to provide the requested post-conviction relief on appeal.

8.4 FORENSIC EVIDENCE: COMPARISON MARKINGS, STRIATIONS, RIFLING IMPRESSIONS

Part of the credibility of ballistics depends upon fair and scientific comparisons. While it is difficult in most cases to claim identicality of weapons with projectiles- a claim of similarity often being the better route- there are select cases where the rifling of a barrel or other unnatural markings in the operation of the weapon itself, give rise to identical claims. Ballistics is in fact a hard science that aggressively needs and depends upon the physical sciences and even more importantly, the tools of comparative micrography (See Figure 8.3).

Commonwealth v. Best
180 Mass. 492, 62 N.E. 748
Supreme Judicial Court of Massachusetts, Essex
January 6, 1902; February 27, 1902
No Number in Original

This is an indictment for murder upon which the defendant has been found guilty of murder in the first degree. The case is here on exceptions. The first exception is to the overruling of a motion to quash the panel because the clerk in issuing the venires did not "require from each town and city a number of jurors as nearly as may be in proportion to their respective number of inhabitants," as required by Pub. Sts. c. 170, § 11. Commonwealth v. Walsh, 124 Mass. 32, 37. The object of this enactment is not to benefit the prisoner, but is expressly stated to be to "equalize as far as possible the duty of serving as jurors." The exception might, perhaps, be disposed of on the ground that the section gives the prisoner no rights, or, when no prejudice is shown, is merely directory as to him. … But it is unnecessary to consider that answer, because, as was decided in a trial for murder in this court, when there is no ground for suspecting misconduct or partiality, the true construction of the statute is not "such as to require that proportion to be observed in every particular occasion

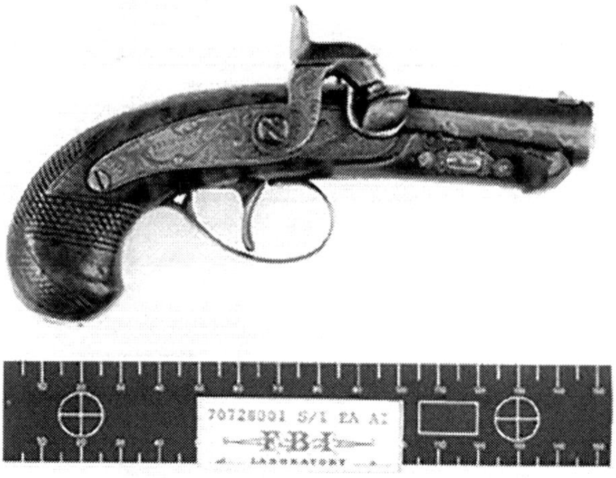

FIGURE 8.3 The pistol used by John Wilkes Booth to assassinate President Abraham Lincoln in April 1865 above an FBI Lab evidence scale. In the late 1990s, the FBI was asked by the National Park Service to authenticate the pistol after it was thought to be stolen from Ford's Theatre years' earlier. See https://archives.fbi.gov/archives/news/stories/2004/march/lincoln_031804 for details. Visit https://www.loc.gov/resource/highsm.04710/?r=-0.059,-0.015,1.123,0.791,0 for a high resolution picture of the pistol.

Source: FBI Media at https://multimedia.fbi.gov/item?type=image&id=6748.

when a jury is to be summoned. That, indeed, is practically impossible. The statute relates equally to civil cases and to criminal cases. In many counties of this Commonwealth the number of towns is much in excess of the number of jurors that are ordinarily required, and it would be impossible to observe it in each particular case with accuracy and with literality. But the intention of the statute, as it seems to us, is to give general directions to the clerks in issuing the venires for jurors, that they shall be so apportioned, taking one term of the court with another, with reference to the population of the towns, that the duty of serving on juries shall be equalized as far as possible." Allen, J., Trial of Henry K. Goodwin, 6, 7. This construction of the statute cuts the root of the defendant's argument. …

Bailey, the person alleged to have been murdered, was last seen alive at about a quarter past nine in the evening of October 8, 1900. Two witnesses testified without objection that between nine and ten they heard his milk wagon pass the house in the direction of Breakheart Hill farm, where he was living alone with the defendant. Nine days later his body was found, cut up, in a pond in Lynn, about six miles from his farm. The government contended that Bailey was killed shortly after his supposed return home and was at once cut up and carried to the pond, and it was not disputed that the movements of Bailey's wagon after his return home were material. These two witnesses testified to having heard a vehicle go by their house, in the direction from the farm, between ten and eleven. They both had described Bailey's wagon as having a rattle known to them, and one of them also mentioned a peculiar sound made by the horse's hoofs. Subject to exception they were allowed to state that it was the same team that had passed before, -- Mr. Bailey's team. The objection is only to allowing an identification by the witnesses through hearing alone, instead of leaving that question to the jury, confining the witnesses to stating the similarity of the sounds to those that had been heard before from the wagon going to the farm. (The road was not travelled beyond the farm, so that it might be assumed that any vehicle coming from that direction came from the farm.) It seems not to need argument to show that this evidence was admissible. Commonwealth v. Hayes, 138 Mass. 185. State v. Rainsbarger, 74 Iowa, 196, 203, 204. So as to an exception, not argued, to allowing a witness to testify that two shots heard between half past nine and ten came from the south or southwest, the direction of Breakheart Hill farm as stated by him. Commonwealth v. Sturtivant, 117 Mass. 122, 133.

An exception was taken to testimony of the defendant's brother-in-law as to some incriminating conversation of the defendant. It now is argued that the communication was privileged because made to an agent of the defendant's lawyer. As to this it is enough to say, without considering other questions, that it does not appear that the witness was acting as an agent, or that the communication was made to him in the belief that he was an agent, or in confidence other than reliance upon the witness's personal friendship.

It was shown that the defendant had accounted for Bailey's absence by the suggestion that there was a reward out for Bailey's arrest because of his having left his wife and children at Wiscasset, Lincoln County, Maine, and having come to Massachusetts with another woman, and that the officers were after him. A witness was produced who testified that he was the only deputy sheriff of Lincoln County, and that there was no warrant out for Bailey's arrest or reward offered by the town of Wiscasset. This was excepted to, and thereupon the question was modified by inserting the words "to your knowledge." It is enough to say that the evidence was admissible to show that the officer who naturally would have been after Bailey if the defendant's suggestion had been true knew nothing of any warrant or reward. But even if we were dealing with matters within the jurisdiction, ... it would be going pretty far to hold that a person having personal knowledge of the state of a record from inspection could not testify that it did not contain a certain warrant, without producing the record.

The government contended that Bailey was shot with a Winchester rifle that was in the kitchen. Two bullets were found in his body, and the government was allowed to prove that another bullet of the same calibre had been pushed through the rifle on or shortly after October 24. It then was allowed to put this bullet in evidence, and also photographs from this and the two bullets from the body, in order to show that the marks from the rifle in the two cases coincided so closely as to prove that all three bullets had passed through the same rifle barrel. This evidence was excepted to. The main ground seems to be that the conditions of the experiment did not correspond accurately with those of the date of the shooting, that the forces impelling the different bullets were different in kind, that the rifle barrel might be supposed to have rusted more in the little more than a fortnight that had intervened, and that it was fired three times on October 10, which would have increased the leading of the barrel. We see no other way in which the jury could have learned so intelligently how that gun barrel would have marked a lead bullet fired through it, a question of much importance to the case. Not only was it the best evidence attainable but the sources of error suggested were trifling. The photographs avowedly were arranged to bring out the likeness in the marking of the different bullets and were objected to on this further ground. But the jury could correct them by inspection of the originals, if there were other aspects more favorable to the defence.

With reference to the bullets found in the body an expert was allowed to testify that they were marked by rust in the same way that they would have been if they had been fired through the rifle found at the farm, and that it took at least several months for the rust that he saw in the rifle to form. It is objected that these were not matters for expert testimony. We see no reason to doubt that the testimony was properly admitted.

After giving the weight of the bullets found in the body, a witness was allowed to give the weight of bullets given up by the defendant at the farm. This obviously was a link in the chain of proof, and needs no special remark.

The last exception argued is to the refusal of the court to rule that there was no evidence warranting a verdict of guilty of murder in the first degree. It is argued somewhat as if the question were whether there was any evidence of murder at all. A sufficient answer is a very brief summary. There was evidence, as we have stated, -- with the weight of it we have nothing to do, -- that Bailey went home to the farm after he last was seen alive. Some one shot him, probably between nine and ten that night, and probably with the rifle at the farm. The defendant was there alone. Bailey's body was cut up, therefore he probably was killed by some one who had opportunity for deliberate action. There was some evidence of ill feeling towards Bailey on the defendant's part, as well as of a possible motive in the way of money. The defendant told the whereabouts of a watch which Bailey always wore when alive, and seemingly had concealed it with some money where it was found. He also gave false reasons for Bailey's disappearance. There was some evidence that Bailey's wagon was driven from the farm, as we have stated, between ten and eleven, to the pond in Lynn. The jury were fully warranted in finding that the defendant killed Bailey. We cannot say that they were not warranted in finding that the murder was in the first degree. ... If there was evidence warranting the finding, it must be left to the jury to say whether it produces or ought to produce a sufficiently strong conviction to justify their verdict. Courts do not attempt to measure the intensity of evidence. They stop with the decision that some evidence has been produced.

Questions:

1. Does the Court take the position that the comparison of the Winchester bullets demonstrated identical traits from the bullets extracted from the murder victim and the test bullet?
2. How and what basis does the defense object to this comparison?
3. What type of demonstrative evidence does the prosecution rely upon when displaying the comparison?

> Read the story about the Oklahoma City Body discussing the chemical residue on Tim McVeigh's clothing, as well as fingerprints on receipt for materials purchased to make the explosive at https://www.fbi.gov/news/stories/the-oklahoma-city-bombing-20-years-later.

People v. Mitchell
94 Cal. 550, 29 P. 1106
Supreme Court of California, Department Two
May 23, 1892
No. 20900

The defendant, having been convicted of murder in the second degree, appeals from the judgment and a refusal of a new trial.

It appears that Oscar Crandall was fatally shot, about midnight, April 23, 1891, near Rathja's saloon, in the town of Red Bluff. Crandall, who had been in the saloon, went out upon the back porch, where he was engaged for a time imitating Chinese talk. Two or three minutes after he went out, Rathja heard the report of a pistol, and went upon the back porch, but saw no one. The back yard was surrounded by a high fence. The saloon was at the corner of Main and Hickory streets, facing on Main. It was bright moonlight, and except as obstructed by the fence, Rathja could see both streets.

Two witnesses claim to have been eye-witnesses of the affray; Peterson, who testified for the prosecution, was 224 feet away, westerly on Hickory Street. He was positive that he recognized defendant's voice, but he could distinguish no one by sight. He saw the flash of the pistol, and the person who fired ran off south, along the alley back of the yard which encloses Rathja's back porch.

This is all the direct testimony against defendant, but there was considerable circumstantial evidence.

Defendant claimed and testified, as did several other witnesses in his behalf, that he was at home, about a half-block from Rathja's saloon, administering what he calls a "shot" of morphine to one Boyden, when he heard the report, and was told that some one was shot, and then ran across the street with Boyden, found the wounded man, and he and Boyden took up the body and carried it to the front of the saloon. While they were so carrying the wounded man the people gathered.

Hughes and Boyden both corroborate defendant's story. Hughes says he was across the street from Rathja's back yard, at the corner of Hickory and High streets. He heard the altercation, saw the flash of the pistol, and the assassin run off, he thought, down Main Street, and being but a few steps from defendant's house, informed a woman there that some one was shot; that Mitchell and Boyden ran past him to the saloon, and the body was carried as defendant stated.

To make out the case for the prosecution by the circumstantial evidence, it became important to fix the precise time of the homicide. It was contended on the part of the people that defendant did not go to his house, as he claimed he did, to administer the hypodermic injection of morphine to Boyden; that he came up from the depot on the Tremont hotel bus, after the arrival of the train, which was due at 11:50, p. m.; that the homicide was committed before the departure of the train, the schedule time of which was 12:05.

They had experiments made as to the time it would require to walk where he said he walked after he left the bus, and claimed that he did not have time to go home and do what he said he did before the report of the pistol.

There was much evidence directed to this point on both sides. Among other things, the defense introduced evidence showing that the train did not leave on that night until after 12:10.

The prosecution, against the objections of the defendant, were permitted to show, by the local agent of the railroad and the night-operator, that by the rules and regulations of the company, the trains must not arrive before their schedule time, and that the conductor must register the exact time of arrival and departure; that such register was in fact kept at this station. The register was then introduced, defendant objecting, from which it appears that the train arrived and left on the night in question on schedule time. Neither of the witnesses had any actual knowledge of the times, nor was the conductor who made the record called as a witness. The evidence was clearly incompetent, and was immaterial in the highest degree. If we can presume, as suggested, that the conductor was absent or dead, that fact would make no difference. It was hearsay evidence, and should have been excluded.

It seems that one H. K. Long had been tried for this same offense, and had been acquitted. Nevertheless, the defense attempted to prove that Long committed the homicide.

C. A. Boyden, a witness for the defense, testified, without objection, that Long told him that he had killed Crandall, and that he got the gun from Morin. In rebuttal, the prosecution asked Morin if he loaned Long a pistol, and he was allowed, against the objection of the defendant, to answer that he did not.

This was not competent. It did not tend to show that Long did not kill Crandall, or that Mitchell did, and it did not tend to contradict the witness Boyden, who did not say that Long procured a pistol from Morin. He only testified as to what Long had said, and this evidence did not tend to disprove that fact. …

Vance, a carpenter and gunsmith, was called as an expert, and allowed, against objection, to state his opinion that a certain cartridge had never been in a pistol, because he saw no mark on it. He was then allowed to compare a ball shown him, and assumed to have been from a pistol claimed by the defense to have been used by Long, with the ball taken from the head of deceased, and to state that they were not similar. The balls were there in court, and shown to the jury.

This was not a proper subject for expert testimony. It did not require special skill to determine whether a cartridge had marks upon it which would indicate that it had been once in the chamber of a pistol, nor that it was covered with verdigris, nor that the end of one bullet was feather-edged, and another not so, or was flat or concave at the end. These are all matters of common knowledge, and the jury was as competent to pass upon them as the witness.

The prosecution, in rebuttal, put in evidence statements made by the defendant while testifying as a witness in the case of People v. Long. Though charged with the same offense for which Long was tried, it is not by the same information. It does not appear that defendant objected to being sworn on that trial, and it seems to be conceded that he did not object to answering the particular questions, the answers to which were given in evidence against him. It is contended simply that the statements are not voluntary, because the defendant was duly sworn to answer questions, and was compelled to answer, or to admit that his answers might criminate himself. Bishop on Criminal Procedure, sections 1255–1257, cites numerous authorities, and concludes that the testimony voluntarily given as a witness are admissions or confessions, competent against him in a criminal cause to which they are pertinent.

Unless he objected to answering the particular matter complained of, and was forced nevertheless to testify, it is considered voluntary. Some few cases there are which seem to sustain the appellant, but nearly all the authorities are the other way. …

We cannot see that our code, which allows the defendant to testify or not as he may choose, and to limit his testimony to such matters as he pleases, can make any difference.

The prosecution called a witness, Hopper, to impeach and contradict defendant's witness, C. D. Boyden, whose brother was accused jointly with defendant. Boyden had said that he did not attempt to get money from Hopper for Long to get away with, in order to save his brother. Hopper testified that Boyden tried to get money for Long to get away with, but did not say that it was to save his brother. The district attorney was then allowed, defendant objecting, to impeach him by asking if he had not testified differently on the Long trial, and proceeded to read for that purpose what purported to be his evidence given at the Long trial.

This was error. A witness, whichever party calls him, cannot be impeached unless he has given testimony against the impeaching party. This witness had not testified against the prosecution. He simply failed to testify to a fact which the district attorney thought he could prove by him. The defense had not proven or attempted to prove any statements made by Boyden to Hopper. Boyden's testimony on the subject was brought out on cross-examination.

The impeaching statements were evidently desired as evidence. If such testimony were admissible, it would be easy to manufacture evidence of that kind. If a witness merely fails to testify as expected, that does not authorize the party calling him to prove that the witness had elsewhere made the desired statements. It is only when he has given damaging testimony that he can be impeached. (People v. Jacobs, 49 Cal. 384.)

Many other points are made by appellant, some of which are at least debatable. They are mostly in regard to attempts at impeachment or comparatively unimportant matters, decision of which would not be of importance on a retrial of the cause.

We think the judgment and order should be reversed and a retrial had.

The Court. -- For the reasons given in the foregoing opinion, the judgment and order are reversed and a new trial ordered.

Questions:

1. Explain the comparison conclusions made by the witness relative to a cartridge and an aligned or associated weapon.
2. What was the background and expertise of the supposed ballistics expert?
3. As a result of the alleged ballistics expert and other issues, what does the court subsequently do?

People v. Rodriguez
2017 IL App (1st) 141379, 79 N.E.3d 345
Appellate Court of Illinois, First District, First Division
May 8, 2017, Decided
No. 1-14-1379

[*P1] Fifteen-year-old Sebastian Rodriguez was charged with first degree murder in connection with the shooting of thirteen-year-old Sameere Conn on October 1, 2008. At the time of the offense, 15-year-olds charged with first degree murder were automatically excluded from juvenile court jurisdiction. Sebastian was accordingly tried, convicted, and sentenced as an adult. Following his jury trial, the circuit court sentenced Sebastian to 50 years in prison: 25 years for the murder and 25 additional years pursuant to a mandatory firearm enhancement. In this direct appeal, Sebastian argues that (1) the circuit court erroneously denied his motion to suppress evidence found during a search of his home,

(2) expert testimony identifying a revolver found in his home as the murder weapon was improperly admitted without a hearing to determine if it was based on generally accepted scientific methodologies, and (3) the imposition of a 50-year sentence on an offender who was 15 years old at the time of his offense is unconstitutional.

[*P2] Shortly after Sebastian filed his notice of appeal, the Illinois legislature raised the age of automatic transfer from juvenile court to criminal court for an individual charged with first degree murder from 15 to 16 years of age and adopted additional sentencing guidelines for defendants who were under the age of 18 at the time of their offenses, including making firearm enhancements discretionary, rather than mandatory, for such individuals. In supplemental briefing, Sebastian argues that these amendments should apply to his case.

[*P3] For the reasons that follow, we affirm Sebastian's conviction for first degree murder, vacate his sentence, and remand this matter to the juvenile court.

[*P4] BACKGROUND

[*P5] A. Pretrial Proceedings

[*P6] Nine days after Sameere Conn's death, Chicago police obtained a warrant to search Sebastian Rodriguez's home for evidence related to the shooting. In the complaint for the search warrant, Detective Ricky Bean identified two eyewitnesses who testified before a grand jury that they knew Sebastian and saw him, dressed in a hooded sweatshirt, fire shots into the convenience store where Sameere was killed, as well as a third eyewitness who identified Sebastian as the individual he saw looking through the glass window of the store's door just before shots were fired through that window. According to the complaint, officers also learned from two other witnesses that Sebastian was known to possess a "kill list" of potential victims that included Sameere. Finally, the complaint alleged that, in connection with prior arrests, Sebastian had given the address 10744 South Hoxie Avenue in Chicago as his home address.

[*P7] Finding this sufficient to establish probable cause, the circuit court issued a warrant to search Sebastian's home for "[o]ne dark colored or grey hooded sweat shirt, [o]ne document containing a list of individual names, [a]nd one handgun." Officers executed the warrant on October 11, 2008, retrieving a revolver from under a floorboard in the bathroom and a number of hooded sweatshirts from elsewhere in the home.

[*P8] Sebastian was charged by grand jury indictment with first degree murder.

[*P9] In his motion to suppress filed on April 26, 2010, Sebastian argued that evidence recovered during the October 11, 2008, search should be excluded because, even if officers had probable cause to arrest him, they had no reason to believe that specific evidence would be found in his home 10 days after the shooting.

[*P10] Although an evidentiary hearing was held on Sebastian's motion to suppress, the testimony offered related only to the scope of the search and the manner in which it was conducted, issues that are not raised in this appeal. The circuit court denied Sebastian's motion, explaining that, in its view, when officers have "a strong identification of a suspected shooter and that person's home," then "it is not beyond logic, nor *** beyond the law, to have probable cause to see if in that person's place of residence, the place they call home, the place in which they keep their items, that there might be evidence of the crime there."

[*P11] On May 9, 2013, Sebastian moved for an evidentiary hearing, pursuant to Frye v. United States, 293 F. 1013 (D.C. Cir. 1923), on the admissibility of expert testimony he expected the State to introduce linking the gun found in his home to a bullet recovered from the scene of the crime. Although he acknowledged that such testimony had historically been admitted by courts, he insisted a Frye hearing was needed because the reliability of the methodologies employed by ballistics experts had recently been questioned in the scientific community.

[*P12] The circuit court disagreed and denied Sebastian's motion. Noting that it was aware of no published opinion of any court concluding that firearm identification evidence was not generally accepted in the scientific community, the court concluded that Sebastian's concerns went to the weight and not to the admissibility of the evidence.

[*P13] B. Trial

[*P14] A four-day trial in this case began on February 4, 2014. Because Sebastian does not contest the sufficiency of the evidence to support his conviction, we include only a brief summary of the trial testimony, with a fuller recitation of the firearms identification testimony, to provide context for the evidentiary issues raised on appeal.

...

[*P19] The physical evidence in this case consisted of

(1) a medium caliber lead bullet fragment recovered from Sameere's body; (2) a fired bullet recovered from a shelf inside Hook's Finer Foods on October 1, 2008; (3) a gunshot residue collection kit consisting of swabs of each of Sebastian's hands plus a control swab, which was administered by police officers shortly after midnight on October 2, 2008; (4) a blue steel .357 Dan Wesson revolver containing six .357 magnum caliber unfired cartridge cases, retrieved from under the floorboards of the bathroom during the October 11, 2008, search of the home at 10744 South Hoxie Avenue in Chicago; and (5) two gray and five black "hoody jackets" also recovered during that search.

[*P20] Brian Mayland, a pattern evidence program manager for the Illinois State Police forensic sciences command, testified as an expert in the field of toolmark and firearm identification. Mr. Mayland previously worked for 17 years as a forensic scientist in firearms and toolmark identification and, for just over one year, as a laboratory director. Although his undergraduate degree was in business, Mr. Mayland testified that he had completed specialized training in the field of firearms identification, including a two-year training program conducted by the Illinois State Police, and had testified as an expert in the field approximately 80 times.

[*P21] Mr. Mayland explained that a cartridge consists of four basic components: the case; the powder inside the case; the bullet, which is seated inside of the case; and the primer, a pressure-sensitive chemical compound located in the head of the case. When a gun is fired, the primer is struck, the resulting spark ignites the powder, gasses from the burning powder create pressure, and the pressure forces the bullet from the mouth of the cartridge down the barrel where rifling—raised and lowered areas known as "lands" and "grooves"—form a twisting pattern along the inside of the barrel that causes the bullet to spin. Mr. Mayland testified that, as a firearm analyst, he uses a comparison microscope to examine two bullets or cartridge cases and compare the marks that are left behind on those items as a result of the firing process. Certain identifying features—like the caliber of the bullet; the number and width of the grooves in the rifling; and the direction of the twist—are known as "class characteristics"; they are present at the time of manufacture and common to an entire class of firearms. Other marks are created by imperfections that develop in a gun over time, as it is fired, and can be unique to a particular gun.

[*P22] In this case, Mr. Mayland examined the fired bullet recovered from the scene of the crime and determined that it was a .38-caliber bullet jacket with six lands, six grooves, and a right-hand twist. He concluded that the metal fragment recovered from Sameere's body was too mutilated to be suitable for comparison. Mr. Mayland then test fired the revolver recovered from Sebastian's house, shooting four bullets into a tank of water, which slows the bullets without damaging them. He compared the test shots to each other to determine if he "could identify test shot with test shot," something he acknowledged is not always possible. In this case he determined that it was. He then compared the test shots side by side with the fired bullet under a comparison microscope. It was Mr. Mayland's opinion "that the fired bullet jacket was fired in that firearm."

[*P23] Defense counsel objected to Mr. Mayland providing this conclusion without elaborating on the specific similarities or differences between the compared specimens that he relied upon as the basis for his opinion. The court sustained the objection, pending further inquiry. When asked to elaborate, Mr. Mayland stated that he "saw a sufficiently similar pattern of individual characteristics that allowed [him] to form an opinion." Specifically, "[t]here were striated marks that lined up when [he] was doing the comparison from the evidence bullet to the test fired bullet." Defense counsel again objected, but this time the trial court overruled the objection.

[*P24] On cross-examination, Mr. Mayland acknowledged that six is the most common number of lands and grooves and it is "very common" for a revolver to have six lands and grooves with a right-hand twist. Based on Mr. Mayland's experience, he believed that hundreds of guns in Chicago could have those same characteristics, noting, however, that he could not be more specific because gun manufacturers "are very close" with such information.

[*P25] Mr. Mayland also noted that the bullet jacket he analyzed was "badly mutilated," consistent with it having struck something. "Based on the condition of the bullet jacket," he said he measured at least two and "probably three" lands and grooves, although he did not know that for certain and did not document his measurements in his notes. Mr. Mayland acknowledged that none of the test shots matched the fired bullet casing exactly. However, he also stated that "no two test shots will ever look exactly the same." Mr. Mayland insisted that, in this case, "there was a sufficiently similar pattern" between the test shots and the fired bullet case for him to form his opinion. Mr. Mayland agreed both that there is no nationally recognized standard to determine that the patterns were close enough to have been generated by the same gun and that his opinion was a subjective one, not capable of verification by objective testing.

[*P26] On redirect examination, Mr. Mayland reiterated that he has compared tens of thousands of bullets and cartridge cases over his career, that he followed all Illinois State Police lab protocols, and that he used methods commonly accepted in the field of firearms identification. Mr. Mayland confirmed that nothing he was asked during cross-examination affected his opinion that the bullet he analyzed was fired from the revolver found in Sebastian's home.

[*P27] Mary Wong, a forensic scientist with the Illinois State Police forensic sciences division, testified as an expert in the field of gunshot residue analysis. Ms. Wong tested the swabs from the residue collection kit administered to Sebastian and the hooded sweatshirts retrieved from his home. None of the items tested positive for gunshot residue. Although Ms. Wong found two "tricomponent particles" on the sample taken from Sebastian's left hand and one on the sample taken from his right hand, she explained that at least three particles from the same sample are required to make a positive identification. All Ms. Wong could conclude from her analysis was that Sebastian "may not have discharged the firearm with either hand" and, "if he did, then the particles were either removed by activity or not deposited or not detected by the procedure." Although tricomponent particles are found in fireworks and car airbags in addition to gunshot residue, Ms. Wong stated that other particles one would expect to find following contact with those items were not present in the samples she tested. However, she acknowledged that gunshot residue particles may be transferred to a person who touches a surface in a room where a gun was fired or who comes in contact with someone who recently fired a gun.

…

[*P33] In its closing argument, the State urged the jury to believe the eyewitness testimony identifying Sebastian as the shooter, which was corroborated by the particles of gunshot residue found on Sebastian's hands and Mr. Mayland's testimony that the gun found in Sebastian's home was the murder weapon. Defense counsel responded by pointing out that there were innocent explanations for a few particles of gunshot residue to be on a person's hands and attacked Mr. Mayland's conclusions as not being based on objective standards or specific measurements. Defense counsel argued that, following the shooting, Sameere's friends heard a rumor that Sebastian killed Sameere and were willing to lie about what they saw to make sure he was convicted. According to defense counsel, it was more likely that some unidentified shooter intending to shoot Anthony Ray, a former gang member who was carrying drugs at the time, had inadvertently shot Sameere.

[*P34] The jury found Sebastian guilty of first degree murder and the circuit court denied his motion requesting a new trial, in which he argued that the circuit court erred when it denied both his motion to suppress and his motion for a Frye hearing. …

[*P37] ANALYSIS

[*P38] On appeal, Sebastian argues that his conviction for first degree murder should be reversed, both because the circuit court erroneously denied his motion to suppress the evidence resulting from the search of his residence for a lack of probable cause and because the circuit court should have conducted a Frye hearing before admitting the testimony of the State's expert on toolmark and firearms identification.

[*P39] Sebastian also challenges his sentence, arguing both that it was unconstitutional when it was imposed and that an amendment to the exclusive jurisdiction statute changing the age from 15 to 16 for the automatic transfer to criminal court of cases involving certain crimes should be applied retroactively to his case. Pursuant to that amendment, Sebastian asks us to vacate his sentence and remand this matter to juvenile court, where the State may seek a discretionary transfer hearing if it chooses. Sebastian alternatively argues that he is entitled to a new sentencing hearing conducted pursuant to amended sentencing guidelines for individuals who were under the age of 18 at the time of their offenses.

[*P40] We address each argument in turn.

…

[*P48] B. Motion for a Frye Hearing

[*P49] Sebastian also argues that the circuit court erred in denying his motion to either exclude the State's toolmark and firearm identification evidence or to hold a Frye hearing to determine the admissibility of that evidence. In support of his contention, both in the circuit court and on appeal, that such evidence is not generally accepted in the scientific community, Sebastian relies primarily on a 2009 report authored by the National Research Council of the National Academy of Sciences (NRC) entitled Strengthening Forensic Science in the United States: A Path Forward. In that report, the NRC noted that toolmark identification has "never been exposed to stringent scientific scrutiny," involves "subjective qualitative judgments by examiners," is "based on unarticulated standards," and lacks any "statistical foundation for estimation of error rates." The NRC concluded that, although there is some benefit to be derived from this testimony, additional studies are needed to address these concerns.

[*P50] The circuit court denied Sebastian's motion for a Frye hearing, concluding that the criticisms raised in the NRC's report go to the weight, and not the admissibility, of toolmark and firearm identification evidence. The court also noted that there are no published opinions holding that such evidence is not generally accepted in the relevant scientific community.

[*P51] In Illinois, "new" or "novel" scientific evidence is only admissible if it meets the standard set out in Frye v. United States, 293 F. 1013 (D.C. Cir. 1923). People v. McKown, 226 Ill. 2d 245, 254, 257, 875 N.E.2d 1029, 314 Ill. Dec. 742 (2007). "[T]he methodology or scientific principle upon which the opinion is based [must be] sufficiently established to have gained general acceptance in the particular field in which it belongs." (Internal quotation marks omitted.) Id. at 254. A court may determine whether a methodology or principle is generally accepted either by conducting an evidentiary hearing or "by taking judicial notice of unequivocal and undisputed prior judicial decisions or technical writings on the subject." Id. A scientific methodology need not be universally accepted or even accepted by a majority of experts in the field; "[i]nstead, it is sufficient that the underlying method used to generate an expert's opinion is reasonably relied upon by experts in the relevant field." In re Commitment of Simons, 213 Ill. 2d 523, 530, 821 N.E.2d 1184, 290 Ill. Dec. 610 (2004). Although it is within the circuit court's discretion to decide both whether a particular witness is qualified to testify as an expert in a particular field and whether the testimony that witness will offer is relevant, we review de novo the circuit court's determination of whether the methodology used by the witness meets Frye's "general acceptance" standard. People v. Nelson, 235 Ill. 2d 386, 430–31, 922 N.E.2d
1056, 337 Ill. Dec. 479 (2009).

[*P52] We first consider whether toolmark and firearm identification evidence is "new" or "novel." The State contends that it is decidedly not, noting that courts have allowed such evidence since at least 1930, when our supreme court held in People v. Fisher, 340 Ill. 216, 240–41, 172 N.E. 743 (1930) that, while a jury is not bound to accept it as true, firearm identification evidence "is competent expert testimony on a subject properly one for expert knowledge." In the decades since Fisher, firearms experts have regularly testified in Illinois courts, for both the prosecution and the defense.

[*P53] Sebastian does not dispute this, but insists that, pursuant to our supreme court's analysis in McKown, firearm identification evidence is nevertheless novel because "there is no record that there has ever been a Frye hearing in Illinois to determine whether generally accepted scientific principles support [it]." The court in McKown held that the horizontal gaze nystagmus (HGN) test, a field sobriety test frequently used by police officers, was a novel methodology subject to the Frye standard. McKown, 226 Ill. 2d at 258. The court explained that its holding was based on "the history of legal challenges to the admissibility of HGN test evidence, and the fact that a Frye hearing ha[d] never been held in Illinois." Id. However, as the court noted, the HGN test was "repeatedly challenged in court, with varying degrees of success," both in Illinois and in other states, and this court had issued "divergent opinions on the topic," such that the general acceptance of the test "remain[ed] unsettled." (Internal quotation marks omitted.) Id. at 257.

[*P54] This case is distinguishable from McKown because the admissibility of firearms identification evidence is not similarly "unsettled" in Illinois. The circuit court noted that it was unaware of any published opinion of any court stating that firearms evidence was not generally accepted in the scientific community, and Sebastian has cited none on appeal. The few out-of- state cases Sebastian cites—in which courts have raised concerns about the reliability of such evidence but have nonetheless held the methodology to be sufficiently reliable to be admitted, at least in some qualified form—do not create the same situation the McKown court was presented with, where legal challenges were resolved both for and against admissibility of the HGN test and the law was truly unsettled. …

[*P55] Similarly unhelpful are cases involving testimony based on scientific methodologies that, although sometimes deemed admissible, never achieved the same sort of widespread acceptance as ballistics evidence … .

[*P56] Although we understand the concerns raised by other courts and by the NCR in its report regarding the subjectivity of firearm identification testimony and the inability to test its accuracy, we cannot say that the circuit court erred in denying Sebastian's motion for a Frye hearing. Toolmark and firearm identification evidence is not new or novel, either pursuant to the plain meaning of those words or in accordance with the analysis employed by our supreme court in McKown. Far from being unsettled, the law in Illinois is consistent in its admission of such evidence. …

[*P57] Nor do we find that the NCR's report so undermines the reliability of ballistics evidence that it has ceased to be "generally accepted" in the scientific community. We agree with the circuit court that the report's concerns go to the weight and not to the admissibility of such evidence. Indeed, our review of the record in this case indicates that—in connection with the his objection that some of Mr. Mayland's testimony lacked foundation, the denial of which Sebastian chose not to contest on appeal—during cross-examination defense counsel explored at length the limitations of Mr. Mayland's conclusions.

…

[*P65] CONCLUSION

[*P66] For the foregoing reasons, we vacate Sebastian's sentence, affirm the judgment of the circuit court in all other respects, and remand this case to the juvenile court.

[*P67]Affirmed in part; sentence vacated; cause remanded.

Questions:

1. Why did the ballistics carry out a comparative control test using fired projectiles from the retrieved weapon into a water tank?
2. What are parts of a projectile cartridge- all of which may have unique identifying characteristic?
3. Does the expert conclude that the test projectile and that discovered at the scene of the crime were unequivocally and undeniably from the same gun? Why or why not?

Use your favorite legal research resource, locate the case below, read the case and answer the questions that follow.

Ricks v. Pauch, United States District Court for the Eastern District of Michigan, Southern Division, March 23, 2020, Decided; March 23, 2020, Filed, Case No. 17–12784.

Questions:

1. The Court's opinion, after conducting an assessment of the test bullet with the bullet extracted from the victim, relied on a "twist" within the weapon to prove similarity. How precisely?
2. Defense counsel claims that the bulltets removed from the victim's body were simply too "pristine." What is the tactic that underlies this observation?
3. Both defense and prosecution examined the "rifling" impressions of the victim's bullet with test bullets fired from the weapon seized. Prove True or False: Prosecution claims that the comparison produced identical results which proves a match.

9 Forensic Pathology

9.1 INTRODUCTION

The role of forensic science in the operation of pathology, both theoretical and applied, is a strong relationship. Indeed, the study of pathology is inherently relevant to the field of forensic science. Stated another way, forensic science comprises the tools, tests and methods that lead to the pathological conclusions, such as: How did this person die? What was the cause of death? What time were injuries inflicted? What type of trauma took place during the crime? What types of instruments were employed to inflict injury? In general, the field of forensic pathology considers the time, manner and cause of death. The forensic pathologist is an expert in determining cause and manner of death and is specially trained to perform autopsies to determine the presence or absence of disease, injury or poisoning; to evaluate historical and law-enforcement investigative information relating to manner of death; to collect medical evidence, such as trace evidence and secretions, to document sexual assault; and to reconstruct how a person received injuries (see Figure 9.1).

FIGURE 9.1 A view of human remains found at a mass gravesite near Mosul, Iraq, during Operation IRAQI FREEDOM. The U.S. Army (USA) Criminal Investigation Command, along with Department of Defense Forensic Pathologists, are in Iraq investigating a mass grave crime scene to gather evidence for possible war crimes trials.

Source: National Archives NAID: 6626605.

DOI: 10.4324/9781003267126-9

9.2 FORENSIC PATHOLOGY AND THE TIME OF DEATH

Dunn v. Jess
981 F.3d 582
United States Court of Appeals for the Seventh Circuit
September 23, 2020, Argued; November 24, 2020, Decided
No. 20-1168

...

A. Postconviction Proceedings

Dunn filed a motion for postconviction relief in the state trial court, contending that his counsel was ineffective for failing to investigate and present evidence of a no-causation defense. To succeed on this claim, under Strickland v. Washington, 466 U.S. 668, 104 S. Ct. 2052, 80 L. Ed. 2d 674 (1984), Dunn needed to show that his counsel's performance was deficient and that he was prejudiced as a result. The court held an evidentiary hearing. Dunn called two witnesses: his trial counsel and a forensic pathologist, Dr. Peter Stephens. The state called the medical examiner who had testified at trial, Dr. Biedryzycki.

Dunn's trial counsel testified that the cause of Schuck-man's death was an important issue in the case. He had consulted with a forensic pathologist, Dr. Robert Corliss, twice before trial in order understand the medical examiner's report on the cause of death. At the initial consultation, Dr. Corliss believed Schuckman had died immediately after receiving the head trauma based on the information in the medical examiner's report. Dunn's counsel erroneously believed that the medical examiner's report stated this conclusion. He memorialized this mistaken belief in two emails to the prosecutors, months apart, in which he stated that the medical examiner had concluded that Schuckman had died immediately from head trauma. He recognized, and pointed out to opposing counsel, that this fact undercut the state's theory of the case. The bartender had witnessed Schuckman seemingly unharmed after Dunn's slap, suggesting it could not have been the cause of Schuckman's death.

In the month leading up to trial, Dunn's counsel consulted with Dr. Corliss again. In that conversation, they focused on the fact that Schuckman sustained multiple head injuries, beyond what would be explained by Dunn's slap and Schuck-man's fall. He planned to cross-examine the medical examiner on this point. At that time, Dunn's counsel was still operating under the erroneous belief that the medical examiner was going to testify that Schuckman's death was immediate from the fatal blow to the back of his head. He expressed this belief just ten days before trial in response to an email from the prosecutor. The prosecutor emailed him and stated that Crochet's attorney had retained a blood spatter expert and a pathologist, who were going to produce reports purportedly establishing that Schuckman had stood up in the area where he was found and then fell and hit his head. Dunn's trial counsel did not follow up with the prosecutor or contact Cro-chet's attorney to inquire about reports' contents. Rather, he wrote back to the prosecutor an hour later, stating that:

Our theory of defense has always been that Mr. Dunn slapped Mr. Schuckman in self-defense; that he didn't die immediately because people heard him talking, mumbling, grumbling after hitting the parking lot, and that the medical examiner will say he died immediately. Therefore, there had to be a second time that his head hit the ground after Mr. Dunn's slap that caused death. I believe that the testimony of the medical examiner, combined with the testimony of [the bartender], supports that there has to be another hit irrespective of what [Crochet's attorney's] pathologist says.

Trial counsel did not speak with Dunn before sending this email, though he testified that Dunn had been clear about not wanting a continuance. He testified that he told Dunn about the prosecutor's email sometime prior to trial.

The expert reports referenced in the email were completed a week after Dunn's trial concluded. There were three expert reports. First, a forensic pathologist wrote a report concluding that Schuckman suffered non-fatal injuries in the parking lot and he died due to injuries he received after falling backwards on the bar's back patio. Second, a blood analyst concluded that Schuckman was not bleeding in the parking lot and he was never upright after sustaining the injury to the back of his head. Third, a neurologist and mechanical engineer calculated that Schuckman could have generated enough velocity falling backwards on the back patio to fracture his skull, and extreme intoxication can result in such an "unbroken" fall.

Trial counsel stated that by the time Dunn's trial started, he understood that the medical examiner would not testify that death was immediate. Given his email to the prosecutor, this understanding must have occurred sometime in the ten days before trial. Dunn's counsel testified that he had no regret over not calling Dr. Corliss as a witness; he planned to draw out the fact that there were multiple injuries when cross-examining the medical examiner. He believed eliciting that testimony from a state's witness was more advantageous for his client than calling his own witness. He further testified that Dr. Corliss told him that the medical examiner's conclusions in the autopsy report were well-reasoned.

During the postconviction hearing, Dunn also presented the testimony of Dr. Stephens, a forensic pathologist. Dr. Stephens opined that the laceration and skull fracture in the back of Schuckman's head were the substantial cause of his death because he likely would have survived the other head injuries. He further opined that Schuckman sustained the skull fracture on the bar's back patio—not the parking lot where Dunn had slapped him. In reaching this conclusion, he relied on the nature of the injury and the lack of blood in the parking lot and on Schuckman's shirt. He testified that the nature of the impact causing the skull fracture would cause an immediate loss of consciousness, Schuckman would have been unlikely to be able to sit up and talk afterward, and he would have died within a matter of minutes. Additionally, the amount of blood found underneath Schuckman's head and the lack of blood in the parking lot supported his conclusion that Schuckman received the skull fracture and laceration on the back patio. If Schuckman had lived longer than a few minutes, the pool of blood would have been bigger. Further, the laceration's size on the back of Schuckman's head indicates it had the potential to bleed significantly and profusely because the scalp is highly vascular. The wound would have bled immediately. So if Schuckman had sustained it in the parking lot, one would expect to find blood both there and on Schuckman's shirt because the bartender and bar patron saw Schuckman upright in the parking lot several minutes after Dunn slapped him. Thus, Dr. Stephens opined that Schuck-man was never upright after sustaining the laceration and skull fracture, and so these injuries must have occurred on the back patio.

Dr. Biedryzycki testified for the state at the hearing. She opined that the laceration was relatively small and not gaping, so it may or may not have bled a lot. The nature of Schuckman's head injury did not lead her to assume that blood had to be present immediately after being injured. Further, she opined that the amount of blood under Schuckman's head could not be used to determine how long he lived after receiving the injuries because blood could have continued to pool around his head even after death due to hydrostatic pressure.

She acknowledged that there was nothing in her autopsy report that was inconsistent with Schuckman sustaining the skull fracture and laceration on the back patio.

The trial court held that Dunn's ineffective assistance of counsel claim failed Strickland's prejudice prong. The Wisconsin Court of Appeals affirmed the denial of Dunn's postconviction motion. State v. Dunn, 2016 WI App 88, 372 Wis. 2d 458, 888 N.W.2d 246 (Wis. Ct. App. 2016) (per curiam). It "was not persuaded that trial counsel performed deficiently in his investigation of Schuckman's death and presentation of a no-causation defense." Id. ¶ 13. The state appellate court noted that Dunn's trial counsel consulted with a forensic pathologist. Id. It further determined that "trial counsel's testimony provides a reasonable explanation for his approach," because he had strategic and practical reasons for not calling Dr. Corliss as a witness. Id. ¶¶ 15-16. Strategically, he believed he could get the medical examiner to testify about the presence of multiple injuries on cross-examination, and practically, Dr. Corliss told counsel that he thought the medical examiner's report was well-reasoned. Id. ¶ 16. The state appellate court concluded that "[w]hile another lawyer may have handled the matter differently, that is not the standard for judging whether counsel's representation was incompetent." Id. The Wisconsin Court of Appeals did not reach Strickland's prejudice prong.

Dunn then filed a petition for a writ of habeas corpus under 28 U.S.C. § 2254 in the United States District Court for the Eastern District of Wisconsin. The district court granted Dunn's habeas petition, concluding that he had received ineffective assistance of counsel. Dunn v. Smith, 430 F. Supp. 3d 568 (E.D. Wis. 2019). It reasoned that Dunn's counsel's investigation of Schuckman's death and of a no-cause defense was insufficient under any reasonable application of Strickland, because he failed to interview the medical examiner before trial, call a forensic pathologist, or investigate the exculpatory reports. Thus, the district court concluded that the state appellate court unreasonably applied Strickland's performance prong.

Because the state appellate court did not address Strickland's prejudice prong, the district court reviewed the prong de novo. The court concluded that "Dunn has demonstrated a reasonable probability that the result would have been different had Dunn's counsel adequately presented the no-causation defense and the jury heard expert testimony of the sort presented by Dr. Stephens." Id. at 579. Because Dr. Stephens's testimony "substantially undermines one's confidence in the result" of the trial, while "the jury may have reached the same conclusion even with such evidence," Dunn had provided sufficient evidence "to establish prejudice under Strickland." Id. at 580. The district court thus granted Dunn's habeas petition, and the state now asks us to reverse.

...

B. Deficient Performance

...

The Wisconsin Court of Appeals concluded that Dunn's trial counsel did not perform deficiently in failing to investigate and present a no-causation defense because he (1) consulted with a pathologist and (2) provided strategic and practical explanations for his approach. Upon closely examining the facts of this case, we find that the Wisconsin Court of Appeals unreasonably applied Strickland because Dunn's counsel was poorly informed and based his strategic decisions on a complete misunderstanding of a key piece of evidence—namely, the medical examiner's opinion on the immediacy of death.

It was unreasonable for the state appellate court to find that trial counsel did not perform deficiently because he consulted with a forensic pathologist. Indeed, at the initial consultation, Dr. Corliss provided trial counsel with a crucial piece of evidence supporting Dunn's no-causation defense: he thought Schuckman died immediately. Whether Dunn caused Schuckman's death was a critical issue in the case. The state's main theory of the case depended on connecting two events: Dunn slapping Schuckman in the bar's parking lot, and Schuckman being found dead hours later on the bar's back patio. Evidence that Schuckman would have died immediately after receiving his fatal head injuries would have lent strong support that these two events were not causally connected because multiple witnesses saw Schuckman alive, apparently unharmed, and upright after Dunn slapped him. Yet Dunn's counsel seemingly misattributed Dr. Corliss's opinion that death was immediate to the medical examiner, and erroneously believed that the medical examiner's report meant she would testify to that at trial. He failed to confirm that the medical examiner would introduce this key piece of evidence at trial or otherwise investigate the immediacy of death. His mistaken belief infected his trial strategy to such an extent that his approach to investigating and presenting a no-causation defense cannot be reasonably viewed as strategic, even with the "heavy measure of deference" afforded him under Strickland. 466 U.S. at 691. His "failure to investigate thoroughly resulted from inattention, not reasoned strategic judgment." Wiggins v. Smith, 539 U.S. 510, 526, 123 S. Ct. 2527, 156 L. Ed. 2d 471 (2003). It is thus unreasonable to view trial counsel's decision to consult with a forensic pathologist as satisfactory performance, considering trial counsel did not use the crucial information gleaned from that consultation.

It was also unreasonable for the state appellate court to find that trial counsel's explanation for his approach contained reasonable strategic and practical reasons. Dunn's counsel explained he did not call Dr. Corliss as a witness because he thought the evidence would be more persuasive if elicited from the medical examiner on cross-examination. The state appellate court found this explanation demonstrated that trial counsel was acting strategically. This overlooks the fact that trial counsel was not able to elicit a crucial piece of evidence from the medical examiner on cross-examination: that Schuckman died immediately. "A court adjudicating a Strickland claim can't just label a decision 'strategic' and thereby immunize it from constitutional scrutiny," and the state court's determination that trial counsel was acting strategically here is not reasonable. Jones, 842 F.3d at 464. In fact, his causation strategy was primarily based on the improbable assumption that the state's medical examiner would testify that the state's causation theory was medically impossible. And when that unlikely strategy blew up, counsel had no Plan B.

Trial counsel and the state appellate court seemingly view the evidence about multiple injuries—which trial counsel was able to elicit on cross-examination—as substitute evidence in support of a no-causation defense. But while evidence that Schuckman had more injuries than could be explained by the slap and fall lends some support to the theory that Dunn did not cause his death, it is not nearly as strong as evidence that Schuckman died immediately after the blunt force trauma to the back of his head. The medical examiner testified that all Schuckman's injuries contributed to his death. In order to convict for felony murder, the jury only needed to find that Dunn's conduct was a substantial factor in bringing about Schuckman's death. … So while the presence of multiple injuries suggests that Schuckman may also have been injured by someone or something other than Dunn, given the medical examiner's testimony that all the injuries contributed to his death, the jury could still reasonably find that the Dunn's slap was a substantial factor in his death. Evidence that Schuckman died immediately from one fatal blow, conversely, is much stronger evidence that Dunn did not cause his death. Multiple witnesses saw Schuckman alive, apparently unharmed, and standing several minutes after Dunn slapped him. Given this, trial counsel's explanation for not calling his forensic pathologist—that he could elicit the relevant information from the medical examiner on cross-examination—cannot be reasonably viewed as strategic. …

Trial counsel's "practical" explanation for why he did not call Dr. Corliss as a witness also fails. Trial counsel testified that he did not call Dr. Corliss because the pathologist told him the medical examiner's conclusions in the report were well-reasoned. But the medical examiner's conclusion that there was an interval of survival was not contained in her report. And when trial counsel spoke with Dr. Corliss in April, trial counsel still had the mistaken belief—memorialized by his email a mere ten days before trial—that the medical examiner would testify that death was immediate. So while it follows that at the time trial counsel spoke with Dr. Corliss there was no contradiction between Dr. Corliss's opinion and the medical examiner's perceived opinion, when trial counsel realized that the medical examiner disagreed on the immediacy of death, his decision to not call his pathologist cannot reasonably be called "practical." For these reasons, the state appellate court unreasonably applied Strickland when it determined that counsel did not perform deficiently because he had strategic and practical reasons for his approach to Dunn's no-causation defense.

…

Sometime in the days before trial, trial counsel figured out that the medical examiner would not testify as he thought she would. Thus, "[e]ven if defense counsel could have initially believed expert testimony unnecessary," finding out the medical examiner was going to testify that Schuckman did not die immediately should "have alerted any reasonable attorney to the need to rebut with a defense expert." Woolley v. Rednour, 702 F.3d 411, 423 (7th Cir. 2012) (finding counsel performed deficiently where he "remained nearly passive in the face of damning" testimony that "effectively hollowed out the core of his client's defense" because he "fail[ed] to retain an expert witness" or "ask for a continuance"). Given that he knew that he no longer could elicit this crucial piece of information on cross-examination, his planned approach to the no-causation defense at trial cannot reasonably be viewed as strategic. Rather than calling Dr. Corliss to testify or requesting a continuance to obtain the exculpatory reports, he decided to rely on his cross-examination of the medical examiner to advance the no-causation defense. But the information he could elicit from her—that Schuckman sustained multiple injuries— only provided weak support that Dunn's slap was not a substantial factor in Schuckman's death. He failed to offer any evidence to rebut the medical examiner's conclusions that there was an interval of survival and all Schuckman's injuries contributed to his death. He did not present any evidence whatsoever that Schuckman died immediately—despite having both a forensic pathologist who had expressed this belief and the information about exculpatory expert reports that suggested that Schuckman died on the back patio and did not stand upright after the fatal injury.

...

C. Prejudice

...

We conclude that there is a reasonable probability that the result would have been different if Dunn's trial counsel had adequately investigated and presented evidence that Dunn did not cause Schuckman's death. Given the evidence that Schuckman was seen alive and apparently unharmed after the slap meant that the medical examiner's testimony at trial was key in linking Dunn's slap to Schuckman's death. Her testimony that all Schuckman's injuries contributed to his death and there was no reason to think he died quickly went unrebutted at trial. Dr. Stephens's testimony cast significant doubt on these conclusions and is sufficient to undermine our confidence in the outcome. ...

Dr. Stephens's testimony contradicted the medical examiner's in two key ways. First, while the medical examiner testified that all Schuckman's injuries contributed to his death, Dr. Stephens opined the skull fracture and laceration to the back of Schuckman's head were the substantial cause of his death and he likely would have survived his other injuries. Second, Dr. Stephens disagreed with the medical examiner about how long Schuckman likely survived after sustaining the skull fracture. The medical examiner testified she had no reason to think that Schuckman would have died immediately from his injuries because they did not affect his vital center; Dr. Stephens testified that Schuckman would have been rendered immediately unconscious, would be unlikely to be able to sit up afterward, and would have died within minutes. An analogous set of facts arose in Rogers v. Israel, 746 F.2d 1288, 1293-94 (7th Cir. 1984). There, like here, a crucial question in the case was whether the victim in a bar fight was immediately incapacitated by his injuries. Trial counsel in both cases failed to rebut the state's pathologist's testimony that incapacitation would not occur immediately, though the counsel could have presented such expert testimony. In Rogers, we found that expert testimony regarding the immediacy of incapacitation "was critical to the presentation of the defense," and so there was a reasonable probability that the jury would have had a reasonable doubt about the defendant's guilt if it had been presented. Id. at 1293. We find the same here.

In addition to rebutting some of the medical examiner's testimony, Dr. Stephens's testimony undermined the state's theory of the case in other ways. He testified that the laceration on the back of Schuckman's head would bleed immediately and profusely. Yet there was no blood found in the parking lot, and Schuckman did not have blood dripping down the back of his head. Dr. Stephens's opinion based on these facts—that Schuckman was injured on the back patio rather than in the parking lot—thus contradicted the state's theory of the case. The medical examiner conceded at the postconvic-tion hearing that nothing in her testimony was inconsistent with Schuckman receiving his injuries on the back patio.

We recognize that the medical examiner's testimony was not the only incriminating evidence in this case. Even at a hypothetical trial where counsel presented testimony like that of Dr. Stephens's, the jury would have learned that Schuckman's cell phone was found at Dunn and Crochet's hotel, Schuck-man's DNA was found on Crochet's pants, and Crochet was spotted walking from the direction where Schuckman was found later that night. "Logically, a verdict weakly supported by the record is more likely to have been affected by errors than one with overwhelming record support." Cook, 948 F.3d at 909. But we do not find this other evidence constitutes overwhelming support that Dunn is guilty of felony murder. Rather, much of this evidence potentially incriminates Crochet. It was Crochet who had Schuckman's blood on his pants, and Crochet who was witnessed leaving the bar, and ten to fifteen minutes later, walking away from the direction where Schuckman was later found. But Dunn's guilt is not necessarily tied with Crochet's—the same witness who testified he saw Crochet walking away from the area where Schuckman was found also testified that he saw Dunn seated in the pickup truck at that time. And notably, despite the forensic and eyewitness testimony implicating Crochet in Schuck-man's death, Crochet was able to leverage the exculpatory expert reports to negotiate a plea bargain for a time-served sentence.

There is of course a chance that even if the jury had heard testimony like that offered by Dr. Stephens, it would have nonetheless convicted Dunn for felony murder. It may have still found that Dunn's slap caused the skull fracture, or that Dunn was liable as a party to a crime based on the evidence against Crochet. But the fact that no evidence was offered about the timing and location of Schuckman's death substantially undermines our confidence in this result. This is sufficient to establish prejudice under Strickland.

II. Conclusion

The Wisconsin Court of Appeals unreasonably applied Strickland when it determined that Dunn's trial counsel did not perform deficiently. Dunn was prejudiced as a result of his trial counsel's errors, and so the district court's grant of habeas relief is
AFFIRMED.

Questions:

1. What time of death dispute consumes the Court's attention?
2. Explain how the defendant's pathological claim so radically differs from the prosecution's argument regarding time of death.
3. What other major determination regarding defense counsel does the court issue and with what result?

Read the DOJ's press release regarding two autopsies that failed to find evidence of wrist injuries in possible civil rights charges at https://www.justice.gov/opa/pr/federal-officials-decline-prosecution-death-jamar-clark.

Locate the case below using your favorite legal research resource, read the case and answer the questions below.

United States v. Moya, __ F.4th __, United States Court of Appeals for the Tenth Circuit July 27, 2021, Filed, No. 20-2006

Questions:

1. Timing of death is often related to the activity of the victim. In this case, how is timing gauged or measured?
2. In a more precise way, the defendant seeks to show the timing of heroin usage and the timing of death could not be correlated. How so?
3. What is the "but-for" standard the Court finds central to a finding of criminal agency and responsibility for death?

9.3 FORENSIC PATHOLOGY: CAUSE AND MANNER OF DEATH

The pathologist's role in criminal and civil litigation has extraordinary impact although none is more compelling than proof in the "cause of death." That correlation is central to criminal agency and damage claims. What causes death is inexorably linked to pathological determinations (See Figure 9.2).

Hall v. Flannery
840 F.3d 922
United States Court of Appeals for the Seventh Circuit
February 17, 2016, Argued; November 4, 2016, Decided
No. 15-2602

Chelsea Weekley suffered a skull fracture as an infant and underwent surgery 17 years later to fix it. She died several days after the surgery, and her mother, Sandra Hall, sued the hospital and

FIGURE 9.2 A forensic pathologist at work.

Source: Maricopa County Medical Examiner at https://www.maricopa.gov/468/Medical-Examiner.

the surgeons. Hall argued that the surgery caused Weekley to suffer a seizure and die, and that the surgeons should have prescribed anti-seizure medication. But the defendants argued that no seizure had occurred and that a heart-related ailment was the likely cause of death. A jury trial was held and the jury found in the defendants' favor.

On appeal, Hall argues that the district judge erroneously permitted three of the defendants' experts to opine about Weekley's likely cause of death. We conclude that Hall forfeited her arguments as to two of these experts by making perfunctory and underdeveloped arguments concerning the experts' testimony, qualifications, and methodology. However, we find that the third expert lacked the requisite qualifications to opine that Weekley's heart ailment was the likely cause of death. Because there is a significant chance that the erroneous admission of this expert testimony affected the outcome of the trial, we vacate the district court's judgment and remand for further proceedings.

I. BACKGROUND

A. Weekley's Surgery and Death

When Chelsea Weekley was approximately five months old, she was dropped and suffered a skull fracture. As the fracture expanded over time, a cyst formed in the area. The fracture and cyst did not become a cause for concern until, at the age of 17, she was hit in the head and suffered a loss of consciousness, blurred vision, and dizziness. After CT and MRI scans confirmed the extent of the fracture and the cyst, Weekley underwent a "cranioplasty" surgery to repair the fracture and the area of the dura (the thick membrane surrounding the brain) where the cyst had formed. The surgery was performed at SSM Cardinal Glennon Children's Hospital ("Hospital") by Dr. Ann Flannery, a neurosurgeon, and Dr. Raghuram Sampath, a neurosurgical resident.

Weekley was discharged a day after the surgery and was found dead in her bed three days later. The coroner observed that Weekley was in a "normal resting position," that her legs were straight and her ankles crossed, and that her right arm was bent near her head "as if using it to lay on." The coroner also observed that her hands were "cramped up," that her bladder was empty, and that her feet were near the headboard while her head was near the foot of the bed.

Weekley's autopsy was performed by Dr. Raj Nanduri, a board-certified forensic pathologist. According to Dr. Nanduri, forensic pathology concerns the effect of sudden, accidental, and suicidal death on the human body. After performing a physical, microscopic, and toxicological examination of Weekley's body, Dr. Nanduri was unable to identify a cause of death. So she asked Dr. Mary Case, a neuropathologist, to examine Weekley's brain. Dr. Case found that the surgery damaged the dura and surface of Weekley's brain. Based on this finding, Dr. Nanduri concluded that Weekley had died from a seizure brought about by the surgical damage. Neither Dr. Nanduri nor Dr. Case was aware of or had reviewed the pre-surgery CT and MRI scans when they made their findings.

...

II. ANALYSIS

...

B. Reversible Error Involving Dr. Ruge

1. Rule 702/Daubert Framework Should Have Been Applied

Hall argues that the district judge erred in permitting Dr. Ruge to testify that Weekley's cause of death was not attributable to seizure, and that a heart-related issue was the likelier explanation. Both Hall and the defendants contend that we should review the admission of this testimony for abuse of discretion. But that overlooks the fact that the abuse of discretion standard is available only when the district judge actually applied the Rule 702/Daubert framework, which did not occur here.

In MIL #49, Hall argued that Dr. Ruge offered cause-of-death opinions without being "qualified through education or experience" and without the requisite "scientific explanation." With that challenge to Dr. Ruge's credentials, the district judge should have conducted a Daubert inquiry, even though Hall did not expressly reference Daubert or Rule 702 by name. And when Hall objected at trial that "there's been absolutely no foundation laid, no qualifications, nothing that would suggest that this gentleman is qualified to give ... opinions as to cause of death," the district judge was squarely faced with a need to determine if Dr. Ruge was qualified as an expert in this area, even if the word "Daubert" was not spoken.

But the objections did not prompt the judge to examine Dr. Ruge's qualifications and methodology or to apply the Daubert test. Instead, the judge focused solely on whether the challenged opinions had been previously disclosed. Perhaps even more curiously, neither party made any effort to clarify that Hall had in fact invoked Rule 702 and Daubert, and that the district judge should therefore comment on more than whether the opinions had been previously disclosed. The parties' failure to do so, however, did not extinguish the need to apply the Rule 702/Daubert framework to Dr. Ruge's opinions. Because that application never occurred, our review of the admission of these opinions is de novo.

2. Adequate Qualifications and Methodology for Dr. Ruge's Seizure Opinion

At trial, Dr. Ruge opined that Weekley had not suffered a seizure before she died. He noted that she had no documented history of seizures, and that her body and bed did not display the typical signs of seizure such as violent convulsions, tongue damage, and urinary incontinence. He also stated that seizures typically do not result from cranioplasty (the surgical procedure Weekley underwent), and that the cranioplasty here had not damaged Weekley's brain.

This opinion was not erroneously admitted. For one, Dr. Ruge possessed sufficient qualifications to opine on the operation performed and the possible occurrence of a seizure afterward. At the time of trial, Dr. Ruge had practiced pediatric neurosurgery for approximately 25 years and was serving as the chief of pediatric neurosurgery for the Advocate Health Care system, which encompassed two major children's hospitals and approximately ten other hospitals. In addition, he was certified by the American Board of Neurological Surgery, was affiliated with multiple neurological and medical societies, and was a former editorial board member for the publications Critical Reviews in Neurology and Child's Nervous System. And perhaps most notably, he had operated on growing skulls fractures like Weekley's, and had published articles on various pediatric neurosurgery topics including epilepsy, cranial cysts, and severe head injury.

In addition, Dr. Ruge's seizure-related opinions were based on sufficiently reliable methodology. Dr. Ruge arrived at his conclusions based on his review of the autopsy report; Weekley's medical records, including the MRI and CT scans taken before surgery, Dr. Flannery's operative report, and Dr. Flannery and Dr. Sampath's post-surgery progress notes; and deposition testimony. ... In addition, Dr. Ruge relied on his medical experience, which as discussed above is substantial, particularly with regard to pediatric neurosurgery. ...

Hall contends that Dr. Ruge lacked the requisite qualifications because pathology is not his area of professional focus. However, this argument ignores the fact that "[o]rdinarily, courts impose no requirement that an expert be a specialist in a given field." Gayton, 593 F.3d at 618 (citation omitted ... The fact that an expert may not be a specialist in the field that concerns her opinion typically goes to the weight to be placed on that opinion, not its admissibility. ...

Here, the issue of whether a seizure occurred shortly before Weekley's death did not concern knowledge that is held solely by pathologists. Dr. Nanduri, the pathologist who examined Weekley after her death, testified that a seizure-related cause-of-death finding is made by excluding all other apparent causes, and that she arrived at her finding here based on the damage to Weekley's dura and cortex—damage that a non-pathologist could have observed. Moreover, Dr. Ruge stated that he "routinely cared for patients who … have seizures as part of their neurosurgical condition," witnessed numerous seizures himself, and operated on individuals with head injuries comparable to Weekley's. So Dr. Ruge possessed the requisite qualifications to testify about seizures. …

3. Inadequate Qualifications for Dr. Ruge's Heart Opinion

While Dr. Ruge was qualified to opine on Weekley's surgery and the possibility of seizure, this qualification does not extend to his opinion that Weekley's heart-related issue was the likelier cause of death. …

Neither Dr. Ruge's trial testimony nor his expert report and curriculum vitae indicate that he possesses any specialized education, knowledge, experience, or skill concerning focal interstitial chronic inflammation specifically, or more broadly cardiology. Indeed, Dr. Ruge acknowledged at trial that when he read the inflammation finding in Weekley's autopsy report, he "didn't know what that was exactly." So he conducted a Google search and found several papers explaining that it is "a finding in young athletes who die suddenly of cardiac arrhythmias," even though their "hearts can look very normal." Based on this research, he opined that focal interstitial chronic inflammation "makes more sense [as a cause of death] based on the whole story, the whole picture."

We do not doubt that Dr. Ruge is an intelligent doctor who possesses considerable knowledge about surgery, pediatrics, and neurology. However, the record lacks sufficient evidence demonstrating that this knowledge and the related experiences render Dr. Ruge qualified to opine about Weekley's heart. … Because Dr. Ruge did not possess the requisite qualifications to opine about the impact of Weekley's heart on her cause of death, we need not address his methodology.

4. Hall's Substantial Rights Were Affected

The fact that Dr. Ruge's heart-related cause-of-death testimony was erroneously admitted does not automatically entitle Hall to a new trial. Rather, she must show that this error had a "substantial influence over the jury." Farfaras, 433 F.3d at 564. We conclude that it did.

It is undisputed that Weekley's cause of death was a critical issue at trial. Hall pointed to seizure—a relatively difficult fact to prove since (as both parties agree) such a finding is made only by ruling out all other possible alternatives. It is no surprise that the defendants chose not only to argue that a seizure had not occurred, but also to offer an alternative cause—heart troubles. And Dr. Ruge played a critical role supporting this alternative cause, testifying about multiple studies he had read that linked the heart condition that Weekley had to the circumstances under which she passed away.

Critically, the import of Dr. Ruge's testimony cannot be minimized on the ground that it was merely "cumulative" of testimony provided by other experts. … To be sure, Dr. Miller and Dr. Rothman briefly referenced the effect a heart attack can have on cause-of-death findings generally, while Dr. Nanduri and Dr. Case acknowledged that heart disturbances can result in death for a healthy young adult. However, Dr. Ruge was the only expert to opine about the purportedly "numerous papers" that identify young athletes who died suddenly of heart-related illness. We cannot ignore the distinct possibility that Dr. Ruge's discussion of these publications played a key role in the jury's verdict, given the thresh-old nature of the cause of death inquiry. …

The defendants contend that reversal would be improper because "there is no basis to conclude that admission of the Defendants' experts' testimony influenced the jury's verdict." They argue that even if the jury had not heard any testimony from the defendants' experts regarding alternative causes of death, the jury could have found for the defendants on the grounds that they did not breach their standard of care, or that any such breach did not cause Weekley's death. In support, the defendants cite several Illinois state court decisions that found a new trial to be unnecessary where it was possible that the jury's verdict rested on an issue that was not subject to appeal. …

But state law is not applicable to this inquiry. "In this circuit the harmlessness standard is treated as procedural and therefore in a diversity case is governed by federal law and [Rule] 61." Sokol Crystal Prods., Inc. v. DSC Commc'ns Corp., 15 F.3d 1427, 1435 (7th Cir. 1994) (citation and footnote omitted); ... Because the defendants have not satisfied Rule 61's harmless-error standard, a new trial is warranted.

III. CONCLUSION

We REVERSE the district judge's decision to admit Dr. Ruge's heart-related cause-of-death testimony, VACATE the judgment, and REMAND the case for proceedings consistent with this opinion.

Questions:

1. In this civil litigation, the pathological determination of cause of death depends upon whether or not the defendant hospital and doctors engaged in malpractice. Exactly what is being alleged by plaintiff in this case?
2. In order to demonstrate that death was caused by a weak heart or some level of heart dysfunction, the plaintiff called 3 expert witnesses. The third witness in the case was typed as not even meeting Daubert standards. Why so?
3. The impact of an unqualified expert witness is not a simple mistake or harmless error in this case. What impact does the Court indicated on the overall proceedings?

Horn v. Kansas
United States Court of Appeals for the Tenth Circuit
September 6, 2019, Filed
No. 18-3258

Petitioner Theodore Vincent Horn II, a Kansas state prisoner appearing pro se, seeks a certificate of appealability ("COA") to challenge the district court's denial of his 28 U.S.C. § 2254 petition for post-conviction relief. We deny Petitioner's request for a COA.

I.

A jury convicted Petitioner in Kansas state court of first-degree murder. The trial judge sentenced Petitioner to life without the possibility of parole for 50 years (a "Hard 50"). Petitioner filed a direct appeal. The Kansas Supreme Court affirmed his conviction and sentence. Petitioner filed five motions for post-conviction relief under Kan. Stat. Ann. § 60-1507. Kansas courts denied each motion.

Petitioner then filed a pro se writ of habeas corpus pursuant to 28 U.S.C. § 2254 in federal district court raising thirty-two claims for relief. The district court denied the petition on all thirty-two claims. Petitioner now seeks a certificate of appealability ("COA") on eleven of those claims.

II.

A COA is a jurisdictional prerequisite to our review of a habeas application. 28 U.S.C. § 2253(c)(1)(A); Miller-El v. Cockrell, 537 U.S. 322, 336, 123 S. Ct. 1029, 154 L. Ed. 2d 931 (2003). Under the Antiterrorism and Effective Death Penalty Act ("AEDPA"), "[w]e will issue a COA 'only if the applicant has made a substantial showing of the denial of a constitutional right.'" Allen v. Zavaras, 568 F.3d 1197, 1199 (10th Cir. 2009) (quoting 28 U.S.C. § 2253(c)(2)). Under this standard, "the applicant must show 'that reasonable jurists could debate whether (or, for that matter, agree that) the petition should have been resolved in a different manner or that the issues presented were adequate to deserve encouragement to proceed further.'" United States v. Taylor,

454 F.3d 1075, 1078 (10th Cir. 2006) (quoting Slack v. McDaniel, 529 U.S. 473, 484, 120 S. Ct. 1595, 146 L. Ed. 2d 542 (2000)). Our "inquiry does not require full consideration of the factual or legal bases adduced in support of the claims," but rather "an overview of the claims" and "a general assessment of their merits." Miller-El, 537 U.S. at 336.

Under AEDPA,

> a state prisoner will be entitled to federal habeas corpus relief only if he can establish that a claim adjudicated by the state courts "resulted in a decision that was contrary to, or involved an unreasonable application of, clearly established Federal law, as determined by the Supreme Court of the United States," or "resulted in a decision that was based on an unreasonable determination of the facts in light of the evidence presented in the State court proceeding." [28 U.S.C. § 2254(d).] Further, "a determination of a factual issue made by a State court shall be presumed to be correct." [Id.] § 2254(e)(1). That presumption of correctness is rebuttable only "by clear and convincing evidence." Id.

Boyd v. Ward, 179 F.3d 904, 911-12 (10th Cir. 1999).

III.

The facts of the case, as determined by the Kansas Supreme Court in Petitioner's direct appeal, are as follows:

> Theodore V. Horn, II, had been an alcoholic for years. When his girlfriend kicked him out of her residence because he had resumed his drinking, he moved in with his grandmother, Tina Weaver, in May 2002. Weaver had raised Horn, and he called her "Mom." She disapproved of his drinking and had called the police when he had previously been drinking or had been drunk at her house. She had once obtained a protection from abuse order against him, though it had expired in June 2000.
>
> On Friday, May 17, 2002—within a week or two of Horn's moving in with Weaver—Krystal Kern, who lived across the street, left for work around noon. She saw Weaver standing inside Weaver's house staring out through the screen door. Kern waved, but Weaver did not acknowledge her, which was unusual because Weaver would always wave, tell her good morning, and wish her a nice day. Kern then heard a man's loud, angry voice coming from inside the house, but did not see Weaver react in any way. Kern never saw her again.
>
> Another neighbor, Greg Fisher, last saw Weaver on May 17 between 5 and 5:30 p.m. when she stuck her head outside her door. The next morning, May 18, he stopped at Weaver's to drop off a bag of fruit and vegetables. He rang the bell, but received no answer, so he left the bag on the doorknob. That afternoon at 2:15 he received a call from his daughter advising him that another neighbor, Betty Roux, was requesting his assistance in checking on Weaver's welfare. When he later approached Weaver's door, he noticed the bag of fruit and vegetables still hanging outside. He received no response to the buzzer, and when he attempted to open the door, a safety chain only allowed it to open several inches. He detected a commotion in the dark living room and Horn called out to say they had overslept but everything was ok. As Fisher walked away, Roux advised that she had already called 911 to come and check on Weaver.
>
> Officer Gregory Johnson of the Wichita police was dispatched that day to 422 North Custer for a check-welfare call. He had been told that he was going to check the welfare of an 89-year-old female whom the neighbors said they had not seen in a couple of days. Officer Michelle Woodrow arrived at the same time, i.e., approximately 3 p.m. Emergency medical services personnel and ambulances also arrived.
>
> The officers spoke to Roux and Roux's daughter. Roux told them that she had not seen Weaver for a couple of days and that she was concerned because she was used to Weaver being at her front door collecting mail. Roux was concerned because the mail was still in Weaver's mail slot, which was unusual. According to Roux, Weaver knew exactly when the mail was supposed to arrive, and if it was a few minutes late, she would call her neighbors. Roux told Officer Johnson that she had knocked on Weaver's door but Horn refused to open it. Horn told Roux that Weaver was okay, but Roux thought his response was unusual so she called 911.

After the conversation with Roux and her daughter in Roux's front yard, Officer Johnson and Officer Woodrow walked next door to 422 North Custer. Johnson confirmed the mail was still in the slot next to the door. He opened the screen door and knocked on the inner door. Johnson heard a male voice from inside ask who it was, and Johnson identified himself as a police officer. For 4-5 minutes, Johnson continued knocking and calling for Horn by name. Officer Woodrow checked to see if the door was locked and pushed the door open. The door opened 3 to 4 inches, but a security chain prevented it from opening further. Through the opening they were able to see what appeared to be the shape of a body nearby under a blanket.

When Horn eventually came to the door, he told the officers he and Weaver were asleep and he was naked. Johnson observed Horn was naked and directed him to get dressed because of the presence of a female officer. He also told Horn to open the door so they could check on his mother. At that point Horn stated, "She's dead. I killed her." He was again directed to put on pants and open the door, to which he stated several times, "This is murder. I killed my mom. I love my mom." After he put on his pants and unchained the door, he was placed in handcuffs. Three additional times he repeated, "This is murder. I killed my mom. I love my mom." Horn was later taken to a hospital in an ambulance because he had passed out in the patrol car.

At the hospital, Lieutenant Kenneth Landwehr observed a blood smear on Horn's right forearm and removed the blood smear with a set of swabs. He observed Horn's hands and detected no sign of injury. Landwehr also took a swab of Horn's penis before medical personnel cleaned it.

Mark Alexander was on duty as a detention deputy at the Sedgwick County jail at 7 p.m. when Horn was brought in. According to Alexander, Horn was screaming, saying that Weaver had been "bitching" at him and he got mad and twisted her neck and took a saw and started cutting her head off. Horn told the officers that she had been dead since Thursday, May 16.

Police found Weaver's body in the living room covered with a green blanket. They observed a massive trauma to her head and a massive amount of pooling of blood underneath it. A wooden carpenter's level, broken and blood-spotted, was nearby. They also observed she was partially decapitated and a carpenter's handsaw was underneath her body. The saw contained hair and blood, consistent with a pattern of blood drops near Weaver's head. In front of the couch and 3 feet from Weaver's feet was a pornographic magazine entitled "Live Young Girls" folded in half lengthwise. The magazine contained Horn's semen and one of his fingerprints.

The forensic pathologist who investigated the cause of Weaver's death testified that based on the receding of rigor mortis in the body, she had probably been dead since 10:15 p.m. on Friday, May 17. He concluded she had died from multiple blunt and sharp force injuries. The neck injuries were consistent with the handsaw and reflected five separate starting points, with one major cut entirely through the vertebrae and ending at the trachea. The pathologist concluded that the major cut alone took between 30 minutes and 2 hours to inflict and that Weaver was still alive when the sawing occurred. Weaver's body was partially clad, with her upper garments pushed over a shoulder and her slacks pulled down to the midpoint on her buttocks. Her genital area also contained small lacerations and abrasions, but no seminal fluid was found on swabs taken from her or her clothing. Horn's DNA was not found on Weaver's fingernails, her clothing, or the bloodstains on the tools used to kill her. Weaver's DNA was not found on Horn's jeans, his penile swab, or his forearm swab.

At trial, Horn testified that on Friday, May 17, he had called his employer around 5 a.m. to see if he would be working that day. The employer had no work for him, so he began to drink. He remembered waking up after passing out on the couch and Weaver griping at him. He got up and ate something in the kitchen. Horn remembered the program Wheel of Fortune was on the television, which typically begins at 6:30 p.m. He returned to the couch and passed out again. It was dark when he woke up again and he learned he had urinated on himself. He got up, but did not remember what he did before he fell asleep again.

Horn testified that he first saw Weaver's body when he retrieved the newspaper. He started crying, got drunk again, and put a spread over her. He had no recollection of killing her.

State v. Horn, 278 Kan. 24, 91 P.3d 517, 520-22 (Kan. 2004).

IV.

...

A. Evidence of pornographic magazine

Petitioner claims that he was deprived of due process rights when prosecutors admitted evidence at trial of the pornographic magazine that was found close to the victim's body. He argues that prosecutors did not charge him with sexual assault and, therefore, the evidence of the pornographic magazine was irrelevant and unfairly prejudicial. Notably, when Petitioner presented these arguments to the state court, he relied only on state law, not on constitutional due process. And as the district court noted, state court evidentiary decisions are matters of state law, and we do not "reexamine state-court determinations on state-law questions" on federal habeas review. Estelle, 502 U.S. at 68.

But even if Petitioner had properly preserved this claim, he would not be entitled to relief. The Kansas Supreme Court determined that the evidence was relevant to "supporting the State's theory of sexual assault as a motive for the murder," and therefore was properly admitted. Horn, 91 P.3d at 527. The court further held that Petitioner had not demonstrated that the trial court's admission of that evidence was an abuse of discretion. Id. at 528.

"[F]ederal courts may not interfere with state evidentiary rulings unless the rulings in question rendered 'the trial so fundamentally unfair as to constitute a denial of federal constitutional rights.'" Moore v. Marr, 254 F.3d 1235, 1246 (10th Cir. 2001) (quoting Tucker v. Makowski, 883 F.2d 877, 881 (10th Cir. 1989)). The state court's decision was not contrary to established United States Supreme Court precedent. Nor did it involve an unreasonable application of United States Supreme Court precedent. Accordingly, reasonable jurists could not debate the resolution of this claim. Petitioner is not entitled to a COA on this issue.

...

B. Ineffective assistance claims

...

a. Expert witness

Petitioner asserts that his trial counsel provided ineffective assistance by selecting an inappropriate expert witness for the defense and for failing to present evidence of Petitioner's blood alcohol content ("BAC") to the jury.

The Kansas Court of Appeals addressed this issue and applied the proper standard. Horn v. State, 2007 Kan. App. Unpub. LEXIS 1115, 2007 WL 1964958, at*2 (Kan. Ct. App. 2007). The court explained:

> [a]t trial forensic toxicologist John Robert Zettl ... testif[ied] for the defense. Zettl, who specializes in drugs and alcohol and their effects upon humans, testified about the typical reactions individuals have to various levels of alcohol in their bodies. Zettl testified that some individuals who have been non-violent alcoholics for a number of years may have episodes where they commit violent acts as a result of intoxication. Zettl also testified that alcohol consumption can make people behave in ways that they never would have sober and that it can cause hallucinations. Based on Horn's blood alcohol concentration level of .516 and a videotape of Horn several hours after he was taken into custody, Zettl stated that Horn was highly intoxicated. In fact, according to Zettl, most people cannot survive at that blood alcohol level. Zettl further testified that high levels of intoxication can lower a person's ability to remember events and can distort a person's ability to gauge time.
>
> Zettl's testimony specifically rebuts Horn's allegation that his counsel failed to introduce sufficient testimony relevant to the effects of intoxication on Horn's psychological state. Nothing was presented in Horn's motion to establish that a second witness would have been able to testify to anything beyond Zettl's testimony or to further support Horn's defense. See State v. Lewis, 33 Kan.App.2d 634, 653, 111 P.3d 636 (2003) (failing to call a witness offering only cumulative testimony is not ineffective assistance). Horn's claim that his Strategic decisions, like which witnesses to call and counsel was

ineffective for failing to call a psychiatrist or psychologist to testify as to the effects of intoxication on Horn's mental state is meritless.

Id.

Strategic decisions, like which witnesses to call andwhat evidence to present at trial, are left to counsel's discretion. Hoxsie v. Kerby, 108 F.3d 1239, 1246 (10th Cir. 1997). This includes expert testimony. Turner v. Calderon, 281 F.3d 851, 876 (9th Cir. 2002) ("The choice of what type of expert to use is one of trial strategy and deserves 'a heavy measure of deference.'") As the district court explained, "[c]laims of ineffective assistance based on calling one expert over another consistently fail." Horn, 2018 U.S. Dist. LEXIS 206600, 2018 WL 6436065, at *7; … Indeed, a habeas petitioner is not entitled "to the testimony of every qualified expert who might possess some specialized knowledge regarding the petitioner's case." Tibbetts, 633 F.3d at 443. Accordingly, the state court's resolution of this claim was not contrary to, or an unreasonable application of, clearly established Federal law. See 28 U.S.C. § 2254(d)(1). We deny Petitioner a COA on this claim.

b. Failure to object to coroner's testimony at trial

Petitioner also alleges that his trial counsel provided ineffective assistance by failing to challenge the testimony of the coroner at trial. Specifically, the coroner testified that bruising appeared on the victim after the autopsy and approximately four days after the victim's death. This, Petitioner claims, is a scientific impossibility, because bruising is caused by blood flow and the victim was deceased at this time. Further, the coroner testified that the victim's neck injury took anywhere from thirty minutes to two hours to achieve, and that the autopsy showed that Petitioner cut the victim's neck while she was still alive. Petitioner claims that is untrue, and that the victim's neck injury took less time than the coroner testified. Accordingly, he claims his trial counsel's failure to challenge the coroner's testimony was ineffective assistance.

Notably, the Kansas Court of Appeals found this claim to be procedurally barred. … And the federal district court concluded that the procedural bar applied in federal court as well. Horn, 2018 U.S. Dist. LEXIS 206600, 2018 WL 6436065, at *10.

Petitioner fails to argue cause and prejudice to overcome the procedural default. See Magar v. Parker, 490 F.3d 816, 819 (10th Cir. 2007). Reasonable jurists, therefore, could not debate the resolution of this claim. Accordingly, we deny Petitioner's request for a COA on this ground.

…

c. Failure to pursue newly discovered evidence claim

Petitioner also asserts that his state appellate counsel denied him effective assistance by not pursuing his claim of newly discovered evidence. Like the preceding claim, the Kansas Court of Appeals concluded Petitioner procedurally defaulted this claim, and the federal district court agreed. Again, Petitioner fails to argue cause and prejudice to overcome the procedural default. Thus, we are procedurally barred from considering this claim for relief.

…

V.

We DENY Petitioner's request for a COA.

Questions:

1. The Horn decision may be unrivaled in its fantastical approach to proving the defendant did not inflict the injuries that caused the death of the victim. Name 2 approaches taken by defendant to claim a lack of agency.

2. Why would the prosecution tie sexual activity with the deceased into a case of horrific as these facts are?
3. The cause of death as testified to by the state's forensic pathologist, was due to multiple forces. Explain.

Locate the following case using your favorite legal research resource, read the case and answer the question below.

Newmiller v. Raemisch, 877 F.3d 1178 United States Court of Appeals for the Tenth Circuit December 18, 2017, Filed No. 16-1396

Questions:

1. This is a case filled with experts and a variety of distinguishing claims. What do all the experts agree as the primary cause of death?
2. One of the disputed conclusions between this bevy of experts is whether or not this sort of would results in immediate death. What facts in this case make that conclusion difficult?
3. Pathologists in this case debated and argued whether a cardiac tamponade had occurred. Point out the variation of opinion.

9.4 FORENSIC PATHOLOGY AND MANNER OF DEATH

In manner of death analysis, the pathologist seeks to discern how the death was inflicted and in what manner the attacker carried out the deed. How a person dies is the "cause." What steps were taken to get to that "cause" is properly labeled "manner."

Read the DOJ's press release regarding a case where injuries sustained prior to death discovered in autopsy at https://www.justice.gov/opa/pr/former-commander-naval-station-guantanamo-bay-convicted-obstructing-justice-connection.

United States v. McKee
752 Fed. Appx. 462 United States Court of Appeals for the Ninth Circuit
October 16, 2018, Argued and Submitted, San Francisco, California; November 20, 2018, Filed
No. 17-10244

Nelson Ray McKee appeals his jury conviction for voluntary manslaughter within Indian Country in the stabbing of his wife, Cheryl Jackson McKee ("Jackson"), on December 31, 2014. At the outset, the district court denied in part McKee's motion to suppress certain evidence as having been obtained in violation of his Fourth Amendment rights. In the course of the trial, the district court also excluded portions of the testimony of a witness, Lekeisha Crutcher, and McKee's expert, Lawrence Lee Renner. McKee received a 120-month sentence, one year above the range indicated in the Guidelines issued by the United States Sentencing Commission. On appeal, he challenges these three rulings and the upward variance in his sentence. We have jurisdiction under 18 U.S.C § 1153, and we affirm.

1. Denial of Motion to Suppress Evidence

The government has the burden of proving that consent was given to a warrantless search. United States v. Arreguin, 735 F.3d 1168, 1174 (9th Cir. 2013) (citing United States v. Welch, 4 F.3d 761, 764 (9th Cir. 1993)). "On appeal, evidence regarding the question of consent must be viewed in the light most favorable to the fact-finder's decision." United States v. Patayan Soriano, 361 F.3d 494, 501 (9th Cir. 2004) (quoting United States v. Kaplan, 895 F.2d 618, 622 (9th Cir.1990)).

We first review the district court's underlying factual findings for clear error. United States v. Lara, 815 F.3d 605, 608-09 (9th Cir. 2016) (citing United States v. Mayer, 560 F.3d 948, 956 (9th Cir. 2009)). The district court concluded that McKee "likely orally consented" to the entry. This finding is not clearly erroneous.

We review de novo the denial of a motion to suppress evidence. Lara, 815 F.3d at 608-09 (citing Mayer, 560 F.3d at 956). Consent to a warrantless search and seizure must be "unequivocal and specific and freely and intelligently given. There must be convincing evidence that [a] defendant has waived his rights. There must be clear and positive testimony." United States v. Shaibu, 920 F.2d 1423, 1426 (9th Cir. 1990) (internal quotation marks omitted) (quoting United States v. Page, 302 F.2d 81, 83-84 (9th Cir. 1962)). The finding that McKee orally consented to the entry meets this threshold. The court therefore did not err in denying the motion.

2. Exclusion of Present Sense Impression Testimony

At trial, Lekeisha Crutcher ("Lekeisha") was prevented from testifying that she heard Brittany Abel say, "I see a van leaving up Hazard Road." The parties agree that the district court erred in excluding this testimony. We therefore review for harmless error. United States v. Morales, 108 F.3d 1031, 1040 (9th Cir. 1997) (en banc) (citing United States v. Rahm, 993 F.2d 1405, 1414 (9th Cir. 1993)). We reverse unless "it is more probable than not that the error did not materially affect the verdict." Id. (citing United States v. Crosby, 75 F.3d 1343, 1349 (9th Cir.1996)). We affirm, because the excluded testimony came in later through Lekeisha's husband, Chad Crutcher. He testified that he heard Abel say, "Well, who's that? Who's that leaving … down the road?" He also testified that he saw the van himself.

3. Exclusion of Expert Testimony

We review the admission or exclusion of expert testimony for an abuse of discretion. Earp v. Cullen, 623 F.3d 1065, 1075 (9th Cir. 2010) (citing United States v. W.R. Grace, 504 F.3d 745, 759 (9th Cir. 2007)).

First, Renner was asked to analyze Jackson's injuries, and the use of a knife to inflict them, as they pertained to the likelihood of a suicide versus a homicide. The court interrupted, stating that the testimony had ventured into a manner-of-death determination. This exclusion was not an abuse of discretion. On voir dire, Renner testified that he had never acted as a forensic pathologist or a coroner and had no experience with manner-of-death determinations. He was accepted as an expert only for crime scene interpretation and reconstruction and blood pattern identification.

Second, Renner was asked whether "the location of the [two] knives at the same spot as the notes" was significant. He responded, "it's less likely that you had two individuals who would assault somebody and then lay the knives both down close together on a table. It would be more consistent with a suicide-type situation." This is the answer that the district court struck, stating that an appropriate foundation as to the location of the knives had not yet been established, that Renner was not offered as an expert on manner of death determinations, and that Renner was being impermissibly asked whether the scene was more consistent with a suicide versus a homicide. The district court then clarified that Renner could be directed to testify regarding where the knives were found, but that he could not be asked whether that made a suicide less likely.

This was an error. Expert opinions on ultimate issues are generally permissible. Fed. R. Evid. 704(a); Morales, 108 F.3d at 1035. And the location of the two bloody knives falls within the scope of Renner's expertise in crime scene reconstruction.

The error, however, was harmless. The defense's alternative theory concerning multiple unknown assailants was never substantially advanced. Whether there was one assailant or two does not necessarily affect the assessment of McKee's guilt. Moreover, the extensive remainder of Renner's testimony on the likelihood of a suicide was admitted.

4. Upward Variance in Sentence

We review the imposition of a sentence for an abuse of discretion in two steps. First, we determine whether the district court committed procedural error. Second, we consider the substantive reasonableness of the sentence. United States v. Carty, 520 F.3d 984, 993 (9th Cir. 2008) (citing Gall v. United States, 552 U.S. 38, 51, 128 S. Ct. 586, 169 L. Ed. 2d 445 (2007). There was no procedural error. McKee's Fifth Amendment rights were not violated. The district court was entitled to take the lack of remorse evidenced by McKee's statement into consideration in imposing the sentence. United States v. Hull, 792 F.2d 941, 943 (9th Cir. 1986) (per curiam). The sentence was not imposed to promote McKee's rehabilitation. Tapia v. United States, 564 U.S. 319, 334, 131 S. Ct. 2382, 180 L. Ed. 2d 357 (2011). In the totality of the circumstances, the sentence was substantively reasonable. Carty, 520 F.3d at 993 (citing Gall, 552 U.S. at 51).

AFFIRMED.

Questions:

1. What was the error the appeals court identifies, and how does it label the error regarding the outcome of the Court case?
2. As to the manner of death, what were the two choices under consideration?
3. Defendant posed a theory that he was not the assailant because the crime scene painted a differing picture. How and with what success?

Using your favorite legal research resource locate the following case, read the case and answer the questions below.

State v. McGuire, 419 N.J. Super. 88, 16 A.3d 411, Superior Court of New Jersey, Appellate Division October 27, 2010, Argued; March 16, 2011, Decided DOCKET NO. A-6576-06T4

Questions:

1. The manner of death in this case is quite disturbing. Why so?
2. The prosecution had to connect the dismembered body to the victim's spouse. By examining different types of evidence, the manner of inflicting death becomes provable and more obvious. Of all the evidence considered in this case, name the two that tend to prove the method and manner of inflicting death.
3. Proof of motive often intersect with the manner of death. Explain the prosecution's various attempts to prove motive and intent in this case.

In the Philips decision, the court grapples with a common scenario that involves the manner of death-namely the question whether there is an intentional homicide or an act of self-defense. Pathological determinations and evidence play a large role in resolving these dilemmas.

Phillips v. Valentine
United States Court of Appeals for the Sixth Circuit
September 1, 2020, Filed
File Name: 20a0512n.06 No. 18-6184

Johnny Phillips shot and killed Phil Glodo in 2007 with a shotgun loaded with birdshot after a day-long argument in which Glodo had been the belligerent party. In 2009, a Kentucky jury convicted Phillips of wanton murder and sentenced him to thirty years in prison. The key evidence at trial was autopsy photographs and medical testimony showing that Glodo had been three feet away or more when the fatal shot had been fired and that he was shot directly in the back of the head between the ears. But in 2013, Phillips discovered an X-ray of the deceased Glodo's skull that was taken by the medical examiner's office and not turned over in the course of Brady discovery. He filed a petition for a writ of habeas corpus on this basis. Phillips argues that this X-ray shows, or could be used in conjunction with expert testimony to show, that the fatal blast was fired at an angle rather than straight-on from behind, which he further argued suggested either that there had been a struggle and that he had indeed been acting in self-defense or, alternatively, that he had not acted wantonly by pointing the gun squarely at the back of Glodo's head and pulling the trigger. The district court denied Phillips's habeas motion, and Phillips appealed.

Since the state concedes that that the X-ray was suppressed, the appeal turns on whether the X-ray is favorable and material. ... Most of the evidence at Phillips's trial was equivocal; the physical evidence that purported to show that he had shot Glodo squarely in the back of the head, and from far enough away not to have been in a close physical struggle, was crucial. Had the X-ray been made available to Phillips at the time of his trial, that trial could have been turned into a "battle of the experts." This is different enough from what actually happened to "undermine[] confidence in the outcome of the trial." Kyles v. Whitley, 514 U.S. 419, 434, 115 S. Ct. 1555, 131 L. Ed. 2d 490 (1995) (quoting United States v. Bagley, 473 U.S. 667, 678, 105 S. Ct. 3375, 87 L. Ed. 2d 481 (1985)). Accordingly, we reverse.

...

I. ANALYSIS

A. The X-Ray Was Suppressed

To take the easiest question first, there is no question that the X-ray was suppressed. The government itself acknowledges "the prosecution's inadvertent suppression of the X-ray of the victim's head[.]" Brady does not require that the suppression be deliberate. Strickler, 527 U.S. at 281-82; ... Nor does Brady require that the prosecutor's office have had knowledge or possession of the material at the time of trial. See id. at 281. (In our case, the Commonwealth maintains that prosecutors denied the existence of the X-ray before the courts in 2013 because it never knew that the X-ray existed. The Commonwealth's current attorney admitted at the evidentiary hearing, however, that this was only speculation.

The Commonwealth admits on appeal that, "Phillips has established cause for the default in that the X-ray was not turned over to Phillips until an open records request was made." As we have seen, "cause for the default" and the Brady suppression inquiry are the same. Brooks, 626 F.3d at 890-91. Therefore, the government has conceded the suppression element of the Brady test.

B. The X-Ray Was Favorable

Phillips's argument is that the X-ray shows that there was far less shot in Glodo's skull than one would expect had the fatal blast been fired straight-on from behind, as the prosecution alleged at trial. The blast, in other words, must have been at an angle, so that only some of the shot emitted from the shotgun entered his skull. This in turn, Phillips argues, shows—or would have allowed him to argue at trial—that the prosecution's theory of the case was incorrect, and that physical evidence corroborated Phillips's tale of self-defense. The evidence could have been used for impeachment of prosecution witnesses (Dr. Schott and forensic examiner Deskins) as well as to serve as the basis for expert testimony in the plaintiff's case-in-chief. The X-ray evidence and the argument built from it, finally, would have played a role in mitigation, as well as exculpation. Without it, jurors convicted Phillips of the second-most-serious crime charged; with it, even if he were not acquitted, perhaps they would have found him guilty only of a lesser included offense.

We evaluate the argument just outlined through two discrete parts of the Brady test. The first, which asks whether the X-ray was "favorable" "asks only which party the evidence favors." Clark v. Warden, 934 F.3d 483, 492 (6th Cir. 2019). Evidence providing an alternative explanation of how the decedent died or undermining the prosecution's theory of the case is favorable. … Moreover, because the favorability inquiry "asks only which party the evidence favors," Clark, 934 F.3d at 492 (emphasis added), credibility inquiries are not appropriate at this stage. Ibid. Normally, the question of credibility arises in the context of witness testimony. See, e.g., ibid. In our case, however, we have an X-ray that does not 'speak for itself,' particularly to those not trained to read one. For this exact reason, the district court had originally ordered an evidentiary hearing, observing that "the Court lacks a medical degree and cannot on its own determine what [the X-ray] shows." As we have seen, at that evidentiary hearing, the Commonwealth produced Dr. Schott as its sole witness, while Phillips produced forensic scientist Dehus. In making its determinations, the district court appears to have equated the value of the X-ray with the credibility of the expert witnesses. The court then used the question of which direction the X-ray was taken from (about which the court questioned both witnesses) as a touchstone for which witness was more credible.

Dehus's mistake as to the direction of the X-ray is damaging, as is his insistent doubling-down in the face of medical testimony to the contrary. However, when stripped of matters we cannot now take into account (such as the wadding issue, or the lack of marring of Glodo's face) and of matters on which Dr. Schott rebutted him (his over-confident statements regarding the direction of the shot, given her points about inability to trace this in a 2-D image), the heart of his testimony is: 1) There were 170 to 220 pieces of shot in the shell that was fired. 2) Had the shell been fired straight-on at the distance indicated, these should all have wound up in Glodo's skull. 3) The X-ray does not show anywhere near that number in Glodo's skull. 4) Therefore, the shot must have been fired at an angle. It is upon the third of these points—whether or not the X-ray shows what one would expect to see—that the question of favorability turns. Here, it seems noteworthy that Dr. Schott did not rebut this point directly. She testified that she could not tell "from the x-ray how many pellets are … in the head[.]" But elsewhere, she stated that X-rays of this type "give me a general idea of the distribution of the pellets or of the bullet."

Given how X-rays work, there would be a significant and visible difference between an X-ray of a skull with a full load of 170-220 metal pellets in it (or nearly that many) and one with far fewer than that. That observation was the heart of Dehus's argument: "It's not possible from the x-ray to individually count each pellet. But it doesn't seem --it doesn't appear to be anywhere near that number." As Dr. Schott herself admitted (indeed, argued), the X-ray is a two-dimensional photograph. Therefore, Dehus's error as to whether it had been taken from the front or the back does not seem to affect the basic ability to argue such points as the density of shot shown. Even with Dehus's skill somewhat in question, we conclude that we cannot dismiss his basic conclusion that there was not nearly enough shot in the skull to support the Commonwealth's theory. Or to put it another way, the X-ray provides some support for Phillips's theory. We particularly cannot discount this conclusion in view of the Commonwealth's and Dr. Schott's inability to rebut it directly.

Phillips also developed—better on appeal than at the evidentiary hearing—a second, somewhat weaker argument: It is possible to use the X-ray to tell if the shot was fired at an angle or straight-on based solely on the distribution of the shot in the X-ray in terms of being on or off-center. This is subject to similar analysis as that above about the sheer quantity of shot. Dr. Schott demonstrated at the evidentiary hearing why the 2-D image could not be used to trace the directionality of any given pellet. But if the mass of the pellets is to one side, that would suggest that they were fired at an angle. As Phillips points out, the mistake over whether the X-ray was taken back-to-front or front-to-back would not matter for this inquiry, because either way, off-centeredness would remain. But Dehus's testimony at the evidentiary hearing was less well developed on whether this was so than in the case of the sheer-quantity argument above.

As to the district court's equating Dehus's credibility with the value of the X-ray, had Dehus's testimony itself been the Brady material, this kind of credibility judgement would have been inappropriate. Cf. Clark, 934 F.3d at 492. Here, however, we are in a somewhat anomalous situation: The Brady material is the X-ray itself, and as an inanimate object, it does not present a "credibility" question in the usual sense—but on the other hand, most judges cannot evaluate it without an expert intermediary. Thus, as compared to Clark, this situation poses a slightly different question: what is the role of the court in evaluating the credibility of the experts who will help it evaluate the meaning (rather than the credibility) of the Brady material?

On the one hand, courts perform a screening role constantly as to expert credibility, both before testimony is given at trial and in evidentiary hearings. Such a role, to some extent, seems necessary here: obviously, it would be a recipe for chaos (and injustice) if courts were obligated to accept as true any testimony, no matter how blatantly incorrect and self-serving, regarding the value of a supposed piece of Brady evidence. On the other hand, however, we have recognized "the Supreme Court's directive that 'the criminal trial … [be preserved] as the chosen forum for ascertaining the truth about criminal accusations.'" Clark, 934 F.3d at 492 (quoting Gumm v. Mitchell, 775 F.3d 345, 364 (6th Cir. 2014) (quoting United States v. Jernigan, 492 F.3d 1050, 1056-57 (9th Cir. 2007) (en banc) (quoting Kyles, 514 U.S. at 440) (brackets in original))). If the evidence suffices to create a battle of the experts, such a battle should be waged at trial, not simply before a judge in a post-hoc setting.

There are good reasons to think that the threshold for the favorability inquiry should be fairly low. Most of the key cases treat it only glancingly or even omit to discuss it. (We note, however, that this is likely due to the fact that the Brady material in question is usually either testimony or police notes that, if believed, are obviously favorable.) The physical evidence here obviously does not have an innate credibility problem, and nor was the testimony surrounding it blatantly self-serving or dishonest. To the contrary, despite an unfortunate misstep or two, it made out the "favorability," in the Brady sense, of the X-ray evidence to Phillips. In short, we hold that the evidence was favorable, and thus we move on to materiality.

C. The X-Ray Was Material

The materiality inquiry asks if there is a "'reasonable probability' of a different result. The question is not whether the defendant would more likely than not have received a different verdict with the evidence, but whether in its absence he received a fair trial, understood as a trial resulting in a verdict worthy of confidence." Kyles, 514 U.S. at 434 (emphasis added); … "A 'reasonable probability' of a different result" is said to have been demonstrated "when the government's evidentiary suppression 'undermines confidence in the outcome of the trial.'" Kyles, 514 U.S. at 434 (quoting Bagley, 473 U.S. at 678). We are warned, moreover, not to conflate materiality with the sufficiency-of-the-evidence inquiry. Kyles, 514 U.S. at 434-45; Bies v. Sheldon, 775 F.3d 386, 399 (6th Cir. 2014). In particular, "A defendant need not demonstrate that after discounting the inculpatory evidence in light of the undisclosed evidence, there would not have been enough left to convict." Bies, 775 F.3d at 399 (quoting Kyles, 514 U.S. at 434-45).

In assessing this part of the Brady analysis, it is important to note how equivocal much of the evidence at Phillips's trial was. For most of the two days, including from many of the prosecution witnesses, the jury heard at length about Glodo's anger and aggression that day. But Dr. Schott's testimony, and her explanation of the photographs of the back of Glodo's skull, appeared to show as a matter of uncontroverted fact that Glodo had been shot squarely from behind (and from a distance of over three feet away). These, combined with Phillips's inconsistent explanations on the day of the shooting—characterizing the shooting on the 911 call and in a police interview as both an accident and self-defense—were, on our reading of the trial transcript, the key turning points. Here was the indisputable (or so it seemed) evidence: Phillips had raised a shotgun, aimed it squarely at the back of Glodo's head when there was a distance of at least three feet between them, and pulled the trigger. No matter what had come before, that is not self-defense, and it is very hard to square with accident. As we have seen, the prosecution strongly emphasized this key point in both its opening and closing arguments.

Viewed in this light, it is easy to see how different the trial could have been had Phillips had the X-ray and expert(s)—firearms and, perhaps, medical—of his own. Instead of an equivocal case swayed by one piece of indisputable evidence, the jury would have seen an equivocal case plus a battle of the experts. Relying on the X-ray, Phillips could have produced an expert to testify that the physical evidence was consistent with his account of a glancing shot, fired in the course of a struggle. Phillips may also have called a medical examiner, so as to have covered some of the weaknesses that Dehus, alone, clearly created.

His counsel could also have used the X-ray for impeachment purposes. Dr. Schott testified at the trial that, based on the pellets she recovered from inside Glodo's head, "the direction [of the pellet's track] was back to front." Phillips's counsel could have impeached this conclusion on the basis of the X-ray. Moreover, Dr. Schott testified at the evidentiary hearing that, during the autopsy and after her technician had taken the X-ray, she recovered only a sample (19, out of an unknown total) of the pellets in the head so that these could be submitted to the police lab for examination. As we have seen, she strongly disclaimed the ability to track the direction of the shot on the basis of the X-ray, saying that it had "no bearing on the direction of the injury." Introducing the X-ray would have allowed the defense to cross-examine Dr. Schott as to why she was so confident that this representative recovery, conducted to obtain material for ballistic analysis, allowed her to reconstruct the path of the shot, while finding no use at all for the X-ray in this regard. Similarly, the X-ray could have been used to weaken Dr. Schott's testimony regarding the photograph, showing the central location of the wound: the argument presumably would be that though the entry wound was in the middle of the back of the head, the X-ray showed that the shot had been at an angle, not straight-on.

The Commonwealth also called a firearms expert at trial for the purpose of showing that the trigger and safety of the shotgun worked normally. Phillips outlines to us the line of questioning on cross-examination of this witness that the X-ray would have made possible:

> Apprised of the x-ray of Glodo's skull, counsel could have elicited compelling testimony from Deskins that Glodo had been struck at a significant angle. Counsel could have asked Deskins to count the number of pellets in the x-ray, and compare that number to the number in the shell. Counsel could have asked Deskins about the pattern of pellets in the x-ray, and compare that to the normal pattern of distribution of shotgun pellets. Finally, counsel could have asked the ultimate question—whether the x-ray indicated a direct shot or an angled shot.

Brief of Appellant at 34. Moreover, as Phillips now points out, the availability of these lines of cross-examination liberates the value of the X-ray (to some degree) from the credibility of Dehus: with the X-ray alone, such lines of cross-examination would have been possible. After all, it is our confidence in the trial, not our confidence in Dehus's testimony, that is the question before us today.

There is also the role this evidence might have played in mitigation. Brady evidence need only be material to either guilt or punishment. Strickler, 527 U.S. at 280. Phillips was charged with intentional murder, wanton murder, second-degree manslaughter, and reckless homicide. (See supra, p. 9 & nn.6-9.) The two lesser-included charges, below that of which Phillips was convicted, have self-defense elements.15 So if the jurors believed that the X-ray indicated the shot was at an angle, and agreed that such a fact indicated that there had been a struggle, then even if they were not convinced that this evidence was wholly exculpatory, they may well have decided that it showed imperfect self-defense and so "mitigated" down to a lesser charge. This would be "mitigation" in only the colloquial sense: really, the jury would be finding him innocent of one charge (wanton murder) and guilty of another, lesser charge.

But it is quite conceivable, alternatively, that they might have mitigated in the technically accurate sense. As Phillips argues:

> Evidence tending to prove that Phillips did not shoot Glodo directly in the back of the head, but rather at a significant angle, would have mitigated Phillips's conduct and likely convinced jurors to impose a more lenient sentence at or near the twenty-year mandatory minimum sentence for the crime of wanton murder.

This argument also undermines confidence in the sentencing decision. When the jury convicted Phillips of wanton murder, they likely found that the discharge was indeed accidental, but that raising a loaded gun and pointing it squarely at the back of a man's head was wanton behavior. For that reason, evidence indicating that the gun was not leveled at Glodo might have gone a long way to mitigate the penalty. A gun held in his general direction, and a nervous slip of the finger, would be very different than taking dead-aim at the back of Glodo's head. Thus, a jury that was provided with the X-ray might well have sentenced Phillips to fewer than the thirty years chosen by the jury.

We have not, we must note, been able to find a case in which Strickler's instruction that Brady is material to punishment as well as guilt, 527 U.S. at 280, has been applied outside the capital context. Were we ruling solely on this basis, that might give us more pause. However, as pointed out above, this is not the sole basis for our ruling; moreover, the Supreme Court is clear on this rule, even if it has not been broadly applied.

The state argues that notwithstanding the X-ray, the evidence showed that Phillips returned to Capps's house, where Glodo was to be found about four nights a week, with a loaded gun in his car; that he got out of his car in the parking lot and that shortly before the fatal shots were fired, he was using a loaded gun to fend off Glodo, and that the gun must have been at least relatively level since holes, arguably from some of the shot, were found in Glodo's windshield. Cf. Phillips, 2012 Ky. App. Unpub. This, the argument then goes, was enough for the jury to find him guilty of wanton murder, defined as "wantonly engaging in conduct which created a grave risk of danger to another and [shooting him] … under circumstances manifesting in extreme indifference to human life," regardless of the X-ray. The district court found this argument persuasive, concluding that: "Even assuming … that Dehus's opinion is correct and that Glodo was shot from behind at an angle and at a distance, that does not undermine confidence in the jury's finding that Phillips committed wanton murder."

The problem, however, is that this is precisely the form a sufficiency-of-the-evidence analysis takes, and the Supreme Court has told us explicitly that the Brady inquiry is not to be equated to such a test. The district court framed the question as whether it could reconcile the new evidence with jury's verdict, rather than whether, in light of the new evidence, it still had confidence in "the outcome of the trial." Kyles, 514 U.S. at 434. Compare the statement of the district court that "[e] vidence that Phillips shot Glodo from behind at a distance is not inconsistent with the jury's verdict," … It was error for the district court to rule on this basis.

The dissent ably presents the case for how a good prosecutor could still have obtained a conviction for wanton murder and a 30-year sentence. The proper legal question, however, is whether the X-ray is "material" in the sense that it undermines our confidence in the course of the trial, not that it requires a conclusion that the outcome would necessarily have been different, or even that it is 50.1% likely that there would have been a different outcome. If there is to be a battle of the documents from the record, the reader is invited to consider whether, as a defense attorney, you would rather go to the jury with only the dissent's photo (Dissent Appendix A), and accompanying testimony; or with the jury also having access to the X-ray (Appendix A to this opinion), and the best accompanying testimony, cross-examination, and argument that could be based on the X-ray ("Where are all the pellets?"). The answer would seem obvious. And as a prosecutor, would you feel that you still could obtain the same conviction and sentence, but you would prefer that the defense not have the X-ray? Again, the answer is clear.

The jury's verdict required it to find that Phillips was not privileged to act in self-protection, which is easier to find when there is a full-on shot to the back of the head. It is easier to find reasonable doubt on that point if Phillips's theory of a struggle and pushing Glodo away is buttressed by evidence that the shot was at a slanting angle. You may consider it a judgment call as to how much better it would be for the defense attorney, and how much harder for the prosecutor, but there is no gainsaying the direction of effect. It is our task to make the call of how much it takes to "undermine confidence." Under the facts and law, as shown above, our confidence is not destroyed but it is undermined.

As a final word, we also find convincing our recent case Carusone v. Warden, N. Cent. Corr. Inst., 966 F.3d 474 (6th Cir. 2020). Carusone had been in a fight over money with another man, whom he admittedly stabbed at least once in the arm and once in the chest; the other man died. Id. at 476. The state tried and convicted Carusone "exclusively on the theory—as the prosecution repeatedly emphasized in its closing argument at trial—that Carusone 'plunged [a] knife into the victim's heart.'" Id. at 475. Later, "Carusone discovered that the State had withheld five pages of the hospital's records (including the ER doctor's report and a chest x-ray) ... ." Id. at 476-77 (emphasis added). This Brady evidence showed that the hole in the side of the victim's heart was consistent with a needle used at the hospital to drain blood from around his heart ("pericardiocentesis"). Ibid. The victim had actually died of a heart attack induced by drugs, stress, and the fight with the Carusone. Id. at 477. In state post-conviction relief proceedings, Carusone produced somewhat better expert testimony than we have in this case, and the state court acknowledged that he had "plainly discredit[ed]" the theory that the state had relied upon at trial. Id. at 475 (brackets in original); see id. at 477. But the state appeals court applied Brady as though it were a sufficiency-of-the-evidence test: It upheld the conviction on an alternate theory that the struggle with Carusone had led, in part, to the fatal heart attack. Id. at 478-79. We granted Carusone a writ of habeas corpus, noting that "the test for materiality under Brady 'is not a sufficiency of evidence test'" but rather that "the suppressed evidence 'undermine[s] confidence in the verdict.'" Id. at 479 (quoting Kyles, 514 U.S. at 434, 435) (brackets in original).

The parallels to our case are plain. Kentucky relied "exclusively on the theory"—which "the prosecution repeatedly emphasized in its closing argument at trial"— that Phillips had shot Glodo "directly from his back into his head." Carusone, 966 F.3d at 474. Like Carusone, Phillips later discovered an X-ray that called this theory into serious question. And the Ohio courts in Carusone's case erroneously treated the Brady inquiry as a sufficiency-of-the-evidence inquiry, just as the district court did here. But that is not the correct inquiry. Kyles, 514 U.S. at 434. If the prosecution suppressed evidence, inadvertently or otherwise, that calls into serious question the theory of death in a murder trial—as it did here and in Carusone—we cannot simply hand-wave that away. We hold that there has been a material Brady violation.

We are reluctant to undo a jury verdict finding someone guilty of such a serious crime. That is particularly true in a situation like this one, where so much of the case turned on the jury's judgment as to what was in someone's mind and, to a lesser extent, on an understanding of local conditions and culture. But we are convinced that, had Phillips had the X-ray to rely on, the course of his trial would likely have been quite different. We are not, in short, "confiden[t] in the outcome of the [original] trial." Ibid.

II. CONCLUSION

For the foregoing reasons, we REVERSE the judgment of the court below and REMAND with instructions to grant a conditional writ of habeas corpus, ordering the state to either retry or release Phillips.

Questions:

1. Experts provide completely opposite views in this case. How so?
2. What pathological evidence appears to have been suppressed by the government, in violation of the Brady doctrine? And in what manner does that evidence seem favorable to a self-defense argument on behalf of the defendant.
3. The Court's opinion also includes a copy of the X-Ray and one Photographic Representation of the Gunshot wound to the back of the head. How does the prosecution characterize these images-as proving what?

Ferris v. City of Cadillac
United States Court of Appeals for the Sixth Circuit March 30, 2018, Filed
File Name: 18a0168n.06 No. 17-1866

BACKGROUND

A. Factual Background

1. The Death

On January 16, 2013, Plaintiff was babysitting his girlfriend's eighteen-month-old daughter, Kalla Fisher, when she purportedly fell down some basement stairs and struck her head on cement. Later that night, Jesse Fisher, Kalla's mother and Plaintiff's then-girlfriend, returned home, and she and Plaintiff checked on Kalla throughout the night. The next morning, Jesse noticed some slight swelling and bruising on the child, but took Kalla to daycare. The daycare provider later contacted Jesse to tell her that the bruising and swelling had worsened. Jesse picked up Kalla and took her to the emergency room. Medical records show that a CT-scan revealed no skeletal fractures or dislocations. As mandatory reporters, emergency room personnel reported the incident to the Grayling Police Department and Children's Protective Services ("CPS").

The next day, Kalla's pediatrician, Dr. Joanna Nigrelli, diagnosed Kalla with "a head contusion and the bruising or [ecchymosis] related to that contusion," but did not see a need to hospitalize her. (R. 197-2, Preliminary Exam., PageID # 5749.) Dr. Nigrelli reported to CPS that she did not think child abuse caused the injury. Based in part on Dr. Nigrelli's examination, CPS determined that the "injury appear[ed] to be an accident." (R. 197-2, Grayling Police Record, PageID # 5674.) CPS and Grayling Police both dismissed the complaint of possible child abuse.

On the night of February 15, 2013, Jesse and Kalla stayed overnight at Plaintiff's home. Jesse put Kalla to bed around 8:30 p.m. Plaintiff later explained that around 11:15 p.m. that evening, he had allowed Kalla and his seven-year-old daughter to join him in the room where he was watching television. And sometime around midnight, Kalla vomited. After cleaning it up, Plaintiff put Kalla and his daughter back to bed and went to bed himself. Jesse found Kalla unresponsive around 9:30 a.m. the next morning. Plaintiff performed CPR until emergency responders arrived, and Kalla was pronounced dead at Mercy Hospital at 10:13 a.m.

The Wexford County Medical Examiner, Dr. Fred Wreford, assigned responsibility for Kalla's autopsy to Defendant Sparrow Health System ("Sparrow"). Defendant Dr. Joyce de Jong, a forensic pathologist employed by Sparrow, was assigned the autopsy. Dr. de Jong performed the autopsy on Sunday, February 17, 2013, accompanied by two autopsy assistants and Cadillac Police Department Detective Todd Golnick. Detective Golnick spoke with Dr. de Jong before the autopsy and told her what he knew about the situation, the history of Jesse and Plaintiff's relationship, and who had been around Kalla prior to her death.

Dr. de Jong's Autopsy Report issued on April 26, 2013. Her findings revealed acute head trauma and multiple contusions and cutaneous injuries. The Autopsy Report notes that the skull was free of fractures and that the brain appeared swollen. In a section titled "Opinions," Dr. de Jong identified the cause of death as "Traumatic Head Injuries" and the manner of death as "Homicide." (R. 210-2, Autopsy Report, PageID # 9510.)

An addendum report was prepared by Defendant Dr. Rudolph J. Castellani, a pathologist, which summarized the results of a neuropathology consultation he conducted. The addendum details Dr. Castellani's "final neuropathologic diagnoses" as (1) acute head trauma, with bilateral subdural hemorrhage, patchy acute subarachnoid hemorrhage, bilateral intraretinal and optic nerve sheath hemorrhages, and bilateral optic nerve sheath hemorrhage; (2) cerebral edema and tonsillar herniation secondary to acute head trauma. Castellani concluded his addendum with the comment that "the findings indicate inflicted head trauma resulting in death." (R. 210-2, Castellani Addendum, PageID # 9518.)

On May 1, 2013, Dr. Fred Wreford issued Kalla's death certificate, which relied exclusively on Doctors de Jong and Castellani's opinions. The death certificate specifies the causes of her death as cerebellar herniation occurring minutes before death, bilateral subdural hematomas and multiple blunt force traumas occurring hours before death, and child abuse occurring weeks before death. The certificate further identifies the manner of death as homicide.

...

3. The Charges

Upon learning that Dr. Schmidt concurred with Doctors de Jong and Castellani about the cause of Kalla's death, Prosecutor Badovinac decided to charge Plaintiff with open murder. At the Preliminary Examination, Dr. de Jong was asked how Kalla died. She replied, "I can say that she sustained blunt force injuries and lethal head injuries." (R. 210-16, Preliminary Exam., PageID # 9901.) She testified that the cause of death was "traumatic head injuries" that were "recent injuries," occurring minutes to hours before Kalla's death. (Id. at PageID # 9898-99.) Dr. de Jong further testified that she "believe[d] [Kalla's] injuries were intentionally inflicted." (Id. at PageID # 9962.) Asked whether a seven-year-old (presumably referring to Plaintiff's daughter) could have caused the lethal impact, Dr. de Jong responded, "I don't believe that" (Id. at PageID # 9964), and, when asked again, "I don't believe so." (Id. at PageID # 9965.)

Dr. Ormsby also testified at the Preliminary Examination. He opined that Kalla "had a major trauma when she fell down the stairs in January … [and that] what actually happened on that night [a month later] is she had a massive intracranial bleed that led to an increased intracranial pressure that led to her demise." (Id. at PageID # 9985.) At the close of the Preliminary Examination, the court determined that there was enough evidence to bind the case over for trial. (Id. at PageID # 10006.)

Plaintiff's counsel in the criminal case requested that Dr. Ljubisa J. Dragovic, a forensic pathologist and neuropathologist, review the Autopsy Report and other medical records. Dr. Dragovic found that "the toddler Kalla Fisher died of bronchopneumonia complicating her head trauma.4 What remains unclear is when the toddler sustained the head trauma." (R. 210-14,

Dragovic Report, PageID # 9848.) Dr. Dragovic saw no evidence of an intentional infliction of head trauma in the materials he reviewed. He described Dr. Castellani's review as an "incomplete professional diagnostic work-up." (Id. at PageID # 9849.) Wexford County Prosecutors met with Dr. Dragovic in September 2014, and Dr. Dragovic advised them that, in his opinion, an earlier injury caused Kalla's death. At the conclusion of this meeting, the prosecutors decided to reevaluate whether to pursue the criminal case.

Prosecutor Badovinac forwarded Dr. Dragovic's report to Dr. Schmidt and asked for his comments. Dr. Schmidt reiterated his opinion that "this girl was killed. Period. She was beaten to death … . The subdural hemorrhage is textbook child abuse." (R. 210-5, Schmidt-Badovinac Email, PageID # 9540.) After examining the input they had received from all the doctors, prosecutors decided that, in light of "competing medical opinions," they were unlikely to be able to persuade a jury beyond a reasonable doubt of Plaintiff's guilt. (R. 200-9, Trautner Dep., PageID # 7772; R. 200-7, Badovinac Dep., PageID # 7703.) In November 2014, the Wexford County Prosecutor dismissed the charges with prejudice.

…

DISCUSSION

Standard of Review

We review a district court's grant of summary judgment de novo. Dixon v. Univ. of Toledo, 702 F.3d 269, 273 (6th Cir. 2012). "Summary judgment is appropriate if there is no genuine issue of material fact and the movant is entitled to judgment as a matter of law." Ondo v. City of Cleveland, 795 F.3d 597, 603 (6th Cir. 2015). "We review the evidence and draw all inferences in the light most favorable to the nonmoving party." Dixon, 702 F.3d at 273.

Analysis

"[G]overnment officials performing discretionary functions generally are shielded from liability for civil damages insofar as their conduct does not violate clearly established statutory or constitutional rights of which a reasonable person would have known." Harlow v. Fitzgerald, 457 U.S. 800, 818, 102 S. Ct. 2727, 73 L. Ed. 2d 396 (1982) (citations omitted). The doctrine of qualified immunity is based on the concern that "[w]hen officials are threatened with personal liability for acts taken pursuant to their official duties, they may well be induced to act with an excess of caution or otherwise to skew their decisions in ways that result in less than full fidelity to the objective and independent criteria that ought to guide their conduct." Forrester v. White, 484 U.S. 219, 223, 108 S. Ct. 538, 98 L. Ed. 2d 555 (1988).

Therefore, "[q]ualified immunity 'gives ample room for mistaken judgments' by protecting 'all but the plainly incompetent or those who knowingly violate the law.'" Johnson v. Moseley, 790 F.3d 649, 653 (6th Cir. 2015) (quoting Hunter v. Bryant, 502 U.S. 224, 229, 112 S. Ct. 534, 116 L. Ed. 2d 589 (1991) (per curiam)).

This Court follows a two-step inquiry to determine whether to apply qualified immunity. Martin v. City of Broadview Heights, 712 F.3d 951, 957 (6th Cir. 2013). "The first step is to determine if the facts alleged make out a violation of a constitutional right." Id. "The second is to ask if the right at issue was 'clearly established' when the event occurred such that a reasonable officer would have known that his conduct violated it." Id. "Plaintiff bears the burden of showing that defendants are not entitled to qualified immunity." Seales v. City of Detroit, No. 17-1096, 724 Fed. Appx. 356, at *2 (6th Cir. Jan. 31, 2018) (citing Chappell v. City of Cleveland, 585 F.3d 901, 907 (6th Cir. 2009)). At summary judgment, the question is whether the facts, when construed most favorably to Plaintiff, show that Defendants violated Plaintiff's clearly established constitutional rights.

This Court has held that "a forensic expert may be subject to suit under § 1983 for deliberately withholding the existence of exculpatory forensic evidence or fabricating forensic evidence." Moldowan v. City of Warren, 578 F.3d 351, 397 (6th Cir. 2009). We have also said that "[i]t is well established that a person's constitutional rights are violated when evidence is knowingly fabricated and a reasonable likelihood exists that the false evidence would have affected the decision of the jury." Gregory v. City of Louisville, 444 F.3d 725, 737 (6th Cir. 2006) (citing Stemler v. City of Florence, 126 F.3d 856, 872 (6th Cir. 1997)); ... Similarly, to succeed on a Fourth Amendment malicious prosecution claim, a plaintiff must prove that the defendant made "'deliberate or reckless falsehoods,' and 'mere negligence' will not create liability." Caminata v. County of Wexford, 664 F. App'x 496, 500 (6th Cir. 2016) (quoting Johnson, 790 F.3d at 655). We have observed that a criminal defendant's right to be free from prosecution based on fabricated evidence existed at least as early as 1992. Gregory, 444 F.3d at 744 n.8. Thus, the only question remaining for this Court on appeal is whether Plaintiff has introduced sufficient facts to show that Defendants violated his constitutional rights by knowingly or recklessly producing false evidence. We hold that Plaintiff has failed to do so.

Plaintiff asserts that Defendants fabricated their expert opinions on three key points: first, that Kalla's injuries were intentionally inflicted; second, that her injuries occurred within minutes to hours of her death; and third, that given the amount of force necessary, her injuries had to have been inflicted by an adult-sized person. To make his case, Plaintiff relies heavily on Gregory, a case involving a forensic examiner's inaccurate report regarding the number of hairs she recovered from a rape scene. 444 F.3d 725. Although the examiner testified that she had "found five hairs," and that these hairs matched the plaintiff's hair, a later deposition revealed that the examiner had actually recovered seven hairs, two of which were non-matching. Id. at 732, 734. Further, the plaintiff alleged that none of the hairs were in fact a match, and the plaintiff's expert testified that the examiner's findings were "far afield of what any reasonable forensic examiner would find from the evidence[.]" Id. at 744 (emphasis added). The court held that summary judgment was inappropriate, finding that a reasonable jury could conclude that the examiner "deliberately withheld the existence of those two nonmatching hairs" and "fabricated her report." Id.

Plaintiff has latched onto the "far afield" language in Gregory and invokes it at every turn to argue that his expert's contrary opinions show that Defendants fabricated evidence. But, of course, mere assertion of a legal standard is not enough to create a triable issue of fact. ... Moreover, Gregory did not create a new "far afield" standard for fabrication; it simply recognized that whether evidence is "fabricated" will often turn on circumstantial evidence and some expert reports can be so grossly deficient that they raise a triable issue of fact as to whether the explanation for their deficiency is fabrication. That was precisely the situation in Gregory: the Court concluded that the defendant's report was so deficient that a reasonable jury could find that the defendant's conclusions were not just wrong but fabricated.

This case is different. In Gregory, "the allegedly false forensic report and testimony did not arise from a flawed investigation; rather, the examiner admitted knowing there were seven hairs, yet chose to report only five." Caminata, 664 F. App'x at 500 (citing Gregory, 444 F.3d at 744-45). In this case, there is no evidence to suggest that Defendants' reports contained deliberate omissions or reckless falsehoods.

Plaintiff's primary evidence supporting his claim is his own experts' opinions. He relies on the opinions of Doctors Ormsby and Dragovic in his criminal case, as well as expert reports from Doctors Janice Ophoven and Mark J. Shuman, who were retained for his civil case to argue that Defendants were reckless in concluding that Kalla's injuries were intentionally inflicted.

Dr. Ophoven identifies numerous errors in the analysis and conclusions of Defendants. In her expert report, she claims that Dr. de Jong:

> [F]ormulated her conclusions in this case without obtaining the medical records from the prior fall in Grayling, the photos of the bruising from the prior fall in Grayling, or the medical records from Kalla Fisher's treating pediatrician. It is expected that a medical examiner obtain all of these materials before rendering any opinions regarding the cause and manner of death and timing of injury.

(R. 197-7, Ophoven Report, PageID # 6677.) She further claims that Dr. Castellani should not have opined on the manner of death, but instead "should have performed the analysis he was contracted to perform and not reached opinions that were outside his training and experience." (Id.) And as for Dr. Schmidt, Dr. Ophoven asserts that for him to have reached conclusions "as to the cause and manner of death without reviewing the slides [(i.e., the slides that were prepared as part of the autopsy report)] is far afield of what any reasonable forensic pathologist would have done." (Id. at PageID # 6678.) In her opinion:

> [A] full review of the materials that were available at the time of the autopsy should have led to a conclusion that the manner of death was undetermined. To conclude that this was a homicide based upon the materials that were available and the investigation that was performed was far afield of what any reasonable forensic pathologist would conclude from the evidence.

(Id. at PageID # 6677.) However, Dr. Ophoven's report also notes that the medical findings do "raise concerns for head injury associated with abuse." (Id. at PageID # 6674.) And her report recognizes that "the case has findings raising concerns of a suspicious death," though she ultimately concludes that "in the final analysis there is insufficient evidence available to determine what happened to the little girl." (Id. at PageID # 6676.) Similarly, Dr. Shuman concluded that the manner of death could not be determined, but he also found that Kalla's head injury "could have occurred accidentally or could have been inflicted by an adult or child." (R. 210-12, Shuman Report, PageID # 9788.) Significantly, no expert retained in this case has offered the opinion that Kalla's death was, in fact, an accident.

Plaintiff argues that his experts' opinions show that Defendants could not have concluded from the available medical evidence that Kalla's injuries were intentionally inflicted. Simply put, because Doctors Ophoven and Shuman concluded that the injuries could have been caused accidentally, Plaintiff argues that Defendants' contrary conclusions raise the specter of fabrication. (Brief for Appellant at 41 ("Defendants' unequivocal opinions/conclusions that Kalla's injuries were intentionally inflicted were far afield of what reasonable medical examiners/forensic pathologists could conclude.").) But this is not sufficient to show fabrication. The question for this court is not whether a self-interested litigant can find an expert to say the defendants got it wrong, but whether the evidence—including any expert opinions—creates a genuine issue of material fact that Defendants fabricated that evidence. And "[a]lthough [Plaintiff's expert evidence] impugns the quality of [Defendants' investigation, it is insufficient to establish knowing fabrication or deliberate or reckless falsehoods, rather than mere negligence." Caminata, 664 F. App'x at 501.

Dr. de Jong testified that the pattern of bruises seen on Kalla created a suspicion of child abuse and that the absence of a clear explanation for the injuries also supported that conclusion. She also testified that the severity of the injuries convinced her that the death was due to inflicted trauma. She explained that, as a forensic pathologist, she was expected to consider the factual information she was given and determine whether the explanation fit the medical findings. And even Plaintiff's expert, Dr. Shuman, agreed that a forensic pathologist has to interpret findings "in a context of what was going on with the child." (R. 205-15, Shuman Dep., PageID # 8826.) Nonetheless, although Plaintiff's experts acknowledge that the medical evidence could support a conclusion of intentional injury, they believe that it was inappropriate for Dr. de Jong to have reached such a conclusion. Plaintiff's expert evidence may indeed suggest that Dr. de Jong's conclusion was premature, but Plaintiff has introduced no evidence to suggest that this error (if indeed there was one) was intentional or recklessly made. Further, since there was nothing obviously deficient about Dr. de Jong's conclusions, we cannot say that Dr. Castellani or Dr. Schmidt fabricated evidence insofar as they relied on those conclusions or drew parallel ones.

Dr. de Jong testified that the pattern of bruises seen on Kalla created a suspicion of child abuse and that the absence of a clear explanation for the injuries also supported that conclusion. She also testified that the severity of the injuries convinced her that the death was due to inflicted trauma.

She explained that, as a forensic pathologist, she was expected to consider the factual information she was given and determine whether the explanation fit the medical findings. And even Plaintiff's expert, Dr. Shuman, agreed that a forensic pathologist has to interpret findings "in a context of what was going on with the child." (R. 205-15, Shuman Dep., PageID # 8826.) Nonetheless, although Plaintiff's experts acknowledge that the medical evidence could support a conclusion of intentional injury, they believe that it was inappropriate for Dr. de Jong to have reached such a conclusion. Plaintiff's expert evidence may indeed suggest that Dr. de Jong's conclusion was premature, but Plaintiff has introduced no evidence to suggest that this error (if indeed there was one) was intentional or recklessly made. Further, since there was nothing obviously deficient about Dr. de Jong's conclusions, we cannot say that Dr. Castellani or Dr. Schmidt fabricated evidence insofar as they relied on those conclusions or drew parallel ones.

Q. Okay. Your opinion is that the trauma that caused this child's death occurred somewhere … around 12 or 1 a.m. and when she's found unresponsive at 9 in the morning. Correct?
A. I think that's the most likely but it could have been earlier.
Q. It could have been earlier; how much earlier?
A. That's uncertain. Hours. It could be hours earlier.
Q. Would you agree with me that it's your opinion that it couldn't have been more than 24 hours?
A. I don't believe it was more than 24 hours, no.
Q. Is that based upon history or is that based upon pathological findings?
A. Based upon the literature. It's based upon some of it with the history.
Q. In other words, if I, if I ask you to look at this case pathologically with zero history. So the body comes in. You have Kalla Fisher's body. You do the Autopsy Report. You do the autopsy. You do all your slides and your examination and your radiological studies and the histopathology and the brain—you do everything the same but you have no history. Would you be able to time the brain bleed based upon pathological findings in this case?
A. If you're working in a vacuum like that, it, it's probably within24 hours or so, 24, maybe 48?

(R. 197-5, de Jong Dep., PageID # 6402-03 (emphasis added).) Dr. de Jong made clear that when she drew her conclusions, she was not looking solely at the medical data before her, but interpreting that data in the context of her background understanding of the circumstances leading up to Kalla's death. That she used this information to draw a conclusion different from Plaintiff's experts—a conclusion that none of Defendants' experts suggest is foreclosed by the medical evidence—does not suggest fabrication.

Finally, as for whether Kalla's injuries had to have been inflicted by an adult, Plaintiff contends that Defendants "were not qualified to give testimony regarding the amount of force necessary to cause Kalla's injuries." (Brief for Appellant at 38.) He therefore concludes that "there can be no question that any conclusions by the Defendants[] offered to the police and the Prosecutor's Office that Kalla's injuries could only have been inflicted by an adult were made with reckless disregard and were far afield of what a reasonable medical examiner/forensic pathologist could have reasonably concluded given the facts of this case." (Brief for Appellant at 39.) But even if Defendants were not qualified to offer their opinions on this score, their doing so does not suggest that they fabricated those opinions. Indeed, Dr. de Jong explained that she said that Kalla was probably, but not certainly, killed by an adult, and that all the literature supported her belief. And Dr. Castellani testified that it was clear to him that a significant amount of force was used to cause Kalla's injuries and that he was not familiar with any report where injuries of that magnitude had been caused by a child.7 Plaintiff's expert, by contrast, concluded that Kalla's injuries could have been inflicted by either an adult or a child and that there was no scientific way to determine which—child or adult—caused the injuries. This contrary conclusion evinces nothing more than a professional disagreement over what conclusions can properly be drawn from the medical evidence; what the evidence does not do is support the assertion that Defendants fabricated evidence.

In sum, the record supports that Dr. de Jong attempted to conduct a thorough autopsy. "Whether [s]he succeeded is immaterial [to] establishing a Fourth Amendment violation." Caminata, 664 F. App'x at 501. Viewed in the light most favorable to Plaintiff, the record at best reveals that Defendants were negligent in their investigation, "and such a conclusion does not establish a constitutional violation." Id.; see Ahlers v. Schebil, 188 F.3d 365, 373-74 (6th Cir. 1999) ("At best, however, the investigation's lack of thoroughness might support an inference of negligence, but it does not demonstrate knowing or intentional behavior designed to violate [the plaintiff's] constitutional rights."); see also Seigel v. City of Germantown, 25 F. App'x 249, 250 (6th Cir. 2001) ("The facts in this case supported at most a finding of incompetent or negligent investigation, which is insufficient to establish a constitutional violation."). Defendants are thus entitled to qualified immunity on Plaintiff's claims for fabrication of evidence and malicious prosecution.

CONCLUSION

For the reasons set forth above, we AFFIRM the district court's grant of summary judgment in favor of Defendants.

Questions:

1. Whether a child dies as a result of abuse is never an easy question. In most cases, the manner of death is proven by a pattern of long-term behavior and long term physical consequences. In this case, the injuries are typed as "acute." If the pathological exams and autopsy, show acuteness, is this more favorable to the defendant?
2. Despite this finding by the assigned forensic pathologist, the police and prosecutor did not proceed with criminal charges. Part of the reason for this related to contrary expert reports and questions of whether the evidence supported the intentionality required for homicide. What does the Court say about the decision?
3. As a result, the defendant sued the prosecution's pathologists and doctors who deduced that child abuse caused the death of the child. The suit was based on civil rights' grounds. Prove the following True or False: That defendant cannot win that suit unless and until he proved Negligence.

10 Forensic Psychiatry and Psychology

10.1 INTRODUCTION

The impact of behavioral science from the psychological realm and the more medically and neurologically based evidence rooted in psychiatry cannot be fully catalogued. Throughout the justice system, from pleas of incompetence to insanity, from judgments concerning extenuation and mitigating circumstances, and from the evaluation of various defenses posed by defendants that excuse or at least mitigate their culpability, you will encounter the fields of forensic psychiatry and psychology (see Figure 10.1).

One of this nation's greatest hallmarks is the belief that those afflicted with demonstrable mental disease and defect, which is directly correlated to the criminality itself, should not be held accountable. In the certifiably insane, in those suffering from congenital and neurological deformity, the element of mens rea cannot take hold as it does in other offenders. A thinking being, a free willing and free choosing moral agent, is less likely under these conditions to cause harm. A hallmark of Western jurisprudence is its insistence on culpability being tied to mental acuity and competency. To be sure, the public tends to the skeptical in these matters given the many charades the system has witnessed over the last century. That skepticism has evolved into a legislative effort to restrict or even eliminate the defense and replace with alternatives such as Guilty but Mentally Ill.

In 2020, the U.S. Supreme Court ruled the State of Kansas could do just that. Despite this, even the most hardened prosecutors know an insane person, in some cases by casual observation and in others by report and assessment that strike at the very heart of what it means to be a rational being. Some defendants do qualify. Even in the Middle Ages, official authorities hesitated in the condemnation and execution of those who appeared mentally disturbed. Known as the "wild beast" theory, the justice system had empathy for those whose mental faculties were clearly in severe disorder and whose awareness and relationship with reality was genuinely suspect. Foaming at the mouth, speaking with spirits and suffering from extraordinary delusions occurred then as it does now, and a short visit to any state mental hospital confirms the tragedy of mental illness and, even more importantly, its substantiality.

The declaration that an accused suffers from a form of insanity is often associated with legal defense strategy. Stereotypically, the public believes that any defendant laden with a mental disease or defect can avail themselves of this way of getting out of being responsible, but the reality of insanity pleas and adjudications is a very different matter. First, the claim rarely sees the light of day in a courtroom due to its stringent and cumbersome requirements. Less than 1% of American criminals file the pleas and notice of insanity and even fewer succeed in the argument, although this view is often debated with slighter higher numbers of successful pleas. "While rates vary from state to state, on average less than one defendant in 100—0.85 percent—actually raises the insanity defense nationwide."

Next, questions of legal insanity vastly contrast with the medico-psychiatric perspective. For example, a finding of psychiatric conditions in no way guarantees a correlation and connection to a finding of legal insanity. In this way, there really are two forms of insanity: factual-medical and legal. It is the latter category the defendant pleads. Only by the declaration of legal insanity can the defendant escape adjudication and eventual judgment. The distinction is not artificial since the law already understands that some portion of the criminal population will suffer from some type of psycho-psychiatric disorder. Most clinical studies on criminal lifestyles verify that criminal pathology is the product of many forces, including addictions, obsessions, alienations, antisocial

DOI: 10.4324/9781003267126-10

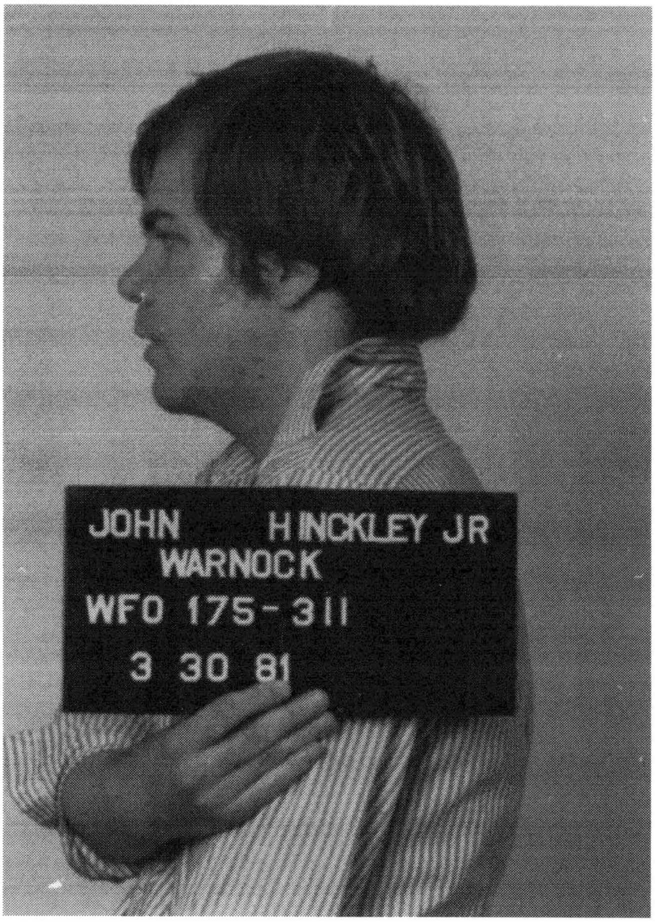

FIGURE 10.1 Mugshot of John Hinckley, Jr., by the Washington Field Office (WFO) on March 30, 1981. Hinckley attempted to assassinate President Ronald Reagan at the Washington Hilton Hotel earlier that day. A jury found Hinckley not guilty by reason of insanity on June 21, 1982. See https://www.fbi.gov/history/artifact-of-the-month/march-2021-limousine-piece-from-reagan-assassination-attempt for details.

Source: FBI Media at https://multimedia.fbi.gov/?q=insanity&perpage=50&page=1&searchType=image.

behaviors and troubled family life. Your run of the mill criminal can always find a psychiatrist who can diagnose some disorder. To allow these general conditions to be an absolute defense in criminal cases would generate not only enormous controversy but also injustice. Legal insanity, therefore, is very narrowly defined and construed. The finding accepts the reality of a mental disease or defect then moves to an intimate correlation that makes it substantially difficult for that defendant to know right from wrong, or to understand or appreciate the criminality of the act in question, or to engage in criminal conduct because of the disorder itself. As a result, legal insanity causes a blindness of understanding in the criminal agent, an inability to know the right and wrong of the conduct chosen or difficulty weighing and evaluating the moral, ethical and legal dimensions of criminal conduct. The insanity defense demands that the defendant respond to two primary questions:

1. Does the accused understand the act to be criminal (good or bad/right from wrong)?
2. Does the offender suffer from some mental disease or defect that directly correlates to this inability to appreciate the fundamental illegality engaged in?

Read the story about a case which involved a defense that ADHD, ADD, ASD and OCD prevented the defendant from having the level of intent necessary for conviction at https://www.justice.gov/usao-sdfl/pr/man-who-cyberstalked-and-threatened-families-parkland-victims-sentenced-prison.

10.2 THE THREE TESTS OF INSANITY

Before exploring how, when and in what way forensic psychiatry and psychology are employed in our legal system and narrowly applied to the adjudicative phase of the processing, we need to examine the three basic tests that have been employed since the early 19th century to the present. All three tests are still utilized, but it will depend on the jurisdiction, for some use M'Naghten and others the ALI test of substantial appreciation or the Durham product rule, which seeks to correlate crime to a mental disease or defect. At the far end of the legal insanity spectrum (even more elastic than substantial appreciation) is the concept of "irresistible impulse," which signifies a free actor who cannot control his or her actions. Few states accept this line of reasoning, and cases which succeed are rare to be sure.

DANIEL M'NAGHTEN'S CASE.
May 26, June 19, 1843

The prisoner had been indicted for that he, on the 20th day of January 1843, at the parish of Saint Martin in the Fields, in the county of Middlesex, and within the jurisdiction of the Central Criminal Court, in and upon one Edward Drummond, feloniously, wilfully, and of his malice aforethought, did make an assault; and that the said Daniel M'Naghten, a certain pistol of the value of 20s., loaded and charged with gunpowder and a leaden bullet (which pistol he in his right hand had and held), to, against and upon the said Edward Drummond, feloniously, wilfully, and of his malice aforethought, did shoot and discharge; and that the said Daniel M'Naghten, with the leaden bullet aforesaid, out of the pistol aforesaid, by force of the gunpowder, etc., the said Edward Drummond, in and upon the back of him the said Edward Drummond, feloniously, etc. did strike, penetrate and wound, giving to the said Edward Drummond, in and upon the back of the said Edward Drummond, one mortal wound, etc., of which mortal wound the said E. Drummond languished until the 25th of April and then died; and that by the means aforesaid, lie the prisoner did kill and murder the said Edward Drummond. The prisoner pleaded Not guilty.

Evidence having been given of the fact of the shooting of Mr .Drummond, and of his death in consequence thereof, witnesses were called on the part of the prisoner, to prove that he was not, at the time of committing the act, in a sound state of mind. The medical evidence was in substance this: That persons of otherwise sound mind, might be affected by morbid delusions: that the prisoner was in that condition: that a person so labouring under a morbid delusion, might have a moral perception of right and wrong, but that in the case of the prisoner it was a delusion which carried him away beyond the power of his own control, and left him no such perception; and that he was not capable of exercising any control over acts which had connexion with his delusion: that it was of the nature of the disease with which the prisoner was affected, to go on gradually until it had reached a climax, when it burst forth with irresistible intensity: that a man might go on for years quietly, though at the same time under its influence, but would all at once break out into the most extravagant and violent paroxysms.

...

Lord Chief Justice Tindal (in his charge): -The question to be determined is, whether at the time the act in question was committed, the prisoner had or had not the use of his understanding, so as to know that he was doing a wrong or wicked act. If the jurors should be of opinion that the prisoner was not sensible, at the time he committed it, that he was violating the laws both of God and man, then he would be entitled to a verdict in his favour: but if, on the contrary, they were of opinion that when he committed the act he was in a sound state of mind, then their verdict must be against him.

Verdict, Not guilty, on the ground of insanity.

...

Your Lordships are pleased to inquire of us, secondly, "What are the proper questions to be submitted to the jury, where a person alleged to be afflicted with insane delusion respecting one or more particular subjects or persons, is charged with the commission of a crime (murder, for example), and insanity is set up as a defence?" And, thirdly, "In what terms ought the question to be left to the jury as to the prisoner's state of mind at the time when the act was committed?" And as these two questions appear to us to be more conveniently answered together, we have to submit our opinion to be. that the jurors ought to be told in all cases that every man is to be presumed to be sane, and to possess a sufficient degree of reason to be responsible for his crimes,until the contrary be proved to their satisfaction; and that to establish a defence on the ground of insanity, it must be clearly proved that, at the time of the committing of the act, the party accused was labouring under such a defect of reason, from disease of the mind, as not to know the nature and quality of the act he was doing; or, if he did know it, that he did not, know he was doing what was wrong. The mode of putting the latter part of the question to the jury on these occasions has generally been, whether the accused at the time of doing the act knew the difference between right and wrong: which mode, though rarely, if ever, leading to any mistake with the jury, is not, as we conceive, so accurate when put generally and in the abstract, as when put with reference to the party's knowledge of right and wrong in respect to the very act with which he is charged. If the question were to be put as to the knowledge of the accused solely and exclusively with reference to the law of the land, it might tend to confound the jury, by inducing them to believe that an actual knowledge of the law of the land was essential in order to lead to a conviction; whereas the law is administered upon the principle that every one must be taken conclusively to know it, without proof that he does know it. If the accused was conscious that the act was one which he ought not to do, and if that act was at the same time contrary to the law of the land, he is punishable; and the usual course therefore has been to leave the question to the jury, whether the party accused had a sufficient degree of reason to know that he was doing an act that was wrong: and this course we think is correct, accompanied with such observations and explanations as the circumstances of each particular case may require.

Questions:

1. Although an old case, from 1843, the influence of M'Naghten is still strong in nearly two dozens American states. Why do you think this be so? What about this test seems timeless and always applicable?
2. What does the court mean when it references the defendant as "carried him away beyond the power of his own control."
3. Prove this statement True or False: The Court concluded that the defendant lacked the necessary understanding to meet the requirement of mens rea.

A New York State decision from 1847, just as few years after the issuance of the English M'Naghten Rule, already shows the impact of the decision in an American jurisdiction.

FREEMAN v. PEOPLE
4 Denio 9 Supreme Court of Judicature of New York January, 1847, Decided

The prisoner was tried ... and found guilty of the crime of murder, upon which verdict sentence of death was pronounced. ...

When the prisoner was brought before the Court ... a plea that he was then insane was interposed by counsel on his behalf, which, being denied by the public prosecutor, a jury was impaneled to try the issue so joined. On the trial of this issue, various objections were made and exceptions taken by the prisoner's counsel, and the first question to be decided is, whether these exceptions can be re-examined on a writ of error.

The statute declares that "No insane person can be tried, sentenced to any punishment, or punished for any crime or offense, while he continues in that state." 2 R. S., 697, sec. 2. This, although new as a legislative enactment in this State, 3 Id., 832, was not introductory of a new rule, for it is in strict conformity with the common law on the subject. "If a man," says Sir William Blackstone, "in his sound memory commits a capital offense, and before arraignment for it he becomes mad, he ought not to be arraigned for it; because he is not able to plead to it with that advice and caution that he ought. And if, after he has pleaded, the prisoner becomes mad, he shall not be tried; for how can he make his defense? If, after he be tried and found guilty, he loses his senses before judgment, judgment shall not be pronounced; and if, after judgment, he becomes of non-sane memory, execution shall be stayed, for peradventure, says the humanity of the English law, had the prisoner been of sound memory, he might have alleged something in stay of judgment or execution. Indeed," it is added, "in the bloody reign of Henry the eighth, a statute was made which enacted, that if a person being compos mentis, should commit high treason, and after fall into madness, he might be tried in his absence, and should suffer death, as if he were of perfect memory. But this savage and inhuman law was repealed by the Statute of 1 and 2 Ph. and M., ch. 10. For, as is observed by Sir Edward Coke, 'the execution of an offender is for example, ut poena ad paucos, metus ad omnes perveneat; but so it is not when a madman is executed; but should be a miserable spectacle, both against law, and of extreme inhumanity and cruelty, and can be no example to others.'" 4 Bl. Com., 24. The true reason why an insane person should not be tried, is, that he is disabled by an act of God to make a just defense if he have one.

...

With these lights before us, the construction of the statute, which forbids the trial of any insane person, cannot be attended with much difficulty. A state of general insanity, the mental powers being wholly perverted or obliterated would, necessarily, preclude a trial; for a being in that deplorable condition can make no defense whatever. Not so, however, where the disease is partial, and confined to some subject, other than the imputed crime and the contemplated trial. A person in this condition may be fully competent to understand his situation in respect to the alleged offense, and to conduct his defense with discretion and reason. Of this the jury must judge; and they should be instructed that if such is found to be his condition, it will be their duty to pronounce him sane. In the case at bar the court professed to furnish a single criterion of sanity, that is, a capacity to distinguish between right and wrong. This, as a test of insanity, is by no means invariably correct; for while a person has a very just perception of the moral qualities of most actions, he may, at the same time, as to some one in particular, be absolutely insane and, consequently, as to this, be incapable of judging accurately between right and wrong. If the delusion extends to the alleged crime, or the contemplated trial, the party manifestly, is not in a fit condition to make his defense, however sound his mind may in other respects be. Still, the insanity of such a person being only partial, not general, a jury under a charge like that given in this case, might find the prisoner sane;

for in most respects he would be capable of distinguishing between right and wrong. Had the instruction been that the prisoner was to be deemed sane, if he had a knowledge of right and wrong in respect to the crime with which he stood charged, there would have been but little fear that the jury could be misled; for a person who justly apprehends the nature of a charge made against him, can hardly be supposed incapable of defending himself in regard to it in a rational way. At the same time it would be well to impress distinctly on the minds of jurors, that they are to gauge the mental capacity of the prisoner, in order to determine whether he is so far sane as to be competent in mind to make his defense, if he has one; for, unless his faculties are equal to that task, he is not in a fit condition to be put on his trial. For the purpose of such a question, the law regards a person thus disabled by disease, as non compos mentis, and he should be pronounced unhesitatingly, to be insane within the true intent and meaning of this statute.

Where insanity is interposed as a defense to an indictment for an alleged crime, the inquiry is always brought down to the single question of a capacity to distinguish between right and wrong at the time when the act was done. In such cases the jury should be instructed that "It must be clearly proved that, at the time of committing the act, the party accused was laboring under such a defect of reason from disease of the mind, as not to know the nature and quality of the act he was doing, or, if he did know it, that he did not know he was doing what was wrong. The mode of putting the latter part of the question to the jury on these occasions has generally been, whether the accused, at the time of doing the act, knew the difference between right and wrong; which mode, though rarely, if ever, leading to any mistake with the jury, is not deemed so accurate when put generally and in the abstract, as when put with reference to the party's knowledge of right and wrong in respect to the very act with which he is charged." 2 Greenl. Ev., sec. 373.

...

But to return to the trial of the preliminary question in the present case. The jury found, not as the issue required them to do, that the prisoner was or was not insane, but that he was "sufficiently sane in mind and memory to distinguish between right and wrong." This verdict was defective: it did not directly find anything, and certainly not the point in issue, but evaded it by an argumentative finding. At the utmost the jury only made an approach towards the point to be decided, but failed to reach it. Their should have been required to pass directly on the question of insanity, and should not have been allowed to evade it by an argumentative verdict of any sort. Such a finding as this would be objectionable in a civil proceeding, and in a criminal case should not be allowed. In the Matter of Morgan, a lunatic, 7 Paige 236.

...

I proceed to another class of questions made by the prisoner's counsel.

The verdict on the preliminary issue was rendered July 6. In the course of the trial, and shortly after the 15th of that month, several medical witnesses were sworn and examined on the part of the prisoner, with a view to establish his insanity the preceding March, when the alleged murder was perpetrated. One of these witnesses, Doctor Van Epps, had known the prisoner from his childhood, and had visited and examined him with a view to ascertain his mental condition, both before and after July 6. The others had never seen the prisoner until July 15; but they also had examined him, on and after that day, in order to be prepared to express an opinion on the question of his sanity or insanity.

That part of the bill of exceptions which states the questions made and exceptions taken, in regard to these witnesses, is, perhaps, liable to some misapprehension, and it may be that I have not rightly understood what was intended to be decided by the court. I have read this part of the bill of exceptions repeatedly, with an anxious desire to collect its true meaning; and although I would not affirm positively, that its meaning may not have been misapprehended, I still think no error has been fallen into in regard to the views of the court. As I understand the bill of exceptions, the court held that it was competent for these, or other medical witnesses, to express an opinion upon the question of the insanity of the prisoner at the time of the alleged murder, but that such opinion must be formed upon facts and circumstances which occurred, or observations made, before July 6, when the verdict on the preliminary issue was rendered, and could, in no degree, rest upon anything observed in the appearance, manner or condition of the prisoner since that time; and that the witnesses could not, with a view to fortify the conclusion of insanity at the time of the homicide, be allowed to express an opinion that he was insane at the trial, or had been at any time since July 6. Nor was it even allowable to say they had examined the prisoner, since that time, with a view to ascertain his mental condition. These restrictions were deemed proper by the court, as I gather from the bill of exceptions, on the ground that the verdict, on the preliminary issue, had conclusively established, for all purposes connected with this trial, the sanity of the prisoner at the time when that verdict was rendered.

I cannot adopt the suggestion, made on the argument, that July 6 may have been taken as a reasonable time by which to bound the inquiries made of these witnesses; on the contrary, I think it quite clear that the court regarded the preliminary verdict as decisive of the question of present insanity and, therefore, limited the witnesses to the time when that verdict was rendered. In giving reasons for his opinion that the prisoner was insane, Doctor Van Epps spoke of an interview with him since July 6, when he "was stopped by the court, who then remarked (an objection having been made by the counsel for the people) that the question of present sanity had been tried and a verdict rendered on the 6th of July instant, and that the question of the present sanity could not then be again retried;" that the question now was as to the sanity of the prisoner when the deed was done, the preceding March, "and that the evidence of insanity must be confined to facts before and at the time of committing the act, and up to the 6th of July instant, when the verdict of sanity was rendered." Doctor Hun, another of these witnesses, had first seen the prisoner July 15. The prisoner's counsel "proposed to prove by this witness that, in his opinion, the prisoner is and was insane at the time of the commission of the crime. This was objected to by the counsel for the people, on the ground that the verdict on the preliminary issue, rendered on the sixth day of July instant, was and is conclusive that the prisoner was sane on that day; and that the same cannot be contradicted by evidence." The court did not pass directly upon this offer and objection; although the ground stated by the counsel for the people is understood to have been precisely that which the court acted upon. This witness was asked if he had made a personal examination of the prisoner, since his arrival at the court, which was July 15, "with reference to the state of his mind." To this the counsel for the people objected, and the court refused to allow the witness to give an answer. He was then asked if it was his opinion, founded upon personal examination since July 6, that the prisoner was insane March 12, when the homicide was perpetrated. This was objected to by the counsel for the people, and the court sustained the objection. The witness was then asked his opinion, founded on such examination, as to the prisoner being insane at the time when the question was put. This was also objected to and excluded by the court. Doctor McNaughton was examined under like restrictions. "The court decided that the witness should not testify as to any examination made by him of the prisoner since the sixth day of July instant," and particularly "instructed the witness that he should, in any future testimony to be given by him upon the trial of this cause, exclude all knowledge or information gained by him of or about the prisoner by or from any personal examination made by him of said prisoner, since the said sixth day of July instant." These references to points decided and views expressed by the court, clearly show, in my opinion, that the court regarded the preliminary verdict as absolutely conclusive for all purposes, in this case, that the prisoner on and after the sixth of July was in a sane state.

The views of the court upon this part of the case were, in my opinion, clearly erroneous. In strictness the verdict on the preliminary issue was not before the court and jury on the trial of the issue of guilty or not guilty, nor was it, in any respect, material to such trial. But if it should be regarded as a fact in the case, of which the court and jury, while engaged in the trial of the main issue, might take notice, no such consequence as that deduced by the court would follow from it.

The only object of the preliminary trial was to ascertain the mental condition of the prisoner, in order to determine whether he should then be tried on the indictment. This, I repeat, was the only object of that trial, and the result at which the first jury arrived could have no possible bearing or just influence upon the trial of the main issue. The indictment was not to be tried piece meal, but at one time, and by a single jury. If, therefore, the opinion sought to be obtained from these medical witnesses, was otherwise competent and proper, and that seems to have been conceded, it is perfectly clear that the preliminary verdict constituted no obstacle to its reception.

I am not about to inquire how far, or under what circumstances, the opinion of medical witnesses may be admissible on the question of insanity, although, in general, nothing is better settled than that such evidence is competent. 1 Phil. Ev., 290; Shelf. Lun., 62, 67-73; 1 Greenl. Ev., sec. 440. And I entertain no doubt that such a witness should be allowed to express an opinion in regard to the mental condition of a person alleged to be insane in the month of March, although the opinion may have been founded solely on an examination made in the succeeding July. In most cases, undoubtedly, the opinion would be more satisfactory and convincing, when based on observations made at or about the time to which the inquiry relates. But this is not decisive against the reception of such evidence, though founded on examinations made at a later period. The competency of the testimony is one question, and its effect another. The first is for the court and the latter for the jury. It will sometimes, undoubtedly, be found, and perhaps not unfrequently, that the mental malady is such that an examination would disclose beyond all peradventure to a skillful physician, what must have been the condition of the patient for months or years before. The lateness of the time when the examination was made, as well as the character of the malady, are certainly to be considered in determining the degree of consequence which should be given to the opinion of the witness, but unless the intervening time is much greater than from March to July, that can furnish no solid objection to the admissibility of the evidence. If I could, therefore, adopt the suggestion that July 6 was taken by the court as a reasonable limitation to inquiries of this description, I should still be unable to agree that the court had a right to impose any such restriction upon the witnesses. It was competent for such witnesses to state what their opinions were, whether founded on examinations before or during the trial; and these opinions might not only extend to the mental condition of the prisoner at the time when the homicide was perpetrated, but they might be brought down to the very time when the witness was speaking. The latter would be admissible, not because the present insanity of the prisoner would necessarily control the verdict, but because it tended to fortify the conclusion that insanity existed in the preceding March. But, although such are my views upon this part of the case, it is not supposed that the court excluded the evidence of the opinion of these witnesses in consequence of the lateness of the period when their examinations had been made. The evidence was shut out, as I understand the case, because the verdict on the preliminary issue was supposed to constitute an insuperable bar to its reception. This, as before said, was in my judgment erroneous. Upon the whole case, therefore, I think the judgment of the court below should be reversed, and a new trial ordered.

Whether the prisoner was or was not insane at the time of the homicide or the trial, is not a question before us on this bill of exceptions, and no opinion on that subject is intended to be expressed or intimated.

Judgment reversed.

Questions:

1. The Court's evaluation of legal insanity depends upon one, singular line of reasoning. Explain.
2. As to expert and lay witness testimony as to the sanity of the defendant, timing is a critical variable. Is it fair to argue that witnesses and the court played loose and fast on the timing of the crime, and the observations of the defendant either being close or remote in nature?
3. Under this ruling, what must insanity deprive the defendant of?

The United States Supreme Court recently ruled on Kansas's insanity law which has long been based on McNaghten.

Locate the case below using your favorite legal research resource, read the case and answer the questions that follow.

Jones v. Ryan, United States Court of Appeals for the Ninth Circuit, February 16, 2021, Argued and Submitted, San Francisco, California; June 28, 2021, Filed No. 18-99005

Questions:

1. Explain what it means to transition from moral incapacity and cognitive incapacity?
2. The Supreme Court seems to adopt a negative slant towards the McNaghten test or right and wrong. Point an example of this in the overall opinion.
3. Defendant in this case makes a curious and very creative legal argument-that the Kansas statute repeals and abolishes the insanity defense. Refute the defendant's assertion.

Durham v. United States
214 F.2d 862 United States Court of Appeals for the District of Columbia Circuit
March 19, 1954, Argued; July 1, 1954, Decided
No. 11859

Monte Durham was convicted of housebreaking, by the District Court sitting without a jury. The only defense asserted at the trial was that Durham was of unsound mind at the time of the offense. We are now urged to reverse the conviction (1) because the trial court did not correctly apply existing rules governing the burden of proof on the defense of insanity, and (2) because existing tests of criminal responsibility are obsolete and should be superseded.

I.

Durham has a long history of imprisonment and hospitalization. In 1945, at the age of 17, he was discharged from the Navy after a psychiatric examination had shown that he suffered 'from a profound personality disorder which renders him unfit for Naval service.' In 1947 he pleaded guilty to violating the National Motor Theft Act and was placed on probation for one to three years. He attempted suicide, was taken to Gallinger Hospital for observation, and was transferred to St. Elizabeth's Hospital, from which he was discharged after two months. In January of 1948, as a result of a conviction in the District of Columbia Municipal Court for passing bad checks, the District Court revoked his probation and he commenced service of his Motor Theft sentence. His conduct within the first few days in jail led to a lunacy inquiry in the Municipal Court where a jury found him to be of unsound mind. Upon commitment to St. Elizabeths, he was diagnosed as suffering from 'psychosis with psychopathic personality.' After 15 months of treatment, he was discharged in July 1949 as 'recovered' and was returned to jail to serve the balance of his sentence. In June 1950 he was conditionally released. He violated the conditions by leaving the District. When he learned of a warrant for his arrest as a parole violator, he fled to the 'South and Midwest obtaining money by passing a number of bad checks.' After he was found and returned to the District, the Parole Board referred him to the District Court for a lunacy inquisition, wherein a jury again found him to be of unsound mind. He was readmitted to St. Elizabeths in February 1951. This time the diagnosis was 'without mental disorder, psychopathic personality.' He was discharged for the third time in May 1951. The housebreaking which is the subject of the present appeal took place two months later, on July 13, 1951.

According to his mother and the psychiatrist who examined him in September 1951, he suffered from hallucinations immediately after his May 1951 discharge from St. Elizabeths. Following the present indictment, in October 1951, he was adjudged of unsound mind in proceedings under Sec. 4244 of Title 18 U.S.C., upon the affidavits of two psychiatrists that he suffered from 'psychosis with psychopathic personality.' He was committed to St. Elizabeths for the fourth time and given subshock insulin therapy. This commitment lasted 6 month --until February 1953 --he was released to the custody of the District Jail on the certificate of Dr. Silk, Acting Superintendent of St. Elizabeths, that he was 'mentally competent to stand trial and * * * able to consult with counsel to properly assist in his own defense.'

He was thereupon brought before the court on the charge involved here. The prosecutor told the court:

'So I take this attitude, in view of the fact that he has been over there (St. Elizabeths) a couple of times and these cases that were charged against him were dropped, I don't think I should take the responsibility of dropping these cases against him; then Saint Elizabeths would let him out on the street, and if that man committed a murder next week then it is my responsibility. So we decided to go to trial on one case, that is the case where we found him right in the house, and let him bring in the defense, if he wants to, of unsound mine at the time the crime was committed, and then Your Honor will find him on that, and in your decision send him back to Saint Elizabeths Hospital, and then if they let him out on the street it is their responsibility.'

Shortly thereafter, when the question arose whether Durham could be considered competent to stand trial merely on the basis of Dr. Silk's ex parte statement, the court said to defense counsel:

'Im am going to ask you this, Mr. Ahern: I have taken the position that if once a person has been found of unsound mind after a lunacy hearing, an ex parte certificate of the superintendent of Saint Elizabeths is not sufficient to set aside that finding and I have held another lunacy hearing. That has been my custom. However, if you want to waive that you may do it, if you admit that he is now of sound mind.'

The court accepted counsel's waiver on behalf of Durham, although it had been informed by the prosecutor that a letter from Durham claimed need of further hospitalization, and by defense counsel that ' * * * the defendant does say that even today he thinks he does need hospitalization; he told me this morning.'

Upon being so informed, the court said, 'Of course, if I hold he is not mentally competent to stand trial I send him back to Saint Elizabeths Hospital and they will send him back again in two or three months.' In this atmosphere Durham's trial commenced.

His conviction followed the trial court's rejection of the defense of insanity in these words:

'I don't think it has been established that the defendant was of unsound mind as of July 13, 1951, in the sense that he didn't know the difference between right and wrong or that even if he did, he was subject to an irresistible impulse by reason of the derangement of mind.

'While, of course, the burden of proof on the issue of mental capacity to commit a crime is upon the Government, just as it is on every other issue, nevertheless, the Court finds that there is not sufficient to contradict the usual presumption of (sic) the usual inference of sanity.

'There is no testimony concerning the mental state of the defendant as of July 13, 1951, and therefore the usual presumption of sanity governs.

'While if there was some testimony as to his mental state as of that date, the burden of proof would be on the Government to overcome it. There has been no such testimony, and the usual presumption of sanity prevails.

'Mr. Ahern, I think you have done very well by your client and defended him very ably, but I think under the circumstances there is nothing that anybody could have done.' [Emphasis supplied.]

We think this reflects error requiring reversal.

In Tatum v. United States we said, 'when lack of mental capacity is raised as a defense to a charge of crime, the law accepts the general experience of mankind and presumes that all people, including those accused of crime, are sane.' So long as this presumption prevails, the prosecution is not required to prove the defendant's sanity, like any other fact, must be proved as part of the prosecution's case beyond a reasonable doubt." Here it appears that the trial judge recognized this rule but failed to find 'some evidence.' We hold that the court erred and that the requirement of 'some evidence' was satisfied.

In Tatum we held that requirement satisfied by considerably less than is present here. Tatum claimed lack of memory concerning the critical events and three lay witnesses testified that he appeared to be in 'more or less of a trance,' or 'abnormal,' but two psychiatrists testified that he was of 'sound mind' both at the time of examination and at the time of the crime. Here, the psychiatric testimony was unequivocal that Durham was of unsound mind at the time of the crime. Dr. Gilbert, the only expert witness heard, 9. so stated at least four times. This crucial testimony is set out in the margin. 10. Intensive questioning by the court failed to produce any retraction of Dr. Gilbert's testimony that the 'period of insanity would have embraced the date July 13, 1951.' And though the prosecution sought unsuccessfully in its cross-and recross-examination of Dr. Gilbert to establish that Durham was a malingerer who feigned insanity whenever he was trapped for his misdeeds, it failed to present any expert testimony to support this theory. In addition to Dr. Gilbert's testimony, there was testimony by Durham's mother to the effect that in the interval between his discharge from St. Eliabeths in May 1951, and the crime 'he seemed afraid of people' and had urged her to put steel bars on his bedroom windows.

Apparently the trial judge regarded this psychiatric testimony as 'no testimony' on two grounds: (1) it did not adequately cover Durham's condition on July 13, 1951, the date of the offense; and (2) it was not directed to Durham's capacity to distinguish between right and wrong. We are unable to agree that for either of these reasons the psychiatric testimony could properly be considered 'no testimony.'

(1) Following Dr. Gilbert's testimony that the condition in which he found Durham on September 3, 1951 was progressive and did not 'arrive overnight,' Dr. Gilbert responded to a series of questions by the court:

Q. (Court). Then is it reasonable to assume that it is not possible to determine how far this state of unsound mind had progressed by July 13th? Isn't that so? A. (Dr. Gilbert). As to the seriousness of the symptoms as compared with them and the time I observed him, that's true, except that his travels were based, according to his statement to me, on certain of the symptoms and his leaving Washington, his giving up his job and work and leaving the work that he had tried to do.

'Q. But you can't tell, can you, how far those symptoms had progressed and become worse by the 13th of July? A. No, now how far they were, that is correct.' (Emphasis supplied.)

Thereafter, when the prosecutor on recross asked Dr. Gilbert whether he would change his opinion concerning Durham's mental condition on July 13, 1951, if he knew that Durham had been released from St. Elizabeth's just two months before as being of sound mind, the court interrupted to say: 'Just a minute. The Doctor testified in answer to my question that he doesn't know and he can't express a definite opinion as to his mental condition on the 13th of July:' This, we think, overlooks the witness' unequivocal testimony on direct and cross-examination, 11. and misconceives what he had said in response to questioning by the court, namely, that certain symptoms of mental disorder antedated the crime, although it was impossible to say how far they had progressed.

Moreover, any conclusion that there was 'no testimony' regarding Durham's mental condition at the time of the crime disregards the testimony of his mother. Her account of his behavior after his discharge from St. Elizabeths in May 1951, was directly pertinent to the issue of his sanity at the time of the crime.

(2) On re-direct examination, Dr. Gilbert was asked whether he would say that Durham 'knew the difference between right and wrong on July 13, 1951; that is, his ability to distinguish between

what was right and what was wrong.' He replied: 'As I have stated before, if the question of the right and wrong were propounded to him he could give you the right answer.' Then the court interrupted to ask:

'The Court. No, I don't think that is the question, Doctor --not whether he could give a right answer to a question, but whether he, himself, knew the difference between right and wrong in connection with governing his own actions. * * * If you are unable to answer, why, you can say so; I mean, if you are unable to form an opinion.

'The Witness. I can only answer this way: That I can't tell how much the abnormal thinking and the abnormal experiences in the form of hallucinations and delusions --delusions of persecution --had to do with his anti-social behavior.

'I don't know how anyone can answer that question categorically, except as one's experience leads him to know that most mental cases can give you a categorical answer of right and wrong, but what influence these symptoms have on abnormal behavior or anti-social behavior --

'The Court. Well, your answer is that you are unable to form an opinion, is that it?

'The Witness. I would say that that is essentially true, for the reasons that I have given.'

Later, when defense counsel sought elaboration from Dr. Gilbert on his answers relating to the 'right and wrong' test, the court cut off the questioning with the admonition that 'you have answered the question, Doctor.'

The inability of the expert to give categorical assurance that Durham was unable to distinguish between right and wrong did not destroy the effect of his previous testimony that the period of Durham's 'insanity' embraced July 13, 1951. It is plain from our decision in Tatum that this previous testimony was adequate to prevent the presumption of sanity from becoming conclusive and to place the burden of proving sanity upon the Government. None of the testimony before the court in Tatum was couched in terms of 'right and wrong.'

Finally, even assuming arguendo that the court, contrary to the plain meaning of its words, recognized that the prosecution had the burden of proving Durham's sanity, there would still be a fatal error. For once the issue of insanity is raised by the introduction of 'some evidence.' so that the presumption of sanity is no longer absolute, it is incumbent upon the trier of fact to weigh and consider 'the whole evidence, including that supplied by the presumption of sanity * * *' on the issue of 'the capacity in law of the accused to commit' the crime. 12. Here, manifestly, the court as the trier of fact did not and could not weigh 'the whole evidence,' for it found there was 'no testimony concerning the mental state' of Durham.

For the foregoing reasons, the judgment is reversed and the case is remanded for a new trial.

II.

It has been ably argued by counsel for Durham that the existing tests in the District of Columbia for determining criminal responsibility, i.e., the so-called right-wrong test supplemented by the irresistible impulse test, are not satisfactory criteria for determining criminal responsibility. We are argued to adopt a different test to be applied on the retrial of this case. This contention has behind it nearly a century of agitation for reform.

A. The right-wrong test, approved in this jurisdiction in 1882, 13. was the exclusive test of criminal responsibility in the District of Columbia until 1929 when we approved the irresistible impulse test as a supplementary test in Smith v. United States. 14. The right-wrong test has its roots in England. There, by the first quarter of the eighteenth century, an accused escaped punishment if he could not distinguish 'good and evil,' i.e., if he 'doth not know what he is doing, no more than * * * a wild beast.' 15. Later in the same century, the 'wild beast' test was abandoned and 'right and wrong' was substituted for 'good and evil.' 16. And toward the middle of the nineteenth century, the House of Lords in the famous M'Naghten case 17. restated what had become the accepted 'right-wrong' test 18. in a form which has since been followed, not only in England 19. but in most American jurisdictions 20. as an exclusive test of criminal responsibility:

'* * * the jurors ought to be told in all cases that every man is to presumed to be sane, and to possess a sufficient degree of reason to be responsible for his crimes, until the contrary be proved to their satisfaction; and that, to establish a defence on the ground of insanity, it must be clearly proved that, at the time of the committing of the act, the party accused was labouring under such a defect of reason, from disease of the mind, as not to know the nature and quality of the act he was doing, or, if he did know it, that he did not know he was doing what was wrong.' 21.

As early as 1838, Isaac Ray, one of the founders of the American Psychiatric Association, in his now classic Medical Jurisprudence of Insanity, called knowledge of right and wrong a 'fallacious' test of criminal responsibility. 22. This view has long since been substantiated by enormous developments in knowledge of mental life. 23. In 1928 Mr. Justice Cardozo said to the New York Academy of Medicine: 'Everyone concedes that the present (legal) definition of insanity has little relation to the truths of mental life.' 24

Medico-legal writers in large number, 25. The Report of the Royal Commission on Capital Punishment 1949-1953, 26. and The Preliminary Report by the Committee on Forensic Psychiatry of the Group for the Advancement of Psychiatry 27. present convincing evidence that the right-and-wrong test is 'based on an entirely obsolete and misleading conception of the nature of insanity.' 28. The science of psychiatry now recognizes that a man is an integrated personality and that reason, which is only one element in that personality, is not the sole determinant of his conduct. The right-wrong test, which considers knowledge or reason alone, is therefore and inadequate guide to mental responsibility for criminal behavior. As Professor Sheldon Glueck of the Harvard Law School points out in discussing the right-wrong tests, which he calls the knowledge tests:

'It is evident that the knowledge tests unscientifically abstract out of the mental make-up but one phase or element of mental life, the cognitive, which, in this era of dynamic psychology, is beginning to be regarded as not the most important factor in conduct and its disorders. In brief, these tests proceed upon the following questionable assumptions of an outworn era in psychiatry: (1) that lack of knowledge of the 'nature or quality' of an act (assuming the meaning of such terms to be clear), or incapacity to know right from wrong, is the sole or even the most important symptom of mental disorder; (2) that such knowledge is the sole instigator and guide of conduct, or at least the most important element therein, and consequently should be the sole criterion of responsibility when insanity is involved; and that the capacity of knowing right from wrong can be completely intact and functioning perfectly even though a defendant is otherwise demonstrably of disordered mind.' 29.

Nine years ago we said:

'The modern science of psychology * * * does not conceive that there is a separate little man in the top of one's head called reason whose function it is to guide another unruly little man called instinct, emotion, or impulse in the way he should go.' 30.

By its misleading emphasis on the cognitive, the right-wrong test requires court and jury to rely upon what is, scientifically speaking, inadequate, and most often, invalid 31. and irrelevant testimony in determining criminal responsibility. 32.

The fundamental objection to the right-wrong test, however, is not that criminal irresponsibility is made to rest upon an inadequate, invalid or indeterminable symptom or manifestation, but that it is made to rest upon any particular symptom. 33. In attempting to define insanity in terms of a symptom, the courts have assumed an impossible role, 34. not merely one for which they have no special competence. 35. As the Royal Commission emphasizes, it is dangerous 'to abstract particular mental faculties, and to lay it down that unless these particular faculties are destroyed or gravely impaired, an accused person, whatever the nature of his mental disease, must be held to be criminally responsible * * *.' 36. In this field of law as in others, the fact finder should be free to consider all information advanced by relevant scientific disciplines. 37.

Despite demands in the name of scientific advances, this court refused to alter the right-wrong test at the turn of the century. But in 1929, we reconsidered in response to 'the cry of scientific experts' and added the irresistible impulse test as a supplementary test for mining criminal responsibility. Without 'hesitation' we declared, in Smith v. United States, 'it to be the law of this District that, in cases where insanity is interposed as a defense, and the facts are sufficient to call for the application of the rule of irresistible impulse, the jury should be so charged.' 39. We said:

'* * * The modern doctrine is that the degree of insanity which will relieve the accused of the consequences of a criminal act must be such as to create in his mind an uncontrollable impulse to commit the offense charged. This impulse must be such as to override the reason and judgment and obliterate the sense of right and wrong to the extent that the accused is deprived of the power to choose between right and wrong. The mere ability to distinguish right from wrong is no longer the correct test either in civil or criminal cases, where the defense of insanity is interposed. The accepted rule in this day and age, with the great advancement in medical science as an enlightening influence on this subject, is that the accused must be capable, not only of distinguishing between right and wrong, but that he was not impelled to do the act by an irresistible impulse, which means before it will justify a verdict of acquittal that his reasoning powers were so far dethroned by his diseased mental condition as to deprive him of the will power to resist the insane impulse to perpetrate the deed, though knowing it to be wrong.' 40.

As we have already indicated, this has since been the test in the District.

Although the Smith case did not abandon the right-wrong test, it did liberate the fact finder from exclusive reliance upon that discredited criterion by allowing the jury to inquire also whether the accused suffered from an undefined 'diseased mental condition (which) deprive(d) him of the will power to resist the insane impulse * * *.' 41. The term 'irresistible impulse,' however, carries the misleading implication that 'diseased mental condition(s)' produce only sudden, momentary or spontaneous inclinations to commit unlawful acts. 42.

As the Royal Commission found:

'* * * In many cases * * * this not true at all. The sufferer from (melancholia, for example) experiences a change of mood which alters the whole of his existence. He may believe, for instance, that a future of such degradation and misery awaits both him and his family that death for all is a less dreadful alternative. Even the thought that the acts he contemplates are murder and suicide pales into insignificance in contrast with what he otherwise expects. The criminal act, in such circumstances, may be the reverse of impulsive. It may be cooly and carefully prepared; yet it is still the act of a madman. This is merely an illustration; similar states of mind are likely to lie behind the criminal act when murders are committed by persons suffering from schizophrenia or paranoid psychoses due to disease of the brain.' 43.

We find that as an exclusive criterion the right-wrong test is inadequate in that (a) it does not take sufficient amount of physic realities and scientific knowledge, and (b) it is based upon one symptom and so cannot validly be applied in all circumstances. We find that the 'irresistible impulse' test is also inadequate in that it gives no recognition to mental illness characterized by brooding and reflection and so relegates acts caused by such illness to the application of the inadequate right-wrong test. We conclude that a broader test should be adopted. 44.

B. In the District of Columbia, the formulation of tests of criminal responsibility is entrusted to the courts and, in adopting a new test, we invoke our inherent power to make the change prospectively.

The rule we now hold must be applied on the retrial of this case and in future cases is not unlike that followed by the New Hampshire court since 1870. 47. It is simply that an accused is not criminally responsible if his unlawful act was the product of mental disease or mental defect. 48.

We use 'disease' in the sense of a condition which is considered capable of either improving or deteriorating. We use 'defect' in the sense of a condition which is not considered capable of either improving or deteriorating and which may be either congenital, or the result of injury, or the residual effect of a physical or mental disease.

Whenever there is 'some evidence' that the accused suffered from a diseased or defective mental condition at the time the unlawful act was committed, the trial court must provide the jury with guides for determining whether the accused can be held criminally responsible. We do not, and indeed could not, formulate an instruction which would be either appropriate or binding in all cases. But under the rule now announced, any instruction should in some way convey to the jury the sense and substance of the following: If you the jury believe beyond a reasonable doubt that the accused was not suffering from a diseased or defective mental condition at the time he committed the criminal act charged, you may find him guilty. If you believe he was suffering from a diseased or defective mental condition when he committed the act, but believe beyond a reasonable doubt that the act was not the product of such mental abnormality, you may find him guilty. Unless you believe beyond a reasonable doubt either that he was not suffering from a diseased or defective mental condition, or that the act was not the product of such abnormality, you must find the accused not guilty by reason of insanity. Thus your task would not be completed upon finding, if you did find, that the accused suffered from a mental disease or defect. He would still be responsible for his unlawful act if there was no causal connection between such mental abnormality and the act. 49. These questions must be determined by you from the facts which you find to be fairly deducible from the testimony and the evidence in this case. 50.

The questions of fact under the test we now lay down are as capable of determination by the jury as, for example, the questions juries must determine upon a claim of total disability under a policy of insurance where the state of medical knowledge concerning the disease involved, and its effects, is obscure or in conflict. In such cases, the jury is not required to depend on arbitrarily selected 'symptoms, phases or manifestations' 51. of the disease as criteria for determining the ultimate questions of fact upon which the claim depends. Similarly, upon a claim of criminal irresponsibility, the jury will not be required to rely on such symptoms as criteria for determining the ultimate question of fact upon which such claim depends. Testimony as to such 'symptoms, phases or manifestations,' along with other relevant evidence, will go to the jury upon the ultimate questions of fact which it alone can finally determine. Whatever the state of psychiatry, the psychiatrist will be permitted to carry out his principal court function which, as we noted in Holloway v. U.S., 'is to inform the jury of the character of (the accused's) mental disease (or defect).' 52. The jury's range of inquiry will not be limited to, but may include, for example, whether an accused, who suffered from a mental disease or defect did not know the difference between right and wrong, acted under the compulsion of an irresistible impulse, or had 'been deprived of or lost the power of his will * * *.' 53.

Finally, in leaving the determination of the ultimate question of fact to the jury, we permit it to perform its traditional function which, as we said in Holloway, is to apply 'our inherited ideas of moral responsibility to individuals prosecuted for crime * * *.' 54. Juries will continue to make moral judgments, still operating under the fundamental precept that 'Our collective conscience does not allow punishment where it cannot impose blame.' 55. But in making such judgments, they will be guided by wider horizons of knowledge concerning mental life. The question will be simply whether the accused acted because of a mental disorder, and not whether he displayed particular symptoms which medical science has long recognized do not necessarily, or even typically, accompany even the most serious mental disorder. 56.

The legal and moral traditions of the western world require that those who, of their own free will and with evil intent (sometimes called mens rea), commit acts which violate the law, shall be criminally responsible for those acts. Our traditions also require that where such acts stem from and are the product of a mental disease or defect as those terms are used herein, moral blame shall not attach, and hence there will not be criminal responsibility. 57. The rule we state in this opinion is designed to meet these requirements. Reversed and remanded for a new trial.

Questions:

1. Does the ruling in Durham uphold long standing law that every defendant is presumed to be sane and that to show otherwise requires the state to overcome that presumption.
2. Why is the Durham case often referred to as the "product rule" of legal insanity?
3. Durham rejects the McNaghten test or right and wrong and does so without charity. The test is essentially labeled as unscientific and utterly out of step with the science and discipline or psychiatry. Describe the court's view on this.

Read the press release from the Tennessee Courts regarding the State's Supreme Court setting a standard of review in insanity defense cases at https://www.tncourts.gov/press/2002/08/29/supreme-court-sets-standard-appellate-review-insanity-defense-cases.

Petric v. People of the Virgin Islands
61 V.I. 401
Supreme Court of the Virgin Islands
June 10, 2014, Argued; October 16, 2014, Filed
S. Ct. Criminal No. 2013-0067

Ivan Petric appeals his convictions for first-degree murder, unauthorized possession of a firearm during the commission of a crime of violence, first-degree animal abuse, and unauthorized possession of ammunition. For the reasons that follow, we reverse Petric's convictions and remand for the Superior Court to enter a judgment of acquittal for unauthorized possession of ammunition and a judgment of not guilty by reason of insanity on the remaining charges.

I. FACTUAL AND PROCEDURAL BACKGROUND

...

On July 27, 2007, the People charged Petric with three counts of first-degree murder, three counts of unauthorized possession of a firearm during the commission of a crime of violence, one count of first-degree animal abuse, and one count of unauthorized possession of ammunition. On April 4, 2009, by order of the Superior Court, Dr. Derek V. Spencer, Chief of Psychiatry at Roy Schneider Hospital on St. Thomas, evaluated Petric. In his report, Dr. Spencer determined that while Petric was capable of assisting in his defense, he likely suffered from an undiagnosed "psychotic disorder." On December 22, 2011, Dr. Spencer updated his report and determined that on the day of the murders, Petric was in a "delusional state" such that Petric "thought that his actions … [were] correct" and "believed that his act was in self-defense." On June 20, 2012, Petric filed a notice of intent to assert an insanity defense.

...

Before the People rested, the Superior Court allowed Petric to present the testimony of Dr. Spencer and Karen Petric in order to accommodate their schedules. Dr. Spencer testified that while Petric was capable of assisting in his own defense, it was apparent that Petric had been suffering from "an untreated psychiatric disorder" for years and that his "actions on [the day of the murders] were driven by delusional beliefs." On cross-examination, the People tried to discredit

Dr. Spencer's testimony that Petric suffered from a mental illness and to demonstrate that Petric's actions on the day of the murders were inconsistent with an individual suffering from a mental illness. Petric then called his mother, Karen Petric, who testified that her family has a history of mental illness, including paranoid schizophrenia, and that she suspected her son was suffering from a similar illness, particularly after he was hospitalized at age fifteen for six weeks due to a psychotic episode. Karen also stated that in 2000, Petric informed her that he was "God" and that people were following him and trying to kill him. She further testified that Petric's father did not believe in psychiatric treatment and would not seek psychiatric care for him.

...

III. DISCUSSION

[1] Petric, as one of several arguments raised on appeal, argues that the Superior Court should have granted his motion for a judgment of acquittal because the People failed to prove beyond a reasonable doubt that Petric was sane at the time of the offenses. Because we agree, we do not reach his other arguments. In addressing the sufficiency of the evidence, "we must view the evidence in the light most favorable to the People, and affirm the conviction[s] if 'any rational trier of fact could have found the essential elements of the crime[s] beyond a reasonable doubt.'" George v. People, 59 V.I. 368, 384 (V.I. 2013) (quoting Mendoza v. People, 55 V.I. 660, 667 (V.I. 2011)).

A. The Insanity Defense in the Virgin Islands

[2] Because the entire trial in this case was conducted using the wrong test for insanity, we must first clarify the insanity defense in the Virgin Islands before addressing Petric's sufficiency argument. At trial both the Superior Court and the parties labored under the mistaken assumption that the applicable test for insanity was whether the defendant had "sufficient reason to know right from wrong" at the time the offenses were committed. This test, commonly referred to as the M'Naghten test, exempts individuals from criminal liability where at the time of committing the offenses the defendant "was labouring under such a defect of reason ... as not to know the nature and quality of the act he was doing; or if he did know it, that he did not know he was doing what was wrong." M'Naghten's Case, 8 Eng. Rep. 718 (H.L. 1843). However, the Virgin Islands Legislature explicitly discarded the M'Naghten test in 1957. 14 V.I.C. § 14(4) (1957 revision note); ... In its place, the Legislature enacted section 14(4) of title 14 of the Virgin Islands Code, which provides that "[a]ll persons are capable of committing crimes and offenses except ... persons who are mentally ill and who committed the act charged against them in consequence of such mental illness." The language of section 14(4) parallels the "product test" for insanity set out by the United States Court of Appeals for the District of Columbia Circuit in 1954. Durham v. United States, 214 F.2d 862, 874-75, 94 U.S. App. D.C. 228 (D.C. Cir. 1954) ("[A]n accused is not criminally responsible if his unlawful act was the product of mental disease or mental defect."), abrogated by United States v. Brawner, 471 F.2d 969, 981-82, 153 U.S. App. D.C. 1 (D.C. Cir. 1972); ...

[3-7] The People argues that the Superior Court did not err in utilizing a test other than that codified in section 14(4) because the United States Court of Appeals for the Third Circuit—sitting in its former capacity as the de facto court of last resort for the Virgin Islands—held in Fredericks that it was not error for a trial court applying Virgin Islands law to instruct the jury on the Third Circuit's test for insanity developed in United States v. Currens, 290 F.2d 751 (3d Cir. 1961). Under the Currens test, a jury must find that "at the time of committing the prohibited act the defendant, as a result of a mental disease or defect, lacked substantial capacity to conform his conduct to the requirements of the law which he is alleged to have violated." Id. at 774. But since the Legislature has statutorily adopted the "product test" for insanity in the Virgin Islands, specifically rejecting other formulations, we cannot endorse the Third Circuit's holding in Fredericks. ... Accordingly, section

14(4) remains the sole test for insanity applicable in the Virgin Islands, to the exclusion of both the M'Naghten test and the Third Circuit's Currens test.

B. Sufficiency of the Evidence

[8] As this Court explained in Nibbs v. People, 52 V.I. 276, 284 (V.I. 2009), to raise the insanity defense a defendant need only introduce "some evidence" tending to show that he was "mentally ill and … committed the act charged against [him] in consequence of such mental illness." Once the defendant introduces "some evidence" of mental illness, the defendant's sanity at the time of the offense becomes an element of the crime, which, like all other elements of the crime, must be proven by the People beyond a reasonable doubt. …

[9] To satisfy the "some evidence" test, the defense need only introduce "a slight quantum of evidence." United States v. McCracken, 488 F.2d 406, 410 (5th Cir. 1974); … In this case, Petric sought to raise the insanity defense by introducing Dr. Spencer's testimony that Petric suffered from a "psychotic disorder" and that his actions in November 2006 were driven by delusional beliefs. In addition, Petric's mother testified that the family had a history of mental illness and described in detail Petric's personal struggles with mental illness. Combined, the testimony of Dr. Spencer and Petric's mother clearly amounted to "some evidence" sufficient to raise the insanity defense and place the burden of proving Petric's sanity on the People. …

[10-12] Although not required, once the defense introduces "some evidence" of mental illness, the first and most obvious way the People can satisfy its burden of proving the defendant's sanity at the time of the offense is by introducing its own expert medical testimony to challenge the defense expert's medical conclusions. … In this case, the People offered no medical testimony that Petric was sane at the time of the murders. Instead, the People sought to discredit Dr. Spencer's testimony that Petric's actions on the day of the murders were driven by delusional beliefs stemming from an undiagnosed "psychotic disorder." In particular, the People challenged Dr. Spencer's ability to conclude that Petric was driven by delusional beliefs by pointing out that Petric was not evaluated until approximately three years after the murders. However, merely pointing out the delay between the acts committed and the psychiatric evaluation does nothing to demonstrate that Petric was sane at the time of the commission of the murders, nor does it undermine any of the bases upon which Dr. Spencer reached his conclusion that Petric's actions at the time were a consequence of his delusional beliefs. … Accordingly, the People's effort to discredit Dr. Spencer's conclusions based on the elapsed time between the commission of the offenses and the psychiatric evaluation, without more, was insufficient to prove beyond a reasonable doubt that Petric committed the murders while sane.

[13] In addition to the People's attempt to discredit Dr. Spencer's conclusion that Petric actions were a consequence of his mental illness, the People presented the lay testimony of the arresting officers as evidence of Petric's mental state at the time of arrest. However, before a non-expert witness may testify to the sanity of the defendant, the party offering the testimony must show a familiarity with the defendant to clearly indicate that the testimony will be of value in determining the defendant's sanity, and the conclusion must be based on the witness's testimony as to specific instances of behavior or conduct near the time of the offense. … The primary testimony of the arresting officers bearing on Petric's mental condition at the time of the murders was only that at the time of his arrest—days after the commission of the crimes—Petric "appeared to be alert" and capable of responding to questions. Significantly, none of the officers testified to having any prior familiarity with Petric, nor were the officers capable of testifying to Petric's mental condition before or at the time of the crimes. … Accordingly, the testimony of the arresting officers indicating that Petric was alert and responsive several days after the murders provided no evidence from which a reasonable juror could conclude beyond a reasonable doubt that Petric was sane at the time of the murders.

[14, 15] The People also attempted to use the observations of the arresting officers at the time of Petric's arrest as circumstantial evidence that Petric was sane at the time of the murders. In particular, the People relied on the testimony of Officer Felix that when Petric was arrested, police

discovered a pair of bloodstained jeans soaking in a wash bucket and two packed suitcases in Petric's room. The officers also observed that the phone cord had been pulled from the wall and that the bodies had been covered up. However, the mere fact that it appeared as though Petric may have been seeking to conceal his actions—by cleaning his clothes, covering the bodies, pulling the phone cord from the wall, and packing suitcases—provides no insight into whether Petric was sane when he committed the murders. Indeed, in order for circumstantial evidence that Petric made efforts to conceal his actions after the murders to be used as evidence of his sanity at the time of the murders, the jury would first have to infer that Petric performed the acts of concealment, then that he did so while sane, and finally—based on these two inferences—that he was sane while committing the murders. … And when there is a complete absence of evidence to show that "the defendant was sane at the critical time, namely, the period when … the killing occurred," the People has failed to produce sufficient evidence to support a finding of sanity …

Requiring the People to produce evidence bearing on a defendant's sanity during the "critical time period"—either immediately before or during the commission of the offense—is particularly important in cases such as this, where the defendant's medical expert testified that the defendant was in a "delusional state" before and during the murders. … Taken together, the People's cross-examination of Dr. Spencer, coupled with the lay testimony of the arresting officers, even when viewed in the light most favorable to the People, was simply insufficient to prove beyond a reasonable doubt that Petric was sane at the time of the offenses.

[16] We turn now to the appropriate remedy. Although this Court has typically held that "'a judgment of acquittal is the appropriate remedy when the evidence is not sufficient to sustain a conviction,'" Farrell v. People, 54 V.I. 600, 619 (V.I. 2011) (quoting Gilbert v. People, 52 V.I. 350, 364 (V.I. 2009)), where the People fails to carry its burden of proving beyond a reasonable doubt that the defendant was sane at the time of the offense, the appropriate remedy is to remand with instructions to enter a judgment of not guilty by reason of insanity. Nibbs, 52 V.I. at 292-93; … Accordingly, due to the absence of sufficient evidence permitting the jury to find beyond a reasonable doubt that Petric was sane at the time of the offenses, we remand this case and direct the Superior Court to enter a judgment of not guilty by reason of insanity on all of Petric's convictions except for unauthorized possession of ammunition.

[17] With regard to that conviction, not only did the People fail to prove his sanity at the time of this offense, but the People also failed to prove that he committed the essential elements of that crime. The People concedes in its brief that it "did not prove that Petric committed this offense" and that Petric's "conviction for unauthorized possession of ammunition should be reversed." Although the parties may not stipulate to the law, Rohn v. People, 57 V.I. 637, 643 (V.I. 2012), we agree that the evidence here was insufficient. As this Court has previously explained, because 14 V.I.C. § 2256(a) (as it existed at the time of Petric's offense) "generally criminalized possession of ammunition 'unless authorized by law' but provided no means to obtain such authorization," the People could not have proven that Petric committed this offense. Nicholas v. People, 56 V.I. 718, 730 (V.I. 2012) (citing Brown v. People, 55 V.I. 496, 500-01 (V.I. 2011)). Accordingly, on remand, the Superior Court must enter a judgment of acquittal for unauthorized possession of ammunition.

IV. CONCLUSION

The evidence was insufficient to support Petric's convictions for first-degree murder, unauthorized possession of a firearm, and first-degree animal abuse because the People failed to carry its burden and introduce sufficient evidence to permit a rational jury to conclude beyond a reasonable doubt that Petric was sane at the time of the offenses. The People also failed to introduce sufficient evidence to support Petric's conviction for unauthorized possession of ammunition. Therefore, we reverse Petric's convictions and remand to the Superior Court to enter a judgment of acquittal for unauthorized possession of ammunition. Further, we direct the Superior Court to enter judgment finding Petric not guilty by reason of insanity on the remaining charges and take further action consistent with that judgment.

Questions:

1. What two tests of insanity does the Court summarily dismiss as not-permissible in this jurisdiction?
2. What does the jurisdictional code mandate as the sole test to be employed when making said determinations?
3. Once the defendant offers evidence of legal insanity, what is the best way to counter the claim on behalf of the Prosecution.

Tests for legal insanity have definitely been liberalized and more reflective of the alleged scientific method of psychiatry and psychology. Gone are the days of moral choice and reflection in those jurisdictions which have picked up on "irresistible impulse" and "lack a substantial appreciation" for the criminality of the acts chosen. In these more modern renditions, the actor's choices are pushed and pulled by determinative forces rather than mere choice. In Brawner the theory of "substantial appreciation receives a thorough vetting.

Locate the following case using your favorite legal research resource, read the case and answer the questions below.

*United States v. Brawner,*471 F.2d 969, United States Court of Appeals for the District of Columbia Circuit, April 12, 1972, Argued En Banc ; June 23, 1972, Decided, No. 22714

Questions:

1. It is clear that this court reject McNaghten nor accept or endorse Durham. In its place the Court adopts a new test outlined by the American Law Institute. What is that test and what suggested definition?
2. The ALI test has often been labeled a "subjective" standard for measuring legal insanity. How so?
3. Brawner also discusses the concept of "Diminished Capacity or Responsibility." Will meeting this standard qualify for a finding of legal insanity?

At the extreme end of the insanity continuum rests the "Irresistable Impulse" Test. Here, the defendant can know right from wrong, can appreciate the criminality of a particular act and can even fail to correlate a mental disease or defect with crime causation yet still be acquitted. The impulse, put simply, cannot be resisted.

Locate the following case using your favorite legal research resource, read the case and answer the questions below.

Parsons v. State, 81 Ala. 577, 2 So. 854 (1887)

Questions:

1. Reading 19th century cases that deal with mental disease or defects, and one quickly discovers an antiquated and less sophisticated view than presently held. For example, provide a few terms that edify this observation.

2. The development of the Irresistable Impulse test was in reaction to the rigidity of the McNaghten Rule. The opinion makes this plain. Point out the role of duress or pressure in pushing a defendant to commit crimes. Target the language of the Court which demonstrates this.
3. The Court provides commentary on "freedom of the will." Why is this relevant to the newly devised test for insanity?

10.3 NEUROLOGICAL EXPLANATIONS AND PROOF OF MENTAL DYSFUNCTION

Elmore v. Holbrook
137 S. Ct. 3, 196 L. Ed. 2d 272 Supreme Court of the United States October 17, 2016, Decided No. 15-7848.

Petitioner Clark Elmore was convicted of murder in 1995 and was sentenced to death. His court-appointed lawyer, who had never tried a capital case before, knew that Elmore had been exposed to toxins as a young adult and that he had a history of impulsive behavior. A more experienced attorney encouraged Elmore's lawyer to investigate whether Elmore had suffered brain damage as a young man. Instead of doing so—indeed, instead of conducting any meaningful investigation into Elmore's life—Elmore's lawyer chose to present a one-hour penalty-phase argument to the jury about the remorse that Elmore felt for his crime. As a result, the jury did not

FIGURE 10.2 Jared Lee Loughner pled guilty to nineteen charges of murder and attempted murder in a shooting in Tucson on January 8, 2011.

hear that Elmore had spent his childhood playing in pesticide-contaminated fields and had spent his service in the Vietnam War repairing Agent Orange pumps. The jury did not hear the testimony of experts who concluded that Elmore was cognitively impaired and unable to control his impulses. The jury heard only from an assortment of local judges that Elmore had looked "dejected" as he pleaded guilty to murder, not from the many independent witnesses who had observed Elmore's searing remorse.

The Constitution demands more. The penalty phase of a capital trial is "a constitutionally indispensable part of the process of inflicting the penalty of death." Woodson v. North Carolina, 428 U.S. 280, 304, 96 S. Ct. 2978, 49 L. Ed. 2d 944 (1976). It ensures that a capital sentencing is "humane and sensible to the uniqueness of the individual." Eddings v. Oklahoma, 455 U.S. 104, 110, 102 S. Ct. 869, 71 L. Ed. 2d 1 (1982).

Elmore's penalty phase fell well below the bare minimum guaranteed by the Constitution. His lawyer acted deficiently in choosing a mitigation strategy without fully exploring the alternatives and in failing to investigate the mitigation strategy that he did choose to present. And had the jury known that Elmore—who had never before been convicted of a crime of violence and felt searing remorse for the heinous act he committed—might be brain damaged, it might have sentenced him to life rather than death.

This Court has not hesitated to summarily reverse in capital cases tainted by egregious constitutional error, particularly where an attorney has rendered constitutionally deficient performance. ... This case plainly meets that standard. For that reason, I respectfully dissent from the denial of certiorari.

I A

Elmore was born in 1951 in central Oregon, where he lived until his teens. Social History, 12 Record 5524-5530. He was exposed to powerful neurotoxins from a young age. Elmore's house in Oregon was located next to an airport from which crop dusters regularly sprayed pesticides. Trial Court Findings of Fact, No. 95-1-00310-1 (Sup. Ct. Whatcom Cty., Wash., Sept. 10, 2004), 14 id., at 6519-6520 (FOF). Decades after Elmore moved away, the state environmental agency took soil samples that showed toxin levels over 4,500 times the maximum amounts allowed by state law. Decl. of Raymond Singer, 11 id., at 5394 (Singer Decl.). Later, Elmore worked on cars and oil pipelines where he regularly melted lead batteries and handled solvents without gloves. FOF, 14 id., at 6521-6522. And when Elmore left home at age 17 to serve in the Vietnam War, he was tasked with repairing Agent Orange pumps without protective equipment. Id., at 6522; Singer Decl., 11 id., at 5395.

Experts who testified at Elmore's postconviction hearing agreed that this exposure placed him at serious risk of brain damage. They conducted neuropsychological tests that revealed mild to moderate cognitive impairments, ..., including a marked inability to control his emotions and impulses, see id., at 7079-7080. Elmore tested in the bottom one percent on tests measuring that characteristic. Id., at 7080. The experts concluded that damage to Elmore's frontal lobe had made him impulsive and susceptible to emotion. ... And they agreed that the murder Elmore later committed was linked to Elmore's cognitive deficits—for instance, by making him unable to "pu[t] on the brakes" when emotional. ...

Elmore was discharged from the Army under honorable conditions in 1972, but found it hard to return to civilian life. 12 id., at 5631. He moved around the United States, taking jobs in hotels, gas stations, farms, and oil fields. Social History, id., at 5532-5538. Elmore was arrested three times—once for stealing checks, once for stealing furniture, and once for stealing appliances from a motel. Reporter's Tr. in No. 95-1-00310-1, 5 id., at 2470-2473 (Trial Tr.); Social History, 12 id., at 5532, 5536. Officers at one prison reported that Elmore was nonviolent and, if anything, was the victim of other inmates' threats. Id., at 5533-5534. After his second conviction, Elmore was incarcerated for two years in Washington state prison, where he was repeatedly raped by another inmate. Id., at 5536. Until the murder for which he was ultimately sentenced to death, and despite his emotional challenges, Elmore was never convicted of a violent crime.

Elmore's death sentence arises out of a murder that he committed in 1995. The crime was horrific. Elmore raped and murdered his stepdaughter, first strangling her with a belt, then driving a sharp object through her ear, and finally bludgeoning her with a hammer. In re Elmore, 162 Wash. 2d 236, 244, 172 P. 3d 335, 340 (2007). Elmore was apparently motivated by fear that the victim would tell the authorities that he had previously sexually abused her. Ibid. After several days of misdirecting the authorities, Elmore turned himself in and confessed. FOF, 14 Record, at 6460. In the wake of the murder, Elmore expressed extreme remorse. A jailhouse minister who visited Elmore in prison later attested that, the day after he arrived, he "was huddled into a ball at the back of the room, shaking uncontrollably." Decl. of Dana Paul Sellars, 11 id., at 5399 (Sellars Decl.). Elmore, he said, "was unlike any prisoner I had counseled before. He was wracked with anguish and dripping with remorse." Id., at 5400. A correctional officer at the prison later testified that Elmore appeared "in a state of disbelief about what he had done" and was "an emotional wreck." Decl. of Donald Pierce, id., at 5404-5405.

B

The jury that sentenced Elmore to death learned about the terrible crime he committed, but heard virtually nothing about his troubling background and cognitive defects. A lawyer named Jon Komorowski was appointed to represent Elmore at trial. Komorowski had never previously worked on a capital case. Decl. of Jon Komorowski, 11 id., at 5325 (Komorowski Decl.). On Komorowski's advice, Elmore pleaded guilty to capital murder without any negotiations with the prosecution. Id., at 5326. Because Elmore pleaded guilty, the trial consisted of only a penalty phase. During that penalty phase, the State presented nine witnesses, all of whom testified regarding the horrific circumstances of the crime. Trial Tr., 5 id., at 2348-2580. The State also presented evidence of Elmore's three criminal convictions, all two decades old. Id., at 2470-2473.

Komorowski's mitigation case for Elmore lasted only an hour. … The theme was remorse: "[T] here are no excuses in this case and none are offered. There is acceptance of responsibility and punishment." Trial Tr., 5 Record 2367; see also id., at 2580-2658 (defense case). The only character witnesses were the three judges who had presided over Elmore's pretrial appearances, who testified that he had sought to plead guilty from the outset. Id., at 2581-2582, 2587-2588, 2590-2592. One described Elmore as "somewhat upset" and "overwhelmed," a second as "dejected." Id., at 2586, 2592. The defense investigator read out a "bare bones" summary of Elmore's biography. Id., at 2306, 2599-2601. Finally, an expert witness testified that Elmore's prior convictions were not violent felonies under Washington's three-strikes law. Id., at 2644-2658.

Years later, in postconviction proceedings, Komorowski acknowledged his error, explaining that the decision not to investigate Elmore's medical history was "the product of … inexperience" and "not a strategic decision." Komorowski Decl., 11 id., at 5329. He admitted that he and the defense team had reviewed Elmore's prison records and some of his hospital records, and had spoken to Elmore's family, who had told them about Elmore's hardships as a child and as a young adult. PRP Tr., 15 id., at 6907-6911. And he consulted with more experienced counsel, including an attorney named Todd Maybrown, who strongly advised Komorowski to investigate indicia of "organic brain disorder" and cautioned that the testimony of a psychologist with no neurology background would not be sufficient. Decl. of Todd Maybrown, 12 id., at 5540-5541 (Maybrown Decl.). Maybrown advised Komorowski that "he might need to hire a medical doctor to try to determine if his client suffered from brain damage." Id., at 5541-5542.

But Komorowski did not hire a neuropsychiatrist, nor did he conduct any further investigation into the possibility of brain damage. Komorowski consulted with three mental health professionals, but none of them tested for any sort of brain damage. PRP Tr., 15 id., at 6985-6986, 7406-7407. The first administered a personality test and concluded that Elmore was not insane, but recommended that Komorowski consult a second expert about whether Elmore was a psychopath. Id., at 7404-7405, 7412. The second concluded that Elmore was not a psychopath: He demonstrated remorse and empathy, and his crime was impulsive and reactive, indicating heightened emotional arousal rather than psychopathy. Id., at 7230-7231. The third was not a licensed psychologist at all. Id., at 6889, 6911, 6924. The two psychologists later agreed that, had Komorowski told them about Elmore's exposure to toxins, they would have recommended neuropsychological testing. Id., at 7422-7423, 7243-7244.

C

Elmore moved for postconviction relief in state court, arguing that Komorowski's representation deprived him of effective assistance of counsel in violation of the Sixth Amendment. But the Washington Supreme Court denied his claim. In re Elmore, 162 Wash. 2d 236, 172 P. 3d 335. "There is no question that the defense team did investigate petitioner's mental health deficiencies," the state court held. Id., at 258, 172 P. 3d, at 347. "Rather, the issue is whether counsel's failure to conduct further evaluations amounted to deficient representation. We believe it did not." Ibid. The Washington Supreme Court ruled that Komorowski did not perform below the constitutional standard and had instead made a "strategic" decision to curtail the investigation. Ibid. According to the state court, Komorowski's strategy was defensible for four reasons: Presenting more mitigation evidence might have opened the door to damaging rebuttal evidence; additional witnesses would have been "cumulative" of the judges who testified; Komorowski worried that if he did not rush to trial, the prosecution might find witnesses who would testify that Elmore's remorse was waning; and Elmore had objected to the presentation of any mitigation case. Id., at 257-258, 263-265, 172 P. 3d, at 346-347, 348-350.

A Federal District Court denied Elmore's habeas petition, and the Ninth Circuit affirmed. Elmore v. Sinclair, 799 F. 3d 1238 (2015). Two judges held that the Washington Supreme Court's decision was not unreasonable. Id., at 1243. The third would have found that the Washington Supreme Court's determination that Komorowski was constitutionally effective was unreasonable, but concurred because he did not believe that the question of prejudice was beyond debate. Id., at 1256-1257 (opinion of Hurwitz, J.). Elmore petitioned for certiorari.

II

...

Many observers, on and off this Court, have questioned the reliability and fairness of the imposition of capital punishment in America. ... Whether our system of capital punishment is inconsistent with the Eighth Amendment, as these critics have charged, is not at issue here. I do believe, however, that whatever flaws do exist in our system can be tolerated only by remaining faithful to our Constitution's procedural safeguards.

All crimes for which defendants are sentenced to death are horrific. ... But not all defendants who commit horrific crimes are sentenced to death. Some are spared by juries. The Constitution guarantees that possibility: It requires that a sentencing jury be able to fully and fairly evaluate "the characteristics of the person who committed the crime." Gregg v. Georgia, 428 U.S. 153, 197, 96 S. Ct. 2909, 49 L. Ed. 2d 859 (1976) (joint opinion of Stewart, Powell, and Stevens, JJ.). That guarantee is a bedrock premise on which our system of capital punishment depends, and it is a guarantee that must be honored—especially for defendants like Elmore, whose lives are marked by extensive mitigating circumstances that might convince a jury to choose life over death. Only upon hearing such facts can a jury fairly make the weighty—and final—decision whether such a person is entitled to mercy.

I respectfully dissent from the denial of certiorari.

Questions:

1. Defendant was convicted of murder and sentenced to death. His crime, even the U.S Supreme Court indicates was "horrific.´ Describe how he killed and who that victim was.
2. Distinguish the two roles of the forensic psychologist and forensic psychiatrist in this case.
3. What was the cause of the neurological damage?

Locate the case below using your favorite legal research resource, read the case and answer the questions below.

Jefferson v. GDCP Warden, 941 F.3d 452, United States Court of Appeals for the Eleventh Circuit, October 17, 2019, Decided No. 17-12160 Non-Argument Calendar

Questions:

1. A claim of ineffective counsel was alleged for defense counsel failing to explore both psychological profiles and even more compellingly neurological damage to the brain. Explain.
2. A claim of brain damage was an issue at sentencing but not trial. What is the story on that and do you consider this weak or strong evidence?
3. A clinical psychologist reviewed the defendant' record as well. In particular, the psychologist highlighted the lower and varying IQ test results. What conclusions are reached?

King v. Westbrooks
847 F.3d 788
United States Court of Appeals for the Sixth Circuit
September 28, 2016
No. 13-6387

This death penalty case arises out of the kidnapping and murder of Diana K. Smith by Petitioner-Appellant Terry King. Following the district court's dismissal of King's petition for a writ of habeas corpus, we granted a certificate of appealability on two issues: whether trial counsel was ineffective for failing to present during the trial testimony about King's intoxication at the time of the murder and whether trial counsel was ineffective for failing to investigate adequately King's mental health and to obtain expert assistance in a timely manner. For the reasons stated below, we AFFIRM the judgment of the district court.

I. BACKGROUND

. . .

In preparation for trial, which began on January 23, 1985, see Post—conviction Tr. ("PCT") Vol. V at 426 (Appellant's 2d Supp. App'x at 796) (Simpson Test.), King's trial counsel, Robert R. Simpson, suspected that King may have had brain damage as a result of a childhood head injury and substance abuse. PCT Vol. IV at 376, 381-82, 384 (Appellant's 2d Supp. App'x at 744, 749-50, 752) (Simpson Test.). In addition to a childhood head injury, in 1982, King—then about nineteen years old—hit his head in a car accident and had double vision for a couple of months afterwards. R. 254-3 at 4 (Gebrow Report at 2) (Page ID #475). From age eight to sixteen, King sniffed gasoline. Id. at 5 (Gebrow Report at 3) (Page ID #476). He also consumed alcohol beginning at age twelve or thirteen and LSD and Quaaludes beginning at age fifteen or sixteen. Id. at 4-5 (Gebrow Report at 2-3) (Page ID #475-76).

On January 15, 1985, Simpson retained a mental-health expert, Martin Gebrow, M.D., to evaluate King. Simpson used private funds from King's family to pay for Dr. Gebrow's services because Simpson was unaware of state law that provided for state funding of an expert. PCT Vol. V at 424, 431-32 (Appellant's 2d Supp. App'x at 794, 801-02) (Simpson Test.). Dr. Gebrow's report indicated that he evaluated King on January 23, 1985.

R. 254-3 (Gebrow Report at 1) (Page ID #474). The report described King's background, including his history of substance abuse. Id. at 1-3 (Page ID #474-76). Dr. Gebrow concluded, "My examination of Mr. King did not reveal any evidence of psychotic thought process. Nor did it reveal any evidence of an organic brain syndrome such as might have been caused by the chronic use of hydrocarbons by inhalation, alcohol, or LSD." Id. at 3-4 (Page ID #476-77). He continued, "This however does not mean that any brain damage does not exist. It would be my recommendation that Mr. King have an electroencephalogram and psychological testing to rule out organicity and/or major thought disorder." Id. at 4 (Page ID #477).

During voir dire, Simpson made an oral motion to "permit the taking of an electroencephalogram" of King, TT Vol. VII at 552-53 (Appellant's App'x at 158-59), which is "a brain wave test that measures the electrical activity of the brain and can ascertain whether or not there are any abnormal electrical discharges which would indicate brain damage," TT Vol. VIII at 642 (Gebrow Test.). In a hearing on the motion, during which Dr. Gebrow testified, the trial court inquired of Dr. Gebrow whether there was "a substantial possibility of damage." Id. at 657. Dr. Gebrow responded, "With the eight year—eight-or-nine-year history of constant hydrocarbon abuse, I think that there would be—could be an excellent chance that this was—that there was some damage" but that it was not a "probability." Id. at 658. Dr. Gebrow also agreed that, based on the examination, King was coherent, his memory appeared to be intact, and that he was able to express himself well. Id. Because Dr. Gebrow "went in cold to do the evaluation," id. at 655, he was not aware of certain conditions that he admitted would affect the evaluation, including prior psychological testing, id. at 648, and evidence of antisocial behavior, id. at 665-66. As a result, the trial court denied King's motion but noted that it would reconsider if Dr. Gebrow reviewed more of King's medical records and decided that an electroencephalogram would still be required. See id. at 670. Having reviewed these records, Dr. Gebrow testified later at trial that an electroencephalogram was not necessary. TT Vol. XII at 383 (Gebrow Test.).

Simpson suggested in his opening statement that King's intoxicated state influenced his actions:

> We think the proof will show that whatever happened to Mrs. Smith, Mr. King's involvement was the product of an incredible quantity of intoxicants. And we think the proof will show that he cannot be held legally responsible for all of his actions to the degree the State would ask you, simply because of the vast quantities of intoxicants that he consumed. And the proof is going to be very clear on that point.

TT Vol. IX at 10 (Appellant's App'x at 161). …

At the penalty phase, King's mother, Billie King, testified that she would find King sniffing gasoline when she came home from work: "Well, you could tell that he had—he had a motorcycle. It was tore up, but it was on the back porch. And he had the gas cap off from the motorcycle. And you could tell that he had been into the gas, and he couldn't hardly sit up. And I whipped him, you know.

He promised me he'd never do it again." TT Vol. XIII at 496. Similarly, King's brother, Gary Edward King, testified that following his father's death, King would sniff gasoline "several times a week." TT Vol. XIV at 509. King also called Robert Booher, M.D., a specialist in "addictionology," to testify about the general effects of LSD, Quaaludes, and alcohol. TT Vol. XVI at 730-37.

Ultimately, the jury found King guilty of first-degree murder and recommended death by electrocution, which the trial court imposed. See State v. King, 718 S.W.2d 241, 243 (Tenn. 1986). Following an unsuccessful direct appeal, see id., King filed a petition for post—conviction relief in state court, in which he raised, among other claims, the same ineffective assistance of counsel claims that are the subject of this appeal. … At post-conviction proceedings, King called a clinical psychologist, Pamela Auble, Ph.D, who had evaluated King and reviewed his medical records. PCT Vols. I-II at 76, 99-100, 106-07 (Appellant's 2d Supp. App'x at 439, 462-63, 471-72) (Auble Test.). She testified, "The psychological testing that I have done and that has been done—the evaluation by Dr. Gebrow that was done prior both raise the question of potential brain damage. This issue still needs to be explored, is not yet conclusive, but is a possible thing that could be explored." PCT Vol. II at 148, 168 (Appellant's 2d Supp. App'x at 513, 533) (Auble Test.). She also testified that she had reviewed a report by a Dr. Kaminski, who performed an EEG on King that "showed negative results."2 Id. at 167 (Appellant's 2d Supp. App'x at 532) (Auble Test.). She further observed that there was no evidence of psychotic thought process. Id.

…

12, 2011). At this stage, King called expert witnesses who stated that in fact King had organic brain damage at the time of Smith's death. An evaluation by James R. Merikangas, M.D., P.C., on June 27, 2000, demonstrated with "a reasonable degree of medical certainty that Mr. King suffers from brain damage and did so at the time of the crime for which he stands convicted." Merikangas Report at 3 (Appellant's App'x at 48). Dr. Merikangas also stated that King does not have antisocial personality disorder, but rather "brain damage which is no fault of his own." Id. at 4 (Appellant's App'x at 49). On March 21, 2001, a psychiatrist named Robert L. Sadoff, M.D., evaluated King and, after reviewing King's previous medical records, including Dr. Merikangas's report, concluded that "[i]t was the combination of all these factors, including his intoxication by several substances at the same time, his brain damage and his personality disorders that substantially impaired his capacity to conform his conduct to the requirements of the law." Sadoff Report at 22 (Appellant's App'x at 79). Finally, a physician named Murray W. Smith, M.D., evaluated King and, in a February 7, 2001 affidavit, stated "Any pre-existing brain damage resulting from the very heavy and frequent use of inhalants from age 8 to age 16, as well as the use of cocaine and amphetamine as found in my evaluation of Mr. King, would further multiply the effects of the alcohol and drugs on the causation of the violent interaction Mr. Terry King had with Ms. Diana Smith." Smith Aff. at 2 (Appellant's App'x at 53). Ultimately, the district court awarded summary judgment against King and dismissed his petition. King has appealed the district court's judgment.

II. DISCUSSION

We granted a certificate of appealability on two questions: (1) "[W]hether trial counsel was ineffective for failing to present testimony about King's intoxication at the time of the murder during the trial" and (2) "[W]hether trial counsel was ineffective for failing to adequately investigate King's mental health and obtain expert assistance in a timely manner." Certificate of Appealability at 2. For the reasons stated below, the answer to both questions is no.

A. Standard of Review

When reviewing a district court's denial of a § 2254 petition, "[t]his court reviews de novo [the] district court's legal conclusions and mixed questions of law and fact and reviews its factual findings for clear error." Moore v. Mitchell, 708 F.3d 760, 774 (6th Cir. 2013). King is entitled to relief only if the Tennessee Supreme Court—which issued "the last reasoned state-court opinion" in this case, Ylst v. Nunnemaker, 501 U.S. 797, 804-05, 111 S. Ct. 2590, 115 L. Ed. 2d 706 (1991)—adjudicated King's ineffective-assistance claims on the merits in a way that:

> 1. resulted in a decision that was contrary to, or involved an unreasonable application of, clearly established Federal law, as determined by the Supreme Court of the United States; or
> 2. resulted in a decision that was based on an unreasonable determination of the facts in light of the evidence presented in the State court proceeding.

28 U.S.C. § 2254(d) (2012).

The Tennessee Supreme Court, applying Strickland and Tennessee ineffective-assistance case law, rejected both of the ineffective-assistance claims that King raises in this appeal. King, 989 S.W.2d at 330-32 (intoxication defense); id. at 332-33 (mental-health investigation). The district court also rejected both claims in its order denying King's § 2254 petition. ...

B. King's Ineffective Assistance of Counsel Claims

To demonstrate that his counsel was constitutionally ineffective under Strickland v. Washington, 466 U.S. 668, 104 S. Ct. 2052, 80 L. Ed. 2d 674 (1984), King must make two showings: "(1) [his] counsel's performance was deficient, or put differently, 'fell below an objective standard of reasonableness'; and (2) the performance prejudiced [King]." United States v. Mahbub, 818 F.3d 213, 230-31 (6th Cir. 2016) (quoting Strickland, 466 U.S. at 687-88). Because the Strickland standard is already "highly deferential," Strickland, 466 U.S. at 689, our review of a state-court decision on a Strickland claim is "doubly deferential" under the Anti-Terrorism and Effective Death Penalty Act of 1996 ("AEDPA"). Cullen v. Pinholster, 563 U.S. 170, 189-90, 131 S. Ct. 1388, 179 L. Ed. 2d 557 (2011) (quoting Knowles v. Mirzayance, 556 U.S. 111, 123, 129 S. Ct. 1411, 173 L. Ed. 2d 251 (2009)). However, "[w]hen a state court relied only on one Strickland prong to adjudicate an ineffective assistance of counsel claim, AEDPA deference does not apply to review of the Strickland prong not relied upon by the state court. The unadjudicated prong is reviewed de novo." Rayner v. Mills, 685 F.3d 631, 638 (6th Cir. 2012).

Although the Tennessee Supreme Court did not specifically state how it ruled on either Strickland prong, it is clear from the substance of its decision that it decided the intoxication defense ineffective-assistance-of-counsel claim based on the deficient-performance prong and the mental-health-expert ineffective-assistance-of-counsel claim based on the prejudice prong. With respect to the intoxication defense ineffective-assistance-of-counsel claim, the court held as follows:

> Although we acknowledge that defense attorneys should strive to present a consistent theory of defense at trial, we must avoid judging the tactical decisions of counsel in hindsight. We have reviewed the circumstances from counsel's perspective at the time and conclude that the change in strategy does not rise to the level of ineffective assistance.

King v. State, 989 S.W.2d at 331-32 (citing Strickland, 466 U.S. at 689; Hellard v. State, 629 S.W.2d 4, 9 (Tenn. 1982)). Because the court focused on Simpson's "tactical decisions" and trial strategy, it appears to have concluded that Simpson's performance was not deficient. And with respect to the mental-health expert ineffective-assistance-of-counsel claim, it held as follows:

> The trial court concluded ... that even if defense counsel had initiated the mental health evaluations earlier, there was no proof that a more favorable report would have been obtained. We find no evidence to preponderate against that finding. Moreover, the record reflects that counsel presented evidence

> through lay witnesses that was remarkably similar to the information provided by Dr. Auble. Appellant's counsel were not ineffective on this issue.

Id. at 333. Because the court focused on the effect mental-health experts would have had on the defense, and not whether Simpson's failure to retain those experts hurt King's defense, it appears to have concluded that Simpson's performance, regardless of its deficiency, did not prejudice King. Nevertheless, we will analyze the deficiency prong of King's intoxication defense ineffective-assistance claim and the prejudice prong of King's mental-health-expert ineffective-assistance claim de novo "because, even under that more liberal standard of review, we conclude that his counsel was not deficient." …

…

2. Mental-Health Expert

King next argues that his counsel was constitutionally ineffective at the guilt and penalty phases for failing to investigate King's mental health on a timely basis and to obtain expert assistance concerning the same. Specifically, King focuses on his attorney's allegedly untimely retention of Dr. Gebrow. As discussed above, we review the deficient performance prong de novo because the state court did not address this prong. We also review the prejudice prong de novo because "even under that more liberal standard of review, we conclude that his counsel was not [ineffective]." …

There is no question that Simpson's delay in retaining Dr. Gebrow fell below an objective standard of reasonableness. "An attorney's ignorance of a point of law that is fundamental to his case combined with his failure to perform basic research on that point is a quintessential example of unreasonable performance under Strickland." Hinton v. Alabama, 134 S. Ct. 1081, 1088-89, 188 L. Ed. 2d 1 (2014) (citing Williams v. Taylor, 529 U.S. 362, 395, 120 S. Ct. 1495, 146 [*798] L. Ed. 2d 389 (2000)). Section 40-14-207 of the Tennessee Code, which has not been amended since the time of King's trial, states in relevant part,

> In capital cases where the defendant has been found to be indigent by the court of record having jurisdiction of the case, the court in an ex parte hearing may, in its discretion, determine that investigative or expert services or other similar services are necessary to ensure that the constitutional rights of the defendant are properly protected.

Tenn. Code Ann. § 40-14-207(b) (West 2002). At a hearing on King's state petition for post-conviction relief, trial counsel stated that he "was unaware of" his ability to obtain state funds in order to hire an expert for King under this section. PCT Vol. V at 424 (Appellant's 2d Supp. App'x at 794) (Simpson Test.). He also indicated that he waited to receive private funds before retaining Dr. Gebrow because he was unaware of this section. Id. at 425 (Appellant's 2d Supp. App'x at 795) (Simpson Test.). King was charged with first-degree murder; his mental state at the time he killed Smith was a critical factor in the jury's determination that he was guilty and that he deserved a death sentence. Such an "inexcusable mistake of law—the unreasonable failure to understand the resources that state law made available to him" constitutes deficient performance. See Hinton, 134 S. Ct. at 1089. Nevertheless, because King has not shown that he was prejudiced by this deficient performance, he has not demonstrated ineffective assistance of counsel.

Even reviewing the prejudice prong de novo, we conclude that habeas relief is not warranted. Fundamentally, King has not shown, with the evidence properly available to us on federal habeas review, that the timely retention of a mental expert would have produced any evidence different from what was already available at the time of trial. Dr. Gebrow testified at a hearing on King's motion to continue the trial that "with the history of gasoline inhalation that there might be a generalized diffused type of brain damage" and that "[y]ou could also find, possibly, some focal point of brain damage." TT Vol. VIII at 644 (Gebrow Test.). He stated, "That is the reason that I requested or recommended that an electroencephalogram and psychological testing be done." Id.

To support his argument that his trial counsel should have obtained a mental expert earlier, King introduced Dr. Auble during post-conviction proceedings. Similarly to Dr. Gebrow, she testified, "The psychological testing that I have done and that has been done—the evaluation by Dr. Gebrow that was done prior both raise the question of potential brain damage. This issue still needs to be explored, is not yet conclusive, but is a possible thing that could be explored." PCT Vol. II at 148 (Appellant's 2d Supp. App'x at 513) (Auble Test.). She also acknowledged that an electro-encephalogram had been performed on King that showed negative results, but that it still had not been determined whether there was evidence of organic brain syndrome. Id. at 167-68 (Appellant's 2d Supp. App'x at 532-33) (Auble Test.). Based on Dr. Auble's nearly identical uncertainty regarding whether King had brain damage, King has not shown that "[t]imely securing the services of an expert would have provided counsel with an expert opinion that related the impact of intoxication and brain damage on King's judgment and behavior at the time of the crime and King's ability to form specific intent for first-degree murder," Appellant's Br. at 75, let alone whether there would be a reasonable probability of a different outcome. See Strickland, 466 U.S. at 695-96. Therefore, King has not shown that he was prejudiced by the delay in retaining a mental-health expert.

To be sure, the findings of the mental-health experts on federal habeas review are troubling. Although the experts presented at trial and during state post—conviction proceedings were never able definitively to determine whether King had brain damage, we now know that King "suffers from brain damage which is no fault of his own." Merikangas Report at 4 (Appellant's App'x at 49). Unfortunately for King, AEDPA does not permit us to consider this evidence. Section 2254(e)(2) controls the admissibility of evidence on federal habeas review:

> If the applicant has failed to develop the factual basis of a claim in State court proceedings, the court shall not hold an evidentiary hearing on the claim unless the applicant shows that—
>
> A. the claim relies on—
>   i. a new rule of constitutional law, made retroactive to cases on collateral review by the Supreme Court, that was previously unavailable; or
>   ii. a factual predicate that could not have been previously discovered through the exercise of due diligence; and
> B. the facts underlying the claim would be sufficient to establish by clear and convincing evidence that but for constitutional error, no reasonable factfinder would have found the applicant guilty of the underlying offense.

28 U.S.C. § 2254(e)(2). This provision controls even if the petitioner seeks relief based on new evidence without an evidentiary hearing. … " Although state prisoners may sometimes submit new evidence in federal court, AEDPA's statutory scheme is designed to strongly discourage them from doing so. Provisions like §§ 2254(d)(1) and (e)(2) ensure that '[f]ederal courts sitting in habeas are not an alternative forum for trying facts and issues which a prisoner made insufficient effort to pursue in state proceedings.'" Pinholster, 563 U.S. at 186 (citation omitted).

The mental-health experts to whom King points at this late stage cannot be considered because they could have been discovered through the exercise of due diligence of post-conviction counsel. … Each of the medical reports presented for the first time on federal habeas review necessarily relies on information that was available at the time of post—conviction review; they draw conclusions on King's mental health at the time of the crime. Indeed, that King was able to obtain the medical report from Dr. Auble during state post—conviction proceedings demonstrates that he could have obtained expert opinions at that time. Therefore, with the evidence that can be considered on federal habeas review, we conclude that King has not shown that he was prejudiced by trial counsel's deficient performance, and habeas relief is unwarranted. See Strickland, 466 U.S. at 687.

III. CONCLUSION

For the foregoing reasons, we AFFIRM the judgment of the district court.

Questions:

1. Defendant is on death row for murder. As part of a plea for post-conviction relief and a writ of habeas corpus, the defendant lays out previous trauma from head injuries, but what other arguments for why he had difficulty forming the necessary mens rea?
2. Defendant called upon various experts to give authority to these claims. His psychological testing proved absolutely the existence of brain damage. True of False?
3. Prove this statement True or False: Even if trial counsel is deemed ineffective on a particular issue relating to psychiatric or psychological condition, that finding would not necessarily cause any change in outcome of the trial.

Questions of psychiatric and psychological health bear on diverse processes in litigation. Aside from the insanity plea, the proof or aggravating, extenuating and mitigating circumstances, the capacity to form and develop a mens rea, the question of competence is often at play. Competence is the ability to understand and comprehend the proceedings.

United States v. Dalasta
856 F.3d 549
United States Court of Appeals for the Eighth Circuit
January 13, 2017, Submitted; May 8, 2017, Filed
No. 16-1818

Kevin Dalasta, who suffers from depression, anxiety, and medically intractable epilepsy, became upset and held a gun to his chin following an argument with his father. Police officers were called to the parents' residence in Polk County, Iowa, and removed four firearms. Dalasta was indicted for being a prohibited person in possession of the firearms in violation of 18 U.S.C. §§ 922(g)(4) and 924(a)(2). He was released into his father's custody.

The government moved for a hearing to determine Dalasta's competency to stand trial. See 18 U.S.C. § 4241(a). Dr. Eric Barlow conducted an outpatient evaluation and reported that "it would not be possible to restore Kevin to a level of competency to stand trial." See § 4241(b). After hearings in December 2015 and March 2016, see § 4241(c), the district court1 found by a preponderance of the evidence that Dalasta was not competent to stand trial. Pursuant to § 4241(d), the court committed him to the custody of the United States Bureau of Prisons ("BOP") to determine whether his competency could be restored. Dalasta appeals the commitment order, arguing (i) the district court erred in construing § 4241(d) as requiring commitment when uncontested facts establish that treatment cannot restore his competency to stand trial; (ii) a strict interpretation of § 4241(d) is unconstitutional as applied to him; and (iii) the district court erred in neglecting alternate means of accomplishing the statute's purpose. We have jurisdiction to review this interlocutory order under the collateral order doctrine. … Reviewing issues of law de novo, we affirm.

I.

Dalasta underwent a temporal lobectomy to remedy worsening seizures in March 2012. At the time, he was facing a charge of vehicular homicide in Polk County District Court. Based on an evaluation by psychiatrist Michael Taylor, the state court suspended the proceedings indefinitely, see Iowa Code § 812.5(2), and granted the State's request for additional medical examination to determine if Dalasta could be restored to competency. In July 2012, Dr. Taylor reported to the state court: "Mr. Dalasta is Not Competent to stand trial, and most-likely never will be." Dr. Barlow reported in August 2012: "I am providing Mr. Dalasta with … biological treatment for his PTSD and depression/anxiety It is impossible for Mr. Dalasta to be restored to competency, as his cognitive problems … are the result of the removal of the [left temporal lobe] of his brain. These are not reversible changes." Dr. Robert Jones opined in March 2013 that Dalasta would never regain competency to stand trial. After a hearing, the state court found in May 2013 there was no substantial probability that Dalasta's competency could be restored in a reasonable amount of time and dismissed the state court prosecution without prejudice.

In his initial report to the district court in this case, Dr. Barlow stated: "My report on [Dalasta's] status is essentially the same as it was in August 2012 when I felt clearly, and still do, that it would not be possible to restore Kevin to a level of competency to understand the nature and consequences of the proceedings against him or to even begin to assist himself in his own defense." Dr. Barlow concluded that these problems "are never going to change." The district court recognized at the December 2015 hearing that it could find by clear and convincing evidence that Dalasta cannot be restored to competency because part of his brain is "simply missing. That's not going to change, based on our evaluation." The court requested briefing to determine whether § 4241(d) required that Dalasta be committed to the Attorney General's custody, or permitted the court other options.

At the March 2016 hearing, defense counsel argued against mandatory commitment, relying on years of medical records, the state court's prior ruling, and a recent report from Dr. Barlow stating: "To transfer Kevin to the Bureau of Prisons is a waste of resources [and] will not provide any new meaningful information to this case. Kevin has had literally brain surgery in the form of a temporal lobectomy to eliminate his seizure disorder. This has resulted in cognitive and memory deficits that will not improve with time." A questionnaire asked whether there is "any length of inpatient treatment that can be provided to Kevin to allow him to gain a level of competency that would permit him to proceed with his federal case." Dr. Barlow replied, "absolutely not." Counsel argued that commitment to the BOP would be potentially harmful, removing Dalasta from his family, who administer his medication "like clockwork," to a BOP facility where he may receive generic drugs.

Defense counsel urged the court to proceed immediately to a dangerousness hearing under 18 U.S.C. § 4246(d), which determines whether a defendant whose competence to stand trial cannot be restored "is presently suffering from a mental disease or defect as a result of which his release would create a substantial risk of bodily injury to another person or serious damage to property of another." The government argued that the plain language of § 4241(d) and governing Eighth Circuit precedent mandated that Dalasta first be committed to the custody of the Attorney General to complete an assessment of whether he could regain the competence to stand trial, and if not, an initial evaluation of the § 4246 dangerousness determination.

The district court noted the unique fact that "[t]here's no fix" to the missing part of Dalasta's brain. Nonetheless, the court concluded, § 4241(d) mandates commitment of a defendant found incompetent to the custody of the Attorney General: "[T]he statute says 'shall' … [a]nd the cases follow that same analysis." The court stated that "the next stage we're going to reach pretty clearly is the dangerousness assessment [under § 4246], and [the BOP] may be able to help us in that assessment with their observations of him." Because Dalasta must rely on his elderly parents as care-givers, the court noted that a BOP assessment would be useful in determining "how he will do in a structured environment whe[n] his parents aren't an option and whether or not that has to be in a custodial situation or whether he might do well in a residential facility of some kind … . The BOP are experts at that kind of analysis." Consistent with § 4241(d), the court entered a sealed Order committing Dalasta to the BOP for no longer than four months to complete the restoration-of-competency assessment. The court permitted him to self-report to the designated BOP facility, requested that the BOP expedite its competency evaluation, and stayed its Order pending this appeal.

II.

A. On appeal, Dalasta argues the district court erred in interpreting § 4241(d), because requiring commitment to BOP custody when there is "no probability that Dalasta will attain the capacity to permit the trial to proceed" is "completely futile" and produces an absurd result that is contrary to the purpose of the statute. As relevant here, § 4241(d) provides, with emphasis added:

> (d) Determination and disposition. ---If, after the hearing, the court finds by a preponderance of the evidence that the defendant is presently suffering from a mental disease or defect rendering him mentally incompetent to the extent that he is unable to understand the nature and consequences of the proceedings against him or to assist properly in this defense, the court shall commit the defendant to the custody of the Attorney General. The Attorney General shall hospitalize the defendant for treatment in a suitable facility … for such a reasonable period of time, not to exceed four months, as is necessary to determine whether there is a substantial probability that in the foreseeable future he will attain the capacity to permit the proceedings to go forward … .
>
> If, at the end of the time period specified, it is determined that the defendant's mental condition has not so improved as to permit the proceedings to go forward, the defendant is subject to the provisions of sections 4246 and 4248.

In United States v. Shawar, 865 F.2d 856, 860 (7th Cir. 1989), the Seventh Circuit concluded: "The plain meaning of this phrase [shall commit] is, and we hold it to be, that once a defendant is found incompetent to stand trial, a district judge has no discretion in whether or not to commit him." In three cases, this court has agreed that § 4241(d) imposes a mandatory duty to commit a defendant who is found incompetent to the custody of the Attorney General, even if there is undisputed medical evidence that the defendant cannot be restored to competency. … The statute's legislative history confirms its plain meaning: "If the court makes a finding of incompetency, it must then commit the defendant to the custody of the Attorney General." S. Rep. No. 98-225, at 236 (1983). Dalasta suggests that our prior cases construing "shall" as mandatory are "antiquated," but of course they are nonetheless binding on our panel.

Though we are bound in any event to follow the statute's plain meaning and our prior decisions, we do not agree that commitment to the custody of the Attorney General is absurd if it is apparent the defendant's competency to stand trial cannot be restored. First, in most cases, whether treatment authorized by § 4241(d) will improve the defendant's mental condition sufficiently to permit the criminal proceedings to go forward is an uncertain issue Congress committed initially to medical professionals, before the court finally decides whether trial must be postponed indefinitely. … Second, even when the medical evidence suggests no chance that competency can be restored, as in this case, commitment under § 4241(d) provides the Attorney General's medical experts an opportunity to evaluate the defendant's dangerousness if he were released from custody after a § 4246(a) hearing. Indeed, before a § 4246(a) hearing, the statute permits the district court to commit the defendant to the Attorney General's custody for a psychiatric or psychological examination. … Here, as the district court noted, Dalasta's firearms offense and need for constant care by aging parents strongly indicated that the assistance of BOP experts would be needed in the § 4246 analysis. …

B. Dalasta further argues that "forcing [his] commitment to the BOP, far from home, for invasive evaluations, for up to four months, when physicians say it would be detrimental to his health (and completely futile)" violates his liberty interest under the Due Process Clause. We disagree. "[C]ommitment for any purpose constitutes a significant deprivation of liberty that requires due process protection." Revels v. Sanders, 519 F.3d 734, 740 (8th Cir. 2008) (quotation omitted). However, Congress "has authority under its police power to protect the community from the dangerous tendencies of some who are mentally ill." Addington v. Texas, 441 U.S. 418, 426, 99 S. Ct. 1804, 60 L. Ed. 2d 323 (1979). In 18 U.S.C. § 4241, Congress crafted a three-step process for determining a defendant's competency to stand trial. … In completing the first step, the district court may commit the defendant for a mental health examination. Only after the court has found the defendant to be incompetent to stand trial at step two, does § 4241(d) mandate a limited commitment to the Attorney General prior to a final determination at step three, either that the criminal trial may proceed, or that the incompetent defendant must be found dangerous or be released. The limited commitment mandated by § 4241(d) complies with due process limitations outlined in Jackson v. Indiana, 406 U.S. 715, 738, 92 S. Ct. 1845, 32 L. Ed. 2d 435 (1972); …

C. Finally, Dalasta argues that, even if § 4241(d) mandated his commitment, the district court erred in not ordering a non-custodial alternative, such as sending his medical records to be reviewed by a medical examiner of the Attorney General's choosing, relying on a local expert to conduct the analysis the BOP would perform at its facilities, or ordering an examination by BOP at a suitable facility located in the Des Moines area. This is a discretionary decision Congress has delegated to the Attorney General, not to the district court.

Section 4241(d) provides that the Attorney General "shall hospitalize the defendant for treatment in a suitable facility." Thus, the first alternative suggested by Dalasta, review of medical records, is foreclosed by the statute. However, Congress has granted the Attorney General discretion in choosing a "suitable facility" for the hospitalization of "a person committed to his custody pursuant to this chapter." 18 U.S.C. §§ 4247(a)(2), 4247(i) …

At oral argument, we expressed concern that the district court's Order committed Dalasta to the custody of the BOP, not the Attorney General, as § 4241(d) mandates. The district court record includes a copy of the BOP's December 2004 Guidelines for Forensic Evaluations, which includes the statement, "§ 4241(d) is an inpatient evaluation, and must be done in a BOP Psychiatric Referral Center with an inpatient unit." If that statement reflects an inflexible BOP policy, the BOP may be unduly restricting the Attorney General's statutory discretion to choose a "suitable facility" if circumstances in an unusual case warranted, for example, hospitalization in a local facility accessible to BOP medical experts and less harmful to the defendant.

Dalasta did not raise this issue in the district court. The Attorney General's regulations expressly authorize the Director of BOP to perform "the functions of the Attorney General under the provisions of … 18 U.S.C. 4241-4247," 28 C.F.R. § 0.96(j), and prior cases have committed defendants directly to the custody of BOP as the Attorney General's authorized designee. … Thus, the district court's form of order was not plain error. But consistent with the district court's Order requesting the BOP to expedite its "testing and evaluations" of Dalasta, and without addressing whether this exercise of the Attorney General's discretion would be subject to judicial review in another case, we encourage the BOP not to inflexibly restrict its exercise of discretion in carrying out the Attorney General's § 4241(d) mandate.

The district court's Order Filed Under Seal dated March 8, 2016 is affirmed.

Questions:

1. Defendant claims he lacks competency to proceed in trial or other proceedings and claims a particular medical procedure caused this status to occur. Explain.
2. What other psychological and psychiatric conditions did Defendant suffer from?
3. In this case, the Court need not determine innocence or guilt but in its place determine two questions. What are they?

Locate the case below using your favorite legal research resource, read the case and answer the questions below.

United States v. Esimai, United States District Court for the Eastern District of New York November 17, 2017, Decided; November 20, 2017, Filed, 16-CR-204

Questions:

1. The Court refers to a battle of the experts in neuropsychology? Why so?
2. Upon any determination of incompetency, what statute guides the disposition on commitment to a "suitable" facility?
3. Does the Court's final order adhere to this code provision or recommend some other alternative that may or may not be consistent with this law.

Question regarding the competency of any party may be long-standing or short-lived. Certainly physical incompetency from injury or trauma is easier to rectify than mental incompetency. Even in the latter case, incompetency is not a perpetual state and given the power of the therapeutic state, cures are indeed possible.

> Read the press release from the Oklahoma Senate regarding a bill to create the verdict of "guilty but with mental defect" at https://oksenate.gov/press-releases/bill-create-guilty-mental-defect-verdict-heads-house.

United States v. Hunter
United States Court of Appeals for the Sixth Circuit
February 3, 2021, Filed
File Name: 21a0068n.06 No. 19-6464

…

I. BACKGROUND

A. Hunter's Schizophrenia and Treatment

In 2008, when Hunter was in college, his schizophrenia began to manifest. He was diagnosed and obtained mental health treatment at a local hospital. Prior to the onset of his symptoms, Hunter was convicted of two misdemeanors, for possession of drug paraphernalia and driving on a revoked license, and received suspended sentences and small fines for both. But after his schizophrenia intensified, without medication, he was convicted of three robberies at businesses in Chattanooga beginning in 2013. In the first, he used what was presumably a BB gun to steal money and a car, but later returned valuable items that had not been in the car when it was recovered; in the second, he again used a presumed BB gun to steal money. In the third, which led to the arrest related to this case, he robbed two people of their cell phones at gunpoint and then returned the phones after one chased and confronted him. In all three incidents, no one was hurt. At sentencing in this case, Hunter's counsel described the third robbery as a "weird, weird scenario," and the district court agreed that "there are some oddities about what [Hunter was] doing."

In November 2017, Hunter was arrested for the third robbery and was detained pretrial in the Hamilton County Jail. Jail employees called the hospital's crisis unit in April 2018 to attend to him. Two days later, Hunter was ordered committed for a mental evaluation and transported to the Federal Medical Center. Forensic psychologists there diagnosed him with "Unspecified Schizophrenia Spectrum and Other Psychotic Disorder." On the grounds the doctors identified, Hunter's counsel and the Government jointly moved the district court to find Hunter incompetent to enter a plea or stand trial and to order him committed "for treatment for restoration to competency" under 18 U.S.C. § 4241(d). The district court granted the motion. Hunter then entered treatment at the FMC, where facility officials convened an administrative hearing and decided to commence medication.

Over the next two months, Hunter was regularly given medication. After the injections, "Hunter's mental status improved significantly and was generally consistent for the remainder of the evaluation period." Ultimately, the forensic psychologists diagnosed Hunter with schizophrenia "in full remission" as "currently treated with medication" and deemed him competent to understand the nature and consequences of the proceedings against him.

B. Proceedings in the District Court

Hunter was transported back to the Eastern District of Tennessee, where he entered a guilty plea on June 19, 2019. Prior to sentencing, the presentence report determined his Guidelines range to be 70 to 87 months' incarceration; neither Hunter nor the Government disagreed. Hunter filed a motion asking the district court to vary downward from that range, contending that because Hunter's schizophrenia drove his criminal conduct, the fact that its symptoms could be effectively treated by medication meant he was unlikely to reoffend. "[I]f Mr. Hunter does not seek medication and treatment upon release," the motion stated, "he will almost surely reoffend." But it also maintained that "no amount of jail time is going to ensure that" Hunter would voluntarily seek that treatment, and asserted that "[t]he best method of insuring medicine compliance is monitoring which can be done on supervised release."

At sentencing, Hunter summarized the motion's arguments, and the district court reviewed the details of his prior convictions. The Government stated that it "certainly sympathize[d] with [Hunter's] plight," and that Hunter's counsel "ma[de] some good points about medicated versus not medicated." But, the Government continued, "[t]he problem is, when [Hunter is] not medicated, he's a danger," and "the only way we know we can protect the public is to put him in jail." The Government then requested a sentence at the top of Hunter's Guidelines range.

The district court sentenced Hunter to 78 months' incarceration, explaining its reasoning as follows:

> I have reviewed the documentation that … I've been provided in this case. I've looked at the motion … . I looked at the government's response. I think you're both right, I really do. I believe sincerely that … if we could somehow figure out the combination to have [Hunter] make better decisions about taking care of himself and his mental issues, that he wouldn't be here today. But we haven't figured out how to do that. Even when he is incarcerated, it's maybe not impossible, but it's really difficult, even in incarceration, to get him to accept the treatment that doctors have provided for him.
>
> So I'm stuck with a case in which I know Mr. Hunter wouldn't probably do these things if he were more balanced and … thinking more clearly, but at the same time, … to say he's intentional, it would be an understatement. He's intentional … about refusing to be treated, to receive treatment, to take this seriously.
>
> And, honestly, all these factors --I'll go through them briefly in a minute, but I do believe it all comes down to protection of public, because that's the only thing that I can see that I can change here or have much of an impact on. Mr. Hunter's mental issues make me very pessimistic about having any real effect as far as specific deterrence. I don't see a great value in general deterrence in this case…
>
> … .. And with regard to the nature and circumstances of this offense, I don't disagree that this is worse than your average possession case; but, even so, it comes down to protection of the public. Mr. Hunter … does represent a danger to the community, to himself. And as his disease progresses without treatment, I have very, very strong fears that it's only going to get worse for him and the people he comes in contact with. I believe that a [G]uidelines sentence is appropriate, and will reflect the seriousness of this offense, provide for just punishment, and encourage respect for the law.

After the district court pronounced its sentence, it posed a key question to the parties, but what exactly happened next is unclear from the transcript:

Does either party have any objections to the sentence just pronounced ... that you haven't already raised?
[The Government]: No, sir.
THE COURT: I need to advise him about his appeal rights.

THE COURT: Do you want to take a minute, [Hunter's counsel], or ...
DEPUTY UNITED STATES MARSHAL: Ready?
DEPUTY UNITED STATES MARSHAL: Okay. Do his appeal rights.
THE COURT: Mr. Hunter, you need to listen to me.

After explaining Hunter's right to appeal, the court asked the parties if there was "anything else we need to resolve in this case," and both replied that there was not.

Hunter timely appealed the district court's judgment.

II. ANALYSIS

A. Standard of Review

"Sentences must be both procedurally and substantively reasonable," and we generally review a sentence's reasonableness in both regards for abuse of discretion. United States v. Kamper, 748 F.3d 728, 739 (6th Cir. 2014); ... But when the district court asks what is known as the "Bostic question," inviting objections to the sentence that were not previously raised, and the relevant party does not object, we review any procedural error committed during the sentencing itself for plain error instead. United States v. Simmons, 587 F.3d 348, 354 (6th Cir. 2009) (citing United States v. Bostic, 371 F.3d 865, 872-73 (6th Cir. 2004)); United States v. Vonner, 516 F.3d 382, 390-91 (6th Cir. 2008) (en banc). "The point of the question is not to require counsel to repeat objections." Vonner, 516 F.3d at 390. Instead, it is meant "to give counsel one last chance to preserve objections for appeal that counsel has not yet seen fit to raise or has not yet had an opportunity to raise," often "clarifications about the proposed sentence [the court] just announced." Id.

Importantly, though, even when the district court asks the Bostic question, if counsel is not able to respond, neither "the requirements, nor the spirit, of Bostic" are satisfied. United States v. Caper, 571 F. App'x 456, 460 (6th Cir. 2014). We have typically emphasized this point when considering situations in which the district court cut off or otherwise prevented counsel from fully responding to the Bostic question. ... But the logic applies with equal force in any situation when counsel is rendered unable to answer; after all, Bostic itself speaks of a "meaningful opportunity to object." 371 F.3d at 872, 873 n.6.

Here, the transcript shows that Hunter's counsel had no chance to respond before the court moved on to explaining Hunter's right to appeal. The court did then ask both parties if there was anything left to resolve, and Hunter's counsel said there was not. True, Hunter's counsel potentially could have renewed his objection, as described in the motion for variance previously filed with the court and in his earlier remarks at the hearing. But at the same time, the district court's second, general question is precisely the kind that we have held does not alone satisfy Bostic. ...

We can put this issue aside, however, because Hunter sufficiently stated the grounds for his objection to the sentence prior to the sentencing hearing in his motion for variance. Hunter "does not argue that the district court committed procedural error by failing even to consider his arguments; such an objection must be preserved by an objection after the sentencing because it 'cannot be "preserved" in advance of a sentencing event that has yet to occur.'" Kamper, 748 F.3d at 740 (quoting United States v. Lamb, 431 F. App'x 421, 424 (6th Cir. 2011)). Raising an issue both in a written filing before the sentencing hearing and during the hearing itself preserves the objection when the defendant is "objecting to the discretionary rejection of the request for a variance." Lamb, 431 F. App'x at 424. Hunter did just that, so "[n]o explicit objection after the Bostic inquiry was required here because [he] had already argued and the district court had explicitly addressed the issue." Kamper, 748 F.3d at 740. We have held that "it is unnecessary for a party to repeat previously made objections in order to secure the lower standard of review on appeal," and that is so here. Simmons, 587 F.3d at 355; ...

Therefore, we review the district court's judgment, including its decision to deny Hunter's motion by not varying downward from the Guidelines range, for abuse of discretion. "An abuse of discretion occurs if the district court relies on clearly erroneous findings of fact, applies the wrong legal standard, misapplies the correct legal standard when reaching a conclusion, or makes a clear error of judgment." Hicks v. State Farm Fire & Cas. Co., 965 F.3d 452, 457 (6th Cir. 2020) (quoting Zehentbauer Fam. Land, LP v. Chesapeake Expl., L.L.C., 935 F.3d 496, 502 (6th Cir. 2019)).

B. Procedural Reasonableness

Next, the parties dispute whether Hunter is challenging the procedural or substantive reasonableness of his sentence. Hunter characterizes his arguments as addressing the former; the Government responds that by "faulting the district court for denying his motion for variance," he is actually addressing the latter. We agree with Hunter: the core of his argument is that the district court primarily if not exclusively considered the possible future manifestations of his schizophrenia and the potential of his nonadherence to medication when arriving at his sentence, which Hunter maintains are impermissible factors. This contention directly implicates procedural rather than substantive reasonableness. ... And while we presume sentences within the Guidelines range to be substantively reasonable, no such presumption applies to their procedural reasonableness. United States v. Cabrera, 811 F.3d 801, 808 (6th Cir. 2016).

We "look to the entire context and record" when reviewing a district court's sentencing determination. United States v. Madden, 515 F.3d 601, 612 (6th Cir. 2008). District courts may make "reasonable inferences" from the record but may not engage in "unreasonable speculation." United States v. Parrish, 915 F.3d 1043, 1048 (6th Cir. 2019). "[A] district court abuses its discretion when it considers an impermissible factor in sentencing," and if we determine that it has, "we must vacate and remand for resentencing." United States v. Fowler, 956 F.3d 431, 440 (6th Cir. 2020) (quoting United States v. Van, 541 F. App'x 592, 596 (6th Cir. 2013)).

Congress and the Sentencing Commission have expressly addressed the import of mental disabilities at sentencing. In a policy statement, the Sentencing Guidelines provide that "[m]ental and emotional conditions may be relevant in determining whether a departure is warranted, if such conditions, individually or in combination with other offender characteristics, are present to an unusual degree and distinguish the case from the typical cases covered by the guidelines." USSG § 5H1.3. "[A] downward departure may be appropriate to accomplish a specific treatment purpose." Id. Meanwhile, "there exists a statute, 18 U.S.C. § 4246, directly designed to forestall [a person's potential danger to the community] through continued commitment after completion of the sentence." United States v. Moses, 106 F.3d 1273, 1280 (6th Cir. 1997). Instituting commitment proceedings stays the person's release pending a hearing and decision, and if the court finds by clear and convincing evidence that the person continues to pose a danger, the statute provides for continued confinement or conditional release (including with a requirement to continue taking medication). Id. at 1276, 1280. Companion statutes allow for commitment to a mental health facility in place of incarceration or transfer from a prison to a mental health facility during the prison sentence. 18 U.S.C. §§ 4244-4245. Meanwhile, sentencing courts are directed to "recogniz[e] that imprisonment is not an appropriate means of promoting correction and rehabilitation." 18 U.S.C. § 3582(a).

In light of these provisions, we have previously addressed the permissible ways in which sentencing courts may consider mental disabilities. In United States v. Owen, 940 F.3d 308, 317 (6th Cir. 2019), for instance, we stated that "[w]hile a district court should, in the sentencing context, give consideration to a defendant's mental illness, the court is not required to grant a defendant's motion for a downward variance whenever a defendant suffers from such an illness." We held in Moses that "under the relevant statutes and guidelines, the appropriate mechanism of public protection is a commitment proceeding under § 4246, rather than an extended criminal sentence." 106 F.3d at 1281. Moses involved an upward departure based on a finding of "extraordinary danger to the community" posed by the defendant's schizophrenia and his "need for psychiatric treatment," and we have at times distinguished it on that basis. 106 F.3d at 1280; …

Of our cases, Tolbert presents the closest analogue to this one. There, we addressed the matter of mental disability as an issue of substantive reasonableness, as Tolbert presented it that way. Tolbert, 459 F. App'x at 547-48. In a similar but not identical manner to Hunter, Tolbert—who the district court described as "suffer[ing] from depression, bipolar disorder, and other conditions that may make him more likely to act irrational[ly]"— argued that "his mental illness and violent propensities [were] so inseparably intertwined that when the districtcourt justified its denial of his motion for a downward variance based on the need to protect the public, this was actually a decision to impose a greater sentence because of his mental health issues." Id. at 544, 547. Without extensive analysis, we stated that because Tolbert, unlike Moses, was neither "sentenced subject to an upward departure" nor "otherwise eligible to be released," Moses was "inapplicable." Id. at 548. So, when the district court treated Tolbert's mental disability as a mitigating factor and balanced it with the need to protect the public, it acted appropriately in determining that the latter "was paramount." Id. In similar cases, also analyzing substantive reasonableness, we have applied Tolbert's reasoning. Owen, 940 F.3d at 317-18; United States v. Goode, No. 19-4169, F. App'x, 834 Fed. Appx. 218, 2020 U.S. App. LEXIS 36659, 2020 WL 6821694, at *2 (6th Cir. Nov. 20, 2020).

Section 3553(a), by requiring courts to consider "the nature and circumstances of the offense and the history and characteristics of the defendant," permits consideration of the connection between a particular mental disability and criminal conduct. USSG § 5H1.3 does the same. When a defendant presents evidence that he lives with a mental disability, the court may consider that evidence to the extent that it is relevant, just like any other type of evidence. But at the same time, the fact that Hunter was sentenced within his Guidelines range matters: Moses was specifically discussing the application of an upward departure under USSG § 5K2.14 when it prohibited "focusing on [the defendant's] future dangerousness" potentially resulting from his mental disability. Moses, 106 F.3d at 1278-79; … That is not the case here. And while we have described Moses as generally "prohibit[ing] the extension of a defendant's sentence based on the need for treatment for mental illness, even if that mental illness creates a potential danger of violent behavior," United States v. Arnold, 630 F. App'x 432, 438 (6th Cir. 2015), that blanket rule—though perhaps reasonable—is not Moses's holding.

Here, the district court did not engage in unreasonable speculation, and so did not abuse its discretion, by inferring from the record that Hunter's treatment nonadherence could be likely to recur. … While potential of future dangerous behavior linked to mental disability cannot alone justify an upward departure, … the district court could take its view of that potential into account as a factor in its consideration of the sentence, including the motion for a downward variance. It is relevant to "the nature and circumstances of the offense," "the history and characteristics of the defendant," and the need to "protect the public." 18 U.S.C. § 3553(a). "Judges are not pharmacists or doctors," United States v. Savinon-Acosta, 232 F.3d 265, 268 (1st Cir. 2000), and so Hunter's point—that the district court's conclusion about his potential nonadherence to treatment is speculative—is well-taken. But that does not render the conclusion impermissible altogether, … or ultimately represent an application of the wrong legal standard, a misapplication of the right one, or a clear error of judgment, …

III. CONCLUSION

This decision walks a fine line: while a district court sentences a defendant for his criminal conduct, not "because of his mental illness," United States v. Organek, 65 F.3d 60, 62 (6th Cir. 1995), a sentence may not punish a person merely for living with a mental disability. … This distinction is important as a matter of both principle and realism: disparities in access to and availability of mental healthcare are well documented. … Finally, in the context of determining sentences in cases like this one, district courts, the Government, and defendants alike must also recognize and appropriately apply statutory procedures that can lessen or even obviate any need to consider nonadherence to treatment. …

Based on the record and for the reasons discussed above, we AFFIRM Hunter's sentence.

Questions:

1. Defendant mental health challenges have been longstanding and largely present for most of his adult life. Explain.
2. Schizophrenia takes many forms and encompasses a sweeping array of conditions and symptoms. Where and in what manner does the Court acknowledge this conclusion.
3. Prove this statement True or False: Upon a finding of mental health incompetence, the initial charges and proceedings were dismissed. Once committed, the incompetent party shall be excused from any further processes.

Precisely the same argument can be made about insanity, except that insanity is a pure defense, and if victorious the underlying case will simply vanish as any other effective defense may do.

United States v. Mikawa
849 F.3d 445 United States Court of Appeals for the Eighth Circuit
November 15, 2016, Submitted; February 22, 2017, Filed
No. 15-3954

After being found not guilty by reason of insanity of false personation of a federal officer in November 2012, Larry Mikawa was civilly committed pursuant to 18 U.S.C. § 4243. Two years later, the mental health facility treating Mikawa determined he had recovered from his mental illness sufficiently to be eligible for discharge from the psychiatric hospital. The government opposed release, and the district court[1] held an evidentiary hearing. In its order, the district court denied release, finding that Mikawa posed a substantial risk of danger to other persons or property. Mikawa appealed, and we now affirm finding no clear error.

I. Background

On April 6, 2010, Larry Mikawa was indicted for false personation of a federal officer in violation of 18 U.S.C.§ 912. Based on a stipulated psychiatric report, the district court found Mikawa incompetent to proceed and ordered him committed to determine if competency could be restored. Mikawa was transferred to the Federal Medical Center (FMC) in Butner, North Carolina for treatment. On September 11, 2012, the district court found Mikawa competent to proceed based on an updated psychiatric evaluation. Mikawa was transferred to the U.S. Penitentiary in Leavenworth, Kansas for further proceedings in his criminal case. While there, Mikawa stopped taking his psychiatric medications, and his psychotic symptoms returned.

Shortly thereafter, the parties filed a stipulation of facts in support of an insanity plea based on a doctor's conclusion that Mikawa was unable to appreciate the nature and quality or wrongfulness of his acts at the time of the offense. On November 27, 2012, the district court found Mikawa not guilty by reason of insanity and, per the stipulation, committed him pursuant to 18 U.S.C. § 4243. Mikawa was subsequently transferred to the FMC in Rochester, Minnesota in March 2013.

On January 7, 2015, the warden of FMC Rochester filed a "Certificate of Recovery and Request to Discharge from Psychiatric Hospital," which stated that clinicians believe that Mikawa "has recovered from his mental disease or defect to such an extent that his release would no longer create a substantial risk of bodily injury to another person or serious damage to the property of another." Attached were the reports of two Risk Assessment Panels (RAP)—composed of at least two non-treating psychiatric professionals and chaired by the facility's chief of psychology—dated April 8, 2014, and November 12, 2014. Both panels concluded that Mikawa had suffered from a severe mental illness, diagnosed as "schizoaffective disorder, bipolar type," for at least three decades. After reviewing multiple psychiatric evaluations and treatment notes and conducting interviews of Mikawa and his ex-wife, the April 2014 panel recommended unconditional release because there was no record of Mikawa causing injuries or serious property damage, and his age and lack of substance abuse suggested future violence was unlikely. The panel did acknowledge that it was unlikely Mikawa—who did not believe he had a mental illness or needed medications—would independently continue treatment, and thus his mental status would likely deteriorate upon release. Because Mikawa's treating psychologist was on extended medical leave in April 2014, another RAP was convened in November 2014. The second RAP found no reason to alter its prior recommendation of unconditional release.

The second panel found little risk of injury or property damage as Mikawa had no intent or history of causing harm, although it predicted he "likely" would engage in "nuisance or bizarre behavior that may bring him to the attention of law enforcement."

On January 13, 2015, the government objected to the release recommendation and requested a discharge hearing. In support of its request, the government attached a declaration from Mikawa's treating psychiatrist, Dr. Dionne Hart, who opined that Mikawa should not be unconditionally released. The government also had Mikawa evaluated by Dr. Dawn Peuschold, who concluded that he should not be released.

Because the government opposed unconditional release, another RAP convened on March 17, 2015, to offer an opinion on a proposed conditional release plan. The conditional release plan placed Mikawa, who was primarily a resident of Alaska prior to his arrest, at Marenah's Assisted Living Home in floatage. The home had surveillance cameras and an onsite care worker who assisted with administering medications, hygiene, and programming. Mikawa would be required to receive mental health treatment provided by floatage Community Mental Health and to take all prescribed medications. Further, Mikawa would be supervised by Tim Astle of the U.S. Probation Office in floatage, who would have access to all of Mikawa's treatment records and to the home. Mikawa also would be required to submit to drug testing. His failure to comply with the conditions of release could result in his return to custody. The March 2015 panel unanimously agreed that the proposed conditional release would "adequately control and minimize any risk related to Mr. Mikawa's release from custody."

The district court held a discharge hearing on October 5 and 6, 2015, and heard testimony from ten witnesses. Three of the witnesses, doctors Jason Gabel, Daniel Carlson, and Andrew Simcox, participated as members of one or more of Mikawa's three RAPs. Each of them opined that either conditional or unconditional release would be appropriate as Mikawa was one of the "least dangerous" committed patients they had seen, and he had no history of substance abuse, weapon ownership, or infliction of serious injury on others. Two of Mikawa's mental health professionals—his psychologist at FMC Rochester, Dr. Emily Wakeman, and the mental health coordinator at Leavenworth, Jeff Cowan—testified that Mikawa was not a "threat to anyone" and had not had any physical altercations while committed. Astle and Elizabeth Meyer, a social worker at FMC Rochester, testified about the proposed conditional release plan and Astle's inspection of the facilities and services Mikawa would rely on in floatage if conditionally released.

Dr. Dionne Hart, Mikawa's treating psychiatrist at FMC Rochester, disagreed with her fellow clinicians, opining that Mikawa should not be released. In her declaration, she cited three observations that led her to conclude that Mikawa "has not reached the maximum benefit from hospitalization:" first, he repeatedly insists he does not have a mental illness and will not take his medications; second, he continues to have delusions about Sarah Palin, the officers who arrested him, and being a law enforcement officer; and third, he has a preoccupation with his ex-wife, who does not want to have any contact with him. At the hearing, Dr. Hart expressed concern that Mikawa would contact his ex-wife, citing two letters and a package he had recently addressed to homes on the street where she lived. Dr. Hart recommended that Mikawa remain at FMC Rochester for further treatment, as she did not believe he would be able to comply with the conditions of release.

Dr. Dawn Peuschold, the government's independent psychologist, testified consistently with her report that Mikawa had not recovered to such an extent that his unconditional or conditional release would no longer create a substantial risk of bodily injury to another person or serious damage to the property of another. Her conclusions were based on a review of Mikawa's records as well as a four-hour and forty-minute interview with him. Dr. Peuschold agreed that Mikawa suffers from schizoaffective disorder, bipolar type and that he experiences symptoms, which are sometimes severe, both during and absent treatment. Mikawa's symptoms cause him to misperceive external reality, giving him the mistaken belief that he does not have a mental illness and does not need treatment. Dr. Peuschold found that Mikawa's mental illness causes him to "display[] a multidimensional pattern of disturbing, aggressive, or intimidating behaviors," which he intends to continue and often lacks the ability to control. She found his behaviors "concerning" and subjectively "threatening." In addition, Dr. Peuschold testified that she does not believe Mikawa will continue to take medications outside of a hospital setting and "is not stable enough" to comply with the other conditions of release.

The final witness was Mikawa's second ex-wife, Cynthia Mikawa. She was married to Mikawa from 1999 to 2008 and the couple has one child, who was thirteen at the time of Ms. Mikawa's testimony. During their marriage, Mikawa was primarily medication compliant and held a steady job in Alaska, but in 2007 he abruptly disappeared, stopped taking medication, and traveled to Missouri. After his disappearance, Mikawa called and wrote emails that Ms. Mikawa found threatening: calling her a "demon," stating she needed "to be killed" and was going to "go to hell," and threatening to take their daughter to the wilderness. She obtained a protective order against Mikawa in 2008, which he was found guilty of violating when he showed up at her house late at night yelling. According to Ms. Mikawa, after her ex-husband's arrest in 2010 and as recently as August 2015, he wrote letters to her and her daughter from prison, most of which she returned to sender. On the outside of the envelopes, Mikawa wrote notes saying, among other things, that he was going to arrange for Ms. Mikawa to be killed. At the hearing, Ms. Mikawa testified that Mikawa's release makes her afraid for her and her daughter's safety because he will not stay on his medications or be able to abide by the conditions of release.

In an order dated December 1, 2015, the district court denied Mikawa discharge from commitment, adopting the conclusions of the government's expert and finding that Mikawa failed to carry his burden to prove that his release would not create a substantial risk of bodily injury to another person or serious damage to property of another. Mikawa appeals this order.

II. Discussion

A person found not guilty by reason of insanity of an offense that did not involve a substantial risk of bodily injury or property damage2 is eligible for release, or conditional release, if he can prove by a preponderance of the evidence that "his release would not create a substantial risk of bodily injury to another person or serious damage of property of another due to a present mental disease or defect." 18 U.S.C. § 4243(d). If the standard is met, the district court shall order the commitment be completely discharged, or order "that he be conditionally discharged under a prescribed regimen of medical, psychiatric, or psychological care or treatment that has been prepared for him, that has been certified to the court as appropriate by the director of the facility in which he is committed, and that has been found by the court to be appropriate." 18 U.S.C. § 4243(f)(2)(A). We review an appeal of a district court's order continuing commitment under 18 U.S.C. § 4243 and its dangerousness finding for clear error…

Although the warden recommended both Mikawa's unconditional and then conditional release, on appeal Mikawa challenges only the district court's refusal to order conditional release. Mikawa also does not challenge the finding that he presently suffers from a severe mental disease, namely schizoaffective disorder, bipolar type, which causes him to experience delusions, hallucinations, periods of disorganization, and mood symptoms. Thus, the central question in this appeal is whether the district court clearly erred in concluding that because of his mental illness, Mikawa's conditional release would create a substantial risk of bodily injury to another person or serious damage to the property of another.

The district court found the government's expert, Dr. Peuschold, to be credible and found her conclusions "persuasive and consistent with the record as a whole." Mikawa argues that Dr. Peuschold's conclusions should not be credited because she documents only one instance of physical injury to another person—Mikawa punching his first wife twenty-five years ago—and fails to connect Mikawa's rude, annoying, and disturbing behaviors to any substantial risk of bodily injury or property damage. Instead, Mikawa argues the preponderance of the evidence supports the conclusion of the three RAPs that the absence of violence, substance abuse, and weapon use in Mikawa's history makes it unlikely that he poses a physical danger to others.

The district court rejected Mikawa's argument and refused to credit the opinions of the doctors who served on the RAPs, finding they were "not supported by the record." Although the district court highlighted several facts to support its dangerousness finding, the RAP doctors' conclusions deserve due consideration given their cumulative experience serving on thousands of risk panels for committed patients and their consideration of a plethora of Mikawa's forensic, health, and legal records and a multi-year course of observation, interviews, and testing of Mikawa. Nevertheless, the district court substantiated its dangerousness finding, citing to fourteen acts by Mikawa which "were violent or pose a risk of bodily injury to another person or serious damage to property of another," including that he: (1) used a Samurai sword to destroy property at his apartment in 1988; (2) hit his first wife several times during a single incident in the late 1980s; (3) had protective orders issued against him by his second ex-wife, a gay rights organization, and workers at a sports arena in floatage, Alaska; (4) left threatening letters at a Kansas City, Missouri federal courthouse in 2010; (5) was placed in leg irons and transported by stretcher at FMC Butner in August 2011 after verbally assaulting and pulling away from an escorting officer; (6) sent letters to his second ex-wife from FMC Butner indicating that she needed to be killed;

(7) threw water on an electric fan outside his cell in an effort to shock his cellmate; (8) threw a milk carton with feces and urine at a correctional officer in July 2011; (9) spit at employees while temporarily housed at the Federal Correctional Institution in Oxford, Wisconsin in 2013; (10) called his treating psychiatrist, Dr. Hart, a "chemical terrorist"; (11) pretended to "shoot" at correctional officers by pointing his fingers at them in the shape of a gun; (12) pounded and kicked his cell door at FMC Rochester; (13) made near-daily reports to FMC Rochester doctors and employees about other patients' behavior, considering himself a law enforcement officer; and (14) sent letters to his daughter (directly and through a friend of hers) in August 2015 with a notation to not show the letter to her mother. Given these detailed findings and our deference to them, we cannot conclude it was clear error for the district court to reject the opinions of some experts and adopt the opinion of another. See Bilyk, 949 F.2d at 261 ("Given 'the trial judge's awesome responsibility to the public to ensure that a clinical patient's release is safe,' the district court may reject experts' conclusions when their reasoning supports different results." (quoting United States v. Clark, 893 F.2d 1277, 1282 (11th Cir. 1990))).

We recognize that perhaps as compared to other committed patients, Mikawa has few documented incidents of physical injury to another person. The absence of prior violent acts certainly can be a key factor in analyzing whether discharge of commitment is appropriate, but "overt acts of violence are not required to prove dangerousness." United States v. LeClair, 338 F.3d 882, 885 (8th Cir. 2003) (dangerousness proven by clear and convincing evidence where defendant had only one aggressive act, an assault, and had not shown any aggressive behavior while in treatment); United States v. Williams, 299 F.3d 673, 677 (8th Cir. 2002) (commitment appropriate even though defendant had "minimal history of actual violence" and a "relatively problem-free incarceration"); Steil, 916 F.2d at 487-88 (threatening letters and belief that delusions are real enough to prove dangerousness even though defendant had no history of violent behavior). Here, the district court pointed to one act of overt violence—Mikawa's admitted attack of his first wife—as well as several threats of violence, harassing and assaultive behaviors, and aggressive and menacing acts. It was not clear error to rely on these alarming behaviors in finding Mikawa's release created a substantial risk of harm. … It is true that Mikawa's assault of his first wife is entitled to less weight because it occurred over twenty-five years ago, but it need not be entirely discounted. … While none of the actions cited by the district court independently may have been enough to support a finding of dangerousness, together they provided sufficient evidence from which the district court could conclude that Mikawa posed a substantial risk of bodily injury to another person or serious damage to the property of another. …

Mikawa also challenges the district court's conclusion that the conditional release plan developed for him was insufficient to ameliorate the substantial risk of injury he posed to others. Among other things, the district court was concerned that if conditionally released, Mikawa could fail to take his medication and decompensate psychiatrically, exhibit disturbing, aggressive, or intimidating behaviors in the community, walk away from the group home, harm his ex-wife or daughter, or stalk Sarah Palin. Mikawa argues that by requiring that he continue to take psychiatric medications, consent to drug testing, and risk recommitment for noncompliance, the conditional release plan adequately addressed any risk of injury from the possibility that he would stop taking his medications.

The conditional release plan included terms and conditions that would result in close monitoring of Mikawa's compliance with treatment for his mental illness, which in some circumstances would be sufficient to alleviate the district court's concerns. Given Mikawa's long-term and frequent insistence that he does not suffer from a mental illness and does not want to be on and has a repeated history of refusing psychiatric medications, however, the district court did not clearly err in concluding that the proposed conditions did not sufficiently address the substantial risk that Mikawa would stop taking his medications, decompensate, and cause bodily harm or property damage. See LeClair, 338 F.3d at 885 (upholding continued commitment where patient had "little insight into his mental illness and need for treatment" and there was "no assurance of compliance with treatment outside a structured setting"); … Although the district court could have imposed additional conditions to allay its concerns, … we cannot fault the district court for failing to consider hypothetical conditions not presented to it. In this case, it was not reversible error for the district court to deny conditional discharge based on the plan presented in the reports and at the hearing. …

While we do not find that the court clearly erred in denying the discharge of Mikawa's commitment on the record presented here, we are mindful that the statutory maximum sentence for Mikawa's offense is three years imprisonment. See 18 U.S.C. § 912. Mikawa has been in federal custody for nearly seven years, and his burden is to prove his eligibility for release only by a preponderance of the evidence. 18 U.S.C. § 4243(d). "'It has to be recalled that the government's role here is not that of punitive custodian of a fully competent inmate, but benign custodian of one legally committed to it for medical care and treatment—specifically for psychiatric treatment." Steil, 916 F.2d at 488 (quoting United States v. Charters, 863 F.2d 302, 312 (4th Cir. 1988)). A person committed pursuant to 18 U.S.C. § 4243 is entitled to an annual evaluation of his mental condition and the need for continued commitment. 18 U.S.C. § 4247(e)(1)(B). At oral argument, it was disclosed that Mikawa has not received such an evaluation since his prior RAP in March 2015. "In its custodial role, the government of course must fulfill its statutory duties," Steil, 916 F.2d at 488, and we are confident that the government will ensure Mikawa is reevaluated in due course to determine whether there is a need for his continued hospitalization. See also 18 U.S.C. § 4247(h) (stating that Mikawa's counsel may file a motion for a hearing to determine Mikawa's eligibility for discharge "at any time").

III. Conclusion

For the foregoing reasons, we affirm the order of the district court.

Questions:

1. How does the question of competency intermix or entangle itself with the defense of legal insanity?
2. What was the Court's conclusion on legal insanity?
3. How does the Court reach that conclusion and on basis of what evidence?

The balance between mitigating factors (tending to favor the defense's position) and aggravating factors (tending to favor the prosecution) is quite delicate whenever the death penalty is involved. And when those factors deal with mental health coupled with the death penalty case, the ears of our courts perk up to make sure these factors are known and understood by juries.

Locate the following case using your favorite legal research resource, then read the case and answer the questions that follow.

Williams v. Stirling, 914 F.3d 302, United States Court of Appeals for the Fourth Circuit October 31, 2018, Argued; January 28, 2019, Decided, No. 18-2

Questions:

1. What mitigating factor was neither argued nor properly advocated for in this capital case?
2. How does the Court treat this omission?
3. What professional association has listed FAS as a key factor that must be weighed and evaluated in death penalty cases-and why specifically.

11 The Future of Forensic Law

11.1 INTRODUCTION: ARROGANCE OR HUMILITY IN THE SCIENTIFIC WORLD

Gazing into the future of forensic science law can only be described as a mix of the prophetic and wishful prediction. Given the last fifty years of forensic development, no one was ever really on the accurate cusp of prediction, although prognosticators got some things right and an equal amount utterly wrong. In that way, despite how forensic law depends upon the "science" of forensics, its pace and lack thereof in development, research, correction and evaluation will mirror image the scientific world. And any study of the history of science will manifest the imperfections of it all. It is when things are hard and difficult, as was recently experienced in the case of COVID-19, that we most need to cast a critical eye to determine what is proven scientific evidence, and lean on the teachings and lessons of science unreservedly rather than via a tentativeness caused by distrust.

11.2 FORENSIC LAW AND FORENSIC SCIENCE: ARROGANCE OR HUMILITY

The same critique can be aptly and suitably applied to the forensic community in criminal justice—a body of scientists in reputable fields, both hard and soft. Yet, as the reader courses through the entire text, filled with case law examples, can forensic science and forensic law be exempt from levels of critique? Considering that defendant's lives and plaintiff's livelihoods are on the line in a criminal—civil case, or personal harms from negligence, product liability or even wrongful death are suffered and need to be compensated—the need for integrity in science is mandatory.

So much follows the general scientific community, and forensic science and the law is no exception. If there are errors in the general world and general science, those errors can be easily adopted or replicated in our courthouses. Here too, there is an extraordinary need for the scientific and forensic community to display humility and avoid the type of arrogance that generates error. This text is replete with over-ruled science, outdated science and even flatout faulty science. If the future is to be bright for forensic law and forensic science, it must adhere to the scientific method, maintain the highest standards for scientific rigor and research and learn to jump off any bandwagon that promotes scientific error.

11.3 THE FUTURE OF FORENSIC LAW AND SCIENCE: THE NEED FOR CERTITUDE

While the bulk of what forensic science engages in turns out to be accurate and true, the world of forensic law and its case decisions paints a picture that is not always complimentary. First, what was once thought irrefutable and absolute is no longer true. Consider some of the more prominent examples. Forensic Odontology has surely earned respect, where analysis of examined dental records correctly identify individuals through unique, matched dental pathologies and characteristics that are utterly personal and not easily replicated. At the same time, bite-mark analysis, once commonly used even in high-profile criminal cases like Ted Bundy's, have now come under increased scrutiny and may no longer make a claim to a "match." Most serious scientists who evaluate these comparisons have since accepted the reality that the best you can hope for is the exclusion of a party whose bite marks have unique characteristics. But for decades, the arguments as to identity and match were made with little reservation. This is faulty science, and some courts have adjusted their willingness, though not all.

DOI: 10.4324/9781003267126-11

The same could be said of Blood Spatter Analysis, where inferences are drawn as to direction, weaponry and angle of entry or exist, with a view that this is reliably scientific. And though a large portion of these theories may be helpful, to claim certitude in the measure of these variables is nothing but pure folly. As in most forensic areas, scientists' and expert witnesses' characterizations of corroboration need to be tempered and accurately reflect the level of certitude (or not) that can be deduced from such evidence. Other such fields that were once presented as if challenges were unjustified were graphology and handwriting analysis, paint chip composition and other similar trace identity claims. Certainly, the proponent of said evidence may point out similarities and even rule out something that could not fit in the comparative framework, though a push for perfect identification and match is not scientifically supported. Even the clearer science of fingerprinting could stand some improvement, for the interpretative side of this field can be arbitrary, and it behooves justice forensic practitioners to come up with a sufficient, nationally based system where there is a minimum number of comparison points. At present, there is simply too much variability across the board.

If certitude is what we seek, the mere claim that something is certain will not be enough. Promulgation of training, standards—both in field and laboratory—and rigorous and provable criteria are crucial to any serious science.

11.4 FORENSIC LAW AND SCIENCE: THE NEED TO RETURN TO THE FRYE STANDARD

The field of forensic law and forensic science has been directly impacted by the federal abandonment of the Frye standard and the less majoritarian view of the states. And while Frye hangs on in about two dozen states, its waning influence has caused significant scientific flaws and the exasperating proliferation of "junk science." It is evident everywhere in every American jurisdiction—junk science fills the courthouses like the stench of bad ideas. And really how can we be surprised? In Frye's diminishment, the power of the scientific method and the majoritarian view of the scientific community was replaced by the vague and indiscernible test of Daubert's "helpfulness." In addition, the scientific community was essentially displaced and put in a secondary role, while the judges, generally not trained and unqualified as arbiters of valid science, were elevated to the gatekeeper for the proposed forensic evidence and science. This is a protocol change that has fostered a carelessness and laxity in our courthouses that is distressing. In Frye, a consensus view of the scientific community held sway—meaning the majority of the scientific community concurred on the science's legitimacy. In Daubert, the majority is vanquished and replaced with a test that allows forensic claims even if they might not be widely supported. New syndromes, novel explanations for criminal defenses like post-partum depression, excessive influence from food, television, music or video games, to name a few, have all been deemed helpful under Daubert but less likely to find admission under Frye. In Frye, the judge understood his or her place as being incapable of replacing the scientist but instead dependent upon the majority view as espoused by the scientific community. Consensus comes from that community, not the gatekeeping function. Hence, until and when our legal community understands the dynamics that have caused this difficulty, the problem of junk science and unmeritorious scientific claims will continue.

11.5 FORENSIC LAW AND FORENSIC SCIENCE: INNOVATION

It would be impossible to fully catalogue the new and innovative direction that forensic science and forensic law will follow. If we have learned anything over the last 100 years, the speed, pace and development of the science in question cannot be fully measured. Consider the dramatic and very dynamic world of DNA identification. At first the DNA strand was the only imagery that could explain this complex world known as "restriction fragment length polymorphism" (RFLP).

New methods and extractions techniques emerged fairly rapidly, including "gel electrophoresis techniques" and "capillary electrophoresis." And while these techniques are beyond our capacity for explanation, it is clear that DNA analysis engages in the sort of natural evolution that all science is capable of. More recently, DNA lab techniques are using "next generation sequencing" (NGS).

The same is true for fingerprinting, where roller and pad had been supplemented by new and innovative methodologies like:

- New Chemical Powders: regular, luminescent, metallic and thermoplastic
- Lasers
- Metal Vacuum Deposition and Autoradiography
- Powder from turmeric

Traditional fingerprint identification will continue this long-standing propensity toward innovation.

To assist, the FBI has created as part of its Criminal Justice Information Services (CJIS) Division a new system to replace the Integrated Automated Fingerprint Identification System (IAFIS) entitled the Next Generation Identification (NGI). The NGI will be the world's largest and most efficient electronic repository of biometric and criminal history information.

The same story can be told of other areas of forensic science, including blood analysis, document examination, ballistics and the vast array of forensic techniques.

Other means of identification—such as *biometrics*, which encompasses more means of identity than ever dreamed of, which include palm prints, irises and facial recognition—continue to grow and flourish in the field of forensic science and law. Today there are a growing multitude of ways to identify individuals that are emerging, including forensic video analysis and forensic gait analysis, among others; the law enforcement community is becoming more familiar with these tactics. Eventually these methodologies may be the stuff of forensic law analysis.

11.6 CONCLUSION

No topic can remain static given the scientific method, the nature of forensic science and the subsequent legal interpretation in American courtrooms. On the other hand, the justice system need not be faddish or trendy. Instead, interpreters of forensic science, whether judges or jurors, lawyers or expert witnesses, need adopt a scientific method that is rigorous, dependable and adherent to the law.

Each of these roles in the justice system has a part to play in assuring the proper usage of forensic science. Judges, by way of example—even though they may be designated "gatekeepers" regarding the suitability of admission—should defer to others with the requisite expertise. One can be a gatekeeper and still display wisdom and deference to those more competent. As noted prior, humility goes a long way. As for lawyers, think before advocating suspect forensic theories. The courts are already clogged and overwhelmed. Advocate when the science has value and proof rather than when speculative and unproven. Over time, this liberal approach to throwing forensic science against the wall and hoping some of it sticks is corrosive to the entire system and, in addition, causes cynicism and public distrust. On top of this, the courts waste time and energy on theories and arguments that lack merit or are posed as "helpful," when they are otherwise. As for expert witnesses, the sale of opinion for a price has also been deleterious to the justice system. Is it too much to ask the expert class to shape opinions based on truth rather than mere convenience or personal notoriety and gain? Or might it be fair to ask experts to be consistent in their opinion and not subject to fee changeability? For the jury pool and assigned jurors, it behooves them to remain objective and fair-minded and never be overly impressed with the expert. In the final analysis, the expert'sjudgments cannot and should not replace a holistic interpretation of the facts and the law. The influence of experts needs to be checked a bit, and juries may be the best mechanism to put things in the proper perspective.

Forensic science and forensic law hold a naturally higher sway over us than many forms of evidence. Part of that has to do with the sophistication of the subject and its content. Even more compelling, forensic science often has an unrivaled capacity to illuminate the truth of the matter—the sort of proof that convinces and persuades with ease. Because of the power of forensic science, all those who encounter it must do so with integrity.

Index

S

T

Made in the USA
Monee, IL
29 August 2024

38287e85-9c1f-4a15-b831-5210e04083a8R01